PARTICLE ACCELERATION IN COSMIC PLASMAS

AIP CONFERENCE PROCEEDINGS 264

PARTICLE ACCELERATION IN COSMIC PLASMAS

NEWARK, DE 1991

EDITORS: G. P. ZANK
T. K. GAISSER

BARTOL RESEARCH INSTITUTE

American Institute of Physics New York

Authorization to photocopy items for internal or personal use, beyond the free copying permitted under the 1978 U.S. Copyright Law (see statement below), is granted by the American Institute of Physics for users registered with the Copyright Clearance Center (CCC) Transactional Reporting Service, provided that the base fee of $2.00 per copy is paid directly to CCC, 27 Congress St., Salem, MA 01970. For those organizations that have been granted a photocopy license by CCC, a separate system of payment has been arranged. The fee code for users of the Transactional Reporting Service is: 0094-243X/87 $2.00.

© 1992 American Institute of Physics.

Individual readers of this volume and nonprofit libraries, acting for them, are permitted to make fair use of the material in it, such as copying an article for use in teaching or research. Permission is granted to quote from this volume in scientific work with the customary acknowledgment of the source. To reprint a figure, table, or other excerpt requires the consent of one of the original authors and notification to AIP. Republication or systematic or multiple reproduction of any material in this volume is permitted only under license from AIP. Address inquiries to Series Editor, AIP Conference Proceedings, AIP, 335 East 45th Street, New York, NY 10017-3483.

L.C. Catalog Card No. 92-73316
ISBN 0-88318-948-8
DOE CONF-9112103

Printed in the United States of America.

CONTENTS

PARTICLE ACCELERATION IN THE SUN, HELIOSPHERE & GALAXY

The Sun as a Lab for Particle Acceleration Mechanisms 3
    D. B. Melrose
Particle Acceleration in the Magnetosphere 15
    Jörg Büchner
Particle Acceleration in the Heliosphere 27
    Martin A. Lee
Particle Acceleration on Galactic Scales 45
    W. I. Axford
High Energy Gamma Ray Observations 57
    Carl E. Fichtel

TRANSPORT THEORY

A Review of Transport Theory 71
    Frank C. Jones
Quasi-Linear Theory and Transport Theory 79
    Charles W. Smith
Particle Transport from a Turbulence Perspective 86
    John W. Bieber and William H. Matthaeus
Recent Developments in Particle Transport Theory:
The Wave Viewpoint 92
    Reinhard Schlickeiser
Particle Acceleration and Transport by Strong MHD-Turbulence
and Shock Wave Ensembles 99
    A. M. Bykov
Anomalous Diffusion of Cosmic Rays Across the Magnetic Field 105
    L. G. Chuvilgin and V. S. Ptuskin
Distribution of Particles and Fields in Turbulent Media 112
    A. Z. Dolginov
A New Nonlinear Diffusion Formalism in a Magnetized Plasma:
Application to Space Physics and Astrophysics 118
    H. Karimabadi and D. Krauss-Varban

PARTICLE ACCELERATION AT NON-RELATIVISTIC SHOCKS

Microphysics and Structure of Quasi-Parallel Shocks:
Observations, Theory, and Implications for Particle Acceleration 125
    Manfred Scholer
Diffusive Shock Acceleration: Acceleration Rate, Magnetic-Field
Direction and the Diffusion Limit 137
    J. R. Jokipii
Numerical Simulations of Time-Dependent Cosmic Ray Mediated Shocks 148
    T. W. Jones

Particle Injection and the Structure of Energetic Particle
Modified Shocks 158
    G. P. Zank, G. M. Webb, and D. J. Donohue
Time-Dependent Cosmic Ray Shocks with Injection:
A Progress Report 164
    D. J. Donohue and G. P. Zank
On the Stability of Cosmic Ray Dominated Shocks 169
    Hyesung Kang, Dongsu Ryu, and T. W. Jones
Two-Fluid Modelling of Particle Acceleration in Modified
Non-Relativistic Shocks 173
    Matthew G. Baring
Particle Acceleration in Modified Oblique Non-Relativistic Shocks 177
    Matthew G. Baring, Donald C. Ellison, and Frank C. Jones
Shock Drift Acceleration 183
    R. B. Decker
Particle Acceleration in Supernova Remnants 189
    L. O'C. Drury
Observational Tests of Particle Acceleration Theory—
Shell Supernova Remnants 195
    Martha C. Anderson and Lawrence Rudnick
Cosmic Ray Origin from Supernova Remnants—A Comparison
with External Galaxies 199
    H. J. Völk
On the First Experimental Confirmation of Hypothesis of
Cosmic Ray Generation in Supernova Explosion 205
    Grant E. Kocharov

STOCHASTIC PARTICLE ACCELERATION

Particle Acceleration in Solar Flares: Observations 213
    Donald V. Reams
Stochastic Acceleration in Impulsive Solar Flares 223
    James A. Miller and Reuven Ramaty
Simulations of Second-Order Fermi Acceleration of Electrons:
Solving the Injection Problem 229
    Galen Gisler
Proton Gyroresonance with Parallel Waves in a Low-Beta
Solar Flare Plasma 235
    Jürgen Steinacker and James A. Miller
Acceleration of Particles by an Ensemble of Shock Waves and
High Energy Emission of Solar Flares 239
    L. G. Kocharov and G. A. Kovaltsov
The Runaway of Fast Electrons into Turbulent Plasma of Solar Flares 247
    Yu. E. Charikov and I. V. Kudrjavtsev
Gamma-Rays and Neutrons as a Probe of the Proton Spectrum During
the Solar Flare of 1988 December 16 253
    P. P. Dunphy and E. L. Chupp
Are Coronal Type II Shocks Piston Driven? 257
    N. Gopalswamy and M. R. Kundu

MHD Turbulence, Reconnection, and Test-Particle Acceleration 261
    Perry C. Gray and William H. Matthaeus
Particle Acceleration at Comets 267
    Tamas I. Gombosi
Pickup Ions Observed at Comet Halley 273
    M. Neugebauer
Particle Acceleration in (by) Accretion Discs 279
    J. I. Katz

PARTICLE ACCELERATION IN RELATIVISTIC FLOWS AND SHOCKS

Energetic Particle Transport in Relativistic Flows 287
    G. M. Webb
Relativistic, Cosmic-Ray Modified Shocks 294
    J. G. Kirk
Particle Acceleration in Disordered Magnetic Fields 298
    A. F. Heavens
Shock Acceleration in a Radiation Field 304
    A. P. Szabo and R. J. Protheroe
Relativistic Shock Waves and the Excitation of Plerions 313
    Jonathan Arons

SIMULATIONS

Monte Carlo Simulation of Electron Acceleration in
Modified Relativistic Shocks 331
    Donald C. Ellison
Ion Acceleration at Collisionless Shock Interactions 342
    Peter J. Cargill
Acceleration of Diffuse Ions Upstream from Parallel Shocks 348
    Kevin B. Quest
Simulations of Collisionless Shocks: Some Implications for
Particle Acceleration 354
    D. Burgess
Simulations of Ion Acceleration at Parallel Shocks 360
    J. Giacalone, D. Burgess, S. J. Schwartz, and D. C. Ellison
Acceleration of Diffuse Ions at Quasi-Parallel Shocks: Simulations 364
    H. Kucharek and M. Scholer

COMPOSITION AND SOURCES OF HIGH ENERGY COSMIC RAYS

The UHE Cosmic Ray Spectrum 373
    Pierre Sokolsky
Cosmic Ray Composition at the Knee 379
    Todor Stanev
The Arrival Directions and Mass Composition of Cosmic Rays
Above 10^{18} eV 386
    A. A. Watson

Ultrahigh-Energy Cosmic Rays from Fanaroff Riley
Class II Radio Galaxies 393
  Jörg Rachen and Peter L. Biermann
Particle Acceleration up to 10^{20} eV 400
  W.-H. Ip and W. I. Axford

ELECTRON ACCELERATION

Electron Acceleration in the Galaxy: Observation and Theory 409
  S. P. Reynolds
Cosmic Rays and Shock Waves in Active Galaxies 417
  Andrew S. Wilson
The Quest for Particle Acceleration in Extragalactic Sources 424
  Lawrence Rudnick, Debora Katz-Stone, and Martha Anderson
Electron Acceleration by Young Supernova Remnant Blast Waves 430
  R. D. Blandford
Diffusive Acceleration of Electrons in Supernova 1987A 436
  Lewis Ball
Simulation of Electron Acceleration at Collisionless Shocks 445
  D. Krauss-Varban
Whistler Waves and Electron-Whistler Scattering in
Astrophysical Plasmas 451
  Matthew G. Baring
Diffusive Transport in Cluster-Center Radio Sources 455
  Jean A. Eilek
Jovian Electron Transport to the Polar Heliosphere: An Analogy
to Magnetospheric Recirculation 461
  J. F. Cooper and D. N. Baker
Direct Electric Field Heating and Acceleration of Electrons in Solar Flares 470
  Gordon D. Holman and Stephen G. Benka
Constraints on Electron Acceleration in the Crab Nebula 471
  A. K. Harding and O. C. DeJager
A Model for the Radio Flare from SN 1987A 477
  J. G. Kirk and M. Wassmann

OBSERVATION PRIORITIES AND PERSPECTIVES

Ground Based Experiments-Observational Priorities 485
  J. A. Goodman
Perspectives of Future Observations on Particle Acceleration
in the Heliosphere 491
  Eberhard Möbius

PREFACE

Particle acceleration is ubiquitous, occurring on all scales in all stellar, space, and galactic environments. After several decades of development, it has become clear that simple theories of particle acceleration are reasonably well understood. However, further significant progress requires the resolution of a number of key, fundamental issues. To address these issues requires the close collaboration of plasma, space, particle, and astrophysicists. In an effort to foster multidisciplinary collaboration in this exciting field, a three-day workshop was held at the Bartol Research Institute from 4–6 December, 1991. Given the BRI's long involvement in cosmic-ray physics and its present mix of plasma, space, particle, and astrophysicists, it seemed particularly appropriate that Bartol organize and host this workshop.

The proceedings reflects the diversity and scope of the meeting. The book is divided into sections corresponding to outstanding issues in particle acceleration, rather than focusing simply on various mechanisms. In this way, we hope to stress the unity of many ideas which find application to diverse environments. Thus, the first section provides a fairly comprehensive review of the sun, magnetosphere, heliosphere, and galaxy as laboratories for particle acceleration. We trust this collection of papers will be valuable to both researchers and students.

The successful organization of any meeting results from the unselfish help of many individuals and the PACP Workshop is no exception. In this regard, we thank Frank Jones (NASA, Goddard Space Flight Center) and Don Ellison (North Carolina State University) for their considerable help in acting as co-organizers of the meeting. The International Scientific Advisory Committee— W. I. Axford, L. O'C. Drury, J. R. Jokipii, C. F. Kennel, M. A. Lee, E. N. Parker and R. Ramaty— provided valuable suggestions and help. We are indebted to the Local Organizers—J. W. Bieber, P. Evenson, D. Martinez, W. H. Matthaeus, N. F. Ness, S. Oughton, C. W. Smith, T. Stanev—for their untiring efforts and to Mrs. L̈aura Rossano for her organizational skills. The PACP workshop was supported by the BRI, NASA, and the NSF under grant number ATM-9016497. Finally, we thank the participants for making the Bartol PACP Workshop an exciting, interesting, and challenging meeting.

G. P. Zank
T. K. Gaisser

Particle Acceleration in the Sun, Heliosphere & Galaxy

THE SUN AS A LAB FOR PARTICLE ACCELERATION MECHANISMS

D.B. Melrose

UNIVERSITY OF SYDNEY, NSW 2006, AUSTRALIA.

ABSTRACT

Mechanisms for the acceleration of fast particles in the solar corona are reviewed briefly. Three specific problems related to acceleration on the Sun are discussed: (1) bulk energization of electrons in solar flares, (2) injection of electrons into diffusive shock acceleration, and (3) the relative abundances of accelerated ions.

1. INTRODUCTION

The acceleration of fast particles in astrophysical plasmas[1-4] is the oldest branch of what is now called plasma astrophysics, and studies of acceleration in the atmosphere of the Sun,[5-9] have contributed in a major and ongoing way to our present understanding. In this paper, after reviewing different possible acceleration mechanisms in section 2, three specific unsolved problems are discussed: bulk energization of electrons in solar flares and elsewhere (section 3), preacceleration of electrons for diffusive shock acceleration (section 4), and the relative abundances of accelerated ions (section 5).

2. REVIEW OF ACCELERATION MECHANISMS

The following eight acceleration mechanisms need to be considered when discussing acceleration of fast particles in the solar corona.

1. Stochastic Acceleration: In modern day terminology, *stochastic acceleration* and *Fermi acceleration* are used synonymously to refer to acceleration by MHD turbulence. Three ingredients are involved. First, is the energy change of individual particles, due to the betatron effect,[10] reflection off moving magnetic inhomogeneities[11] or transit damping.[12] Second, effective scattering is required and this is attributed to wave-particle interactions when the gyroresonant condition is satisfied:

$$\omega - s\Omega - k_\| v_\| = 0, \tag{1}$$

where ω is the wave frequency, $k_\|$ is the component of its wave vector parallel to $\boldsymbol{B}$, s is the harmonic number, $\Omega = |q|B/\gamma m$ is the gyrofrequency of the particle with charge q, mass m, Lorentz factor γ and velocity $v_\|$ parallel to $\boldsymbol{B}$. Resonant scattering, in its simplest form, is due to interactions at $s = \pm 1$ in the resonance condition (1). The relevant resonances are for $\omega \ll |k_\| v_\||$, so that (1) implies $\Omega \approx |k_\| v_\||$, leading to the *threshold conditions*[13]: resonant scattering of ions with Alfvén waves is possible only for $|v_\|| > v_A$ and of electrons by whistlers only for $|v_\|| > 43 v_A$. The third ingredient in the theory is a statistical treatment, usually based either on a Fokker-Planck approach or on a quasilinear approach implying that stochastic acceleration may be regarded as a diffusion in momentum space. Achterberg[14] clarified this further by showing Fermi-type acceleration to be a consequence of Landau damping ($s = 0$ in (1) above) of MHD turbulence in the presence of effective scattering. The diffusive equation is[14-16]

$$\frac{\partial}{\partial t}\langle f\rangle(p) = \frac{1}{p^2}\frac{\partial}{\partial p}\left[p^2 D_{pp}(p)\frac{\partial}{\partial p}\right]\langle f\rangle(p), \qquad (2)$$

where $\langle f\rangle(p)$ is the particle distribution function averaged over pitch angle. For acceleration by MHD turbulence, the diffusion coefficient is of the form

$$D_{pp}(p) = \nu_A \frac{cp^2}{4v}\left(1 - \frac{v_A^2}{v^2}\right)^2, \quad \nu_A = \frac{\pi}{4}\bar{\omega}\left(\frac{\delta B}{B}\right)^2, \qquad (3)$$

where ν_A is the acceleration rate, δB is the magnetic amplitude in the waves and $\bar{\omega}$ is their mean frequency. Stochastic acceleration and diffusive shock acceleration are favored mechanisms for solar cosmic rays.

2. Diffusive Shock Acceleration: Shock acceleration[1-4] involves two quite different mechanisms: *diffusive shock acceleration* and *shock drift acceleration*. Diffusive shock acceleration involves multiple shock crossings, with the particles experiencing first order Fermi acceleration when reflected off scattering centers after each crossing. A major success of the theory is that the predicted energy spectrum of accelerated particles has a power law form of just the type observed in cosmic rays and synchrotron sources. Scattering centers both upstream and downstream of the shock are required. As discussed in sections 4 and 5, problems remain in understanding preacceleration to above the threshold for resonant scattering.

3. Shock Drift Acceleration: Shock drift acceleration occurs in a single reflection or transmission at a shock front,[1,17] and is due to a drift velocity in the front such that the electric field in the front does work on the particle. The magnitude of the energy change is related to the strength of the shock and is roughly independent of the initial energy of the particle. Such changes are important only when the initial energy of the particle is comparable with the change in energy. Shock drift acceleration is important for particles at IPM shocks and at bow shocks.

4. Resonant Acceleration by MHD Waves: An alternative to stochastic acceleration by MHD turbulence is resonant acceleration.[18-20] Formally, resonant acceleration involves high harmonics, $s \gg 1$ in (1), whereas stochastic acceleration involves $s = 0$.[14]

5. Resonant Acceleration by Longitudinal Waves: Acceleration by Langmuir waves seems a favorable mechanism for acceleration of nonrelativistic

electrons.[21] This and the remaining mechanisms mentioned below have all been considered as possible candidates for bulk energization of electrons (section 3). A difficulty with the application of resonant acceleration by Langmuir waves is in identifying an adequate source for Langmuir waves. Resonant acceleration by ion sound or lower hybrid waves, rather than Langmuir waves, is of interest because these waves can be generated through a current instability, and so can be driven by macroscopic motions. However, acceleration by such waves affects virtually all the particles, and is thought to lead to heating rather than acceleration.

6. Runaway Acceleration: The simplest conceivable mechanism for the acceleration of fast particles is by a static electric field, $E_\parallel$, parallel to the magnetic field. Although plasmas are highly conducting and tend to short any $E_\parallel$, an $E_\parallel \neq 0$ can occur under a variety of conditions, notably in the presence of some form of enhanced resistivity. Acceleration by an $E_\parallel$ is usually treated in terms of a runaway process.[22–25] Self-inductance places a severe limitation on runaway acceleration, arising from the fact that the runaway electrons constitute an electric current, as discussed in section 3.

7. Acceleration by Potential Double Layers: One specific type of $E_\parallel$ occurs in a potential double layers.[26–28] Double-layer like structures called electrostatic shocks are a common feature in the Earth's auroral zones, where they are associated with fluxes of accelerated particles.[29] The possibility that acceleration of electrons in the solar corona might be related to acceleration of auroral electrons is discussed in section 3 below.

8. Acceleration during Magnetic Reconnection: Acceleration of fast particles during reconnection, due to the electric field induced by the changing magnetic field, is a further possibility for particle acceleration in the solar corona,[30,31] but as yet there is no adequate statistical theory for the mechanism. From a single-particle viewpoint, it seems that only a small fraction of all particles is accelerated.[7] From a fluid viewpoint, there is evidence from magnetic merging in the Earth's magnetotail that plasma is ejected along the magnetic separatrices at about the Alfvén speed,[32] implying that much more energy goes into ions than electrons.

3. BULK ENERGIZATION OF ELECTRONS

An outstanding unsolved problem is the detailed mechanism, or mechanisms, involved in the primary energy release in solar flares. The accepted interpretation is that the energy goes into electrons with energies $\gtrsim 10\,\mathrm{keV}$, although alternatives are not entirely ruled out.[33,34] The efficiency is thought to be at least 20 percent.[35]

Data on Bulk Energization : Data relating to bulk energization of electrons in the impulsive phase of solar flares comes from the interpretation of hard X-ray bursts,[36] from type III and spike radiobursts,[37] from microwave bursts, and from observation in the IPM of the electrons that generate type III bursts.[38] The hard X-ray data require that $\gtrsim 10\,\text{keV}$ electrons precipitate at a rate up to about $10^{36}\,\text{s}^{-1}$.[39] This rate integrated over the impulsive phase can be so large (up to $\approx 10^{39}$) that the electron population needs to be resupplied by a return current drawing electrons from the chromosphere.[33] A small fraction of the energized electrons escapes from the corona in the form of electron beams, and these beams generate type III radiobursts in the corona and in the IPM, where they can be observed directly by spacecraft.

Bulk energization involves many localized bursts of energy release.[40,37] The microwave data suggest that individual episodes of energy release occur on the shortest timescales ($\lesssim 100\,\text{ms}$) that can be resolved.[41] Shorter time scales are observed in spike bursts, which are suggestive of "microflares" in which the energy release occurs in a region of dimensions 200 km or smaller.[42,43] Bulk energization seems to occur under several different conditions in addition to the impulsive phase of flares. Electron beams form high in corona due to high altitude flares,[44] in type I-III storms,[45] in the herringbone structure of type II bursts[46,47] and in SA events.[48]

Mechanisms for Bulk Energization : Theoretical ideas on how bulk energization occurs include resonant acceleration by plasma turbulence, by parallel electric fields, and in neutral sheets or reconnecting regions. Resonant acceleration of electrons by Langmuir turbulence is very effective in principle and could account for bulk energization[49,50] provided that the postulated Langmuir turbulence is present. However, there is no adequate source for the Langmuir turbulence. Upconversion from ion sound turbulence has been considered,[51] but has been rejected on both practical[6] and fundamental[52] grounds.

Giovanelli[53] proposed a discharge theory for solar flares, which invoked what is now called runaway acceleration.[22-25] Some runaway acceleration is unavoidable in a plasma if there is any parallel electric field. Moreover, one can argue that there must be large ($\approx 10^{10}$ V) potential drops available in solar flux tubes,[10,54-56] and only a small fraction of this available potential need appear along field lines to provide the acceleration. Giovanelli's model was criticized by Cowling[57] on two grounds, one of which is that any process tending to change a current is strongly opposed by the plasma, as implied by Lenz' law.[54] In circuit language, the inductive time scale for the circuit is too long to allow the current to change substantially.[6,24,47,57] This suggests that while runaway acceleration may account for a high energy electron tail, it cannot account for bulk energization.

Acceleration of particles in neutral sheets[58] is favored in some models that invoke magnetic reconnection.[59] However, a difficulty with this suggestion is that most of the energy should go into ions and not electrons. Conversely, this may be regarded as an argument for neutral beams rather

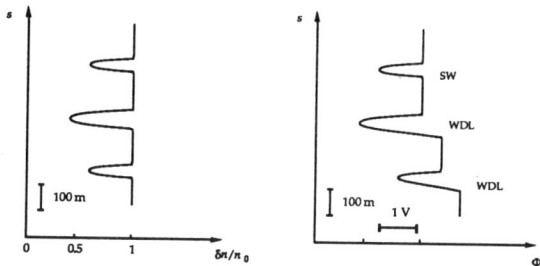

Fig. 1. A schematic description of variation of the relative density ($\delta n/n_0$) and potential (Φ) with height s as observed above the auroral zones from the Viking spacecraft; SW denotes a solitary wave, and WDL denotes a weak double layer.[63,64]

than electrons beams in the primary energy release.[34]

Analogy with Auroral Acceleration : One possible analogy for bulk energization is acceleration during reconnection in the plasma sheet,[34] but this is thought to lead to acceleration of neutral beams[32] rather than of electrons. Another possible analogy is with the electrons accelerated in the auroral zones. This suggests a model involving acceleration by multiple weak double layers (hereafter WDLs).[60]

Acceleration of auroral electrons is attributed to multiple WDLs,[61] but there are alternative opinions.[62] Observations from the Viking spacecraft[63] confirmed earlier data[65] on the presence of structures interpreted as WDLs and solitary waves (hereafter SWs), cf. Figure 1. Associated with the WDLs there is an average electric field directed upward; the total $\Phi \approx$ few kV appears to arise from the potentials across many WDLs in series.[66] The electric field has a fluctuating component with an amplitude much larger than the DC component in the field, and with characteristic frequency of a few hertz, such that electrons are affected by the fluctuating component but ions are not.[67,61] This accounts for two observations that otherwise appear inconsistent with acceleration by WDLs: field-aligned counterstreaming electrons beams,[68,69] and field-aligned energetic electrons and ions propagating in the same direction.[70]

There are two additional features of these models that may be relevant to a possible analogy with bulk energization of electrons in the solar corona. First, the counterstreaming electrons can have a fluence that is large enough to reduce the supply of electrons in ≈ 10 s, implying bulk energization rather than the production of a runaway tail,[68] as required in the solar application. Second, in both the auroral zones and in the solar corona, the actual current, I, is much smaller than the current, $-e\dot{N}$, implied by electrons escaping at a rate $\dot{N}$. This suggests that acceleration by a parallel electric field can produce escaping electrons at a rate that greatly exceeds the rate I/e. In the solar case the hard X-ray data imply $\dot{N} \gtrsim 10^{36}\,\text{s}^{-1}$ and the limit on the current implies $I/e \lesssim 10^{32}\,\text{s}^{-1}$, so that counterstreaming is required with the opposing electron flow rates balanced to one part in $\approx 10^4$.

4. INJECTION OF ELECTRONS INTO DIFFUSIVE SHOCK ACCELERATION

It is widely believed that relativistic electrons in most synchrotron sources are accelerated by the diffusive shock mechanism.[1–4] Diffusive shock acceleration requires efficient scattering, and scattering by waves is effective only if (i) the relevant waves are present at an appropriate level, and (ii) the particles exceed the threshold for resonant scattering. An outstanding problem is how electrons are preaccelerated to above the threshold for their resonant scattering.

The Injection Problem for Electrons : The injection problem for electrons in diffusive shock acceleration[71,72] arises as follows. Diffusive shock acceleration requires that the particles remain near the shock front and cross it many times. An ion of mass m_i can be resonantly scattered by Alfvén waves provided that its momentum, p, exceeds the threshold $m_i v_A$, where v_A is the Alfvén speed. There are two obvious possibilities for resonant scattering of electrons. First, electrons are scattered by the same Alfvén waves as the ions. The threshold is then $p > m_i v_A$, which requires $v > (m_i/m_e) v_A$ for electrons. Second, electrons are scattered by whistlers, in which case the threshold is reduced to $v > 43 v_A$.[13] In the following discussion it is assumed that the latter threshold applies. One suggested preacceleration mechanism is that the electrons are heated through the damping of longitudinal waves excited by ions reflected from a strong, quasi-perpendicular shock.[73–75] It is of interest to ask whether observations of electron heating and acceleration in the solar context support this or suggest any alternative possibility.

Electron Heating at Interplanetary Shocks : Ions and electrons are observed to be heated and accelerated at IPM shocks and at bow shocks. Electrons are heated less efficiently than ions.[76] The electrons that are observed tend to be strongly field aligned,[77] which is consistent with shock drift acceleration rather than diffusive shock acceleration. Whereas there is both a field-aligned and a diffusive component of the ions,[78] the latter implying that ions can be scattered efficiently, there is no corresponding evidence for effective scattering of the electrons. Thus IPM shocks provide no evidence for diffusive shock acceleration of electrons. This is not surprising because such shocks are relatively weak, with the exception of some planetary bow shocks,[79] and diffusive shock acceleration even for ions is expected to be efficient only for Mach numbers $M \gtrsim 5$.[80,81]

Acceleration at Type II Shocks and in SA Events : Although most IPM shocks seem to be relatively inefficient accelerators of electrons, those shocks that produce type II radio bursts[46,47] and SA events[48] are copious accelerators of electrons. In those type II burst that exhibit herringbone structure and in SA events, there is clear evidence of electron beams escaping from the shock front. How these beams are formed is poorly understood. One possibility[82] is an analogy with the Earth's electron foreshock, where the electrons are accelerated by reflection from the shock which is a form of shock drift acceleration.

Acceleration of Relativistic Solar Electrons : The acceleration of relativistic electrons has long been regarded as a characteristic feature of second phase acceleration in flares.[83] The idea is that acceleration occurs in two stages with the first stage producing the mildly relativistic electrons implied by hard X-ray bursts, type III bursts and microwave bursts, and the second phase producing both ions and relativistic electrons; the first phase electrons act as a seed population for stochastic or shock acceleration in the second stage. One of the assumptions implicit in the two-stage model for acceleration is a significant time delay between the first and second phases. More recent observations show that the delay can be as short as about a second,[84] and this has led to the two phases of acceleration being reinterpreted in terms of two steps of acceleration with only a short time delay between them.

Gamma-ray observations imply that relativistic electrons, as well as deca-MeV ions, are accelerated promptly in the impulsive phase. Gamma rays from solar flares include both line and continuum components: the lines are due to promptly accelerated of deca-MeV ions, and the continuum component requires that relativistic electrons also be accelerated within a second or so of the onset of the impulsive phase of a flare. The acceleration of the ions has been discussed by many authors.[84-88] Less attention has been given to the acceleration of relativistic electrons.[89-91] Shock acceleration would require that shocks form on the prompt time scale. Observations[92-94] suggest that the prompt acceleration is not associated with the shocks that produce type II bursts. Stochastic acceleration or resonant acceleration by MHD turbulence seems more plausible.

A possible preacceleration mechanism suggested by the solar observations involves the production of electron beams at shocks, as in type II bursts and SA events. If a strong, quasi-parallel shock transition consists of a region involving subshocks and shocklets,[95] then multiple reflections of such electron beams should lead to energies above the threshold for scattering, when the beams are destroyed and the individual particles become available for diffusive shock acceleration. However, before such speculations can be formulated into a model, further observational evidence is required to identify the particular properties of those shocks that produce electrons beams, which seems to be restricted to regions close to the Sun[96] where in situ data are not available.

5. RELATIVE ABUNDANCES OF ACCELERATED IONS

The abundances of accelerated ions relative to the abundances in the ambient plasma provides an important test for acceleration mechanisms. The solar data have now reached a level of sophistication that imposes severe constraints on acceptable mechanisms.

The Observed Abundances : Data on the abundances come from direct measurement from spacecraft of energetic ions in the IPM plasma and also from the relative intensity of gamma-ray lines observed in some impulsive flares.[5,93] Qualitatively, the relative abundances are close to photospheric values, with the differences from such values referred to as anomalies. Three ratios are useful in characterizing events: the electron to proton ratio, e/p, the ^{3}He/^{4}He ratio and the Fe/O ratio.

The most startling anomaly involves overabundances of ^{3}He often exceeding 10^3, such the ^{3}He/^{4}He ratio can exceed unity. There is a correlation between the ^{3}He-rich events and impulsive electron events.[97] It seems that a high ^{3}He/^{4}He ratio is characteristic of particles accelerated in the impulsive phase of solar flares. Such events have a high e/p ratio and a heavy ion composition characteristic of the flare plasma. Data on the Fe/O ratio can be used to distinguish ions accelerated in impulsive flares, with Fe/O ≈ 1, from ions accelerated at shocks in the IPM plasma, Fe/O ≈ 0.1.[93] The observed abundances raise two general questions. One concerns the implications of the abundances being so close to normal, and the other concerns how one accounts for the specific anomalies observed.

Predicted Anomalies for Various Preacceleration Mechanisms : Stochastic acceleration involves a seed population of already suprathermal particles being accelerated. The following argument suggests that the formation of a seed population through any stochastic preacceleration mechanism should lead to extreme anomalies, inconsistent with the observed abundances. Suppose the seed population is attributed to an acceleration process drawing ions into a suprathermal tail. The nonthermal tail forms through a balance between an acceleration mechanism and collisions (or a collision-like process) tending to maintain the Maxwellian distribution[98]; both processes may be described by a diffusion equation in momentum space of the form (2).[98] It is convenient to define a critical momentum, p_c, where systematic acceleration and slowing down balance. For $p \ll p_c$ the collisions build up a flux of particles in momentum space to resupply the depleted tail and the flux approaches an asymptotic value for $p \gg p_c$, as illustrated in Figure 2. This flux implies the rate, γ_i, at which ions of species i become suprathermal. For ions of charge $q_i = Z_i e$, mass $m_i = A_i m_p$, thermal speed V_i and collision frequency ν_{0i}, subject to stochastic acceleration of the form (2) with $\nu_{Ai} = A_i \nu_{Ap}$ ($i = p$ denotes protons), the rate of production of suprathermal ions depends on Z_i

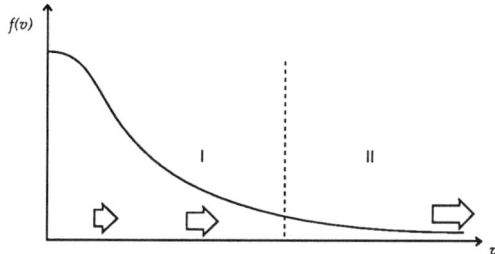

Fig. 2. In region I collisions build up a flux to supply a depleted tail; in region II acceleration removes particles to higher velocities.

and A_i according to[99]

$$\gamma_i = \left(\frac{2}{\pi}\right)^{1/2} Z_i^2 A_i^{1/2} \nu_p e^{-A_i} \exp\left[-\tfrac{1}{2} Z_i \beta (\tfrac{1}{2}\pi - \arctan 2 A_i \beta^{-1})\right], \quad (4)$$

with $\beta = (\nu_{0p} V_p / \nu_{Ap} c)^{1/2}$. The important qualitative point is that this rate is strongly dependent on the species, contrary to observation. This qualitative conclusion follows for any preacceleration mechanism involving particles being drawn out a tail on a Maxwellian distribution. Preacceleration of ions is obviously required for stochastic acceleration, and the foregoing discussion might be taken as an argument against this mechanism and in favor of diffusive shock acceleration. Preacceleration also seems to be required for diffusive shock acceleration[81], although not all authors agree.[71] Galeev *et al.* [81] proposed a specific preacceleration due to gyroresonance absorption of high-k Alfvén waves, and the foregoing arguments suggests that this mechanism should also result in gross anomalies.

Implications for Preacceleration of Ions : An acceptable preacceleration mechanism must be only weakly dependent on Z_i and A_i (except for ^{3}He). If only a small fraction of the ions are preaccelerated, then it seems that this must occur through all the ions in highly localized regions being preaccelerated, e.g., through the formation of localized mass flows at $v > v_A$. One possibility is in reconnection regions where plasma jets with $v \gtrsim v_A$ form along the separatrices, as observed in the Earth's magnetotail.[32] Once such a jet is formed, it propagates until scattering destroys it, and the resulting scattered particles form an appropriate seed population for further acceleration. Another idea involves acceleration of ions in plasma upflows from the chromosphere in response to the precipitation of electrons during a flare.[100]

6. DISCUSSION AND CONCLUSIONS

Observations of solar energetic particles and models for their interpretation have had a major impact on the development of theories for particle acceleration, as summarized in section 2. However, major unsolved problems remain, and three of these are discussed here.

(1) Perhaps the most outstanding unsolved problem concerning particle acceleration in the solar corona is the bulk energization of electrons in the impulsive phase of solar flares. A suggestion made here is that the acceleration may be analogous to that of electrons in the auroral zones, assumed here to be due to multiple WDLs. (2) Preacceleration is required for electrons in both stochastic acceleration and diffusive shock acceleration. One possible preacceleration mechanism suggested by the solar observations is the production of electron beams at shocks, as in type II bursts and SA events. If a strong shock transition consists of a region involving subshocks and reverse shocks then multiple reflections of such electrons beams should lead to energies above the threshold for scattering. (3) The evidence that energetic ions accelerated at IPM shocks have abundances characteristic of the ambient plasma[93] provides strong support for the suggestion that galactic cosmic rays are accelerated from the interstellar medium by supernova shocks.[101,72]

In conclusion, it is worth emphasizing that interchange of ideas between magnetospheric and space physics, solar physics and nonsolar astrophysics has been profitable in the past and is likely to continue to be profitable. These three fields provided complementary information, with the solar case being intermediate between the other two, both in terms of the information available and in terms of the energies of the particles. The Sun plays an important role as a lab to test how acceleration processes scale from conditions that can be probed directly in the heliosphere to large scale astrophysical phenomena for which theories cannote be tested directly.

REFERENCES

1. Toptygin, I.N., *Space Sci. Rev.* **26**, 157 (1980).
2. Axford, W.I., in G. Setti, G. Spada and A.W. Wolfendale (eds) *Origin of Cosmic Rays.* IAU Symposium No. 94, D. Reidel (Dordrecht), p. 339 (1981).
3. Drury, L.O'C., *Rep. Prog. Phys.* **46**, 973 (1983).
4. Jones, F.C., and Ellison, D.C., *Space Sci. Rev.* **59**, 259 (1991).
5. Ramaty, R. *et al.*, in P.A. Sturrock (ed.) *Solar Flares*, Colorado Associated Press, p. 117 (1980).
6. Heyvaerts, J., in E.R. Priest (ed.) *Solar Flare Magnetohydrodynamics*, Gordon and Breach (New York), p. 429 (1981).
7. Forman, M.A., and Ramaty, R., and Zweibel, E. G., in P.A. Sturrock, T.E. Holtzer, D.M. Mihalas and R.K. Ulrich (eds) *The Physics of the Sun, Volume II*, D. Reidel (Dordrecht), p. 249 (1985).
8. Benz, A.O., *Solar Phys.* **111**, 1 (1987).
9. Melrose, D.B., *Aust. J. Phys.* **43**, 703 (1990).
10. Swann, W.F.G., *Phys. Rev.* **43**, 217 (1933).
11. Fermi, E., *Phys. Rev.* **75**, 1169 (1949).
12. Shen, C.S., *Astrophys. J.* **141**, 1091 (1965).
13. Melrose, D.B., *Solar Phys.* **37**, 353 (1974).
14. Achterberg, A., *Astron. Astrophys.* **97**, 259 (1981).
15. Tverskoï, B.A., *Sov. Phys. JETP* **25**, 317 (1967).
16. Kulsrud, R.M., and Ferrari, A., *Astrophys. Space Sci.* **12**, 302 (1971).
17. Pesses, M.E., *J. Geophys. Res.* **86**, 150 (1981).
18. Lacombe, C., *Astron. Astrophys.* **54**, 1 (1977).
19. Barbosa, D.D., *Astrophys. J.* **233**, 383 (1979).

20. Ellison, D.C., and Ramaty, R., *Astrophys. J.* **298**, 400 (1985).
21. Tsytovich, V.N., *Soviet Phys. Usp.* **9**, 370 (1966).
22. Norman, C.A., and Smith, R.A., *Astron. Astrophys.* **68**, 145 (1978).
23. Kuijpers, J., van der Post, P., and Slottje, K., *Astron. Astrophys.* **103**, 331 (1981).
24. Holman, G.D., *Astrophys. J.* **293**, 584 (1985).
25. Takakura, T., *Solar Phys.* **115**, 149 (1988).
26. Carlqvist, P., *Cosmic Electrodynamics* **3**, 377 (1972).
27. Block, L.P., *Astrophys. Space Sci.* **55**, 59 (1978).
28. Raadu, M.A., *Phys. Rep.* **178**, 27 (1989).
29. Mozer, F.S., Cattell, C.A., Hudson, M.K., Lysak, R.L., Temerin, M., and Torbert, R.B., *Space Sci. Rev.* **27**, 155 (1980).
30. Friedman, M., and Hamberger, S.M., *Astrophys. J.* **152**, 667 (1968).
31. Bulanov, S.V., and Sasarov, P.V., *Soviet Astron.* **19**, 464 (1976).
32. Lin, R.P., Anderson, K.A., McCoy, J.E., and Russell, C.T., *J. Geophys. Res.* **82**, 2761 (1977).
33. Brown, J.C., Karlicky, M., MacKinnon, A.L., and van den Oord, G.H.J., *Astrophys. J. Suppl.* **73**, 343 (1990).
34. Martens, P.C.H., and Young, A., *Astrophys. J. Suppl.* **73**, 333 (1990).
35. Duijveman, A., Hoyng, P., and Machado, M.E., *Solar Phys.* **81**, 137 (1982).
36. Dennis, B.R., *Solar Phys.* **100**, 465 (1985).
37. Aschwanden, M.J., Benz, A.O., Schwartz, R.A., Lin, R.P., Pelling, R.M., and Stehling, W., *Solar Phys.* **130**, 39 (1990).
38. Lin, R.P., *Solar Phys.* **100**, 537 (1985).
39. Hoyng, P., Brown, J.C., and van Beek, H.F., *Solar Phys.* **48**, 197 (1976).
40. Sturrock, P.A., Kaufmann, P., Moore, R.L., and Smith, D.F., *Solar Phys.* **94**, 341 (1984).
41. Sturrock, P.A., in B.R. Dennis, L.E. Orwig and A.L. Kiplinger (eds) *Rapid Fluctuations in Solar Flares*, NASA (Washington DC), p. 1 (1987).
42. Benz, A.O., *Solar Phys.* **96**, 357 (1985).
43. Stähli, M., and Magun, A., *Solar Phys.* **104**, 117 (1986).
44. Cliver, E., and Kahler, S., *Astrophys. J.* **366**, L91 (1991).
45. Suzuki, S., and Dulk, G.A., in D.J. McLean and N.R. Labrum (eds), *Solar Radiophysics*, Cambridge University Press, p. 289 (1985).
46. Nelson, G.J., and Melrose, D.B., in D.J. McLean and N.R. Labrum (eds), *Solar Radiophysics*, Cambridge University Press, p. 333 (1985).
47. Cairns, I.H., and Robinson, R.D., *Solar Phys.* **111**, 365 (1987).
48. Cane, H.V., Stone, R.G., Fainberg, J., Stewart, R.T., Steinberg, J.L., and Hoang, S., *Geophys. Res. Lett.* **8**, 1285 (1981).
49. Hoyng, P., *Astron. Astrophys.* **55**, 31 (1977).
50. Smith, D.F., *Astrophys. J.* **217**, 644 (1977).
51. Tsytovich, V.N., Stenflo, L., and Wilhelmsson, H., *Physica Scripta* **11**, 251 (1975).
52. Melrose, D.B., and Kuijpers, J., *Astrophys. J.* **323**, 338 (1987).
53. Giovanelli, R.G., *Mon. Not. R. Astron. Soc.* **108**, 163 (1948).
54. Dungey, J. W., in B. Lehnert (ed.) *Electromagnetic Phenomena in Cosmical Plasma*, IAU Symposium No. 6, North Holland (Amsterdam), p. 135 (1958).
55. Sweet, P.A., in B. Lehnert (ed.) *Electromagnetic Phenomena in Cosmical Plasma*, IAU Symposium No. 6, North Holland (Amsterdam), p. 123 (1958).
56. Colgate, S.A., *Astrophys. J.* **221**, 1068 (1978).
57. Cowling, T.G., in G.P. Kuiper (ed.) *The Sun*, Chicago University Press, p. 532 (1953).
58. Spicer, D.S., *Adv. Space Res.* **2**, 135 (1983).
59. Sakai, J.-I., and Ohsawa, Y., *Space Sci. Rev.* **46**, 113 (1987).
60. Khan, J.I., *Proc. Astron. Soc. Australia* **8**, 29 (1989).
61. Block, L.P., and Fälthammar, C.-G., *J. Geophys. Res.* **95**, 5877 (1990).
62. Bryant, D.A., *Physica Scripta* **T30**, 215 (1990).
63. Boström, R., Holback, B., Holmgren, G., and Koskinen, H., *Physica Scripta* **33**, 523 (1989).

64. Mälkki, A., Koskinen, H., Boström, R., and Holback, B., *Physica Scripta* **39**, 787 (1989).
65. Temerin, M., Cerny, K., Lotko, W., and Mozer, F. S., *Phys. Rev. Lett.* **48**, 1175 (1982).
66. Boström, R., Gustafsson, G., Holback, B., Holmgren, G., Koskinen, H., and Kintner, P., *Phys. Rev. Lett.* **61**, 82 (1988).
67. Hultqvist, B., *J. Geophys. Res.* **93**, 9777 (1988).
68. Sharp, R.D., Shelley, E.G., Johnson, R.G., and Ghielmetti, A.G., *J. Geophys. Res.* **85**, 92 (1980).
69. Lin, C.S., Burch, J.L., Winningham, J.D., Menietti, J.D., and Hoffman, R.A., *Geophys. Res. Lett.* **9**, 925 (1982).
70. Hultqvist, B., Lundin, R., Stasiewicz, K., Block, L., Lindqvist, P.-A., Gustafsson, G., Koskinen, H., Bahnsen, A., Potemra, T.A., and Zanetti, L.J., *J. Geophys. Res.* **93**, 9765 (1988).
71. Eichler, D., *Astrophys. J.* **229**, 419 (1979).
72. Blandford, R.D., in Roger and Landecker (1988), p. 309 (1988).
73. Moses, S.L., Coroniti, F.V., Kennel, C.F., and Scarf, F.L., *Geophys. Res. Lett.* **12**, 609 (1985).
74. Papadopoulos, K., *Astrophys. Space Sci.* **144**, 535 (1988).
75. Cargill, P.J., and Papadopoulos, K., *Astrophys. J.* **329**, L29 (1988).
76. Feldman, W.C., in B.T. Tsurutani and R.G. Stone (eds) *Collisionless Shocks in the Heliosphere: Review of Current Research*, American Geophysical Union, Washington DC, p. 195 (1985).
77. Potter, D.W., *J. Geophys. Res.* **86**, 11,111 (1981).
78. Cornwall, J.M., in T. Chang (ed.) *Ion Acceleration in the Magnetosphere and Ionosphere*, American Geophysical Union, Washington DC, p. 3 (1986).
79. Bagenal, F., Belcher, J.W., Sittler, E.C., Jr, and Lepping, R.P., *J. Geophys. Res.* **92**, 8603 (1987).
80. Völk, H.J., in J. Audouze and J. Tran Thanh Van (eds) *High Energy Astrophysics*, Editions Frotiere, p. 281 (1985).
81. Galeev, A.A., Sagdeev, R.Z., and Shapiro, V.D., in Proceedings of the Joint Varenna-Abastumani International School and Workshop *Plasma Astrophysics*, European Space Agency, ESA SP-251, p. 297 (1986).
82. Holman, G.D., and Pesses, M.E., *Astrophys. J.* **267**, 837 (1983).
83. Wild, J.P., Smerd, S.F., and Weiss, A.A., *Ann. Rev. Astron. Astrophys.* **1**, 291 (1963).
84. Bai, T., Hudson, H.S., Pelling, R.M., Lin, R.P., Schwartz, R.A., and von Rosenvinge, T.T., *Astrophys. J.* **267**, 433 (1983).
85. Melrose, D.B., *Solar Phys.* **89**, 149 (1983).
86. Decker, R.B., and Vlahos, L., *Astrophys. J.* **306**, 710 (1986).
87. Smith, D.F., and Brecht, S.H., *Solar Phys.* **115**, 133 (1988).
88. Sakai, J.-I., *Astrophys. J. Suppl.* **73**, 321 (1990).
89. Ohsawa, Y., and Sakai, J.-I., *Astrophys. J.* **332**, 439 (1988).
90. Miller, J.A., and Ramaty, R., *Solar Phys.* **113**, 195 (1987).
91. Steinacker, J., Dröge, W., and Schlickeiser, R., *Solar Phys.* **115**, 313 (1989).
92. Cane, H.V., and Reames, D.V., *Astrophys. J. Suppl.* **73**, 253 (1990).
93. Reames, D.V., *Astrophys. J. Suppl.* **73**, 235 (1990).
94. Nakajima, H., Kawashima, S., Shinohara, N., Shiomi, Y., Enome, S., and Rieger, E., *Astrophys. J. Suppl.* **73**, 177 (1990).
95. Zank, G.P., Axford, W.I., and McKenzie, J.F., *Astron. Astrophys.* **233**, 275 (1990).
96. Bougeret, J.-L., in B.T. Tsurutani and R.G. Stone (eds) *Collisionless Shocks in the Heliosphere: Review of Current Research*, American Geophysical Union, Washington DC, p. 13 (1985).
97. Reames, D.V., von Rosenvinge, T.T., and Lin, R.P., *Astrophys. J.* **292**, 716 (1985).
98. Gurevich, A.V., *Sov. Phys. JETP* **11**, 1150 (1960).
99. Melrose, D.B., *Can. J. Phys.* **46**, S638 (1968).
100. Winglee, R.M., *Astrophys. J.* **343**, 511 (1989).
101. Meyer, J.-P., *Astrophys. J. Suppl.* **57**, 173 (1985).

PARTICLE ACCELERATION IN THE MAGNETOSPHERE

Jörg Büchner
Max–Planck–Institut für extraterrestrische Physik
Außenstelle Berlin, D–O–1199 Berlin, Germany

ABSTRACT

Observations in the Earth's magnetosphere provide the unique opportunity of investigating particle acceleration *in situ*. In comparison with measurements in the tail of the Earth's magnetosphere we discuss the scaling laws of acceleration in current sheets and due to reconnection. We show how different processes provide energy gains at all observed energy levels, up to several MeV. Extrapolating the obtained scaling laws to astronomical scales we demonstrate that reconnection should be relevant for cosmic particle acceleration in general.

1. INTRODUCTION

Since first satellites were launched, energetic particles have been extensively investigated in the Earth's magnetosphere [1-5]. The observed energies range up to hundreds of MeV in the radiation belts, bursts of several MeV ions have been seen in the magnetotail. These particle energies exceed by far the thermal energy of the source plasmas in ionosphere ($\approx eV$) and solar wind ($\approx 100 eV$). Although they are still much lower than those of cosmic rays, our point is that universal acceleration processes exist and that the magnetosphere can serve as an *in situ* test site for cosmic particle acceleration. Our aim here is to discuss scaling laws of current sheet related acceleration processes in order to verify their validity by magnetospheric observations and to extrapolate them to cosmic scales.

Searching for acceleration to high energies, we are interested mainly in B.) type processes, creating non–Maxwellian features like bumps on tail or power law distribution functions instead of heating or accelerating the whole plasma bulk (A.) type processes) [6]. As far as the source of energy is concerned, rising evidence exist that magnetic fields are universal in space. In cosmic acceleration sites the plasma–β, i.e. the ratio of plasma pressure to magnetic pressure, is usually (much) smaller than unity, i.e. the magnetic energy per particle is (much) greater than the average particle energy. The question is, how to transfer magnetic energy to particles efficiently? The high energies of radiation belts can be reached only by long term pumping of adiabatic particles. The origin of multi–MeV electrons during magnetically quiet times [7] might be the Jupiter rather than the local magnetosphere [4,8]. Auroral acceleration is an universal cosmic energization process, which is at the same time *in situ* observable in the magnetosphere [6]. It is an A.) type process, enhancing the energy of the electron bulk motion toward the ionosphere up to the order of 10 keV. Auroral acceleration is caused by non–MHD field–aligned electric fields at altitudes between 1000 and 6000 km. Different mechanisms of generating parallel electric fields and accelerating particles were proposed in the past. They are all well described and we refer the interested reader to comprehensive current reviews [1,9,10].

In contrast, diffusive shock acceleration is a typical B.) type process [11,12]. The ISEE mission has given convincing evidence for shock acceleration at the bow shock wave of the Earth's magnetosphere, but mainly in the energy range below $200 keV$ [13,14]. Occurring on diffusion time scales, shock acceleration cannot explain bursts of MeV particles, often observed in the magnetotail of the Earth [2,3]. In the tail a significant amount of the diamagnetic depression of the local magnetic field is caused by energetic ions above $50 keV$, at times counting for up to one quarter of the local magnetic field decrease [15]. This reminds estimates that energetic cosmic rays comprise about one third of the energy density of the interstellar medium! At the same time ions stream persistently at high speeds along the plasmasheet boundary layer (PSBL) toward and away from the Earth [16,17]. As far as the the central plasmasheet (CPS) is concerned, recent high–time–resolution measurements on board of AMPTE led to the discovery of ion flow bursts perpendicular to the local magnetic field ($u_i > 4 \cdot 10^5 m/s$) [18]. Lyons and Speiser [19] suggested that current sheet acceleration forms the PSBL beams by energizing plasma mantle ions, reflected in the central plasma sheet backward to the Earth. In section 2. we discuss opportunities and limits of current sheet acceleration, including consequences of the non–linear dynamics of chaotic particle motion, which may cause the formation of spatial structures of the PSBL ion flows (beamlets) and velocity space structures in the CPS. They also provide a mechanism of thermalizing the accelerated plasma, making current sheet acceleration an A.) type process rather than a B.) type one. Larger electric fields can be obtained by magnetic reconnection and the vanishing magnetic field near neutral lines allows an effective, almost run–away acceleration. In sections 3.–5. we discuss different reconnection models in order to estimate the scaling laws of electric field generation and maximum obtainable energy. We compare the derived scaling laws with measurements in the magnetospheric tail. In the conclusion (section 6.) we extrapolate these results to astrophysical situations.

2. CURRENT SHEET ACCELERATION

A simple current sheet magnetic field is given by

$$\mathbf{B} = B_o \tanh\left(\frac{Z}{L_z}\right) \cdot \mathbf{e}_x + B_z(X) \cdot \mathbf{e}_z \quad (1)$$

and the electric field ($\mathbf{E} = E_y \cdot \mathbf{e}_y$)

$$E_y = E_{CS} = \frac{\eta \cdot B_o}{\mu_o \cdot L_z} = \frac{1}{R_m} \cdot v_A \cdot B_o \quad (2)$$

where η is the resistivity, $R_m = \mu_o v_A L_z/\eta$ the magnetic Reynolds number and $v_A = B_o/\sqrt{\mu_o n m_i}$ the Alfvén speed at $|Z| = L_z$ (m_i = ion mass). For the Earth's magnetotail current sheet with a number density $n = 10^6 m^{-3}$, a magnetic field at the PSBL of $B_o = 20 nT$, i.e. $v_A \cong 4 \cdot 10^5 m/s$, and a typical $R_m \approx 1000$ the electric field is less than $10^{-2} mV/m$. Due to the different response of electrons and ions also charge separation electric fields may act [20]. However, a charge separation field larger than the resistive field (2) would need significant plasma outflows in the current direction, which were not observed in the magnetotail. On the other hand, the highly conducting magnetospheric plasma enables a mapping of ionospheric electric fields along the equipotential magnetic lines of force. A typical cross-polar potential of about $50kV$ [21] results in an average cross-tail electric field strength of the order of a few tenth of mV/m. In contrast to the case of magnetic-field-aligned auroral acceleration current sheet electric fields are directed perpendicular to the ambient magnetic field. Hence a run-away along the electric field is prohibited by the Lorentz-force $e \cdot [\mathbf{v} \times \mathbf{B}] = ev_y B_z \mathbf{e}_x$, expelling the particles out of the electric field direction. The obtainable energy is limited by the maximum particle shift along the electric field, in weak electric fields $\Delta Y \cong 2\rho_{nj}$, where $\rho_{nj} = v_{oj}/\Omega_{nj}$ is the maximum possible Larmor-radius in the normal magnetic field $B_z \mathbf{e}_z$ ($\Omega_{nj} = b_n \Omega_{oj}$, $\Omega_{oj} = e_j B_o / m_j$ for the specie $j = e, i, p$, i.e. electrons, ions in general or protons, respectively and $b_n = B_z/B_o$). In one-dimensional current sheets, where $B_z = B_n = $ const., the acceleration law can also be obtained by considering the particle motion in the electric-field-free deHofmann-Teller frame, moving with the $\mathbf{E} \times \mathbf{B}$ speed E_y/B_z in the X-direction [19]. Thus in the Earth's magnetotail every PSBL particle gains an additional speed of about $2\ E_y/B_z$ or an additional energy of about $\Delta W \approx 2keV$, if the above-mentioned parameters are used. Recently it has been shown that ions in the keV energy range and very energetic electrons in the current sheet of the Earth's magnetotail are chaotic [22,23]. For them

$$0.1 < \kappa_j = \sqrt{R_{min}/\rho_{max,j}} < 1, \qquad (3)$$

where $R_{min} = L_z \cdot B_z/B_o$ is the minimum curvature radius of the magnetic field lines and $\rho_{max,j}$ the maximum Larmor-radius $\rho_{max,j} = \rho_{nj}$. In this parameter range the chaotic scattering at the current sheet is modulated in dependence on κ_j, i.e. its initial energy [24,25,26]. Plasma mantle particles are entering the plasma sheet due to the velocity filter effect at different distances from the Earth along the tail. Due to the $\kappa-$ modulated scattering they become either scattered back to the PSBL, as in the Lyons and Speiser model [19], or they become trapped on cucumber orbits inside the CPS [27]. Which way they go depends on their κ value at the moment they enter the neutral sheet. As a result PSBL beams may exhibit spatial structures, beamlets [28]. In the CPS, on the other hand, the lack of particles accelerated to the PSBL at special velocities, creates ring-type structures in the distribution function, observed by satellites [24]. The transition between scattering to PSBL and to CPS should cause interrupted high speed streams, which seem to correspond to the observations of flow bursts in the magnetotail [18]. Notice that the chaotic nature of current sheet scattering leads to a thermalization of particles, trapped on cucumber orbits. Therefore, an enhancement of the particle speed by about E_y/B_z is transformed to thermal energy of the CPS ions. Consequently current sheet acceleration is also an A.) type process. At the same time it explains the formation of the keV component of the PSBL beams and heating of the CPS ions to keV energies. Current sheet acceleration is, however, not very appropriate to accelerate particles to high energies. Stationary plasma flow through a current sheet is a weak generator of electric fields and an inefficient accelerator as well.

3. STATIONARY RECONNECTION

Reconnection was originally suggested as a process inducing electric fields by merging magnetic flux in the solar atmosphere [29]. In the magnetosphere it merges solar wind and magnetospheric plasma at the magnetopause, in the magnetotail it re–connects the merged plasma and magnetic flux back to magnetosphere and solar wind [30]. The latter seems to occur at a distance of about 100 R_E downstream the tail. The Parker–Sweet model describes theoretically stationary reconnection through an elongated thin, resistive diffusion layer (half–thickness: L_z) [31,32]. The strength of the induced electric field is proportional to the reconnection rate, i.e. the Alfvénic Mach number in the inflow (upstream) region $M_o = u_o/v_A = u_e \cdot B_e/(v_A \cdot B_o)$, where u_o, B_o, v_A denote flow velocity, magnetic field strength and local Alfvén speed upstream and B_e, u_e the corresponding parameters downstream the reconnection region. The Parker–Sweet reconnection rate is small, $M_o \cong 1/\sqrt{R_m}$ [31,32], and the resulting electric field $E_{PS} = u_o \cdot B_o = v_A \cdot B_o/\sqrt{R_m}$ is, for the magnetotail current sheet parameters used in section 2., as small as $0.02 mV/m$. Over a tail width of 36 R_E this corresponds to a voltage drop of 7 kV, which is much less than the cross–tail potential mapped from the ionosphere.

In the Petchek model of stationary reconnection [33] the diffusion layer is much smaller than the whole reconnection region, which now contains two pairs of slow mode shocks. The Alfvénic Mach number in the inflow region far from the diffusion layer of Petchek reconnection is $M_o = \pi/8 \ln R_m$ [34]. The electric field strength in the Petchek model is, therefore,

$$E_y = E_P = u_o \cdot B_o = u_e \cdot B_e = \frac{\pi}{8 \ln R_m} \cdot v_A \cdot B_o = \Gamma \cdot v_{Ao} \cdot B_o. \qquad (4)$$

We will further use the typical value $\Gamma \approx \pi/8 \ln R_m \approx 0.1$. The field strength in the tail lobes is $B_o \cong 10 nT$, but the plasma density is very low there, just about $10^4 m^{-3}$. Hence the Alfvén speed is as high as $3 \cdot 10^6 m/s$ and $E_P = 3\ mV/m$. The corresponding total cross tail potential drop would be larger than 1 MeV! It is not obvious, however that particles can be accelerated along the whole X-line, i.e. over the whole potential drop. Acceleration in this potential drop is limited by the real particle shift in the electric field direction. In contrast to parallel electric fields in the case of auroral acceleration, efficient, almost run—away acceleration in the electric field of reconnection is possible only near neutral lines. Taylor-expanding the magnetic field near a X-line one obtains

$$\mathbf{B}(X,Z) = B_o \frac{Z}{L_z} \cdot \mathbf{e}_x + B_n \frac{X}{L_x} \cdot \mathbf{e}_z \qquad (5)$$

where $L_A = (\partial \ln |\mathbf{B}|/\partial A)^{-1}$ $(A = X, Z)$ are characteristic scale lengths of the magnetic field inhomogeneity near the X-line. The ratio $\tan \alpha = L_z/L_x$ determines the separatrix angle α. The constant electric field $E_y \mathbf{e}_y$ for Petchek-reconnection, given by Eq. (4), causes a $[\mathbf{E} \times \mathbf{B}]$ drift of the magnetized plasma. For $X = 0$ plasma approaches the neutral plane along the Z direction at a drift speed $u_z = E_y L_z/B_x(Z)$, which due to $B_x \propto Z$ increases with the decrease of $Z \to 0$. This is true until the plasma magnetization and drift approach break down. Let us call the region of non–adiabatic motion around the X–line, where particles can be efficiently accelerated in E_y, the acceleration region, which we further call AR. Near the X–line particle motion in the Z–direction is stable, but in the X–direction it is unstable, expelling particles to the edge of the AR. This ejection from the AR terminates the acceleration and particles start to drift again [35]. In order to determine the moment of ejection, it is appropriate to neglect the Z–oscillations in the equations of motion. In this limit and assuming $v_y(t=0) = 0$ (because we are looking for acceleration to suprathermal energies) the equations of motion in the $X - Y$ plane can be reduced to a linear acceleration law in the Y–direction and and an Airy–type differential equation, describing the ejection from the X–line region [36–38]:

$$v_y = \frac{dY}{dt} = \frac{e_j}{m_j} \cdot E_y \cdot t = \Gamma \cdot \Omega_{oj} \cdot v_A \cdot t \tag{7a}$$

$$\frac{d^2 X}{dt^2} = b_n \cdot \Omega_{oj}^2 \cdot \frac{X}{L_x} \cdot \frac{E_y}{B_o} \cdot t = \frac{X \cdot t}{\tau_o^3}, \tag{7b}$$

$$\tau_o = \sqrt[3]{\frac{L_x \cdot B_o}{b_n \cdot \Omega_{oj}^2 \cdot E_y}} = \sqrt[3]{\frac{L_x}{b_n \cdot \Omega_{oj}^2 \cdot \Gamma \cdot v_A}}, \tag{7c}$$

where $b_n = B_n/B_o$. For $t > \tau_o$ the solution of Eq. (7b) is growing exponentially fast ($X \propto \exp\{t/\tau_o\}$), i.e. at about τ_o particles become ejected from the AR. Effective acceleration in the AR is possible, therefore, only at times $t < \tau_o$. From Eqs. (7) one can estimate the maximum possible velocity gain in the AR

$$v_y|_{max} \cong \frac{e_j}{m_j} \cdot E_y \cdot \tau_o = \left(\frac{\Gamma}{b_n}\right)^{2/3} \sqrt[3]{v_A^2 \cdot \Omega_{oj} \cdot L_z} \tag{8}$$

and the resulting maximum non–relativistic energy gain is

$$\Delta W_j \cong \frac{m_j}{2} b_n^{-4/3} \cdot (\Omega_{oj} L_z)^{2/3} \left(\frac{E_y}{B_o}\right)^{4/3} = \frac{m_j}{2} \left(\frac{\Gamma}{b_n}\right)^{4/3} \cdot \sqrt[3]{v_A^2 \Omega_{oj} L_z}^2 \tag{9}$$

The scaling of $\Delta W_j \propto E_y^{4/3}$ has recently been verified by particle simulations [37]. The scaling law of peak energies obtainable by stationary reconnection (Eq. (9)) reveals a maximum possible energy gain at a distant tail neutral line of about 100 keV, if $v_A = 3 \cdot 10^6 m/s$, $b_n = \Gamma = 0.1$, $\Omega_{oi} = 1 rad/s$, $L_z = 6 \cdot 10^6 m$. This is less than the MV total cross–tail potential would allow, but it is still more than current sheet acceleration can provide (cf. section 2.). Also it is the correct order of higher ion energies, observed in the PSBL flows. Stationary reconnection at a distant neutral line and current sheet acceleration along the whole tail together are thus able to accelerate all components of the PSBL flows.

4. NON–STATIONARY LAMINAR RECONNECTION

Stationary reconnection and current sheet acceleration are, however, unable to explain the bursty appearance of MeV ion flows, observed in the tail [2,3]. At the same time there is a theoretical discussion going on whether fast steady reconnection at high Reynolds numbers is possible at all [39]. In fact, numerical simulations revealed a tendency that from an initial Petchek reconnection long current sheets develop, which later switch to a highly dynamic or turbulent regime [40]. This is consistent with the observation of reconnection in form of flux transfer events at the magnetopause or the sudden detachment of plasmoids in the tail at distances between 10 and 30 R_E downstream. Non–stationary reconnection might induce larger electric fields and test particle calculations indeed indicate that non–stationary reconnection may accelerate particles to high energies.

Test particle calculations in prescribed, time dependent magnetotail–like reconnection fields have been carried out by Pellinen and Heikkila [41], imposing a constant rate of the changing magnetic field (about $10 nT/min$). In their model protons gained a maximum energy of 60 keV after being accelerated by moving almost parallel to the X–line. Proton acceleration near the O – type neutral line, however, was restricted by gyration in the normal field, deflecting and trapping particles near the neutral line. Electrons left the X– and the O– line regions quickly after reaching an energy of 5 - 15 keV. In order to obtain a more realistic impression about the rate of change of the magnetic field and acceleration, test particle orbits in simulation models of non–stationary magnetotail reconnection have been carried out [42-46]. They revealed that protons at a distance less than a gyroradius from the X–type neutral line can gain energies up to 200 keV [42]. In three–dimensional MHD reconnection models particles may turn back, being accelerated a second time [43]. In a two–dimensional MHD simulation of bursty reconnection test protons gained energies up to 300 keV after a single X–line has split into a pair of two during plasmoid formation [44]. Most energetic protons drifted over large distances parallel to the X–line and became later trapped within the forming plasmoid. The resulting spectrum depends strongly on the source plasma distribution [45]. Energetic protons flowing along the boundary of plasmoids have a power law spectrum $n(W_p) \propto W_p^{-4}$ if the initial distribution in the tail lobe was $n(W_p) \propto W_p^{-3}$. Comparing test particle calculations with energetic particle measurements on board of AMPTE/IRM, the observed ratio between H^+, He^+ and O^+ fluxes could be understood for a cross tail potential along the X–line of 200 kV [46]. A $2\frac{1}{2}$ dimensional particle code has been used to simulate acceleration at single and multiple X–lines self-consistently [47]. Power law distribution functions were obtained, for protons $n(W_p) \propto W_p^{-4.4}$ and for electrons $n(W_e) \propto W_e^{-3.8}$.

All simulations have shown that, as in stationary reconnection, the most efficient acceleration by non–stationary reconnection takes place near an X–line. Thus we consider scaling laws of the near X–line acceleration of non–stationary reconnection. As a time dependend perturbation we choose a tearing mode wave form which modifies the magnetic field near the neutral plane as follows

$$\mathbf{B} = B_o \frac{Z}{L_z} \cdot \mathbf{e}_x + B_1(t) \cdot \sin(k \cdot X) \cdot \mathbf{e}_z \qquad (10)$$

Like in the case of stationary reconnection we neglect Z-oscillations near the forming X-line and obtain the reduced equations of motion in X, Y

$$\frac{d^2 X}{dt^2} = \Omega_{oj} \cdot \frac{B_1(t)}{B_o} \cdot kX \cdot v_y \qquad (11a)$$

$$v_y = \frac{dY}{dt} = \frac{\Omega_{oj}}{k} \frac{B_1(t) - B_1(t=0)}{B_o} + v_{yo} \qquad (11b)$$

As one can see particle ejection from the neutral line (Eq. 11a) and acceleration in the Y-direction (Eq. 11b) depend directly proportional on the final value of the magnetic field perturbation $B_1(t)$.

In the magnetosphere a linear ion tearing mode instability was proposed to pinch the current sheet during geomagnetically disturbed times at a down-tail distance between 10 and 30 R_E [48]. During the linear growth phase of the instability the time dependence of the magnetic field perturbation is given by

$$B_1(t) = B_{1oo} \cdot \exp\left\{\frac{t}{t_{lin}}\right\}, \quad \text{where} \quad t_{lin}^{-1} \cong \Omega_{oj} \left(\frac{\rho_{oi}}{L_z}\right)^{5/2} \cdot \left(1 + \frac{T_e}{T_i}\right)(1 - k^2 L_z^2) \qquad (12)$$

and B_{1oo} is a small initial perturbation. The maximum obtainable particle energy depends on the value of $B_1(t)$ at the end of the linear instability growth phase. The linear instability growth stops, when the half-width of a magnetic island $\Delta Z = 2\sqrt{B_1/(B_o k L_z)}$ exceeds the extent of the ion interaction region $|Z| \leq \sqrt{L_z \cdot \rho_{oi}}$ [49]. This happens, when

$$B_1(t) = B_1(t_o) = k\rho_{0i} \cdot B_o = B_{1o}, \qquad (13)$$

and determines the transition to a nonlinear stage of the instability [50]. The Y-velocity at the end of the linear stage of the instability ($t = t_o$) can easily be found from Eq. (11b), where the initial perturbation $B_1(t=0) = B_{1oo}$ is negligibly small compared to $B_1(t_o) = B_{1o}$. As one easily can see the energy gain at the end of the linear instability growth is of the order of the initial energy. Hence, the linear tearing mode instability is not an efficient accelerator at all.

Different possibilities exist to continue the tearing mode growth nonlinearly up to its final saturation. Let us consider the consequences of an explosive nonlinear tearing mode growth after $B_1(t) = B_1(t_o) = B_{1o}$ is reached [49-51]. The growth of the field perturbation $B_1(t)$ of the explosive tearing mode is given by

$$B_1(t) = \frac{B_{1o}}{1 - (t - t_0)/\tau_{non}}, \qquad (14a)$$

$$\frac{1}{\tau_{non}} = \Omega_{0i} \cdot \frac{B_{1o}}{B_0} \frac{1 - k^2 L_z^2}{k L_z} \cdot \left(\frac{\rho_{0i}}{L}\right)^{3/2} \cdot \left(1 + \frac{T_e}{T_i}\right), \qquad (14b)$$

where t_o is now the starting moment of the explosive, nonlinear tearing mode instability. For typical magnetotail parameters one finds $\tau_{non} \cong 10 \ldots 100s$, which agrees well with the duration of the observed particle bursts. Combining Eqs. (11a,b) and (14a,b) one obtains a differential equation describing the ejection from the X-line:

$$\frac{d^2 X}{dt^2} = \Omega_{oj}^2 \cdot \left(\frac{B_{1o}}{B_0}\right)^2 \cdot \frac{X}{(1 - t/\tau_{non})} \cdot \left[\frac{1}{(1 - t/\tau_{non})} - \frac{1}{(1 - t_0/\tau_{non})}\right] \qquad (15)$$

Eq. (15) bears different solutions for electrons and ions [52].

Electrons become accelerated almost immediately to relativistic energies and they leave the AR on a time scale t_{acc} much faster than the mode growth time ($t_{acc} \cong L_y/c \cong 4\pi/(kc) \ll \tau_{non}$). During their acceleration they see, therefore, an almost constant electric field. This field is maximum close to the moment of saturation of the nonlinear instability $t = t_{sat}$. An appropriate criterion for the saturation of instability is $B_1(t_{sat}) = B_o$. For $kL_z = 0.33$, $\rho_{oi}/L_z = 0.33$ i.e. $k\rho_{oi} = 0.1$ and $B_{1o}/B_o = k\rho_{oi} = 0.1$ one finds $t_{sat} = 0.9\tau_{non}$. Using the smallness of the acceleration time $t_{acc} \ll t_{non}$, the maximum (relativistic) energy gain of electrons in the nonlinear tearing mode can be written as [52]

$$\Delta W_e = P_y \cdot c \cong 4\pi m_e \left(\frac{\Omega_{oe}}{k}\right)^2 \frac{1 - k^2 L_z^2}{k^2 L_z^2} \left(\frac{\rho_{oi}}{L_z}\right)^{3/2} \left(1 + \frac{T_e}{T_i}\right) \qquad (16)$$

For the typical magnetospheric parameters $L_z = 6 \cdot 10^6 m$ and $B_o = 20 nT$ one finds a maximum obtainable electron energy of several MeV. Taktakishvili and Zelenyi [52] obtained a power law $n(W_e) \propto W_e^{-4}$ for the electron distribution below the peak energy.

In contrast to electrons protons have to spend all the mode growth time in the AR in order to obtain significant energy. The maximum energy gain for (non-relativistic) protons is, if the initial velocity $v_y(t_o)$ is neglected, given by

$$\Delta W_p = \frac{m_p}{2} \cdot v_y^2(t_{sat}) \cong \frac{m_p}{2} \cdot \left(\frac{\Omega_{op}}{k}\right)^2 \cdot \left(1 - k^2 \rho_{op}^2\right)^2 \qquad (17)$$

For parameters typical for the Earth's magnetospheric tail ($B_o = 20 nT$, $\tau_{non} = 30 s$ and $k^{-1} = 10^7 m$) the resulting maximum proton energies are of the order of 3 MeV [52]. The spectrum of the accelerated ions in this model was found to be a power law with spectral exponents between 2 and 7 in the energy interval 0.3 - 1 MeV and 4 - 7 for energies larger than 1 MeV [53]. It is worth mentioning that the ion acceleration near a X-line creates an inverse velocity dispersion, i.e. ions ejected first leave the AR at lower energies than those, ejected close to t_{sat}. Such inverse velocity dispersion was indeed observed in the Earth's magnetotail presumably close to a reconnection event [54].

5. TURBULENT RECONNECTION

Since 1979 Parker [55] argues in favor of rapid reconnection on the scale of $\tau_A = L/v_A$, the Alfvén transit time through the region of maximum inhomogeneity with a spatial scale L. Calculations, carried out by Matthaeus, Goldstein and Ambrosiano [56-58], indicate that a rapid evolution of turbulent MHD reconnection through stages of reconnection and magnetic island coalescence indeed may accelerate particles to high energies. In their model acceleration takes place while particles are trapped in bubbles, formed by reconnection. As a result of test particle computations these authors came to the conclusion that the acceleration scales as $\Omega_o \tau_A$ [56-58]. This is a large number in space plasma, where the gyroperiod usually is much smaller than the Alfvén transit time. The resulting scaling law of peak energies is

[56] $\Delta W_j = 0.5 m_j v_A^2 \Gamma^2 \Omega_{oj}^2 \tau_A^2$ and numerically a power law spectrum $n(W) \propto W^{-2.7}$ was obtained. However, in the two-dimensional turbulent MHD model the test particle motion was not limited in the third ($Y-$) direction. In real situations the particle motion will be either bounded by a finite Y-extent of the acceleration site or by an acceleration time, effectively shorter than τ_A. For $\Gamma\Omega_o\tau_A > 1$ one finds $t_{acc} = \sqrt{0.5\tau_A/\Gamma\Omega_o}$ [57]. Taking into account a possible difference in mass m_j and charge state Z_j of test particles and background plasma ions (m_o, Z_o), respectively, the scaling law in the non-relativistic case is, finally,

$$W_j = \frac{m_j}{2} \cdot v_A^2 \cdot \left[\Gamma \cdot \frac{\omega_{pj} L_y}{c} \frac{Z_j}{Z_o} \frac{m_o}{m_j}\right]^2 \cong \frac{m_j}{2} \cdot \Gamma^2 \cdot \Omega_{oj}^2 \cdot L_y^2 \qquad (18)$$

The second expression is a simplification obtained from the first by putting $Z_o = Z_j$ and $m_o = m_j$. Notice that this expression corresponds to the acceleration law (7a) for $t = t_{acc} = \tau_A$, where τ_A is the Alfvén transit time in the Y-direction. Notice that the upper estimate for the peak energy, obtainable by turbulent reconnection, is of the same order as the estimate of Eq. (17), obtained for explosive reconnection, if one takes into account that the length of the neutral line is of the order of $L_y \approx 2\lambda = 2\pi/k$ [50]. The scaling of the maximum obtainable energy from turbulent reconnection would coincide with that of stationary Petchek reconnection (combining Eqs. (4) and (7)), if particles could freely move along the X-line over a distance of L_y. As we have seen in section 3., however, in stationary X-line reconnection particles are expelled out of the AR after $t_{acc} \cong \tau_o$ (cf. Eq. (7c)). Turbulent reconnection, instead, seems to allow an accumulation of elementary acts of acceleration by reconnection in bubbles. If so, however, it might be more probable that this process does not extract the total available potential drop but just some fraction of it. Hence the maximum obtainable energy gain by turbulent reconnection in the magnetotail is probably less than a MeV [57], i.e. less than an explosive stage of the tearing mode instability could provide.

6. CONCLUSION

Comparing the scaling laws, derived for current sheet acceleration and reconnection, and observations in the Earth's magnetotail one have seen that current sheet acceleration, stationary and non-stationary reconnection including turbulent one accelerate particles in the magnetosphere to different energy levels. In dependence on their efficiency as an electric field generator and as a particle accelerator they contribute to the formation of different parts of the energy distribution. Non-stationary laminar and turbulent reconnection are most appropriate to explain the peak energies in the spectrum of observed energetic particle flows and also their often observed bursty appearance. Due to propagation effects (dispersion) the event analysis of spectrum formation is, unfortunately, a more complicated undertaking. Multi-satellite measurements, planned with the projects INTERBOL and CLUSTER [59] will be appropriate to verify theoretical models of acceleration by means of *in situ* observations.

The successful explanation of particle acceleration in the magnetospheric tail makes it reasonably to apply the obtained scaling laws to other cosmic acceleration sites.

First, let us follow the original proposal of reconnection as a mechanism of solar particle acceleration [29]. A typical value of the solar magnetic field in potential reconnection regions is $B_o \cong 0.1T$, a typical length scale of the acceleration region is $10^5 - 10^6 m$. For an assumed Alfvén speed of $v_A \cong 10^6 m/s$ (for $n \cong 10^{16} m^{-3}$) the reconnection electric field can become as large as $E \cong \Gamma \cdot v_A \cdot B_o \cong 10^4 V/m$ and the potential $10^9 - 10^{10} V$. In this field protons can be accelerated up to energies of the order of several $1 Gev$, while the energy of relativistic electrons is limited by synchrotron losses to $1 - 2 GeV$ [60].

Flare-like energy releases have been observed at red dwarfs. For an estimated magnetic field strength of $(0.1 - 1)T$ [61], an Alfvén speed probably of the same order as in solar active regions and a field inhomogeneity scale of about $10^5 - 10^6 m$, the resulting energy of the fastest particles would be $10^{12} - 10^{14} eV$. These magnitudes fit well the observed ones [62]. $10^{17} eV$ cosmic rays could be accelerated in galactic magnetic fields, if the reconnection scale length would be of the order of $10^{14} m$.

ACKNOWLEDGMENTS

The author gratefully thanks I. Axford, J. Borovsky, M. L. Goldstein, G. Haerendel, E. Möbius, M. Scholer and L. M. Zelenyi for meaningful discussions and remarks. Part of this research was supported by the Deutsche Agentur für Raumfahrtangelegenheiten, project 50 QN 9102 3.

REFERENCES

1. G. Haerendel, in *Plasma Astrophysics* ESA publications division, **SP-285**, Vol. I 37 (1989).
2. S. M. Krimigis, in American Institute of Physics conference proceedings, **No. 56**, 179 (1979).
3. M. Scholer, in *Magnetic Reconnection in Space and Laboratory Plasmas*, ed. E. W. Hones Jr., American Geophysical Union, Wahington D.C., 216 (1984).
4. D. N. Baker, et al., J. Geophys. Res., **91**, 10771 (1986).
5. E. Möbius, J. Atmosph. and Terrestr. Phys. **52**, (1990).
6. G. Haerendel, ESA Journal, **4**, 197 (1980).
7. D. N. Baker et al., Geophys. Res. Lett., **6**, 531 (1979).
8. E. N. Paulikas, G. A., and J. B. Blake, in *Quantitative Modeling of Magnetospheric Processes, Geophys. Monogr. Ser.* ed. by W. P. Olson, AGU, Washington DC, **21**, 180 (1979).
9. J. E. Borovsky, Los Alamos National Laboratory Preprint No. LA-UR-91-3269 (submitted to: J. Geophys. Res., **97**, 1992).
10. J. L. Burch in *Auroral Physics*, eds. C. I. Meng, M. J. Rycroft, and L. A. Frank, Cambridge University Press, Cambridge, 97 (1991).
11. L. Drury, Rep. Prog. Phys., **46**, 973 (1990).
12. D. C. Ellison, J. C. Jones, and S. P. Reynolds, Ap. J. **360**, (1990).
13. M. Scholer, F. M. Ipavich, G. Gloeckler, and D. Hovestadt, J. Geophys. Res., **85**, 4602 (1980).
14. E. Möbius et al., Geophys. Res. Lett. **14**, 681 (1981).

15. E. Kirsch, S. M. Krimigis, E. T. Sarris, R. P. Lepping, and T. P. Armstrong, Geophys. Res. Lett. **4,** 137, (1977).
16. A. T. Y. Lui et al., J. Geophys. Res. **82,** 1235 (1977).
17. K. Takahashi, and E. W. Hones Jr, J. Geophys. Res. **93,** 8558 (1988).
18. W. Baumjohann, G. Paschmann, and H. Lühr, J. Geophys. Res., **95,** 3801 (1990).
19. L. R. Lyons, and T. W. Speiser, J. Geophys. Res., **87,** 2276 (1982).
20. H. Alfvén, J. Geophys. Res. **73,** 4379 (1968).
21. J. P. Heppner, J. Geophys. Res., **82,** 1115 (1977).
22. J. Büchner, Astron. Nachr. **307,** 191 (1986).
23. J. Chen, and P. J. Palmadesso, J. Geophys. Res. **91,** 1499 (1986).
24. J. Chen, G. Burkhart, and C. Huang, Geophys. Res. Lett. **17,** 2237 (1990).
25. G. Burkhart, and J. Chen, J. Geophys. Res. **96,** 1499 (1991).
26. J. Büchner, Geophys. Res. Lett., **18,** 1595 (1991).
27. J. Büchner, and L. M. Zelenyi, J. Geophys. Res. **93,** 11,821 (1989).
28. M. Ashour-Abdalla, J. Berchem, J. Büchner, and L. M. Zelenyi, Geophys. Res. Lett. **18,** 1603 (1991).
29. R. G. Giovanelli, Mon. Not. R. Astron. Soc., **107,** 338 (1947).
30. J. W. Dungey, Phil. Mag., **44,** 725 (1953).
31. E. N. Parker, Astrophys. J. Suppl., **8,** 177 (1963).
32. P. A. Sweet, in *Electromagnetic phenomena in Cosmical Physics,* ed. B. Lehnert, Cambridge University Press, New York, 123 (1958).
33. H. E. Petchek, *NASA Spec. Publ.* **SP-50,** 425 (1964).
34. V. M. Vasyliunas, Rev. Geophys. **13,** 303 (1975).
35. B. U. Ö. Sonnerup, J. Geophys. Res. **76,** 8211 (1971).
36. S. V. Bulanov, and P. V. Sasorov, Sov. Astron. **19,** 464 (1976).
37. G. R. Burkhart, J. F. Drake, and J. Chen, J. Geophys. Res. **95,** 18,833 (1990).
38. H.-J. Deeg, J. E. Borovsky, and N. Duric, Phys. Fluids B **3,** 2660 (1991).
39. D. Biskamp, Phys. Lett., **105A,** 124 (1986).
40. M. Scholer, Geophys. Astrophys. Fluid Dynamics, **62,** 51, (1991).
41. R. J. Pellinen, and W. J. Heikkila, J. Geophys. Res. **83,** 1544 (1978).
42. T. Sato, H. Matsumoto, and K. Nagai, J. Geophys. Res. **87,** 6089 (1982).
43. J. Birn, and E. W. Hones Jr., J. Geophys. Res., **86,** 6802 (1981).
44. M. Scholer, and F. Jamnitzky, J. Geophys. Res. **92,** 12,181 (1987).
45. M. Scholer, and F. Jamnitzky, J. Geophys. Res. **94,** 2459 (1989).
46. D. Sachsenweger, M. Scholer, and E. Möbius, Geophys. Res. Lett. **16,** 1027 (1989).
47. D. C. Ding, and L. C. Lee, Adv. Space. Res. **11,** 117 (1991).
48. K. Schindler, J. Geophys. Res. **79,** 2803 (1974).
49. A. A. Galeev, F. C. Coroniti, M. Ashour-Abdalla, Geophys. Res. Lett. **5,** 707 (1978).
50. A. A. Galeev, Space Sci. Rev., **23,** 411 (1979).
51. T. Terasawa, J. Geophys. Res. **86,** 9007 (1981).

52. A. L. Taktakishvili, and L. M. Zelenyi in *Reconnection in Space Plasma,* Proc. of the International Workshop, September 1988, Potsdam, **ESA SP – 285, Vol. II,** 227 (1989).
53. L. M. Zelenyi, A. S. Lipatov, D. G. Lominadse, and A. L. Taktakishvili, Planet. Space Sci. **32** 313 (1984).
54. D. V. Sarafopoulos, and E. T. Sarris, Planet. Space Sci. **36,** 1181 (1988).
55. E. N. Parker, in *Proc. 22th Int. Cosmic Ray Conf.,* Clarendon, Oxford **5,** 35 (1979).
56. W. H. Matthaeus, J. L. Ambrosiano, and M. L. Goldstein, Phys. Rev. Lett. **53,** 1449 (1984).
57. M. L. Goldstein, W. H. Matthaeus, J. L. Ambrosiano, Geophys. Res. Lett. **53,** 1449 (1984).
58. J. L. Ambrosiano, W. H. Matthaeus, M. L. Goldstein, and D. Plante, J. Geophys. Res. **93,** 14,383 (1988).
59. J. Büchner, in *Proc. Int. Workshop on Space Plasma Physics Investigations by Cluster and Regatta,* Graz, Austria, 2.–22. February 1990, **ESA SP-306** 117 (1990).
60. S. V. Bulanov, and F. Cap, Adv. Space Res. **65,** 837 (1984).
61. D. J. Mullan, Astron. and Astrophys. **40,** 41 (1985).
62. S. V. Bulanov, and F. Cap, Astronomicheski Journal **65,** 837 (1988).

PARTICLE ACCELERATION IN THE HELIOSPHERE

Martin A. Lee,
Institute for the Study of Earth, Oceans and Space,
University of New Hampshire, Durham, NH 03824

ABSTRACT

This review presents an overview of the various populations of energetic particles accelerated in the heliosphere: energetic storm particles, corotating ion events, upstream ions at Earth's bow shock, the cosmic ray anomalous component, and interstellar and cometary pickup ions. In each case the review presents or describes original and/or key observations, the current understanding of its origin, and its particular impact on theoretical progress.

1. INTRODUCTION

The heliosphere is that volume of space dominated by gas from the Sun. From the point of view of energetic particles, it consists of the solar wind (generally neglecting the Sun!), the solar wind termination shock, and the shocked subsonic gas extending to the heliopause. The heliopause is the division between solar and interstellar plasma and magnetic field. Twenty years ago the heliosphere was viewed as a passive domain for energetic particles. Solar energetic particles were known to be injected periodically at its center, galactic cosmic rays continually penetrated from outside, and occasional effluence leaked from Earth's magnetosphere. The heliosphere was considered to be responsible for merely dispersing and cooling these particle populations. It was a degrading receptacle!

However, in the last twenty years our view of the heliosphere has changed dramatically. Space exploration has discovered a great variety of new energetic particle populations accelerated at different sites throughout the heliosphere: "energetic storm particle" (ESP) events associated with traveling shocks throughout the heliosphere (their discovery actually harks back to the earliest era of space discovery); "corotating" ion events associated with the forward and reverse shocks bounding corotating interaction regions (CIR) in the solar wind; ion enhancements associated with "merged interaction regions" (MIR) in the outer heliosphere; the "field-aligned beams" (or "reflected", R), "intermediate" (I) and "diffuse" (D) ions at Earth's and other planetary bow shocks; the cosmic ray "anomalous" component; cometary and interstellar pickup ions (pickup ions are hardly energetic but their acceleration is a topic of current interest). Figure 1 is a schematic diagram of the heliosphere showing these populations of accelerated particles and their approximate acceleration sites. In this review we do not discuss solar flare particles, galactic cosmic rays, and magnetospheric energetic particles, which are generally not considered of "heliospheric" origin. They are the subject of articles in this volume by Melrose, Axford and Büchner, respectively.

28 Particle Acceleration in the Heliosphere

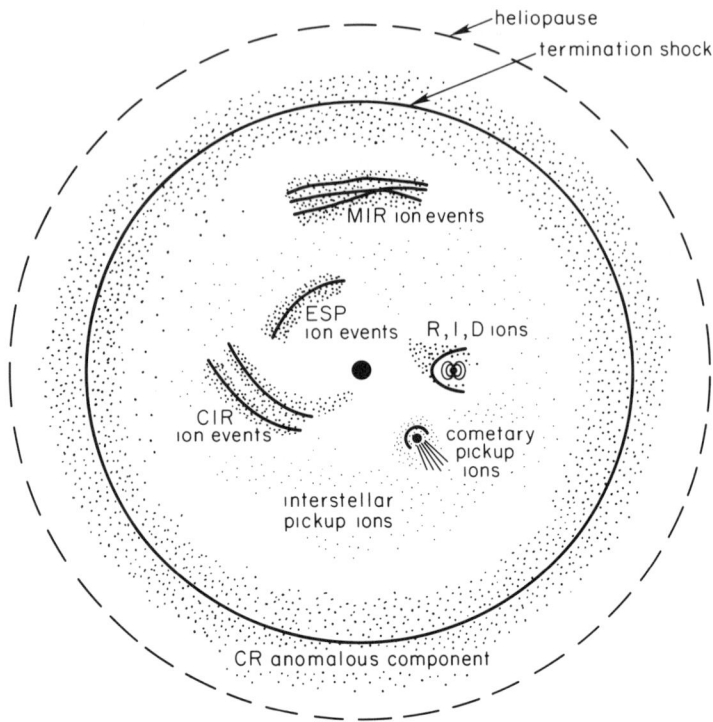

Particle Acceleration in the Heliosphere

Fig. 1. Heliospheric energetic particle populations with the Sun at the center. Solid lines indicate shocks.

Most of these heliospheric energetic particle populations are directly associated with collisionless shock waves. The origin of the cosmic ray anomalous component is thought to be the solar wind termination shock[1]. Thus, with the exception of interstellar pickup ions there is nearly a one-to-one correspondence between shocks and energetic particle populations in the heliosphere. Since the solar wind is super-fast-magnetosonic it is no surprise that it is riddled with shock waves. But originally it was a surprise that the shocks are such prolific particle accelerators. The study of these heliospheric particles moved forward with advances in acceleration and transport theory in the 1970's and 1980's. Thus the heliosphere has provided an effective laboratory for detailed in situ studies of particle acceleration which form a basis for applications throughout the Cosmos.

In this introductory review we shall tour the heliosphere and its energetic particle populations. For each population we shall describe its basic characteristics, highlight its discovery and/or key observations, and emphasize its role in advancing our understanding of acceleration processes.

2. ENERGETIC STORM PARTICLE EVENTS

Figure 2 from Bryant et al.[2] shows an early Explorer 12 measurement of an "energetic storm particle" (ESP) event, so named because of the close temporal correspondence of the event with magnetic storm sudden commencement (SC) at Earth as shown in the lowest panel. Solar flare particles are recorded from the 3+ flare indicated in the lowest panel. The superposed ESP enhancement also occurs at the same time as a Forbush decrease in the cosmic ray count rate recorded by the Deep River neutron monitor. The energy spectrum of the ESP event is clearly softer than that of the flare particles and contains very few protons with energies greater than ~100 MeV.

The origin of ESP events was very early postulated to be a collisionless shock[3], which could also serve as the origin of the sudden commencement and the Forbush decrease. Fisk[4] was the first to provide a quantitative theory for the shock origin of ESP events. He connected solutions of the energetic particle transport equation[5,6] across the shock and derived expressions for the particle differential intensity. If he had not assumed a power-law energy spectrum and not allowed a discontinuity in particle intensity at the shock, he would have discovered the mechanism of diffusive shock acceleration in which particles scatter across the shock many times and are accelerated by being coupled to the shock compression. Scholer and Morfill[7] also came close with their work on ESP events. However, the discovery of that mechanism and its ability to produce power-law energy spectra had to await the work of Axford et al.[8], Krymsky[9], Blandford and Ostriker[10], and Bell[11] on the origin of galactic cosmic rays.

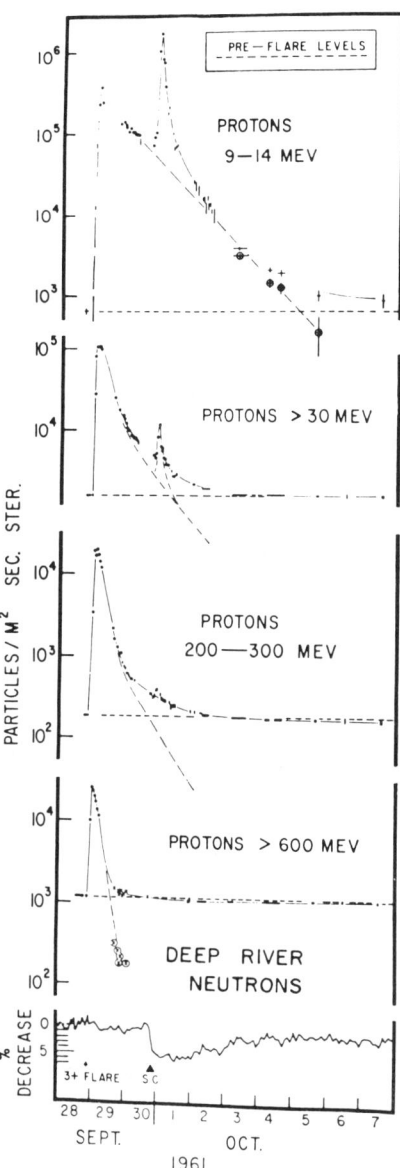

Fig. 2. ESP event of 30 September 1961 superposed on solar flare protons from the 3+ flare of 28 September 1961 (Fig. 18 from Bryant et al.[2]).

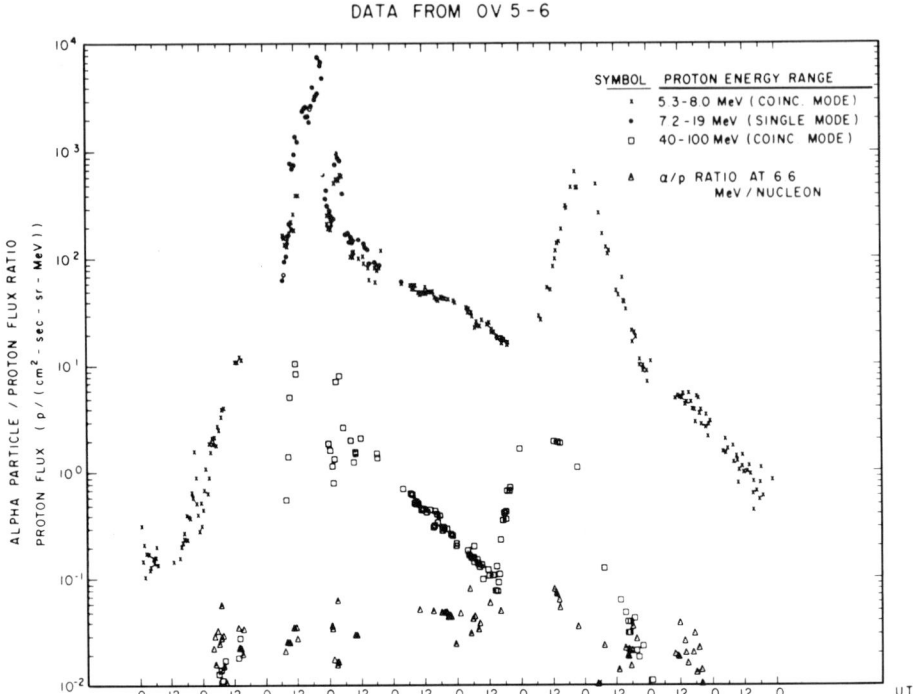

Fig. 3. ESP event of 4 August 1972 (Fig. 1 from Yates et al.[26]).

Fisk actually derived particle scattering mean free paths near the shock which were much too small. The events he chose to address were those later called "shock spike events", which have short durations ≤ 1 hour near Earth. Sarris and Van Allen[12], Pesses et al.[13] and Decker[14] interpreted these increases as transient reflections of solar flare ions from the shock front whenever the shock is locally quasi-perpendicular. Particles are mirrored and reflected from the increased field strength at a "fast" shock and gain energy by drifting parallel to the electric field in the shock frame[15]. This acceleration mechanism became known as "shock drift" acceleration.

For several years there was much controversy between the proponents of the two types of shock acceleration, diffusive and drift, concerning which type properly described ESP events. A major step forward was made by Jokipii[16], who showed that both mechanisms are included in the energetic particle transport equation if the drift term is kept, and that the relative contribution of each is frame dependent. The current understanding, based on extensive measurements of ESP events by ISEE-3[17,18], is that diffusive acceleration of solar wind ions dominates ESP events at quasi-parallel shocks, whereas shock drift acceleration of energetic solar flare ions dominates ESP events at quasi-perpendicular shocks.

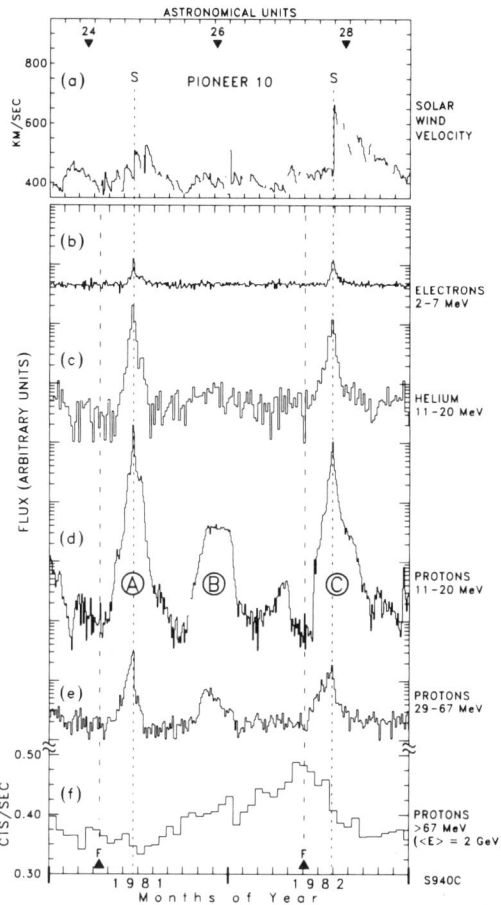

Fig. 4. ESP events associated with shocks in the outer heliosphere as observed by Pioneer 10 (Fig. 1 from Pyle et al.[30]).

Based on the work of Bell[11], Lee[19] developed a comprehensive theory for ESP events at quasi-parallel shocks including diffusive shock acceleration and wave excitation by the accelerated ions so that the ion spatial diffusion coefficient is determined self-consistently. Kennel et al.[20,21,22] performed the most thorough ever study of an interplanetary traveling shock (the famous shock of 11, 12 November 1978) and its associated ESP event; they determined that the energetic particle and magnetic field fluctuation data were generally in good agreement with the theory of Lee[19]. Furthermore, Tan et al.[23] found that spatial diffusion coefficients derived from observed gradients of several ESP events were in very good agreement with those predicted theoretically based on wave excitation by the accelerated ions.

However, some subtle discrepancies between observations of ESP events and current application of the theory of diffusive shock acceleration have been uncovered by Tan et al.[24]. Tan et al. and Sanahuja and Domingo[25] observed in several events that the energetic storm particles downstream of the shock are isotropic in a frame moving with respect to the solar wind plasma along the magnetic field and away from the shock with a speed which correlates with, but is larger than, the local Alfvén speed. It is unclear why this frame is the frame in which the energetic particles scatter to isotropy. Also the compression ratio determined by this frame downstream and the unstable Alfvén waves upstream yields a power-law index for the ESP event energy spectrum which does not agree with observations.

The huge ESP event of 4 August 1972 pictured in Figure 3 from Yates et al.[26] also played a special role in the study of shock acceleration in the heliosphere. Eichler[27] showed that the shock was a "mediated" shock in which the energetic particle pressure is comparable with the thermal plasma pressure and modifies the structure of the shock. "Mediated"

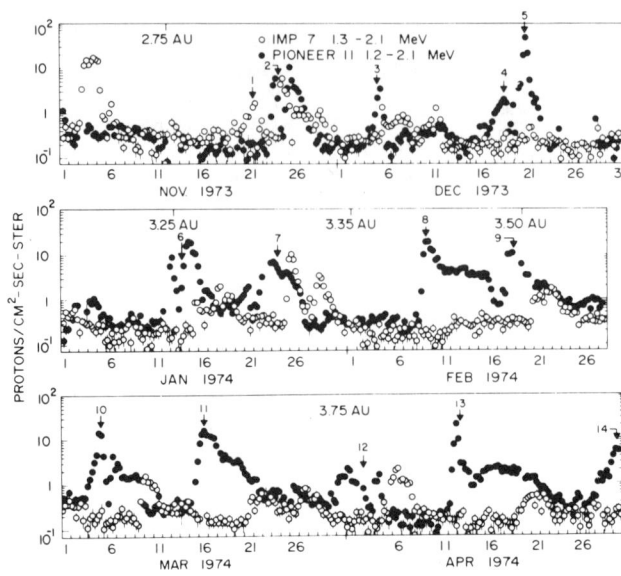

Fig.5. Proton intensities measured by Pioneer 11 (with IMP 7 as a reference at 1 AU) showing the corotating ion events (Fig.2 from McDonald et al.[36]).

shocks are very rare in the heliosphere but are expected to be the origin of galactic cosmic rays in interstellar space[28,29].

ESP events continue to be present in the outer heliosphere although their occurrence rate is reduced due to decay and coalescence of shock waves. Figure 4 from Pyle et al.[30] shows energetic particle events measured by Pioneer 10 between heliocentric radial distances of 24-28 AU. It is clear that periods of solar activity produce a coalescence of flare and/or shock produced energetic particles in the outer heliosphere. These events (A, B and C with a small event between B and C in the Figure) have been called "super events" by Müller-Mellin et al.[31] and exhibit clearly the 154-day periodicity in solar flare activity first noted by Rieger et al.[32] for flare associated γ-ray emission. It is clear that events A and C with an entrained shock wave show dramatic ESP enhancements of protons, helium and electrons.

Interplanetary traveling shocks driven by flares or coronal mass ejections may accelerate particles more efficiently when they are closer to the Sun. Indeed, Mason et al.[33], Lee and Ryan[34], and Cane et al.[35] have all argued from different points of view that coronal shocks are responsible for the acceleration of solar flare particles in so-called "gradual" events.

3. COROTATING ION EVENTS

One of the most surprising of the early results from the Pioneer spacecraft was the discovery that the MeV proton flux often increased with heliocentric radial distance. The expectation had been that protons in this energy range are solar flare particles which decrease in intensity with distance from the Sun. Figure 5 from McDonald et al.[36] compares IMP 7 data at 1AU with Pioneer 11 data at the indicated distances for the period 1 November 1973 to 1 May 1974. The energetic particle events shown exhibit a 26-day periodicity with generally two peaks per period and were called "corotating ion events" implying a spatial structure which corotated with the Sun. At somewhat greater heliocentric radial distances, shown in Figure 6 from Barnes and Simpson[37], the events were found to correlate very closely with corotating interaction regions (CIR) in the solar wind. These regions are formed when a coronal hole extends to the heliographic equator and its fast solar wind creates a pattern of compression and rarefaction in the solar wind over several solar rotations. Figure 6 shows the dominance of solar flare particle events in the temporal profile measured by IMP 8 at 1 AU and the dominance of the corotating events at 4-6 AU. The double-peaked structure of the corotating events occurs because the events peak at the forward and reverse shocks bounding the CIR, a clear signature of shock acceleration.

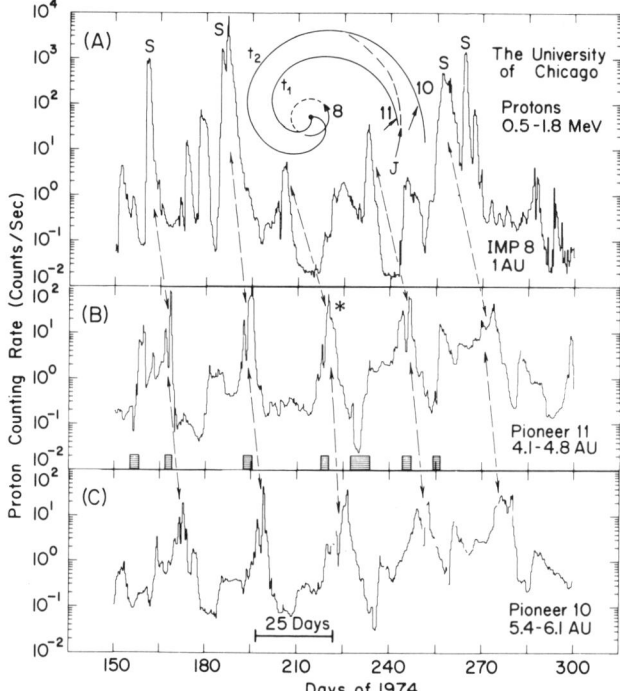

Fig. 6. Corotating events measured by Pioneers 10 and 11 showing ~25 day periodicity and double-peaked structure (Fig. 2 from Barnes and Simpson [37]).

Palmer and Gosling[38] initially proposed a shock origin of the corotating events and interpreted the ion gradients and anisotropies as resulting from a balance between convective and diffusive transport of ions reflected and accelerated at the shock. Fisk and Lee[39] employed the relatively new theory of diffusive shock acceleration to account successfully for the gradients, anisotropies and energy spectra of the events. Rather than the power-law spectra generally produced by shock acceleration, the corotating ion events have spectra which are exponential in particle speed[40,41,42] resulting from a balance between shock acceleration and adiabatic deceleration in the diverging solar wind. Scholer et al.[43] demonstrated that corotating event intensity correlated with solar wind temperature, implying that the shocks accelerate the ions out of the solar wind plasma. The mystery of how relatively weak quasi-perpendicular shocks are able to accelerate ions out of the solar wind remains. It is a mystery we shall return to in §5.

The corotating interaction regions become more complex in the outer heliosphere as the forward and reverse shocks spread and cross each other and parcels of solar wind plasma are multiply shocked. These regions are similar to those associated with the "super events" which result from a coalescence of flare-produced shock waves. Such regions in the outer heliosphere have been dubbed "merged interaction regions" (MIR) by Burlaga and colleagues[44]. These regions affect the modulation of galactic cosmic rays and are likely to accelerate particles at their entrained shocks and turbulence.

4. ENERGETIC PARTICLES UPSTREAM OF EARTH'S BOW SHOCK

Although energetic particles in the energy range 10-100 keV/charge and associated ultra-low-frequency (ULF) waves had been observed prior to 1978 in the vicinity of Earth's bow shock[45,46], it was the International Sun Earth Explorer (ISEE) Mission which was dedicated to understanding their behavior. ISEE-1 and ISEE-2 could measure particle and wave intensities along the sub-solar bow shock under a variety of solar wind conditions and unravel temporal and spatial structure. ISEE-3 could measure upstream particles as they escaped the shock into interplanetary

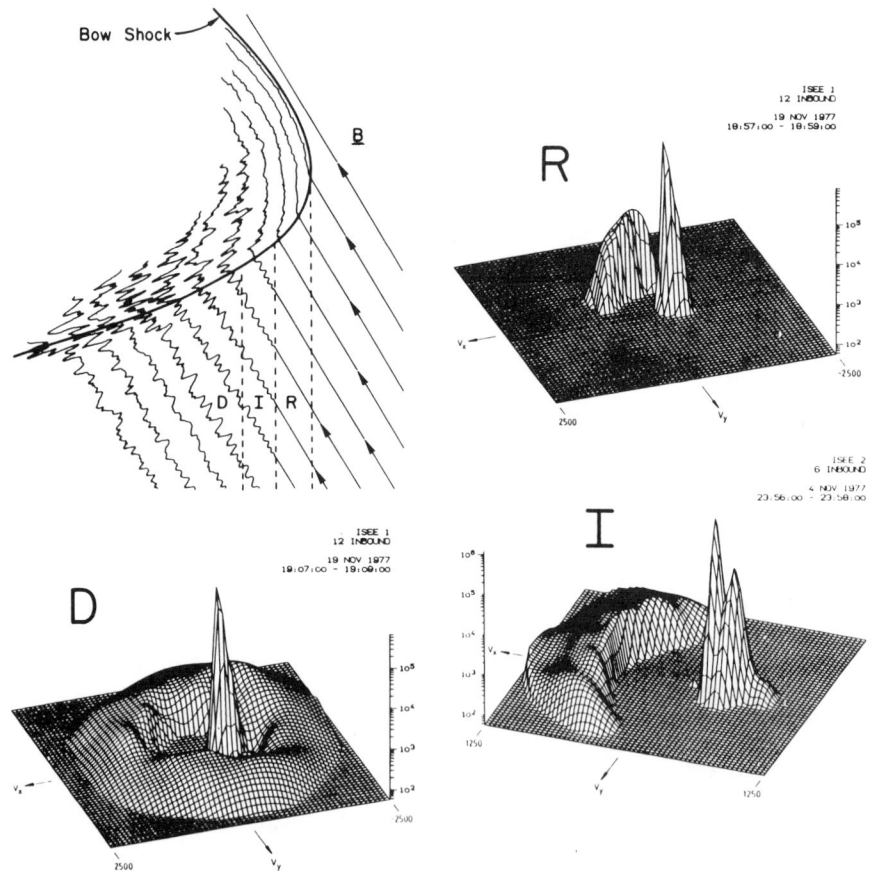

Fig. 7. Reflected (R), intermediate (I) and diffuse (D) ion distributions and their locations upstream of Earth's bow shock (Figs. 4, 13, and 7 from Paschmann et al.[47]).

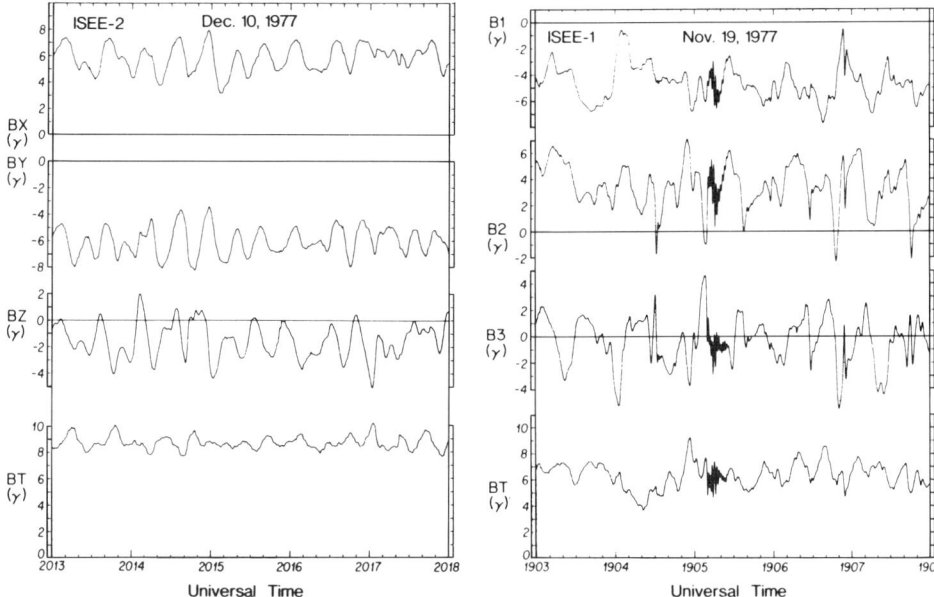

Fig. 8. Fluctuations in **B** associated with "intermediate" ions (Fig. 15 from Hoppe et al.[48]).

Fig. 9. Fluctuations in **B** associated with "diffuse" ions (Fig. 20 from Hoppe et al.[48]).

space. An initial collection of papers entitled "ISEE Upstream Waves and Particles" was published as a special issue of the Journal of Geophysical Research (86, 4317-4536, 1981).

Figure 7 shows the three classes of energetic particle distribution functions (taken from Paschmann et al.[47]) observed and a schematic diagram of the regions of space where they are found. Each class is represented by a contour plot of $v^4 f(v_x, v_y, v_z = 0)$, where $f(\mathbf{v})$ is the distribution function. In each case the major enhancement is the solar wind, in direction $-V_{sw} \hat{e}_x$. The reflected (R) ions are observed to stream along the upstream magnetic field **B** away from the shock at speeds of 1-2 V_{sw}. The reflected ions were later dubbed "field-aligned beams" (FAB) to distinguish them from the "reflected" ions transmitted through quasi-perpendicular supercritical shocks. The intermediate (I) ions are similar to R ions but have undergone some pitch-angle scattering in the forward hemisphere in velocity space. Figure 8 (from Hoppe et al.[48]) shows the large-amplitude transverse wave forms with periods of 10-20s typically associated with the I ions. The diffuse (D) ions are nearly isotropic and extend to energies of ~ 100 keV/charge. Figure 9 (also from Hoppe et al.[48]) shows the wave forms associated with the diffuse ions, which have evolved to reveal complex structures including the so-called "high-frequency wave packets" as shown[48,49]. The composition, spectra and spatial distribution of the upstream ions are described by Ipavich (this volume).

36 Particle Acceleration in the Heliosphere

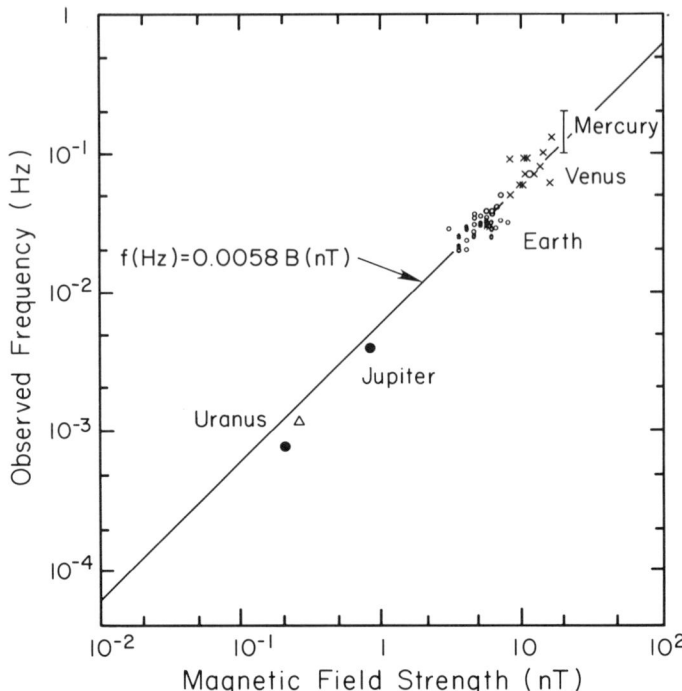

Fig. 10. Dominant frequency of upstream waves plotted versus |B| at various planets (Fig. 8 from Russell et al.[65]).

Based on the ISEE observations the picture which emerged was that these distributions represent temporal phases of shock acceleration as a given flux tube first contacts and then sweeps across the nose of the shock toward the flanks. Field-aligned beams are produced initially as solar wind ions reflect from the quasi-perpendicular shock near the point of field tangency[50]. These ions excite right-hand circularly polarized waves[51,52,53,54] which scatter ions that follow to form the intermediate ion distributions. The waves in Figure 8 are left-hand circularly polarized due to Doppler shift in the spacecraft frame. The intermediate ions then scatter into the backward hemisphere, exciting the opposite wave polarization, returning ions to the shock to initiate the process of diffusive shock acceleration, and thereby producing the diffuse ions. Thus the bow shock provides a laboratory to study shock acceleration (and injection) under conditions of limited geometry and finite acceleration time[55-60].

These temporal and spatial limitations play the role of adiabatic deceleration for the corotating ion events in limiting the acceleration of upstream ions to exponential spectra [in this case in energy/charge[61]]. However, ironically it is not yet understood how the exponential spectrum is formed. Finite time of connection to the bow shock is reasonable but need not yield energy/charge dependence and cannot account

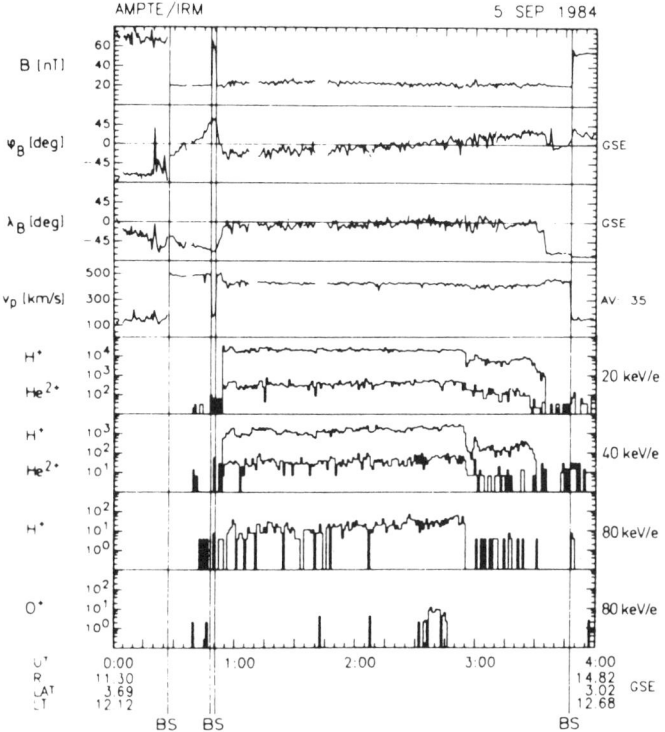

Fig. 11. Magnetic field B, plasma velocity V_p, and upstream ions at Earth's bow shock (Fig. 1 from Möbius et al.[67]).

for the exponential spectra when **B** is radial. Lateral drift transport, diffusion across field lines, or a free-escape boundary upstream are other possibilities but all are unsatisfying from various points of view. Based on escape via perpendicular diffusion, Lee[62] developed a coupled theory for the ion acceleration and wave excitation under conditions of radial **B** which could account for most observed features of the diffuse ions and associated waves.

Ellison[57] pioneered Monte Carlo calculations of ion shock acceleration in application to Earth's bow shock. In contrast with the theory of diffusive shock acceleration, Monte Carlo calculations can treat the behavior of the low energy ions as well. Based on the assumption that all particles scatter elastically in the frame of the local center of mass with a scattering mean free path λ proportional to rigidity, Ellison was able to obtain the entire shock structure as well as the energetic ion spectrum (assuming a free escape boundary). Ellison et al.[63] showed that the predicted composition and spectra are in accord with detailed measurements made by AMPTE/IRM during several bow shock crossings. Although the Monte Carlo technique has shortcomings (it neglects the wave field and the origin of ion scattering) it has helped provide the conceptual breakthrough that quasi-parallel shock structure is due to particle scattering in the local plasma frame and that energetic particles are a natural consequence of this process.

Upstream particles with similar characteristics have also been observed at Jupiter[64] and are likely to occur at all planetary bow shocks. Figure 10 shows the dominant frequency of wave activity observed upstream of several planets (from Russell et al.[65]). Since these frequencies are dominated by Doppler shift, the fact that they are proportional to |**B**| indicates that all waves are predominantly excited by protons with speeds

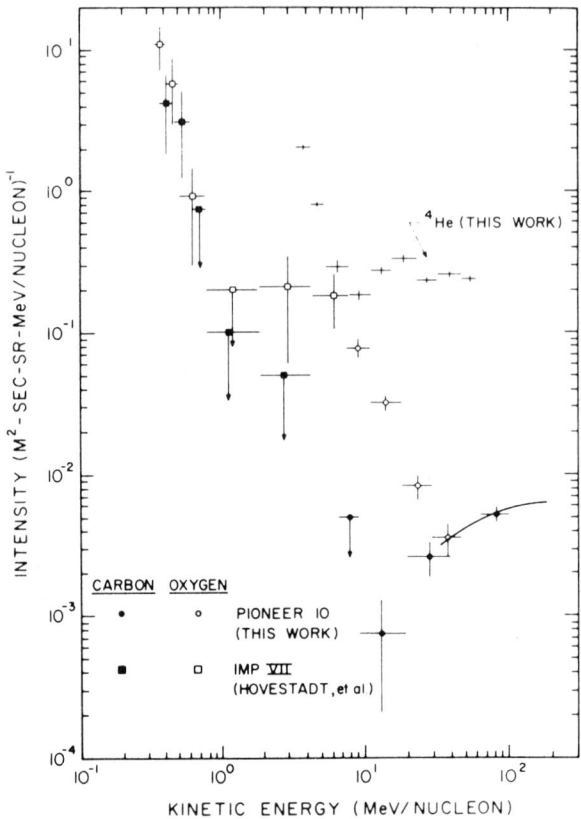

Fig. 12. Energy spectra of He, C and O (Fig.2 from McDonald et al.[71]).

of about 2 V_{SW}, as they are at Earth. These protons presumably dominate the energy density of upstream energetic ion populations at all the planets.

In spite of the evidence for a shock-accelerated origin of the upstream ions, Krimigis and coworkers[66] have argued that they are produced by leakage from the magnetosphere. Recent AMPTE/IRM measurements by Möbius et al.[67] shown in Figure 11 have demonstrated that indeed leaked particles are observed upstream of Earth's bow shock. The 80 keV/charge intermittent O^+ shown in the bottom panel appearing during this upstream ion event (dominated by H^+ and He^{++}) is undoubtedly of magnetospheric origin. Since the magnetospheric ions of solar wind origin have higher charge states and smaller scattering mean free paths, their probability of escape would be even smaller than that of O^+. The completely dominant and continuous H^+ and He^{++} is therefore clearly accelerated from the solar wind at the bow shock.

Electrons with energies up to ~ 100 keV are also observed upstream of Earth's bow shock[68]. They are apparently accelerated out of the solar wind by the "shock drift" mechanism near the point of field tangency[69].

5. THE COSMIC RAY ANOMALOUS COMPONENT

Measurements of cosmic ray energy spectra in the range 1-50 MeV/nucleon by both IMP 7[70] and Pioneer 10[71] revealed a surprise. Inbetween energetic solar flare particles at energies below ~1 MeV/nucleon and galactic cosmic rays at energies above ~ 100 MeV/nucleon, there appeared a separate enhancement in oxygen and helium, but not in carbon. This new component shown in Figure 12 from McDonald et al.[71] was dubbed the "cosmic ray anomalous component". It was recognized almost immediately that if these particles were galactic cosmic rays they would extrapolate to unacceptably high intensities in interstellar space. A heliospheric origin of the anomalous component was suggested by Fisk et al.[72]. They noted that their composition is consistent with an origin as interstellar neutral gas, ionized in the inner heliosphere by photoionization or charge exchange with solar wind protons, picked up by the solar wind, and accelerated somewhere in the outer heliosphere. Carbon is easily ionized and only expected with at most a low density in the inner heliosphere. Since the accelerated particles are then singly-ionized they have a high rigidity and relatively easy access back into the inner heliosphere as observed. Although this origin was soon accepted, a reasonably direct measurement of a charge state of +1 for the anomalous component was made only recently by Adams et al.[73].

Recent measurements by Voyagers 1 and 2 have revealed detailed spectra of the anomalous component in the outer heliosphere where its intensity is higher. Figure 13 from Cummings and Stone[74] shows enhancements in O, N, Ne and Ar as well as a small contribution of C, indicating that neutral carbon does exist in the very local interstellar medium. Evidence for the presence of anomalous component hydrogen in the spectrum of cosmic ray protons was also extracted from the Voyager 1 and 2 measurements by Christian et al.[75].

The origin of Fisk et al.[72] begged the question of the mechanism which could accelerate 1 keV/nucleon pickup ions to energies of 10-100 MeV/nucleon. Fisk[76] himself suggested "transit-time damping" of turbulence in the outer heliosphere, an idea also pursued by Klecker[77]. However, the energy density of the anomalous component now observed is too great to be supplied by solar wind turbulence. The most likely mechanism, first suggested by Pesses et al.[78], is acceleration at the solar wind termination shock. Jokipii[1] showed that the predicted spectra and gradients are consistent with the Voyager observations. Two important features of shock acceleration are apparent in Jokipii's model: Firstly, shock drift acceleration dominates at the quasi-perpendicular termination shock and effectively limits the energy gain of each ion to the potential of ~250 MeV between the heliographic pole and equator. Thus the spectra of the different ion species line up beautifully when plotted versus total energy[79]. Secondly, the interstellar pickup ions are readily accelerated out of the solar wind at the nearly perpendic-

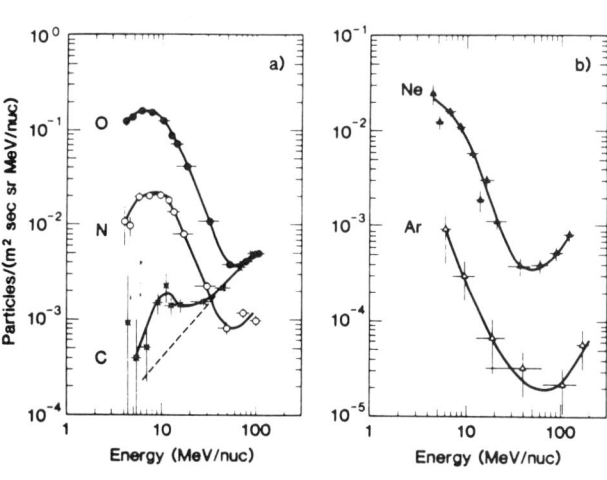

Fig. 13. Voyager energy spectra of anomalous component O, N, C, Ne and Ar (Fig. 1 from Cummings and Stone[74]).

ular termination shock, contrary to conventional intuition. Indeed Jokipii[80] argues that only the rapid acceleration rates at a quasi-perpendicular shock are consistent with the maximum age of the anomalous component inferred from their charge state[81].

A final interesting feature of acceleration of the anomalous component at the termination shock is that the theory places a limit on the distance to the shock. Requiring that the pressure in anomalous component hydrogen not exceed the ram pressure of the solar wind places the shock inside about 80 AU[79], and provides perhaps the best hope that we can look forward to a spacecraft traversal of the shock in the next decade. The large pressure of the anomalous component opens the possibility that the termination shock is mediated[82].

6. INTERSTELLAR AND COMETARY PICKUP IONS

Pickup ions in the solar wind are created when atoms or molecules from comets, planets, artificial clouds or the interstellar gas streaming through the heliosphere are ionized, and picked up and accelerated by the solar wind electric and magnetic fields onto helical trajectories in the frame of the solar wind. The initial ring beam distribution of pickup ions in velocity space is quickly scattered in pitch angle by magnetic fluctuations in the solar wind to form a shell with a radius approximately equal to V_{sw}. The shell is then cooled adiabatically by the diverging solar wind. Pickup ions may also be accelerated.

Figure 14 from Möbius et al.[83] shows the first measurement of interstellar pickup ions. Using instrumentation on the AMPTE/IRM spacecraft they measured the energy spectrum of He^+, which is the only interstellar pickup ion species created within 1 AU. The rapid decline in the energy spectrum beyond 4 E_{sw}, where E_{sw} is the energy of solar wind He^{++} in the spacecraft frame, is the signature of pickup ions, whose spacecraft-frame speed is less than or equal to twice V_{sw}. The rapid decline is an indication that acceleration of the ions in the inner heliosphere is not very important. However, stochastic acceleration of the ions by solar wind turbulence can modify the ion spectrum, particularly in the outer heliosphere where adiabatic deceleration is negligible and a large timescale can compensate for the slow inefficient acceleration process[84]. The fact that the energy density in the turbulence is less than that in the pickup ions prompted Bogdan et al.[85] to develop the first self-consistent theory of second-order Fermi or stochastic acceleration to account for wave damping in response to the acceleration. They also applied the theory to the acceleration of cometary pickup ions.

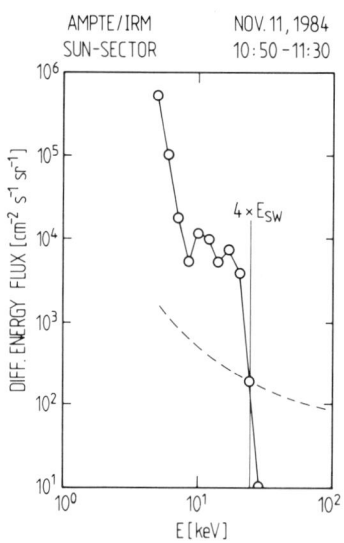

Fig. 14. Energy spectrum of interstellar pickup He+ (Fig. 2 from Möbius et al.[83]).

Of course at the solar wind termination shock the ions are accelerated to form the cosmic ray anomalous component as described in § 5.

The flybys of Comets Giacobini-Zinner in 1985 and Halley in 1986 produced new data on the behavior of cometary pickup ions, which are dominated by protons and ions of the water group: H_2O^+, OH^+ and O^+. Figure 15 taken from Coates et al.[86] shows the intensity of water group ions as a function of speed in the solar wind frame measured by Giotto during the inbound journey toward Comet Halley. The local solar wind speed indicated correlates well with ion speed far from the comet, a direct signature of the pickup process. Closer to the comet there is some evidence of adiabatic acceleration as the solar wind decelerates due to the mass loading by the pickup ions[87]. Within the cometary bow wave, occurring at 19:22 hours, there is evidence for additional acceleration due either to shock acceleration or to stochastic acceleration in the large-amplitude turbulence excited in part by the isotropization of the pickup ions upstream of the bow wave[87,88,85]. The additional acceleration is not large due to the limited timescale and the limited energy density in the turbulence. Acceleration of the pickup ions at the bow shock has also required relaxation of the standard restriction of the analytical theory of diffusive shock acceleration, that particle speeds be much greater than the fluid or shock speeds[89].

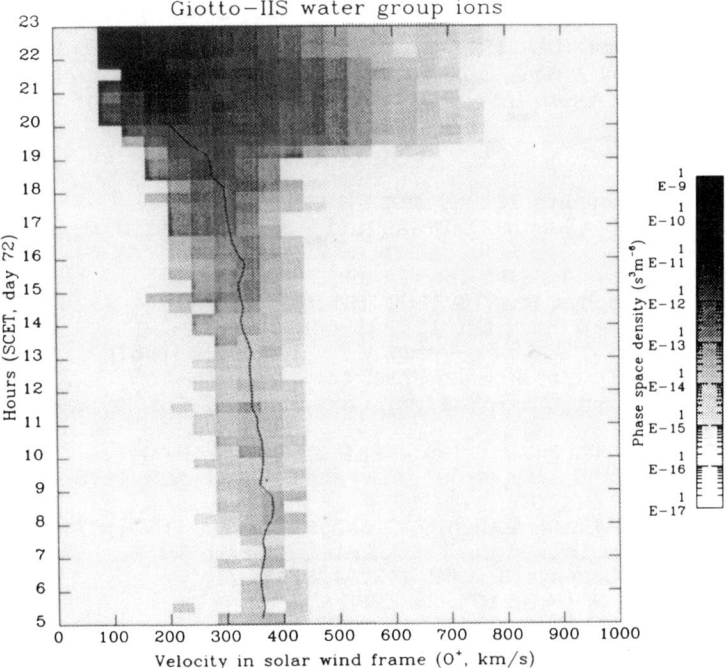

Fig. 15. Omnidirectional distribution of water-group pickup ions measured by Giotto along the inbound trajectory (Fig. 5 from Coates et al.[86]).

7. CONCLUSIONS

We have attempted to provide an overview of particle acceleration in the heliosphere. Besides energetic solar flare particles and galactic cosmic rays, there is a variety of populations of energetic particles produced in the heliosphere. In each case we have attempted to describe its origin and reproduce or describe original observations.

With extensive observations of these energetic particles by a variety of spacecraft over the last 20 years, the heliosphere has become an effective laboratory for the study of particle acceleration in the Cosmos. As a result the development of acceleration theory has advanced together with the study of energetic particle behavior in the heliosphere. We have attempted to emphasize the impact each population has had on theoretical progress.

We fully expect the excitement in this field of research to continue as the Voyager and Pioneer spacecraft approach the termination shock, and as Ulysses, which has already made the first detection of pickup hydrogen in the solar wind[90], probes the unexplored regions of the heliosphere above the solar poles. The heliosphere is indeed an active domain.

Acknowledgements.

The author wishes to thank the conference organizers at Bartol Research Institute for hosting a very successful and stimulating gathering. He also wishes to state his indebtedness to G.P. Zank for his patience in awaiting this typescript. This work was supported, in part, by NSF Grant ATM-8903703, by NASA Space Physics Theory Program Grant NAG5-1479, and by NASA Grants NAGW-1531 and Grant NAG5-1548.

REFERENCES

1. J.R. Jokipii, J. Geophys. Res. **91**, 2929 (1986).
2. D.A. Bryant, T.L. Cline, U.D. Desai, and F.B. McDonald, J. Geophys. Res. **67**, 4983 (1962).
3. T. Gold, J. Geophys. Res. **64**, 1665 (1959).
4. L. A. Fisk, J. Geophys. Res. **76**, 1662 (1971).
5. E.N. Parker, Planet. Space Sci. **13**, 9 (1965).
6. L.J. Gleeson and W.I. Axford, Astrophys. J. **149**, L115 (1967).
7. M. Scholer and G. Morfill, Solar Phys. **45**, 227 (1975).
8. W.I. Axford, E. Leer and G. Skadron, Proc. Int. Conf. Cosmic Rays 15th **11**, 132 (1977).
9. G.F. Krimsky, Doklady Akad. Nauk. SSR **234**, 1306 (1977).
10. R.D. Blandford and J.P. Ostriker, Astrophys. J. **221**, L29 (1978).
11. A.R. Bell, M.N.R.A.S. **182**, 147 (1978).
12. E.T. Sarris and J.A. Van Allen, J. Geophys. Res. **79**, 4157 (1974).
13. M.E. Pesses, R.B. Decker and T.P. Armstrong, Space Sci. Rev. **32**, 185 (1982).
14. R.B. Decker, J. Geophys. Res. **86**, 4537 (1981).
15. P.D. Hudson, M.N.R.A.S., **131**, 23 (1965).
16. J.R. Jokipii, Astrophys. J. **255**, 716 (1982).
17. M. Scholer, F.M. Ipavich, G. Gloeckler and D. Hovestadt, J. Geophys. Res. **88**, 1977 (1983).
18. P. Van Nes, R. Reinhard, T.R. Sanderson, K.-P. Wenzel and R.D. Zwickl, J. Geophys. Res. **89**, 2122 (1984).
19. M.A. Lee, J. Geophys. Res. **88**, 6109 (1983).

20. C.F. Kennel et al., J. Geophys. Res. **89**, 5419 (1984).
21. C.F. Kennel et al., J. Geophys. Res. **89**, 5436 (1984).
22. C.F. Kennel et al., J. Geophys. Res. **91**, 11917 (1986).
23. L.C. Tan, G.M. Mason, G. Gloeckler and F.M. Ipavich, Proc. Int. Conf. Cosmic Rays 21st **5**, 297 (1990).
24. L.C. Tan, G.M. Mason, G. Gloeckler and F.M. Ipavich, J. Geophys. Res. **93** 7225 (1988).
25. B. Sanahuja and V. Domingo, J. Geophys. Res. **92**, 7280 (1987).
26. G.K. Yates, L. Katz, B. Sellers and F.A. Hanser, in "Correlated Interplanetary and Magnetospheric Observations", ed. D.E. Page (D. Reidel, Dordrecht, Holland, 1974), p. 597.
27. D. Eichler, Astrophys. J. **247**, 1089 (1981).
28. L. O'C. Drury and H.J. Völk, Astrophys. J. **248**, 344 (1981).
29. R.D. Blandford and D. Eichler, Phys. Rept. **154**, 1 (1987).
30. K.R. Pyle, J.A. Simpson, A. Barnes and J.D. Mihalov, Astrophys. J. **282**, L107 (1984).
31. R. Müller-Mellin, K. Röhrs and G. Wibberenz, in "The Sun and the Heliosphere in Three Dimensions", ed. R.G. Marsden (D. Reidel, Hingham, MA, 1986), p. 349.
32. E. Rieger, G. H. Share, D.J. Forrest, G. Kanbach, C. Reppin and E.L. Chupp, Nature **312**, 623 (1984).
33. G.M. Mason, G. Gloeckler and D. Hovestadt, Astrophys. J. **280**, 902 (1984).
34. M.A. Lee and J.M. Ryan, Astrophys. J. **303**, 829 (1986).
35. H.V. Cane, D.V. Reames and T.T. von Rosenvinge, Astrophys. J. **93**, 9555 (1988).
36. F.B. McDonald, B.J. Teegarden, J.H. Trainor, T.T. von Rosenvinge and W.R. Webber, Astrophys. J. **203**, L149 (1976).
37. C.W. Barnes and J.A. Simpson, Astrophys. J. **210**, L91 (1976).
38. I.D. Palmer and J.T. Gosling, J. Geophys. Res. **83**, 2037 (1978).
39. L.A. Fisk and M.A. Lee, Astrophys. J. **237**, 620 (1980).
40. M.A.I. Van Hollebeke, F.B. McDonald, J.H. Trainor and T.T. von Rosenvinge, J. Geophys. Res. **83**, 4723 (1978).
41. G. Gloeckler, D. Hovestadt and L.A. Fisk, Astrophys. J. **230**, L191 (1979).
42. R.A. Mewaldt, E.C. Stone and R.E. Vogt, Geophys. Res. Lett. **6**, 589 (1979).
43. M. Scholer, G. Morfill and M.A.I. Van Hollebeke, J. Geophys. Res. **85**, 1743 (1980).
44. L.F. Burlaga, F.B. McDonald, M.L. Goldstein and A.J. Lazarus, J. Geophys. Res. **90**, 12027 (1985).
45. E.W. Greenstadt, I.M. Green, G.T. Inoye, A.J. Hundhausen, S.J. Bame and I.B. Strong, J. Geophys. Res. **73**, 51 (1968).
46. J.R. Asbridge, S.J. Bame and I.B. Strong, J. Geophys. Res. **73**, 5777 (1968).
47. G. Paschmann, N. Sckopke, I. Papamastorakis, J.R. Asbridge, S.J. Bame and J.T. Gosling, J. Geophys. Res. **86**, 4355 (1981).
48. M.M. Hoppe, C.T. Russell, L.A. Frank, T.E. Eastman and E.W. Greenstadt, J. Geophys. Res. **86**, 4471 (1981).
49. T. Hada, C.F. Kennel and T. Terasawa, J. Geophys. Res. **92**, 4423 (1987).
50. M.F. Thomsen, in "Collisionless Shocks in the Heliosphere: Reviews of Current Research", ed. B.T. Tsurutani and R.G. Stone (AGU, Washington, DC, 1985), p. 253.
51. D.H. Fairfield, J. Geophys. Res. **74**, 3541 (1969).
52. A. Barnes, Cosmic Electrodyn. **1**, 90 (1970).
53. S.P. Gary, J.T. Gosling and D.W. Forslund, J. Geophys. Res. **86**, 6691 (1981).
54. D.D. Sentman, J.P. Edmiston and L.A. Frank, J. Geophys. Res. **86**, 7487 (1981).
55. T. Terasawa, Planet. Space Sci. **27**, 365 (1979).
56. M.A. Lee, G. Skadron and L.A. Fisk, Geophys. Res. Lett. **8**, 401 (1981).
57. D.C. Ellison, Geophys. Res. Lett. **8**, 991 (1981).

58. D. Eichler, Astrophys. J. **244**, 711 (1981).
59. G. Skadron and M.A. Lee, J. Geophys. Res. **88**, 9975 (1983).
60. M.A. Lee and G. Skadron, J. Geophys. Res. **90**, 39 (1985).
61. F.M. Ipavich, A.B. Galvin, G. Gloeckler, M. Scholer and D. Hovestadt, J. Geophys. Res. **86**, 4337 (1981).
62. M.A. Lee, J. Geophys. Res. **87**, 5063 (1982).
63. D.C. Ellison, E. Möbius and G. Paschmann, Astrophys. J. **352**, 376 (1990).
64. D.N. Baker, R.D. Zwickl, S.M. Krimigis, J.F. Carbary and M.H. Acuña, J. Geophys. Res. **89**, 3775 (1984).
65. C.T. Russell, R.P. Lepping and C.W. Smith, J. Geophys. Res. **95**, 2273 (1990).
66. S.M. Krimigis, D.G. Sibeck and R.W. McEntire, Geophys. Res. Lett. **13**, 1376 (1986).
67. E. Möbius, D. Hovestadt, B. Klecker, M. Scholer, F.M. Ipavich, C.W. Carlson and R.P. Lin, Geophys. Res. Lett. **13**, 1372 (1986).
68. K.A. Anderson, J. Geophys. Res. **86**, 4445 (1981).
69. C.S. Wu, J. Geophys. Res. **89**, 8857 (1984).
70. D. Hovestadt, O. Vollmer, G. Gloeckler and C.Y. Fan, Phys. Rev. Lett. **31**, 650 (1973).
71. F.B. McDonald, B.J. Teegarden, J.H. Trainor and W.R. Webber, Astrophys. J. **187**, L105 (1974).
72. L.A. Fisk, B. Kozlovsky and R. Ramaty, Astrophys. J. **190**, L35 (1974).
73. J.H. Adams et al., Astrophys. J. **375**, L45 (1991).
74. A.C. Cummings and E.C. Stone, in "Proceedings of the Sixth International Solar Wind Conference", ed. V.J. Pizzo, T.E. Holzer and D.G. Sime (NCAR/TN-306 + Proc, Boulder, CO, 1988), p. 599.
75. E.R. Christian, A.C. Cummings and E.C. Stone, Astrophys. J. **334**, L77 (1988).
76. L.A. Fisk, J. Geophys. Res. **81**, 4633 (1976).
77. B. Klecker, J. Geophys. Res. **82**, 5287 (1977).
78. M.E. Pesses, J.R. Jokipii and D. Eichler, Astrophys. J. **246**, L85 (1981).
79. J.R. Jokipii, in "The Physics of the Outer Heliosphere", ed. S. Grzedzielski and D.E. Page (Pergamon Press, Oxford, 1990), p. 169.
80. J.R. Jokipii, Private Communication (1992).
81. J.H. Adams and M.D. Leising, Proc. Int. Conf. Cosmic Rays 22nd **3**, 304 (1991).
82. M.A. Lee and W.I. Axford, Astron. Astrophys. **194**, 297 (1988).
83. E. Möbius, D. Hovestadt, B. Klecker, M. Scholer, G. Gloeckler and F.M. Ipavich, Nature **318**, 426 (1985).
84. P.A. Isenberg, J. Geophys. Res. **92**, 1067 (1987).
85. T.J. Bogdan, M.A. Lee and P. Schneider, J. Geophys. Res. **96**, 161 (1991).
86. A.J. Coates, B. Wilken, A.D. Johnstone, K. Jockers, K.-H. Glassmeier and D.E. Huddleston, J. Geophys. Res. **95**, 10249 (1990).
87. P.A. Isenberg, J. Geophys. Res. **92**, 8795 (1987).
88. T.I. Gombosi, K. Lorencz and J.R. Jokipii, J. Geophys. Res. **94**, 15011 (1989).
89. T.I. Gombosi, J.R. Jokipii, J. Kota, K. Lorencz and L.I. Williams, "The telegraph equation in charged particle transport", Astrophys. J., submitted (1992).
90. G. Gloeckler et al., EOS **73**, Supplement, 246 (1992).

PARTICLE ACCELERATION ON GALACTIC SCALES

W.I. Axford
Max-Planck-Institut für Aeronomie, D-3411 Katlenburg-Lindau, FRG

ABSTRACT

A brief survey is given of the history and current ideas concerning the origin of cosmic rays in the galaxy and in extragalactic sources. It is argued that shock acceleration in various guises is the essential and conceptually most economical acceleration mechanism.

I. HISTORICAL INTRODUCTION

The first suggestion that there might be an extra-terrestrial source of ionization in the atmosphere was made by Wilson in 1901 (1). However, on performing his experiments underground and finding no change he came to the conclusion that this could not be the case and that the observed ionization is a property of air itself. Various measurements performed during the next decade did not resolve the matter but after a series of balloon flights, particularly one on 7 August 1912, Hess discovered an increase of ionization with altitude which could only be explained as being the result of "the existence of a hitherto unknown and very penetrating radiation, coming mainly from above and being most probably of extra-terrestrial origin" (2). The fact that the effect was observed to be independent of whether it was night or day suggested a source outside the solar system.

Confirmation of the existence of the "Höhenstrahlung" came from further flights to higher altitudes by Kohlhörster (3) and Millikan and Cameron (4). It was natural at a time when the only known sources of ionization were associated with radioactivity that the extra-terrestrial radiation should be assumed to be of the most penetrating type known, namely gamma rays. Millikan, who seems to have had an esoteric bent, made the hypothesis that the "cosmic rays" are in fact gamma rays produced by the synthesis of nuclei, which he described colourfully as the "birth cries of atoms". With a better understanding of the energy losses of gamma rays in matter it became evident that synthesis is energetically inadequate and Jeans made the opposite hypothesis that instead, the supposed gamma rays are the result of nuclear disintegration (the "death of the universe"). It is perhaps not surprising that Rutherford was moved to remark that it was time for "more work and less talk", although for such a distinguished scientist it is surprising that he himself was not able to contribute anything to the development of the subject.

The first indications that particles rather than gamma rays are involved (Skobelzyn (5), Bothe and Kohlhörster (6)) required a new hypothesis not involving small-scale nuclear processes. Noting that energies of the order of at least 1-10 GeV would be required, Bothe made the perceptive remark that the particles "could get these energies in weak but extensive fields of force", presumably in the galaxy or elsewhere. In fact these ground-level observations of particles must have involved secondaries rather than primary cosmic rays and it was not until the discovery of the geomagnetic latitude effect by Clay and others (7) and the east-west effect (8,9) that it became clear that mainly positively charged particles with energies of at least 10 GeV are required. With the discovery of mesons and the development of an understanding of shower theory it became necessary to account for energies as high as 10^{15} eV (10).

The first ideas put forward to explain the origin of cosmic rays as particles rather than gamma rays paid no attention to Bothe's remark but instead tended to invoke a stellar origin as first suggested by Clay ("they must originate in the hot stars of the Milky system"). The first serious theory, advanced by Swann (11), invoked the kind of electric and magnetic fields that might have been expected in a growing sunspot (or starspot): Swann's first model is impossibly singular, requiring the particle energy and magnetic field to be initially zero with the latter increasing linearly with time and the particle moving on a fixed circle. With more realistic initial conditions and with the electric field consistent with the demands of plasma physics, the possible energy gain is then limited by the need for the magnetic moment of the particle to be conserved.

In 1934, Baade and Zwicky (12) advanced the hypothesis that cosmic rays are emitted by supernovae, which "with all reserve" they considered to "represent the transition of an ordinary star into a neutron star". This remarkable paper has its defects from the point of view of our present knowledge: for example, as supernovae were considered to be relatively rare it was accordingly argued that extragalactic events are dominant. Nevertheless it represented the first major step in the development of modern theories of the origin of cosmic rays. Colgate and his colleagues (13) carried this approach further in a series of papers based on the fact that a supernova shock front emerging from the outermost layers of the progenitor star should become highly relativistic and produce cosmic rays directly. The idea, which received much attention in the 1960s, fails however because of the nuclear disintegrations which should occur and, especially, the energy losses associated with the subsequent expansion into the surrounding interstellar medium. We nevertherless accept that supernovae are involved in the acceleration of cosmic rays, but not at the initial stage of the explosion (14).

Attempts to construct theories of cosmic ray acceleration involving the sun and ordinary stars were made in the late 1940s by Teller and Ritchmeyer and Alfvén (15). These ideas were instructive but not convincing. (There are of course "cosmic rays" of solar origin associated with flares and, accordingly to currently-favoured ideas, the termination of the solar wind, but these have nothing to do with the bulk of the cosmic rays nor with the mechanisms considered in these theories.) Nevertheless this activity had the beneficial effect of stimulating Fermi to think about the problem.

Fermi's first attempt to construct a cosmic ray acceleration theory lead to what is now termed his "second order" mechanism (16). It is assumed that particles continually collide with moving magnetic "irregularities" in interstellar space. Although the energy gain associated with a head-on collision is the same as the energy loss associated with an overtaking collision, there is a slight predominance in the rate at which head-on collisions occur which leads to a net acceleration. A balance of this acceleration with absorption due to collisions with interstellar matter leads to a power-law spectrum for the cosmic rays which matches that observed if the "irregularities" are encountered about once per year, with reasonable assumptions concerning the density of interstellar matter and the mean speed of the "irregularities" ($V/c \simeq 10^{-4}$).

This theory was almost immediately recognized to be unacceptable as a result of measurements by Freier and others (17) showing the presence of heavy nuclei in the cosmic rays with a relative absence of light fragmentary nuclei such as Li, Be and B. This implies that the cosmic rays are at most 10^6 - 10^7 years old and that the acceleration time is accordingly short – too short to be accounted for on the basis of Fermi's original theory. In his cogently argued Henry Norris Russell lecture of 1954, Fermi gave a much improved theory which resolved the difficulties satisfactorily, although not in full detail (18). First, he suggested that galactic escape should predominate over collision losses, so that all species behave similarly. Second, he argued that head-on collisions with irregular magnetic fields might be more efficient and also more probable than overtaking collisions (or their equivalent) so that the acceleration is more rapid, giving rise to a "first order" mechanism.

Modern theories of cosmic ray acceleration involving diffusive and drift acceleration in shocks, associated for example with supernova remnants (19,20), can be regarded as extensions of Fermi's revised theory. There are additional aspects associated with storage and leakage of cosmic rays from the galaxy when it comes to determining the equilibrium spectrum. The leakage process, which was originally discussed by Morrison, Olbert and Rossi (21), Ginzburg and Syrovatski (22) and others, still remains in a rather unsatisfactory state and is usually treated on an ad hoc basis. Nevertheless, considerable progress has been made and the present state of our theoretical understanding of the origin of cosmic rays can be regarded with some satisfaction if not complete confidence. An origin in "weak but extensive fields of force" in the galaxy is favoured, as originally proposed by Bothe, and we have had some success in accounting for the spectrum, power requirements and composition of the cosmic rays, as described briefly in the following sections and in other papers in these Proceedings. The possibility that there is a contribution from point sources such as pulsars is still not excluded however, although at the present time it is not at all clear how these could explain the observed properties of cosmic rays (23). Furthermore, there is most probably an extragalactic component which is prevalent at the highest energies, although not very significant as far as the bulk of cosmic rays in the galaxy are concerned. Current theories of the origin of these particles do not differ very much in principle from that of Burbidge and Hoyle (24), although in the interest of economy of hypotheses it is usually assumed that shock acceleration is involved as at lower energies.

II. THE OBSERVED PROPERTIES OF COSMIC RAYS

Because of the difficulty of measuring anything other than total energy per particle, E, at high energies, it is customary to depict the overall cosmic ray spectrum in these terms, as indicated in Figure 1. It should be remembered, however, that any electromagnetic acceleration mechanism tends to produce a spectrum which is better defined in terms of the particle rigidity, P (volts), which is simply E (eV)/Z for very energetic particles.

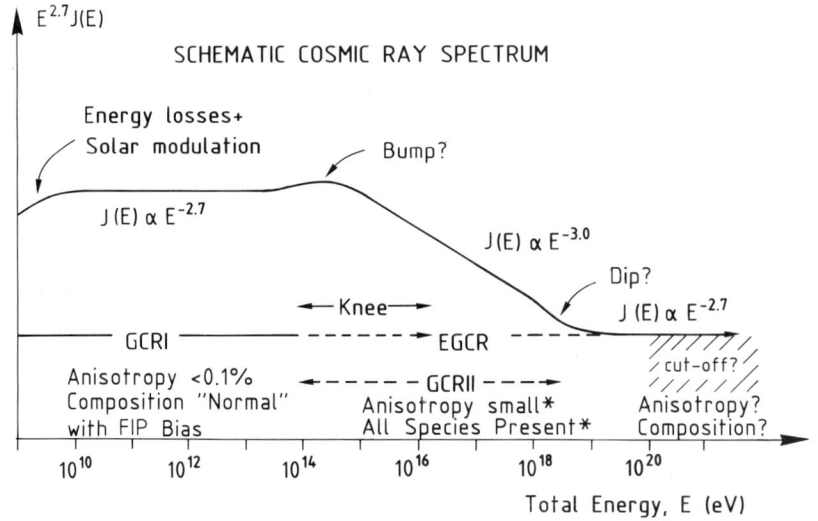

Fig. 1.

In Figure 1 the differential intensity J(E) has been multiplied by E^k, where k = 2.7 is the index of the power law describing the main part of the spectrum in the range 10^{10} - 10^{14} eV. This has the advantage of emphasizing subtle variations in shape of the spectrum which tend to pass unnoticed if one simply depicts J(E).

Below 10^{10} eV the spectrum bends down as a result of propagation losses and the effects of solar modulation. The spectrum is a rather good power law up to about 10^{13} eV where it begins to merge with the "knee", which may in turn be associated with a slight "bump" in the spectrum between 3×10^{13} and 3×10^{15} eV. It is argued that the cosmic ray spectrum below the knee is produced by diffusive shock acceleration in supernova remnants with the knee itself representing a rigidity-dependent cut-off in this process, associated with the finite lifetime of the remnants. These cosmic rays are termed GCRI: they have a rather normal distribution of elements and isotopes with a bias towards species with low FIPs (first ionization potentials) and a contribution due to fragmentation of the primaries. The GCRI show evidence for a small (< 0.1%) anisotropy, which, however, appears to be a local effect of no particular significance.

Above the knee, at around 3×10^{15} eV, a second power law spectrum is observed with a spectral index of about 3.0. Composition measurements are difficult and indirect in this region but it appears that there is a mixture of elements (25) and not a preponderance of protons or of heavy species such as Fe. No significant anisotropy has been detected despite earlier reports, which were somewhat over-interpreted (26). Indeed one should not expect much in the way of anisotropies in this region (GCRII) if the source(s) of the particles are distributed and their gyroradii in the galactic magnetic field small compared with the scale-length of any density variations. The sun is improbably situated almost on the mid-plane of the galaxy at the present epoch and only the second harmonic of the anisotropy should be evident if the intensity falls off to either side of the plane.

Above about 10^{19} eV the spectrum flattens and has a slope something like that of GCRI. Since particles in this energy range have quite large gyroradii and are accordingly difficult to accelerate and contain in a galactic disc with a scale height of 10^2 - 10^3 pc, we presume that they are of extragalactic origin (EGCR). We have no reliable information concerning the anisotropy and composition of these cosmic rays.

Apart from the three components of the cosmic radiation defined here, the spectrum shows three possible features which may have implications concerning the origin of the particles. First, the knee appears to be associated with a slight bump and possibly with some change in composition (27). (More observations are required to determine the reality of the bump since it occurs in a region where high and low energy techniques overlap. It is clear, however, that the overall spectrum is continuous at the knee with a smooth transition from GCRI to GCRII.) The second feature is the "dip", which appears in the range 10^{18} - 10^{19} eV (28); this is a slight but definite feature which appears to indicate a cut-off in the acceleration of GCRII and a transition to a region of dominance of an independent population with a flatter spectrum, namely EGCR. Finally, one may ask whether there is an end to the spectrum at about 10^{20} eV, which would be expected for extragalactic particles due to propagation losses (29). We must wait for the results obtained from much larger arrays at present being built before this can be determined.

The continuity of the spectrum has important implications in itself. First, it is easy to obtain a concave feature such as the transition around 10^{19} eV by combining two independent (e.g. power law) spectra as indicated in Figure 2a. One does not have to assume anything remarkable to get a smooth join in the combined spectrum although the presence of the dip does suggest that in this case the join occurs not far from the point at which the GCRII spectrum cuts off entirely. Second, it is quite another matter to obtain a convex feature such as that

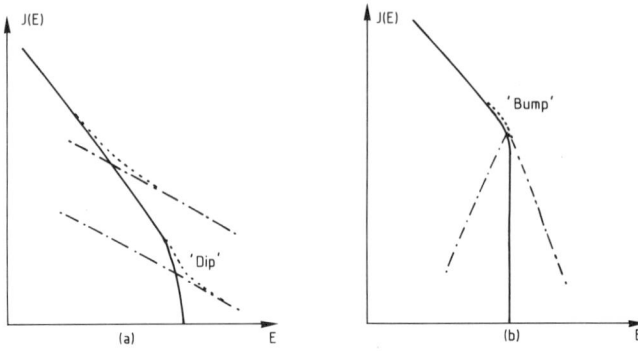

Fig. 2a,b.

found at the knee by combining two independent spectra as indicated in Figure 2b, since one spectrum would have to cut off where the second begins and at the join they would both have to have the same amplitude. Such a coincidence is possible but most unlikely if the sources of the particles comprising the two spectra are really independent. It is more likely that the two spectra "know" about one another and this is possible if the high energy component is obtained by a second stage of acceleration of the low energy component; that is, GCRII are obtained as a result of the post-acceleration of GCRI. There is a further constraint on the post acceleration process in that *all* GCRI in the vicinity of the cut-off must have an opportunity to take part in the post-acceleration, otherwise there would be a jump in the spectrum at this point. This is an argument based on respect for the principle of Occam's razor and does not exclude a two-source model: it simply suggests that it is less likely.

III. THE ACCELERATION OF GCRI

Shock acceleration has become a fashionable subject mainly during the last decade. In fact, the phenomenon and the basic mechanisms have been known for about 30 years. It was first observed as an effect on galactic cosmic rays in association with geomagnetic storm sudden commencements in 1957 (30) and, more clearly, on solar flare particles in 1961-2 (31). The "shock-drift" mechanism was analysed first around 1960 (32), while the more complicated diffusive acceleration mechanism, although well understood in principle in 1963-4 (33), was not completely solved until 1976-8 (19). (Note, however, that the jump conditions and near solutions were obtained in 1967-71 (34)).

Hoyle (35) was the first to propose shock acceleration as a means of producing cosmic rays in his consideration of particle acceleration in strong radio sources. He produced a solution for strong cosmic ray modified shocks but introduced a hypothesis rather than solving a diffusion equation, which was not available at the time. The idea was not forgotten as it was noted in connection with the problem of energy input to the interstellar gas that "..it seems possible that shock heating may provide a significant energy source for the cosmic ray gas" (36).

The subject of diffusive shock acceleration has been well reviewed (20) and it is sufficient to simply state the implications for the acceleration of GCRI here:

1) A simple "non-linear" treatment suggests that the efficiency of energy transfer from the background gas to cosmic rays is O(1), which is essential to account for the power requirements of GCRI on the basis of the properties of supernova remnants.

2) The accelerated particles have a spectrum which is a power law in rigidity with index almost 2 for strong shocks (19). The total spectrum produced by an evolving supernova remnant should also closely approximate a power law in rigidity with spectral index about 2.1 or 2.2 (37).

3) All species are treated in a similar way so that the composition of the accelerated particles should be similar to that of the source plasma (which may itself have a FIP bias).

4) The shocks associated with supernova remnants involve cosmic ray energy densities comparable with the kinetic energy density of the incoming flow and are therefore subject to microscopic (streaming) and macroscopic (Rayleigh-Taylor) instabilities. These instabilities should cause the cosmic ray scattering mean free path to approach the so-called Bohm limit (38).

5) As a consequence of the finite lifetime of the shock the spectrum has a rigidity cut-off at about $P = 10^{14}$ volts for a Sedov blast wave with standard supernova remnant parameters (20,39). This cut-off can be identified with the knee of the overall spectrum.

With the above results as a basis for calculating the source function for GCRI it is a simple matter, after considering the observed spectra of secondaries and primaries, to deduce that the spectrum of GCRI can be accounted for if the galactic escape time has the form of a power law in rigidity, $P^{-0.6}$, at least up to about 10^{13} volts. Beyond this the escape law may turn up to produce the bump in the spectrum beginning at about 3×10^{13} eV. However, there are other possible explanations for the bump, including the escape of high energy particles from the remnant ahead of the shock so that they do not undergo any serious adiabatic deceleration (40). (The highest energy particles tend to be those which stay with the shock for the longest time and therefore suffer least from adiabatic losses.)

This appears to be an adequate basis for an understanding of the origin of GCRI, apart from the cause of the FIP bias, which we take to be an indication of magnetic stripping of easily photo-ionized species from the exterior layers of dense neutral hydrogen clouds by the hot "coronal" gas flowing around them. (Note that Lyman alpha, which should be a prominent component of the radiation field in the coronal medium is capable of photo-ionizing all the low FIP elements but not those considered to be high FIP.)

It is worth noting that the cosmic ray induced turbulence associated with supernova remnant and other shocks should be saturated up to wave lengths long enough to scatter efficiently particles with rigidities near the cut-off of GCRI, but not at longer wavelengths. This turbulence is continually renewed by the passage of more shock waves but, except by cascading to longer wavelengths (i.e. greater than about 0.1 pc), there is no strong source of turbulence capable of efficiently scattering GCRII. We may therefore assume to a first approximation that GCRII propagate almost "scatter-free", although of course this cannot be strictly true.

IV. THE ACCELERATION OF GCRII

Apart from those theories in which the threat of Occam's razor is ignored, most earlier attempts to account for GCRII raise more questions than they answer. It is typically assumed that the acceleration mechanism required for GCRI is effective to energies of the order of 10^{18} eV or more and the high energy spectrum is shaped by additional losses: either interaction with an intense photon background or a change in the escape law. This requires that a large fraction of the power input to cosmic rays goes into GCRII since the source spectrum must be almost P^{-2}. We have criticized these suggestions elsewhere and believe that they do not hold the answer to the problem (41).

A genuine post-acceleration theory has been considered by Jokipii and Morfill (42), making economical use of diffusive shock acceleration at the galactic wind termination shock. This avoids the afore-mentioned threat but suffers from several difficulties, the most important of which is associated with the fact that the second stage of acceleration is not "contiguous" with the first (i.e. diffusive shock acceleration by supernova remnants). In this case the GCRI first lose energy in helping drive the galactic wind and after acceleration are modulated by the same wind as they make their way back to the galaxy. This makes excessive demands on the power available and does not guarantee that the resultant spectrum in the galaxy is smoothly convex at the knee (41,43).

We prefer to take the obvious way out of these difficulties by looking for a post-acceleration mechanism which operates within the galaxy and directly on all the particles that have been released after acceleration by large supernova remnants. The only important acceleration process available in the galaxy as a whole must involve the shocks which are so abundantly generated by supernovae, stellar winds, expanding "bubbles" resulting from supernova/OB associations and their interactions with each other and with dense interstellar clouds. These shocks, which usually accelerate any particle encountering them, also involve a subsequent expansion, which produces a corresponding deceleration. However, as noted as a general possibility by Fermi (18), there is no reason why the two processes should cancel each other and indeed we find that this is not the case in the most simple situation where the particles interact successively with a series of independent Sedov supernova remnants in a uniform ambient magnetic field (44). Energy gains easily exceed energy losses until the gyroradius of the particles is no longer small compared with the size of the remnants. Thus the acceleration is essentially a first order Fermi process; there is also a second order acceleration since there are losses as well as gains but this is slow and unimportant. There are similarities between this scheme and that proposed by Gurevich and Rumyantsev (45).

A detailed description of this post-acceleration process is given in an accompanying paper (46). It is sufficient to note that a reasonable account of the origin of GCRII can be achieved if the mean residence time of these particles in the region of acceleration is about 10^5 years and shock waves are encountered every 100 pc or so. We require that the supernova remnants are numerous and large (i.e. rather old) and hence that they propagate mainly in the hot component of the interstellar medium. In this case it may be possible to match the overall spectrum of GCRII with an ambient magnetic field strength of about 6 microgauss and to obtain the "dip" marking the large gyroradius cut-off in the vicinity of 10^{18} eV. The results are to some extent sensitive to the position of the knee in the spectrum and also to the relative abundances of the major species involved.

In reality the situation must be more complicated than we have assumed in this first model. The supernovae must overlap and their shock waves must interact with each other and with shocks from other sources. However, this can only enhance the acceleration process. We are left with the problem of how to determine the rate of escape of particles from the galaxy, which must be to some extent connected with the interaction of particles with the large scale structures which accelerate them, but of course in this respect ours is in no worse a situation than any other theory.

There are other versions of this post-acceleration approach to explaining GCRII. Bykov and Toptygin (47) concentrate on OB associations as the main location for the second stage, which is quite acceptable as these are probably the source regions of most GCRI. In this case one must introduce a residence time for cosmic rays in the association which may be quite different from that in the galaxy as a whole. Furthermore the post-acceleration involves the source population of GCRI and the spectra of both components would then be shaped by escape from the galaxy. In principle the two schemes are the same but Bykov and Toptygin have made an additional important contribution in that they have provided a new generalized transport/acceleration equation which can be treated analytically rather than using a Monte Carlo procedure. Berezhko (48) has taken a similar approach in that he considers post-acceleration of the ambient rather than the source GCRI population. However, he has invoked the relatively rare interactions involving relativistic shocks which may be associated with pulsar-driven winds and which give rise to correspondingly large energy changes. In this case the main requirement is to show that a typical particle at the knee has a good chance of encountering such a shock so that the resultant spectrum is continuous.

A somewhat different approach has been taken by Bell (49) who invokes the very large electric fields existing in pulsar winds imbedded in the same supernova remnants that are the sources of GCRI to post-acclerate particles to GCRII within these sources. Again, it is necessary to show that every particle at the knee, determined by the first stage of acceleration associated with the external remnant driven shock, has a good chance of being accelerated in the pulsar wind.

Whichever mechanism, or combination of mechanisms, is responsible for the existence of GCRII, if it involves post acceleration of GCRI produced by the currently-favoured supernova remnant shock acceleration process, then the spectra of individual species must be similar and rigidity-dependent. Furthermore, the composition should vary continuously at the knee in the manner originally envisaged by Peters (50). Thus there are quite definite predictions which can be tested by any experiment capable of making precise spectral and composition measurements in this region.

It is noteworthy that post-acceleration by repeated shock encounters must also affect GCRI at much lower energies than at the knee and indeed this has been invoked as an important process by several authors. There is however a difference if the particles propagate diffusively since they are unable to move quickly and freely between shock encounters and on interacting with any given shock are eventually "swallowed" by it and undergo adiabatic cooling in the subsequent expansion. This process has been examined by Blandford and Ostriker (51) and as emphasised by Cesarsky (52), it cannot be very important otherwise it would deform the spectra of secondary species in a manner inconsistent with observations.

V. EXTRAGALACTIC COSMIC RAYS

In order to account for the presumed extragalactic component of the cosmic ray spectrum (EGCR) it is difficult to escape the conclusion that the stong radio galaxies are responsible as originally suggested by Burbidge and Hoyle (24). Various analyses made of this suggestion lead to very much the same conclusions, namely that high energy particles ($> 10^{20}$ eV) can be accelerated in the shocks which must terminate the jets flowing out from the nuclear regions of the parent galaxy (53). There is some uncertainty concerning the strength of the magnetic fields involved however and this affects the processes controlling the energies achievable. We reject the idea that the "minimum energy" field can be assumed since this refers only to the region of maximum radio emission which should in fact be situated at the contact surface between jet material and the shocked extragalactic gas. The magnetic field in the jet should be much smaller and also consistent with the formation of the jet in the centre of the galaxy. Nevertheless it appears to be just possible for particles of sufficiently high energy to be produced which in turn are affected by propagation losses and (time dependent) diffusion before reaching our own galaxy (46).

REFERENCES

1. C.T.R. Wilson, Proc. Camb. Phil. Soc. **11**, 52 (1901).
2. V.F. Hess, Physik Z. **13**, 1084 (1912).
3. W. Kohlhörster, Physik Z. **14**, 1153 (1913).
4. R.A. Millikan and G.H. Cameron, Phys. Rev. **28**, 851 (1926).
5. D. Skobeltzyn, Z. Physik **43**, 354 (1927); **54**, 686 (1929).
6. W. Bothe and W. Kohlhörster, Z. Physik **56**, 751 (1929).
7. J. Clay, Proc. Acad. Sci. Amsterdam **30**, 1115 (1927); **31**, 1091 (1928); A.H. Compton, Phys. Rev. **43**, 387 (1933).
8. B. Rossi, Phys. Rev. **36**, 606 (1930).
9. T.H. Johnson, Phys. Rev. **43**, 381 (1933); Rev. Mod. Phys. **10**, 193 (1938); L. Alvarez and A.H. Compton, Phys. Rev. **43**, 835 (1933); B. Rossi and S. de Benedetti, Phys. Rev. **45**, 214 (1934).
10. P. Auger, Nature **135**, 820 (1935); J. Phys. Radium **10**, 39 (1939); H.J. Bhabha and W. Heitler, Proc. Roy. Soc. **A159**, 432 (1937); J.F. Carlson and J.R. Oppenheimer, Phys. Rev. **51**, 220 (1937).
11. W.F.G. Swann, Phys. Rev. **43**, 217 (1933); J. Franklin Inst. **258**, 383 (1954).
12. W. Baade and F. Zwicky, Proc. Nat. Acad. Sci. **20**, 239 (1934).
13. S. Colgate and M.H. Johnson, Phys. Rev. Lett. **5**, 235 (1960).
14. V.L. Ginzburg, Dokl. Akad. Nauk, SSSR, **92**, 1133 (1953).
15. R.D. Ritchmeyer and E. Teller, Phys. Rev. **75**, 1729 (1949); H. Alfvén, Phys. Rev. **75**, 1732 (1949).
16. E. Fermi, Phys. Rev. **75**, 1169 (1949).
17. P. Freier, E.J. Lofgren, E.P. Ney, and F. Oppenheimer, Phys. Rev. **74**, 1818, and H.L. Bradt and B. Peters, Phys. Rev. **74**, 1828 (1948); **80**, 943 (1950).
18. E. Fermi, Astrophys. J. **119**, 1 (1954).
19. W.I. Axford, E. Leer and G. Skadron, Proc. 15th ICRC **11**, 132 (1977); G.F. Krimsky, Dokl. Akad. Nauk SSSR **234**, 1306 (1977); R.D. Blandford and J.P. Ostriker, Astrophys. J. **221**, L29 (1978); A.R. Bell, MNRAS **182**, 147 and 443 (1978).
20. W.I. Axford, Proc. 17th ICRC **12**, 155 (1981); L.O'C. Drury, Rep. Prog. Phys. **46**, 163 (1983); R.D. Blandford and D. Eichler, Phys. Rep. **154**, 1 (1987).
21. P. Morrison, Handbuch der Physik (Springer, Berlin, 1961), Vol. **46-1**; P. Morrison, S. Olbert and B. Rossi, Phys. Rev. **94**, 440 (1954).
22. V.L. Ginzburg and S.I. Syrovatskii, The Origin of Cosmic Rays (Pergamon, 1964).
23. J. Arons, IAU Symposium No. 94, 175 (1980).
24. G. Burbidge and F. Hoyle, Proc. Phys. Soc. **84**, 141 (1964).
25. G.L. Cassiday et al., Astrophys. J. **356**, 669 (1990).
26. R.W. Clay and A.G.K. Smith, Astrophysical Aspects of the Most Energetic Cosmic Rays (World Scientific, Singapore, 1991), 125.

27. T.K. Gaisser, Astrophysical Aspects of the Most Energetic Cosmic Rays (World Scientific, Singapore, 1991), 146.
28. A.A. Watson, these Proceedings (1991).
29. K. Greisen, Phys. Rev. Lett. **16**, 748 (1966);
    G.T. Zatsepin and V.A. Kuzmin, JETP Lett. **4**, 73 (1966).
30. Y.L. Blokh, L.I. Dorman and N.S. Kaminer, Proc. 6th ICRC **4**, 173 (1959).
31. W.I. Axford and G.C. Reid, J. Geophys. Res. **67**, 1692 (1962); **68**, 1793 (1963);
    D.A. Bryant, T.L. Cline, U.D. Desai and F.B. McDonald, J. Geophys. Res. **67**, 4983 (1962).
32. V.D. Shabansky, Sov. Phys. JETP **41**, 1107 (1961);
    L.I. Dorman and G.L. Freidmann, Problems of MHD and Plasma Phys., 77 (1959).
33. J.A. Van Allen and N.F. Ness, J. Geophys. Res. **72**, 935 (1967);
    J.R. Jokipii and L.R. Davis, Phys. Rev. Lett. **13**, 739 (1964).
34. L.J. Gleeson and W.I. Axford, Astrophys. J. **149**, L115 (1967);
    L.A. Fisk, Ph.D. Thesis, Univ. California, San Diego (1970);
    L.A. Fisk, J. Geophys. Res. **76**, 1662 (1971).
35. F. Hoyle, MNRAS **120**, 338 (1960).
36. R.C. Newman and W.I. Axford, Astrophys. J. **153**, 595 (1968).
37. T.J. Bogdan and H.J. Völk, Astron. Astrophys. **122**, 129 (1983);
    H. Moraal and W.I. Axford, Astron. Astrophys. **125**, 204 (1983).
38. L.O'C. Drury and S.A.E.G. Falle, MNRAS **223**, 353 (1986);
    S.V. Chalov, Sov. Astron. Lett. **14**, 114 (1988);
    G.P. Zank, J.F. McKenzie and W.I. Axford, Astron. Astrophys. **233**, 275 (1990).
39. P.O. Lagage and C.J. Cesarsky, Astron. Astrophys. **125**, 294 (1982);
    L.O'C. Drury, Proc. 21st ICRC **12**, 85 (1990).
40. E.G. Berezhko, Y.K. Yelshin and A.A. Turpanov, Proc. 20th ICRC **2**, 171 (1987).
41. W.I. Axford, Astrophysical Aspects of the Most Energetic Cosmic Rays (World Scientific, Singapore, 1990) 406.
42. J.R. Jokipii and G.E. Morfill, Astrophys. J. **312**, 170 (1987).
43. W.I. Axford, Proc. 20th ICRC **8**, 120 (1987).
44. W.-H. Ip and W.I. Axford, these Proceedings (1991).
45. L.E. Gurevich and A.A. Rumyantsev, Astrophys. Space Sci. **72**, 261 (1980).
46. W.-H. Ip and W.I. Axford, Astrophysical Aspects of the Most Energetic Cosmic Rays (World Scientific, 1991) 273.
47. A.M. Bykov and N. Toptygin, Sov. Phys. JETP **98**, 1255 (1990).
48. E.G. Berezhko, Proc. 22nd ICRC **2**, 436 (1991).
49. A.R. Bell, Proc. 22nd ICRC **2**, 420 (1991).
50. B. Peters, Nuovo Cimento (Suppl.) **14**, 436 (1959).
51. R.D. Blandford and J.P. Ostriker, Astrophys. J. **237**, 793 (1980).
52. C.J. Cesarsky, Proc. 20th ICRC **8**, 87 (1987).
53. P. Biermann, these Proceedings (1991).

HIGH ENERGY GAMMA RAY OBSERVATIONS

Carl E. Fichtel
NASA/Goddard Space Flight Center, Greenbelt, MD 20771

ABSTRACT

Since the trajectories of astrophysical charged particles are bent by magnetic fields and normally curl many, many times before their detection, their origin may not be inferred from their directions as is the case with photons. Fortunately, charged particles reveal their presence through interactions in many instances leading to high energy gamma rays. Bremsstrahlung, Compton, Synchrotron, and curvature radiation all generally have a monotomically decreasing energy spectrum reflecting that of the parent particles, whereas nucleon-nucleon radiation has a maximum at about 70 MeV reflecting the nature of the interaction process. Gamma radiation has been seen coming from neutron stars in pulses with the same period as the radio pulsar. Solar gamma rays also have been seen as have short bursts of gamma rays whose origin remain a mystery. The galactic diffuse gamma radiation which reveals the distribution of cosmic rays in our galaxy has been studied, and new results should add greatly to our understanding of the scale of coupling between cosmic rays and interstellar matter. Beyond our galaxy, active galaxies are seen in gamma rays implying a huge energy in the form of cosmic rays to be present there. The Optically Violent Variable quasar 3C 279 is particularly astounding. During an active state observed by the Energetic Gamma Ray Experiment Telescope (EGRET) on the Compton Gamma Ray Observatory, 3C 279 was seen to be emitting approximately 10^{48} erg s^{-1}, if its radiation is isotropic.

INTRODUCTION

By the very nature of the processes that produce them, astrophysical high energy gamma rays are related directly to high energy particles. As energetic particles interact with matter, photons, and magnetic fields, they produce gamma rays which carry information on the parent electrons and nuclei. These interactions may occur in the environment of a neutron star, a black hole, a supernova remnant, an active galaxy, or other object where relativisic particles are present. They may also occur in interstellar space as the cosmic rays collide with the diffuse matter and photons throughout our galaxy or other galaxies.

Since charged particle directions are quickly changed and scrambled on astronomical scales by magnetic fields, their origins cannot be determined by the directions of arrival; hence, high energy gamma ray astronomy provides the means of studying charged particles where direct measurements are not possible. Gamma rays also are relatively penetrating and hence may arrive from regions that may not be studied by some other forms of electromagnetic radiation. Thus, although the density of high energy gamma rays is low, they have compensating advantages that make their study of considerable value in the study of astrophysics energetic particles.

Until recently the information that existed on high energy gamma rays came mostly from SAS-2 and COS-B, supplemented by some high altitude scientific balloon flights. With the launch of the Arthur Holly Compton Gamma Ray Observatory, significant additional information has become available from the Energetic Gamma Ray Experiment Telescope (EGRET) which observes the high energy (20 MeV--30 GeV) gamma ray range with a sensitivity of from ten to twenty times that of the two earlier satellites just mentioned. Because the Compton Observatory has just recently been launched, the next section will give a brief description of EGRET. The subsequent three sections will describe information on charged particles in galactic sources, our own and other normal galaxies, and active galaxies.

THE EGRET INSTRUMENT

The EGRET instrument has the typical components used in high energy gamma ray telescopes: an anticoincidence dome to discriminate against charged particles, a spark chamber system with interspersed conversion material to convert the photons and determine the trajectories of the secondary electron-positron pairs, a triggering telescope that detects the presence of the pair with the direction of motion, and an energy measuring device, which in the case of EGRET is a NaI(Tl) crystal. A description of the instrument and its general capabilities is given by Kanbach[1]. The results of the instrument calibration, both before and after launch, will be given by Thompson et al.[2]. The telescope covers the energy from about 20 MeV to over 20 GeV. It has an effective area of about 1.5×10^3 cm^2 between 0.5 and 1.0 GeV., decreasing slowly above this energy and falling below this energy first slowly and then more rapidly to the low energy threshold. Strong, hard-spectrum sources may be located to five to ten arcmin. by EGRET. The instrument is designed to be free of internal background, and the calibration results have verified that the instrument background is at least an order of magnitude below the lowest natural level of diffuse background. The timing accuracy is 0.1 millisecond are better. Because of the very low flux levels of astrophysical gamma ray sources, the observing periods are typically 2 weeks in duration, although a few are shorter or longer.

GALACTIC SOURCES

a. Pulsars

Almost immediately after their discovery, pulsars were proposed to be associated with neutron stars[3], and this relationship is now generally accepted. The large release of energy, the very fast period, and the remarkably small variation of the period seem to dictate that the pulsed radiation must be from a massive object of small size.

The highest intensity gamma-ray pulsar is the one in Vela, PSR 9833-45, as observed at the Earth in the gamma-ray region for which $(1.2 \pm 0.2) \times 10^{-5}$ photons (E > 100 MeV) cm^{-2} s^{-2} are seen. The two most striking features of this pulsar, as first seen by SAS-2[4,5] are the two gamma-ray pulses as opposed to the one in the radio region, and the fact that neither gamma-ray pulse is in phase with the radio pulse. These features were confirmed by the data obtained later from the COS-B satellite gamma-ray telescope[6,7]. More recently high energy gamma ray observations of the Vela pulsar have been made by EGRET; the pulse profile is shown in Figure 1. Notice in particular how narrow the first pulse is. Although there is evidence of radiation between the pulses where they are closer together in time, there is none above the diffuse radiation in the region where the pulses are further apart.

The observational picture for the second strongest gamma-ray pulsar PSR 0532+21 in the Crab nebula is much simpler to describe. This pulsar, which is faster than PSR 0833-45 and was the first gamma-ray pulsar reported[9], is seen with the double pulsed structure in the radio, optical, X-ray and gamma-ray regions.

The pulse profile obtained by EGRET is similar to that of the Vela one as shown in Figure 2.

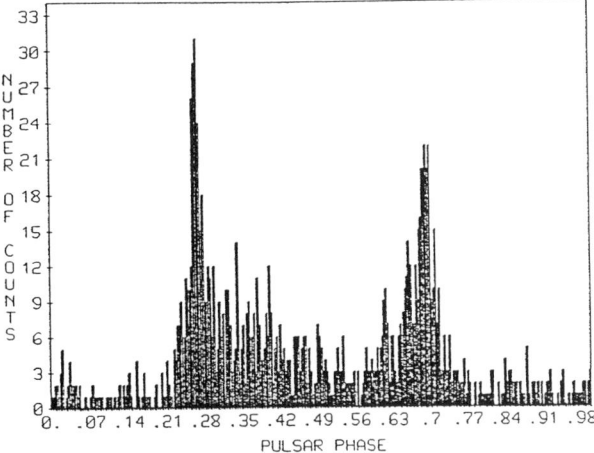

Fig. 1. Fine time resolution phase distribution of the gamma radiation (100 MeV) from PSR 0833-45 as observed by EGRET[8].

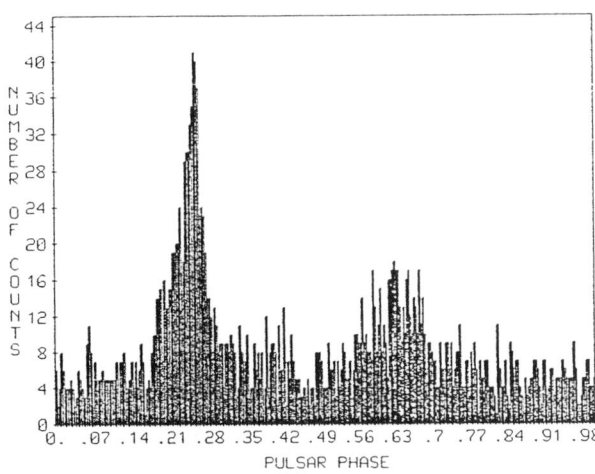

Fig. 2. Fine Time resolution phase distribution of the gamma radiation (> 50 MeV) from PSR 0531 + 21 as observed by EGRET[10].

There are two general classes of theoretical models for pulsars which seem to have evolved. In the outer gap model[11-13], the gamma radiation originates in the pulsar magnetosphere. There are two variations within this model, and it is possible that the two observed gamma-ray pulses are an example of each. In the polar cap model (Daugherty and Harding 1982), the gamma emission comes from the polar cap, arising from both curvature and synchrotron radiation in the strong magnetic fields. The photons formed in this way also interact until the cascade develops to the level where the gamma rays escape.

b. Supernovae

Supernovae have always been objects of special interest, and the detection of gamma ray lines over two decades ago[15] has added new excitement. There are also several mechanisms that could lead to high energy gamma rays from supernovae. For example, since it is generally believed that the cosmic rays below about 10^{15} eV are galactic, and since supernovae appear to be the most energetic events occurring in the galaxy, supernovae have been suggested as the primary sources of these cosmic rays. A continuum emission from the Crab supernova remmant has been again at very high energies (> 0.1 TeV) by Vacanti et al.[10].

Other supernovae have not been seen to have this type of radiation, but it may be only a question of sensitivity. Except in a region of abnormally high density, gamma-rays from cosmic ray interactions would not be expected to be seen from a supernova unless it was close on a galactic scale. Approximately a few times 10^{49} ergs per supernova in the form of relativistic particles are required if they are to be the source of cosmic rays. A straightforward calculation[17] using a density, the same as that locally gives an intensity of about 10^{-7} photons (e > 100 MeV) $cm^{-2} s^{-1}$ for a supernova 1 kpc away. The Vela supernova remnant is closer than the Crab, but in an apparently less dense region. Hence, the matter interaction continuum from the Vela supernova remnant probably would be difficult to observe.

c. Other Galactic Objects

Besides the two pulsars, there is only one other clearly established high energy gamma ray source; this is the one with galactic coordinates (195, + 4 1/2°) sometimes called Geminga. It was seen by both the SAS-2[18,19] and COS-B[20] gamma-ray telescopes. Because it is in a region of the galactic plane with a relatively low diffuse flux and no known large clouds, it is a strong candidate for a point source; however, there is no object at other wavelengths within the solid angle of uncertainty of source position which is generally accepted as the source of the gamma rays although candidates have been suggested.

Several reasonably conventional theories for the origin of the Geminga gamma radiation have been proposed, including a young pulsar whose radio emission is beamed in such a way that it is not seen at the Earth. An intriguing new proposal by Lake[21] is that the origin of the radiation lies in a lump of Weakly Interacting Massive Particles (WIMP's), i.e., annihilating dark matter. He notes, that, although this source is only 4 1/2° from the galactic plane, 9 percent of the celestial sphere is within 5° of the disk. Hence, it is not unreasonable that the source should be in the halo where the lumps of WIMP's would be expected. This proposed explanation might be confirmed with better energy spectral information.

Several localized excesses[22], as well as Cygnus X-3, seen by Galper et al.[23] and SAS-2[24] but not COS-B, remain interesting candidates for study by EGRET.

d. Solar Flares

From the measurements with the gamma-ray spectrometer on the Solar Maximum Mission (SMM), it is known that some cosmic gamma-ray bursts and solar flares have spectra that extend into the high energy (> 10 MeV) region. In solar flares, this hard component is believed to arise from bremssstrahlung and high energy ions that interact in the solar photosphere (π^o decay), whereas for gamma-ray bursts, neutron stars with high magnetic fields and possibly pulsar-like activity may be responsible.

EGRET observed high-energy gamma radiation above 30 MeV from the Sun following an intense flare around 2:00 U.T. on June 11, 1991. Gamma rays were detected up to energies of a GeV. After the decay of most of the X-ray flare, which causes substantial deadtime losses in EGRET, high-energy emission from the Sun was registered in the spark chamber during the interval from about 3:30 UT to at least 10:30 UT. The solar origin of this emission is assured by the time profile of the gamma-ray count rate and by time resolved sky maps, which show a clear maximum at the position of the Sun. A preliminary spectrum averaged over the existing data is shown in Figure 3.

e. Gamma Ray Bursts

The EGRET spectrometer described in Kanbach et al [1] records spectra of bursts up to 140 MeV with an adjustable integration time. In addition, the rate of one of the four sets of anticoincidence photomultiplier tubes is read every 1/4 s. This mode is continual. A burst was observed by both COMPTEL and EGRET on May 3, 1991. This is the first event that has been seen by imaging detectors.

Six individual photons were detected in the EGRET spark chamber, allowing a determination of the burst arrival direction which was $l^{II} = 171.9° \pm 1.3°$, $b^{II} = 5.3° \pm 1.1°$. In addition, three energy spectra were measured by EGRET from 1 to 200 MeV; they were measured during the first second after the BATSE trigger, the next 2 seconds and the subsequent 4 seconds. These spectra are shown in Figure 4. The first two spectra exhibit a similar differential spectra index of about -2.2 with no apparent high-energy cut-off. By the time of the third spectrum, an additional soft component is evident.

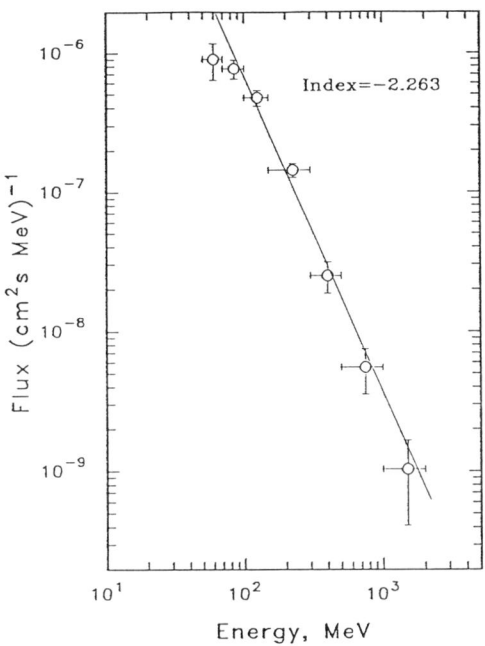

Fig. 3. Differential energy spectrum of the June 11, 1991 solar flare integrated over the event[25].

IV COSMIC RAYS AND THE DIFFUSE GAMMA RADIATION

a. Our Galaxy

It is well known now from the results of OSO-3, SAS-2, and COS-B that the high energy gamma ray sky is dominated by radiation from the galactic plane. The galactic diffuse radiation was well anticipated before it was observed. As early as 1952, Hayakawa[27] noted the importance of meson-producing interactions between cosmic rays and the interstellar gas. In the same year, Hutchinson[28] discussed the production of Bremsstrahlung radiation by the cosmic rays. Even earlier, Feenberg and Primakoff[29] examined the astrophysical significance of the Compton effect in regard to cosmic ray electrons, although it is now realized that Compton radiation is the smallest of the three.

Although the point source contribution to the high energy diffuse gamma radiation cannot be estimated with certainty, several factors suggest that point sources may not be a major contributor. These include the fact that the distribution of the great majority of the high energy gamma radiation along the galactic plane can be reproduced in great detail with calculations based on cosmic ray interactions with the galactic matter

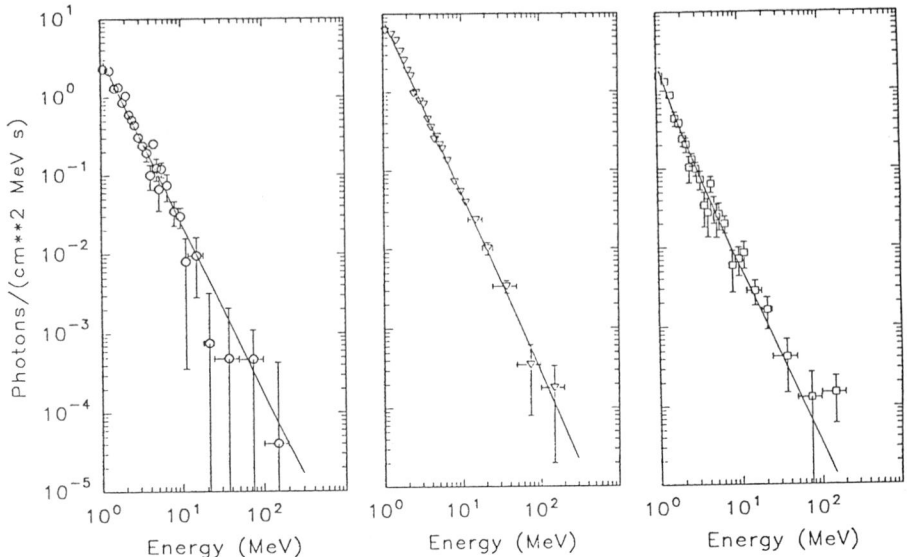

Fig. 4. Energy spectra from the Total Absorption Shower Counter[26]. The three spectra were acquired sequentially with accumulation times of 1, 2, and 4 seconds for (a), (b), and (c), respectively. Power law fits in the form $AE^{-\gamma}$ where E is in MeV were made. For spectrum (a), $\gamma = 2.13 \pm 0.08$ and $A = 3.26 \pm 0.25$. For spectrum (b), $\gamma = 2.24 \pm 0.03$ and $A = 8.71 \pm 0.49$. Spectrum (c) does not fit a single power law. As discussed in the text, the fit is the sum of two power laws where one is constrained to have the average index of the other two, $\gamma = 2.22$, with a fitted value of $A = 0.935 \pm 0.018$. The second power law has index $\gamma = 4.0 \pm 0.8$ and $A = 1.40 \pm 0.03$ and contributes mainly at low energies.

and photons, the energy spectrum is well represented by the results of these same calculations, and the energy spectrum uniformity along the plane. As an example of the spectrum which results from the three interaction processes mentioned above, the calculated energy spectrum[17] of the galactic gamma radiation for a region near the galactic center is shown in Figure 5 and compared to data. The spectral shape elsewhere in the galactic plane is nearly the same with only minor variations resulting from small differences in the relative number of secondary electrons and the relative contribution of the Compton electrons.

Even the earliest SAS-2 results of Kniffen et al.[31] showed the general correlation of the gamma radiation with the galactic structure, and the final results of SAS-2 and COS-B show it quite clearly. Detailed calculations comparing the predictions of the expected galactic high energy gamma ray distribution with that which is observed have been made by many authors. The general approach is outlined by

Fichtel and Trombka[17], as well as elsewhere. In general, the molecular hydrogen normalization is treated as an adjustable parameter as long as the result lies within allowable limits. The Compton radiation is the most uncertain of the three because of the limited knowledge of the photon uncertain of the three because of the distribution, but since it is a small contributor, this situation is not a major concern.

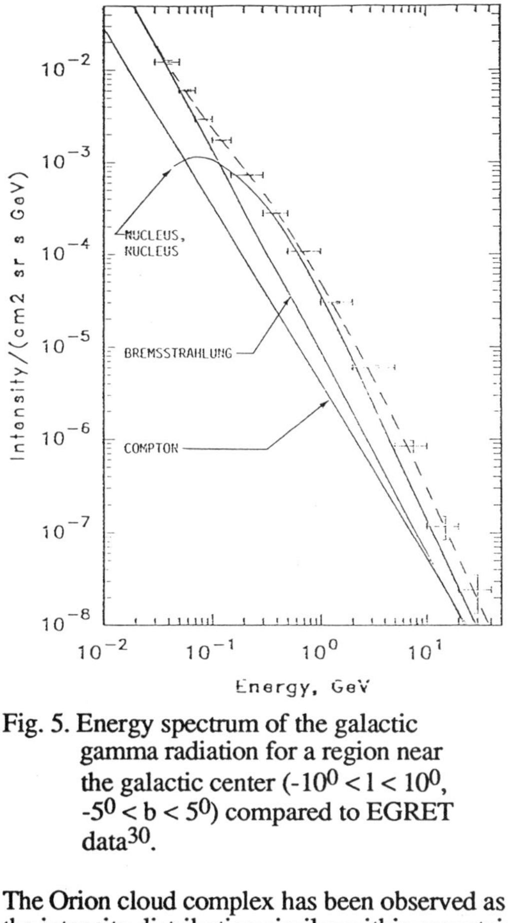

The question of whether the cosmic ray density is correlated with the matter, and, if it is, on what scale has not been resolved, in large part because of the lack of the more detailed data required for a detailed correlation analysis. The most significant complication is that since the molecular hydrogen is mathematically similar to assuming a positive cosmic ray gradient towards the center in terms of the gamma-rays produced. When the full sky survey by GRO is completed, currently estimated to be August 1992, data from EGRET should exist to go far in answering the question of cosmic ray density distribution.

Galactic molecular clouds contain much of the diffuse galactic matter and are generally believed to be the location for the formation of stars. There are, however, many unanswered questions about the nature of these regions and their formative processes. The limited sensitivity and angular resolution of the SAS-2 and COS-B instruments has restricted the gamma-ray information on these objects, however, two positive identifications were made with the COS-B telescope, namely and the Orion cloud complex.

Fig. 5. Energy spectrum of the galactic gamma radiation for a region near the galactic center ($-10^0 < l < 10^0$, $-5^0 < b < 5^0$) compared to EGRET data[30].

The Orion cloud complex has been observed as an extended gamma-ray source[32] with the intensity distribution similar within uncertainties to the estimated matter distribution. The gamma-ray excess[33] with the cloud seems quite probable. There is, at the present, a controversy over whether or not the gamma ray intensity is more than would be expected on the basis of cosmic rays at the local intensity level interacting with the matter in the cloud.

ACTIVE GALAXIES

Prior to the observations of EGRET, only one active galaxy, 3C 273, had been seen in the high energy (E ≥ 30 MeV) gamma ray range. One radio galaxy and two Seyferts had been seen in the low energy gamma ray interval, and many more active galaxies had been observed in the hard X-ray range. For many of these observed

X-ray galaxies, as well as the active galaxies observed in low energy gamma rays, the upper limits derived from SAS-2 and COS-B high energy gamma ray data are substantially (more than an order of magnitude) below an extrapolation of a power law spectra, suggesting that a sharp spectral change in the low energy gamma ray region may be a general feature of these galaxies[34]. (See e.g. Fichtel and Trombka[17])

3C 273 is the brightest X-ray quasar and was the first quasar identified as a source of gamma rays[34]. The differential energy spectrum of 3C 273 steepens from the X-ray range to the gamma ray region, with the slope of the differential energy spectrum changing from 1.4 in the hard X-ray region to 2.7 in the high energy (E > 50 MeV) gamma ray region. The COS-B instrument observed 3C 273 in gamma rays several times, and no significant variation in the flux among the observation was observed[35] (Bignami et al. 1981). The spectrum, shown in Fig. 6, is consistent with several of the massive black hole models, including the synchrotron, self Compton type models, possibly including some degree of subsequent photon-photon interactions and the Penrose pair production and Compton scattering processes, involving infalling protons and electrons.

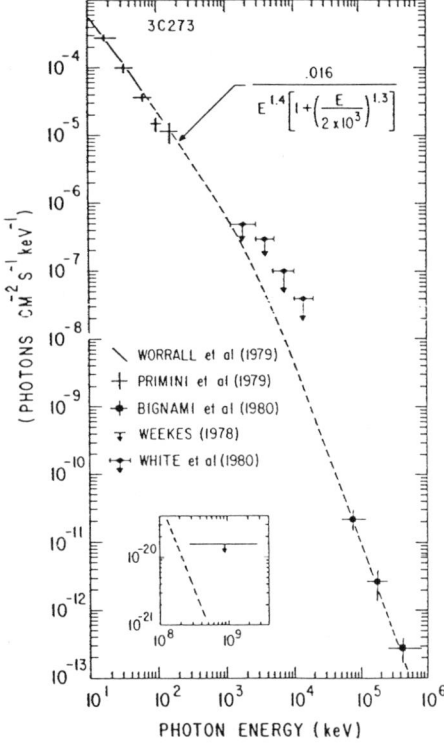

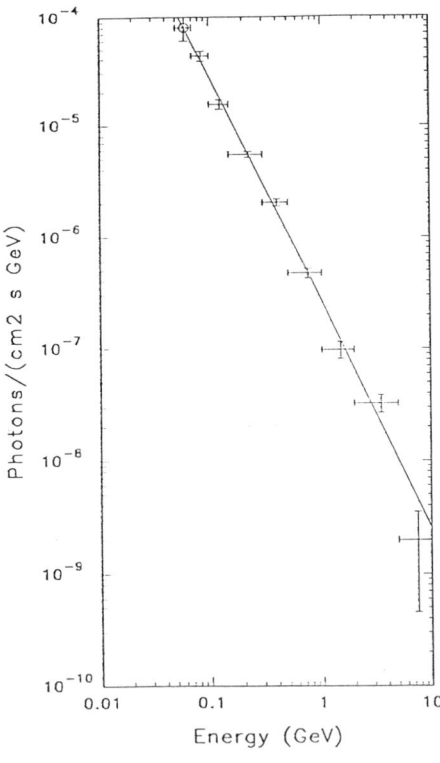

Fig.6. X- and Gamma Ray Spectrum of 3C-273 (Fichtel and Trombka[17])

Fig. 7. A differential high-energy gamma ray spectrum observed for 3C 279 during the period June 15-28, 1991.[36]

EGRET has now observed intense gamma radiation from the direction of the quasar 3C 279 throughout the energy range from 30 MeV to over 5 GeV during the period 1991 June 15-28 [36]. Its spectrum is well represented by a photon differential power law exponent of 2.0 ± 0.1, with a photon intensity above 100 MeV of (2.8 ± 0.4) x 10^{-6} cm^{-2} s^{-1}. 3C 279 was not detected by either of the earlier high energy gamma ray telescopes SAS-2 or COS-B. For E > 100 MeV, the 2 σ upper limits were 1.0 x 10^{-6} cm^{-2} s^{-1} in 1973 from the SAS-2 observations and 0.3 x 10^{-6} cm^{-2} s^{-1} for the combined 1976, 1978, and 1980 COS-B observations. Hence, there has been a large increase in high energy gamma ray intensity relative to the earlier times, as there has been in the radio, infrared, optical, and X-ray ranges. This source is the most distant and by far the most luminous gamma ray source yet detected.

Figure 7 shows the observed spectrum with the best power law from a least squares fit; the 1 σ errors are statistical only. Although the sensitivity analysis is still undergoing some refinement, the systematic errors associated with the instruments sensitivity values are believed to be less than 15% except below 70 MeV.

The Optically Violent Variable (OVV) quasar 3C 279 (z = 0.538) has exhibited several types of activity over the time it has been observed. It was the first quasar in which apparent superluminal motion was deteccted[37], and that type of activity continues (e.g., Unwin et al [38]). Makino et al.[39] have observed a 20% change in X-ray intensity (2-10 keV) on a time scale of less than 1 hour. In 1988 and 1989, its luminosity in the radio, infrared, optical, and X-ray bands increased and reached a maximum level as much as an order of magnitude higher than that seen earlier[40-42]. Since late 1990, the radio brightness of 3C 279 has again increased, to an even higher level than in 1988, and its X-ray intensity has reached a level similar to that in 1988[43]. This unusual level of activity, as well as the general character of 3C 279, suggest that it is a prime target for study in the high energy gamma ray range. It was, for example, one of four quasars listed by Kanbach et al.[1] as likely candidates to be examined by EGRET.

Figure 8 combines the present result with measurements in lower frequency ranges made at earlier times. It is seen that the energy output per logarithmic energy interval observed in the high energy gamma ray band is larger than observed previously in any other frequency band. (Note, however, that the EGRET observations are not simultaneous with those at lower frequencies.) Also, a comparison of the slopes of the spectra in the X-ray and gamma ray ranges suggests that the spectrum must steepen slightly between these two photon energy ranges.

If the emission from 3C 279 is isotropic, its gamma ray luminosity between 100 MeV and 10 GeV is about 1.6 x 10^{51} photons s^{-1}, or approximately 1.1 x 10^{48} erg s^{-1} for H_0 = 75 km s^{-1} Mpc^{-1}. Should this be true and the central object be a black hole, the Eddington limit would imply a mass near 10^{10} $M_\odot$. However, as discussed below, it may very well be that the emission is beamed; for one steradian it would be about 10^{47} erg s^{-1}, for example. For comparison, our own galaxy emits approximately 5 x 10^{38} erg s^{-1} (E > 100 MeV).

Numerous mechanisms have been suggested for producing gamma radiation in association with quasars. A sample of these is summarized by Hartman et al.[36]. The shape and spectral index observed for 3C 279 suggest the presence of an extraordinary source of relativistic particles with an energy spectrum at least as hard as that observed for charged cosmic rays near the Earth. There is no way of knowing from these data whether the spectrum extends across the many additional decades to the extreme energy required by the extragalactic part of the cosmic ray spectrum. However, this evidence for a large concentration of relativistic particles in a quasar, with energies extending to

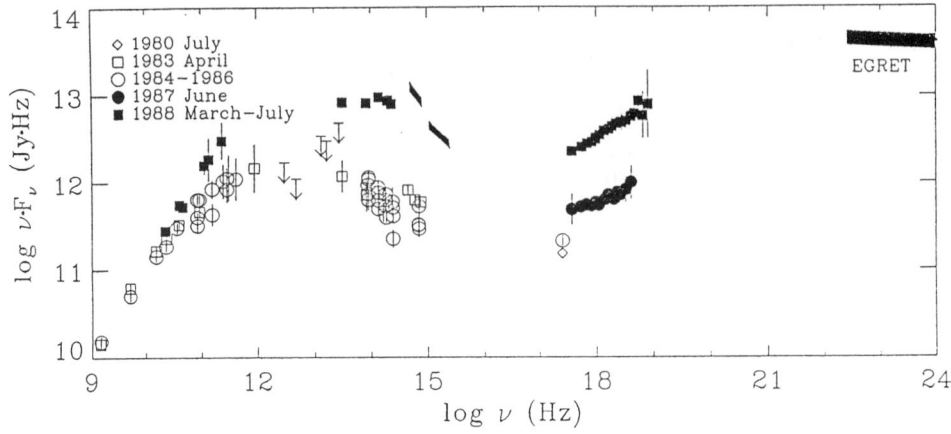

Fig. 8. The representation of the multifrequency spectrum of 3C 279 for the quiescent (open symbols), and enhanced (filled circles) periods are adapted from Makino et al.[39] The quiescent data are from Wilkes and Elvis[44], Landau et al.[45], and Brown et al.[46]. The enhanced data from the period March-July 1988 are from Urry (1988), Peterson, Wagner and Korista (1988), Neugebauer[49], Teräsranto[50] and Matsuo et al.[51]. The data of EGRET are indicated by its name[36].

at least 10 GeV, could provide some support for the concept that quasar jets may be the source of the extragalactic cosmic rays, as suggested by Colgate[52]. Gamma ray observations of 3C 279 in its flaring state by ground-based TeV- and PeV-energy telescopes could prove to be valuable in addressing this question.

POSTLUDE

Two other results of interest to this conference have been reported between the time of the conference and the submission of this paper, January 30, 1992. Both are contained in IAU circular 5431[53] and were presented at the January, 1992 American Astronomical Society meeting[30]. They are summarized below.

Three other quasars have been tentatively identified from the EGRET data, indicating that a class of quasars may be emitting high energy gamma rays in copious quantities at least some of the time. They are 4C 38.41, PKS 0528+134, and PKS 0208-512. All are radio loud, flat spectrum sources. A large fraction of quasars cannot be emitting such great quantities of high energy gamma rays continuously, or their combined radiation would far exceed the observed diffuse radiation. These observations do, however, give some further support to the possibility that quasars may be the source of the extragalactic cosmic rays, discussed above.

The other result is the time variability of the 3C 279 source. During the period from June 15 to 28, the high energy gamma radiation was observed to rise by a factor of about four or five to a maximum on June 25 and then fall sharply over the next three days. Based on the standard arguements involving the speed of light, this result implies a source region of less than about 10^{-3} pc. Combined with the very large total amount of energy released, this result seems to strongly favor the interpretaion of 3C 279 as a supermassive black hole.

REFERENCES

1. Kanbach, G. et al 1988, Space Science Reviews, 49, 669
2. Thompson et al., 1992, to be submitted to the ApJ Suppl.
3. Gold, T. 1968, Nature, 218, 731
4. Thompson, D. J., Fichtel, C. E., Kn iffen, D. A., and Ögelman, H. B., 1975 ApJ Letters, 200, L79
5. Thompson, D. J., Fichtel, C. E., Kniffen, D. A., and Ogelman, H. B. 1977, ApJ Letters, 214, L17
6. Buccheri, R., et al. 1978, A&A, 69, 141
7. Kanbach, G et al. 1980, AA, 90, 163
8. Kanbach et al., 1992a, to be published
9. Browning, R., Ramsden, D., and Wright, P. J. 1971, Nature, 232, 99
10. Nolan et al., 1992, to b e published
11. Cheng, K. S., Ho, C., and Ruderman, M. 1986a, ApJ, 300, 500
12. Cheng, K. S., Ho, C., and Ruderman, M. 1986b, ApJ, 300, 522
13. Ruderman, M. 1990, NASA Conference Publication 3071, Eds. C. Thompson, D. J., Fichtel, C. E., Kniffen, D. A., and Ögelman, H. B. 1975, ApJ Letters, 200, L79
14. Daugherty, J. K. and Harding, A. 1982, ApJ, 252, 337
15. Clayton, D. D., Colgate, S. A., and Fishman, G. J. 1968, ApJ
16. Vacant, G, et al. 1991, ApJ, 377, 467-479
17. Fichtel, C. E. & Trombka, J. I. NASA SP-453 Fichtel, S. Hunter, P. Sreekumar, and F. Stecker, 1989
18. Fichtel, C. E. et al.1975, ApJ, 198, 163
19. Thompson, D. J. et al. 1977, ApJ, 213, 252
20. Masnou, J. L. et al. 1977, ESA SP-124, 33
21. Lake, George, 1990, to be published in Nature
22. Mayer-Hasselwander, H. A. & Simpson, G. 1988, 27th Cospar
23. Galper, A. M.et al. 1975, 14th ICRC, 1, 95
24. Lamb, R. C. et al 1977, ApJ Letters, 212.L63
25. Kanbach et al., 1992b, to be published
26. Schneid et al., 1992 to be published in Ast. and Astrophys
27. Hayakawa, S. 1952, Prog. Theor. Phys, 8, 571
28. Hutchinson, G. W. 1952, Phil. Mag., 43, 847
29. Feenberg, E & Primakoff, H. 1948, Phys. Rev, 73, 449
30. Fichtel, C. E., et al., 1992, American Astronomical Society Meeting January 12-16. 992, 44.01
31. Kniffen, D. A. et al. 1973, ApJ Letters, 186, L105
32. Caraveo et al., 1981, 17th Int. Cosmic Ray Conf. Vol. 1, 139
33. Swanenburg et al., 1981, ApJ Letters 243, L69
34. Swanenburg, B. N. et al. 1978, Nature, 275, 298
35. Bignami, G. F. et al. 1981, A&A, 93, 71
36. Hartman, R. C. et al 1991, to be published in the ApJ Letters
37. Whitney, A. R. et al. 1971, Science, 173, 225
38. Unwin, S. C., Cohen, M. H., Biretta, J. A., Hodges, M. W., & Zensus, J. A. 1989, ApJ, 340, 117

39. Makino, F., et al. 1989, ApJ (Letters), 347, L9
40. Robson, E. L., Smith, M. G., Aycock, J., & Walter, D. M. 1988, IAU Circ. No. 4556
41. Kidger, M. R. & Allan, P. M. 1988, IAU Circ. No. 4595
42. Makino, F. & Ohaski, E., 1989, IAU Circ. No. 4736
43. Makino, F., Fink, H. H., & Clavel, J. 1991, Nagoya Conference Proceedings, 1
44. Wilkes, B. J. & Elvis, M., 1987, ApJ, 323, 243
45. Landau, R. et al. 1986, ApJ, 308, 78
46. Brown, L. M. J. et al 1989, ApJ, 340, 129
47. Urry, C. M. 1984, Ph.D. thesis, The Johns Hopkins University (NASA Technical Memorandum 86103)
48. Peterson, B. M., Wagner, R. M., & Korista, K. T. 1988, IAU Circ. No. 4634
49. Neugebauer, G. 1988, private communication
50. Teräsranta, H. 1988, private communication
51. Matsuo et al., 1989, Pub. Astr Soc. Japan, 41, 865
52. Colgate, S., 1990, Americal Astronomical Society Meeting, June 10-14, 1990, 51.01
53. Kanbach, G., et al., 1992, I. A. U. Circular 5431

Transport Theory

A REVIEW OF TRANSPORT THEORY

Frank C. Jones
Laboratory for High Energy Astrophysics, Code 665
Nasa/Goddard Space Flight Center
Greenbelt, MD 20771
U.S.A.

ABSTRACT

The acceleration of particles in a plasma should be governed by a transport equation that describes how the particles are transported in space and in energy. A complete equation would describe all possible ways in which particles can exchange energy and momentum with the plasma in which they are embedded. Such equations have been developed through the years and today represent most of the processes thought to be important in the acceleration of particles in astrophysical plasmas. We shall discuss how these equations are derived and how the various terms relate to the modes of energy exchange between the particles and plasma.

INTRODUCTION

The theory of the transport of energetic charged particles in a moving plasma is contained, almost in its entirety, in one equation, the so called Diffusion-Convection Equation. This equation was first applied to modulation theory by Parker in 1958[1]. In 1965[2] he modified it to include the energy changes that the solar wind produced in the cosmic ray particles, the main effect of which is the "adiabatic cooling" of the particles during the time that they are in the heliosphere. (For a different approach to deriving this equation see the work of Dolginov and Toptygin[3,4].) This equation was not applied to particle acceleration as such until 1977 when it was used [5,6,7] in the theory of diffusive shock acceleration.

In a certain sense, a complete transport equation will describe all possible particle acceleration processes since it will include transport along the energy axis as well as in three dimensional space. In this talk I will attempt to show how energy change terms arise in the transport equation and what their effects can be. Further, we will show how a more generalized form of the diffusion-convection equation may be derived and see that it leads to an additional mode of energy change for the charged particles.

DERIVATION OF TRANSPORT EQUATION

We first begin with the conservation equation for the energetic particles in a flowing plasma;

$$\frac{\partial U(\vec{x},p)}{\partial t} + \vec{\nabla}\cdot\vec{S}(\vec{x},p) + \frac{\partial S_p(\vec{x},p)}{\partial p} = 0 \qquad (1)$$

where

$$U(\vec{x},p) = p^2 \int d\hat{p} f(\vec{x},\vec{p}) \qquad (2)$$

and

$$\vec{S}(\vec{x},p) = \frac{p^3}{\gamma m}\int d\hat{p}\,\hat{p}\, f(\vec{x},\vec{p}) \qquad (3)$$

(we will discuss $S_p(\vec{x},p)$ shortly).

If we assume that $\overline{f}(\vec{x},\vec{p})$ can be described by its first two moments and that Fick's Law applies in the plasma rest frame we may write

$$\overline{f}(\vec{x},\overline{p}) = \frac{1}{4\pi \overline{p}^2}\left\{\overline{U}(\vec{x},\overline{p}) + \frac{3m\overline{\gamma}}{\overline{p}}\overline{\hat{p}}\cdot\overline{\vec{S}}(\vec{x},\overline{p})\right\} \qquad (4)$$

© 1992 American Institute of Physics

and
$$\overline{\vec{S}(p)} = -\overleftrightarrow{K} \cdot \vec{\nabla}\overline{U}(p) \tag{5}$$
where items with an overbar are in the plasma rest frame.

Since $f(p)$ is an invariant[8]
$$\overline{f}(\overline{p}) \to f(p) = \overline{f}(\overline{p}) \tag{6}$$
and since $\vec{p} = \overline{\vec{p}} + m\overline{\gamma}\vec{V}$ we have
$$f(\vec{p}) = \overline{f}(\vec{p} - m\overline{\gamma}\vec{V}) \approx \overline{f}(\vec{p}) - m\overline{\gamma}\vec{V} \cdot \frac{\partial}{\partial \vec{p}}\overline{f}(\vec{p}) \tag{7}$$

Noting that
$$\frac{\partial}{\partial \vec{p}} = \hat{p}\frac{\partial}{\partial p}$$
we obtain
$$f(\vec{p}) = \frac{1}{4\pi p^2}\left\{\overline{U}(p) - \frac{3m\overline{\gamma}}{p}\hat{p} \cdot \overleftrightarrow{K} \cdot \vec{\nabla}\overline{U}(p)\right\}$$
$$-\frac{m\overline{\gamma}}{4\pi}\frac{\partial}{\partial p}\left\{\vec{V} \cdot \hat{p}\frac{\overline{U}(p)}{p^2} - \frac{3m\overline{\gamma}}{p^3}\vec{V} \cdot \hat{p}\hat{p} \cdot \overleftrightarrow{K} \cdot \vec{\nabla}\overline{U}(p)\right\} \tag{8}$$
where for the sake of consistency we must drop the last term that is a tensor addition to the pressure that is proportional to $\hat{p}\hat{p}$.

Applying the definitions 2 and 3 to equation 8 we obtain
$$U(p) = \overline{U}(p) \tag{9}$$
$$\vec{S}(p) = -\overleftrightarrow{K} \cdot \vec{\nabla}U(p) - \frac{1}{3}\vec{V}p\frac{\partial U(p)}{\partial p} + \frac{2}{3}\vec{V}U(p)$$
$$= -\overleftrightarrow{K} \cdot \vec{\nabla}U(p) - \frac{1}{3}\vec{V}\frac{\partial}{\partial p}(pU(p)) + \vec{V}U(p) \tag{10}$$

We have the first two terms in equation 1 and we must now find the flux in momentum space $S_p(\vec{x}, p)$; to this end we note that the pressure exerted by the energetic particles is given by:
$$P = \frac{p^4}{m\gamma}\int d\hat{p}\hat{p}\hat{p}f(\vec{x}, \vec{p})$$
$$= \frac{1}{3}U(p)pv \tag{11}$$

The force density due to the gradient of this pressure is
$$\vec{F} = -\frac{1}{3}v\vec{\nabla}(U(p)p) \tag{12}$$
and the work done per unit volume by the flowing plasma against this force density is just
$$W = -\vec{V} \cdot \vec{F} = \frac{1}{3}v\vec{V} \cdot \vec{\nabla}(U(p)p) = <\dot{\varepsilon}> \tag{13}$$
where ε is the energy density of the particles with momentum p.

The flux along the momentum axis is just the change in momentum density per unit time so we have
$$S_p = <\dot{p}> = \frac{1}{v}<\dot{\varepsilon}> = \frac{1}{3}\vec{V} \cdot \vec{\nabla}(U(p)p) \tag{14}$$

Inserting the values obtained for $U(p)$, $\vec{S}(p)$ and $S_p(p)$ into equation 8 we obtain

$$\begin{aligned}\frac{\partial U}{\partial t} &= \vec{\nabla} \cdot \left(\overleftrightarrow{K} \cdot \vec{\nabla} U - \vec{V} U + \frac{1}{3}\vec{V}\frac{\partial}{\partial p}(Up)\right) - \frac{1}{3}\frac{\partial}{\partial p}\left(\vec{V} \cdot \vec{\nabla}(Up)\right) \\ &= \vec{\nabla} \cdot (\overleftrightarrow{K} \cdot \vec{\nabla} U - \vec{V} U) + \frac{1}{3}(\vec{\nabla} \cdot \vec{V})\frac{\partial}{\partial p}(Up) = 0 \end{aligned} \quad (15)$$

which is the diffusion-convection equation including the adiabatic heating or cooling depending on the convergence or divergence respectivly of the plasma flow velocity $\vec{V}$.

WHAT IS THE ACCELERATION TERM?

There has been some discussion as to whether the first or second line of equation 15 better represents the true picture of energy change for the energetic particles being carried along by the flowing plasma. The term S_p does represent the local rate with which the plasma and the particles are exchangeing energy (or rather momentum) and it is claimed by some that this is the appropriate term to describe the energization or cooling of the particles. To answer this question it is useful to consider the behavior of some average property of the particle distribution. If $\alpha(\vec{x},p)$ represents some function over the four dimensional phase space, $\vec{x}, p$ its average value is given by

$$\langle\alpha\rangle = \int d^3x dp \alpha(\vec{x},p) U(\vec{x},p) \quad (16)$$

and the rate of change of $\langle\alpha\rangle$ with time is

$$\frac{d\langle\alpha\rangle}{dt} = \int d^3x dp \alpha(\vec{x},p) \frac{\partial U(\vec{x},p)}{\partial t} \quad (17)$$

which after several integrations by parts yields

$$\frac{d\langle\alpha\rangle}{dt} = \int d^3x dp \left[(\vec{V} + \vec{\nabla}\cdot\overleftrightarrow{\kappa})\cdot\vec{\nabla}\alpha + \frac{p}{3}\vec{V}\cdot\vec{\nabla}\frac{\partial\alpha}{\partial p} - \frac{p}{3}\vec{\nabla}\cdot(\vec{V}\frac{\partial\alpha}{\partial p})\right] U. \quad (18)$$

We see that the second term which comes from the second term of the top line of equation 15 and represents the Compton-Getting effect makes no contribution if α is purely a function of $\vec{x}$ or p; in other words it does not move the center of gravity or mean momentum of the particle distribution. Also we note that although there is no diffusive flux unless there is a spatial gradient of U the mean position (given by setting $\alpha = \vec{x}$) is changed only by the gradient of κ.

If we set $\alpha = p$ on the other hand we see that the mean value of the momentum (or energy) is shifted by the term proportional to the divergence of the flow velocity. If we represent a single particle at a position $\vec{x}_s$ and momentum p_s by a delta function distribution function $U(\vec{x},p) = \delta(\vec{x} - \vec{x}_s)\delta(p - p_s)$ we see that the term

$$\dot{p} \equiv -\frac{p_s}{3}\vec{\nabla}\cdot\vec{V} \quad (19)$$

gives the rate of change with time of the momentum of an individual particle and should be considered as the acceleration term.

DERIVATION FROM THE BOLTZMANN EQUATION

In the above derivation of the transport equation no account was taken of the *physics* of the interaction between the plasma and the particles; it was merely assumed that in the plasma rest frame the particles diffused passively with a phenomenological diffusion tensor $\overleftrightarrow{\kappa}$ (it was made a tensor to allow for the possibility that diffusion might proceed at different rates in different directions). A more correct procedure would be to employ the Boltzmann Equation to describe the evolution of the particle distribution function in the presence of large scale electric and magnetic fields and scattering centers that could be in motion with respect to the plasma.

Gleeson and Axford[9] were the first to derive the transport equation from the Boltzmann equation: where the left hand side represents the streaming of the particles through the large scale, average force fields represented by $\vec{F}$ and the right hand side represents the change in the distribution function f due to the stochastic forces or scattering.

In their original work[9] Gleeson and Axford treated the scattering term as arising from hard sphere scattering and they did not include the effects of large scale magnetic or electric fields (*i.e.* they set $\vec{F} = 0$.). If, however, we wish to investigate the full origin of the convective velocity such forces must be included. This may be done by including the force term.

The Boltzmann Equation is

$$\frac{\partial f}{\partial t} + \vec{v} \cdot \vec{\nabla} f + \vec{F} \cdot \vec{\nabla}_p f = \left(\frac{\partial f}{\partial t}\right)_c \qquad (20)$$

where the force field $\vec{F}$ is produced by the electric and magnetic fields. Since the only large scale electric field that can persist in a plasma is that produced by the plasma's flow speed, $\vec{V}_p$ we may write

$$\vec{F} = \frac{\vec{p} \times \vec{\omega}}{\gamma} - m\vec{V}_p \times \vec{\omega} = (\vec{p} - m\gamma\vec{V}_p) \times \left(\frac{\vec{\omega}}{\gamma}\right) \qquad (21)$$

where $\vec{p}$, m, and γ are the particles momentum, mass, and Lorentz factor respectively, $\vec{V}_p$ is the plasma flow velocity and $\vec{\omega} = e\vec{B}/mc$ is the particles local gyrofrequency.

The collision or scattering term is given by

$$\left(\frac{\partial f}{\partial t}\right)_c = \int d^3V d^3p' \{f(\vec{p}',\vec{x})F(\vec{V},\vec{x})\sigma(\vec{V},\vec{p}' \to \vec{p}) \mid \vec{v}' - \vec{V} \mid$$
$$- f(\vec{p},\vec{x})F(\vec{V},\vec{x})\sigma(\vec{V},\vec{p} \to \vec{p}') \mid \vec{v} - \vec{V} \mid\} \qquad (22)$$

From this beginning the derivation of the transport equation is long and quite tedious; the interested reader is referred to the literature.[10] Here we will simply state the assumptions that are required and state the result. First the assumptions are:

• In the rest frame of the scatterer the scattered particles do not change their energy in a single scattering.

• The scattering process is such that it will drive the particle distribution towards isotropy in this frame.

• The momentum transfer, averaged over the scattering cross section, may be written as a tensor acting on the original momentum.

Nothing was assumed about the strength of the scattering so the results are as valid for weak, pitch angle scattering as they are for hard spheres.

The equation that is obtained is:

$$\frac{\partial U}{\partial t} - \vec{\nabla} \cdot \left(\overleftrightarrow{K} \cdot \vec{\nabla} U - \vec{V}_1 U + \frac{1}{3}\vec{V}_1 \frac{\partial}{\partial p}(Up)\right) + \frac{1}{3}\frac{\partial}{\partial p}\left(\vec{V}_2^T \cdot \vec{\nabla}(Up)\right)$$
$$= -\frac{1}{3}\frac{\partial}{\partial p}\left\{(m\gamma)^2 p^2 \frac{\partial}{\partial p}\left(\frac{U}{p^2}\right)\left[\langle \delta \vec{V} \cdot \overleftrightarrow{\nu} \cdot \delta \vec{V} \rangle\right.\right.$$
$$\left.\left. + (\vec{V}_s - \vec{V}_p) \cdot \overleftrightarrow{\nu} \cdot (\overleftrightarrow{\nu} + \overleftrightarrow{\Omega})^{-1} \cdot \overleftrightarrow{\Omega} \cdot (\vec{V}_s - \vec{V}_p)\right]\right\}. \qquad (23)$$

where $\overleftrightarrow{K}$ is the diffusion tensor given by:

$$\overleftrightarrow{K} = -\frac{1}{3}\left(\frac{p}{m\gamma}\right)^2 (\overleftrightarrow{\nu}^T - \overleftrightarrow{\Omega})^{-1}, \qquad (24)$$

and $\overleftrightarrow{\nu}$ is the momentum transfer scattering frequency tensor, $\overleftrightarrow{\Omega}$ is the gyrofrequency tensor defined by $\overleftrightarrow{\Omega} \cdot \vec{A} = \vec{\omega} \times \vec{A}$.

The superscript (T) indicates the transposed matrix or vector. The velocities of the individual scattering centers fluctuate about their mean velocity $\vec{V}_s$ by an ammount $\delta \vec{V}$ and the angle brackets indicate an average over the scattering centers' velocity distribution function.

The left hand side would be identical to that of equation 15 except for the appearance of the two flow velocities $\vec{V}_1$ and $\vec{V}_2$ instead of the single one $\vec{V}$. The right hand side is now non zero and reflects the inclusion of effects that produce momentum diffusion or *stochastic acceleration*.

The two velocities $\vec{V}_1$ and $\vec{V}_2$ are tensor averages of the plasma flow velocity $\vec{V}_p$ and the mean flow velocity of the scattering centers $\vec{V}_s$ given by

$$\vec{V}_1 = (\overleftrightarrow{\nu}^T - \overleftrightarrow{\Omega})^{-1} \cdot (\overleftrightarrow{\nu}^T \cdot \vec{V}_s - \overleftrightarrow{\Omega} \cdot \vec{V}_p) \qquad (25)$$

$$\vec{V}_2 = (\overleftrightarrow{\nu}^T + \overleftrightarrow{\Omega})^{-1} \cdot (\overleftrightarrow{\nu}^T \cdot \vec{V}_s + \overleftrightarrow{\Omega} \cdot \vec{V}_p). \qquad (26)$$

If we examine the definitions of the velocities $\vec{V}_1$ and $\vec{V}_2$ we can see that simplifications occur in certain cases. In particular if the plasma and scattering centers flow with the same velocity we have

$$\vec{V}_1 = \vec{V}_2 = \vec{V}_p = \vec{V}_s$$

and we obtain the original Convection Diffusion equation, with second order Fermi acceleration due to the random component of the scatterers motion. Further, if $\overleftrightarrow{\Omega} \gg \overleftrightarrow{\nu}$ (as is usually the case) and $\vec{V}_p$ has a significant component perpendicular to $\vec{B}$ such that $\overleftrightarrow{\Omega} \cdot \vec{V}_p \gg \overleftrightarrow{\nu}^T \cdot \vec{V}_s$ we obtain again a simpification that may be seen from the following considerations.

With a slight rearrangement we may write $\vec{V}_1$ and $\vec{V}_2$ as

$$\vec{V}_1 = \vec{V}_s - (1 - \overleftrightarrow{\Omega}^{-1} \cdot \overleftrightarrow{\nu}^T)^{-1} \cdot \overleftrightarrow{\Omega}^{-1} \cdot \overleftrightarrow{\Omega}(\vec{V}_s - \vec{V}_p)$$

$$\vec{V}_2 = \vec{V}_s - (1 + \overleftrightarrow{\Omega}^{-1} \cdot \overleftrightarrow{\nu}^T)^{-1} \cdot \overleftrightarrow{\Omega}^{-1} \cdot \overleftrightarrow{\Omega}(\vec{V}_s - \vec{V}_p) \qquad (27)$$

where $\overleftrightarrow{\Omega}^{-1}$ can *not* be regarded as the simple inverse of $\overleftrightarrow{\Omega}$ because $\overleftrightarrow{\Omega}$ has a zero eigenvalue and thus has no proper inverse. Therefore $\overleftrightarrow{\Omega}^{-1} \cdot \overleftrightarrow{\Omega}$ is *not* the unit tensor, rather it is the *projection* operator on the plane perpendicular to the local magnetic field.

Since we have assumed that $\overleftrightarrow{\nu}/\overleftrightarrow{\Omega} \ll 1$ we have

$$\vec{V}_1 = \vec{V}_2 = \vec{V}_s - \vec{V}_{s\perp} + \vec{V}_{p\perp} = \vec{V}_{s\parallel} + \vec{V}_{p\perp}. \qquad (28)$$

Once again we have $\vec{V}_1 = \vec{V}_2$ however both velocities are equal to the perpendicular component of $\vec{V}_p$ in the directions perpendicular to $\vec{B}$ and to the parallel component of $\vec{V}_s$ in the direction parallel to $\vec{B}$. In other words the scatterers drag the particles along the magnetic field and the particles undergo $\vec{E} \times \vec{B}$ drift across the field.

ENERGY CHANGES DUE TO DRIFTS AND MAGNETIZATION

In a moving plasma that contains a magnetic field an electric field $\vec{E}_m$ is produced by the motion of the "frozen in" magnetic field. It has been suggested on various occasions that some of these energy change terms should be added to the transport equation since they are different from the adiabatic term that appears there; we shall see that this is not the case, these modes are contained in the adiabatic term.

The *entire* effect of the electric field in the derivation of equation 23 is contained in the plasma flow velocity, $\vec{V}_p$ (one could say with equal validity that the electric field is *caused* by the flow velocity) and we have seen that the particles move perpendicular to the magnetic field due to the $\vec{E} \times \vec{B}$ particle drift. How can a divergence in this flow cause an energy change in the particle? The answer, as was shown by Kóta[11] and Jokipii[12] lies in the gradient and curvature drifts.

Noting that the particles energy change comes from the dot product of the induced electric field with the particles drift motion $\vec{V}_d$ and with its motion around its gyro orbit we have

$$\vec{E}_m = -\frac{1}{c}\vec{V} \times \vec{B}$$

$$\vec{V}_d = \frac{1}{3}\frac{\gamma mcv^2}{e}\vec{\nabla} \times \frac{\vec{B}}{B^2} \quad (29)$$

$$\frac{d\mathcal{E}}{dt} = e\vec{E}_m \cdot \vec{V}_d + \frac{e\omega}{2\pi}\oint \vec{E}_m \cdot d\vec{l}$$

$$= -\frac{1}{3}\gamma mv^2(\vec{V} \times \vec{B}) \cdot \left(\vec{\nabla} \times \left(\frac{\vec{B}}{B^2}\right)\right) \quad (30)$$

$$+\frac{e\pi r_g^2}{2\pi c}\vec{\omega} \cdot \vec{\nabla} \times (\vec{V} \times \vec{B}). \quad (31)$$

Since $r_g = v_\perp/\omega$ and $\vec{\omega} = e\vec{B}/\gamma mc$ we have

$$\frac{d\mathcal{E}}{dt} = \frac{1}{2}\gamma mv_\perp^2\left\{\frac{\vec{B}}{B^2} \cdot \vec{\nabla} \times (\vec{V} \times \vec{B})\right\}$$

$$-\frac{1}{3}\gamma mv^2\left\{\frac{\vec{B}}{B^2} \cdot \vec{\nabla} \times (\vec{V} \times \vec{B})\right\}$$

$$-\frac{1}{3}\gamma mv^2\vec{\nabla} \cdot (\vec{V} - \hat{b}\hat{b} \cdot \vec{V})$$

In an isotropic particle distribution $v_\perp^2 = \frac{2}{3}v^2$ so the first two terms in the above equation cancel and noting that $\vec{V} - \hat{b}\hat{b} \cdot \vec{V} = \vec{V}_\perp$ we are left with

$$\frac{dp}{dt} = \frac{1}{v}\frac{d\mathcal{E}}{dt} = -\frac{p}{3}\vec{\nabla} \cdot \vec{V}_\perp, \quad (32)$$

which is just the perpendicular component of the adiabatic energy change term found in the transport equation.

We see, therefore, that this term includes all energy change effects that arise from the interaction of the particles with the electric and magnetic fields in the moving plasma. When this equation is applied to shock acceleration we see that the so called "shock drift" acceleration mechanism is already included and is not a fundamentally different process that needs to be added to the equations.

STOCHASTIC ACCELERATION

If we now turn our attention to the right hand side of equation 23 we see the term

$$-\frac{1}{3}\frac{\partial}{\partial p}\left\{(m\gamma)^2 p^2 \frac{\partial}{\partial p}\left(\frac{U}{p^2}\right)\left[\langle\delta\vec{V} \cdot \overset{\leftrightarrow}{\nu} \cdot \delta\vec{V}\rangle + (\vec{V}_s - \vec{V}_p) \cdot \overset{\leftrightarrow}{\nu} \cdot (\overset{\leftrightarrow}{\nu} + \overset{\leftrightarrow}{\Omega})^{-1} \cdot \overset{\leftrightarrow}{\Omega} \cdot (\vec{V}_s - \vec{V}_p)\right]\right\}.$$

This is clearly a momentum diffusion or stochastic acceleration term with the part that is proportional to $\langle\delta\vec{V} \cdot \overset{\leftrightarrow}{\nu} \cdot \delta\vec{V}\rangle$ due to the random motion of the scattering centers; the part proportional to $(\vec{V}_s - \vec{V}_p) \cdot \overset{\leftrightarrow}{\nu} \cdot (\overset{\leftrightarrow}{\nu} + \overset{\leftrightarrow}{\Omega})^{-1} \cdot \overset{\leftrightarrow}{\Omega} \cdot (\vec{V}_s - \vec{V}_p)$ however, is not so obvious. If we make use

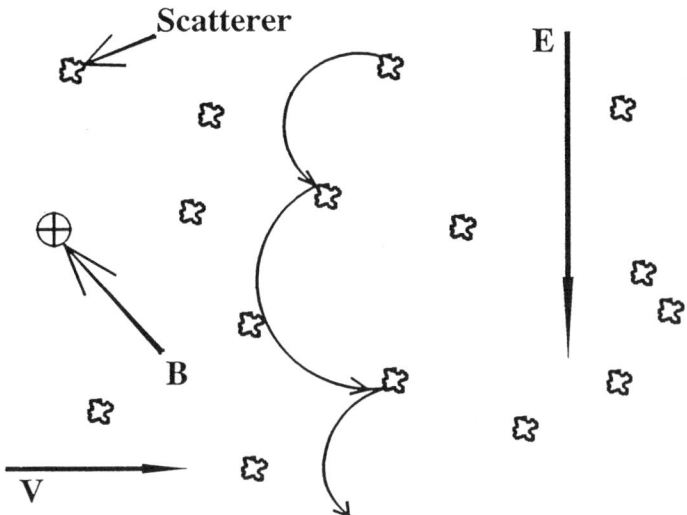

Figure 1: Particle Undergoing Cross Field Diffusion in a Flowing Plasma

of the approximation $\overset{\leftrightarrow}{\Omega} \gg \vec{\nu}$ as before this term may be written as $(\vec{V}_s - \vec{V}_p) \cdot \vec{\nu} \cdot \overset{\leftrightarrow}{\Omega}^{-1} \cdot \overset{\leftrightarrow}{\Omega} \cdot (\vec{V}_s - \vec{V}_p)$ and we see that it is the square of the scattering center velocity relative to the plasma in the direction perpendicular to the magnetic field that enters here rather than a random or stochastic velocity. And yet it *is* stochastic acceleration

This stochastic acceleration is unusual because it is produced by a *steady flow* of the scattering centers. If there were no magnetic field or if the scatterers were moving at the same speed as the plasma the particle population would quickly begin to drift along with the scattering centers and no further energization would be possible. However, with a magnetic field that is not moving with the scatterers the particles are not free to drift along with the scatterers but are bound to the field lines. As they gyrate about the field lines they are struck repeatedly by the scatterers and since *the collisions are not correlated with the gyrophase of the particles* the collisions are sometimes overtaking and sometimes head on, which is just the prescription for stochastic, Fermi acceleration to occur.

We can see that such a term is necessary if we consider the following situation that is sketched in Figure 1 A plasma with embedded magnetic field pointing into the plane of the figure is flowing to the right with velocity $\vec{V}_p$ and thereby inducing an electric field, $\vec{E}$. If particles are diffusing perpendicular to the magnetic field parallel or antiparallel to the electric field it might at first appear that they should also gain or lose energy respectivly. However, it can quickly seen that this is not so if the scatterers are moving with the plasma for in the flow frame of reference there is no electric field and hence no energy change. In the frame in which there *is* an electric field this can be understood[13] if one realizes that all collisions that move a particle in a direction parallel to the electric field are overtaking collisions in which the particles give up energy to the scattering centers and vice versa for the antiparallel direction.

This argument does not hold, of course, if the scatterers are not moving in the frame in which there is an electric field; there is no energy exchange on scattering in this case so the energy gained or lost from moving in the electric potential will be retained by the particles. This diffusive motion in three dimensional space is described by the first term in equation 23 and the corresponding diffusion in energy space is given by the last term. From the first term we can see that

$$\frac{(\delta y)^2}{\delta t} = \frac{1}{3}\frac{v^2}{\omega^2}\nu, \qquad (33)$$

the energy change is given by $\delta\epsilon = eE\delta y$ and since $E = V_p B/c = V_p \gamma m c \omega /ec$ we have

$$\frac{(\delta \epsilon)^2}{\delta t} = \frac{1}{3}(\gamma m V_p v)^2 \nu. \qquad (34)$$

From the last term we have

$$\frac{(\delta p)^2}{\delta t} = \frac{1}{3}(\gamma m V_p)^2 \nu \qquad (35)$$

and since $\delta\epsilon = v\delta p$ the two expressions are equal.

CONCLUSIONS

The above considerations apply whenever we can consider the energetic particle population to be approximately isotropic (adequately described by the first two moments) in any reference frame intrinsic to the problem. This means that the transport equation can not be used to describe the initial acceleration of thermal particles by plasma shocks or relativistic shocks where the energetic particle speeds are never much greater than the flow speeds (however, see Jones and Ellison[14] for an indication of how well it works even when it shouldn't). In most other situations, however, it describes almost any acceleration process that can be caused by a moving plasma. It describes shock acceleration for both parallel shocks and oblique ones (shock drift acceleration is included), and stochastic acceleration by the turbulent motion of the scatterers as well as by their motion across the magnetic field. One class of processes is not properly treated by this equation, however, those in which the interaction is strongly dependant on the particles momentum such as a resonance interaction with a narrow spectrum of plasma waves. Such processes would not be expected to accelerate particles to high energy but they can be very important as a means of "preheating" certain species of particles for injection into the main accelerator. The major problem in studying the acceleration of charged particles in astrophysical plasmas is to identify the primary terms of the transport equation that govern the given situation and discover how to extract a useful solution from the equation.

REFERENCES

1. E. N. Parker, *Phys. Rev.*, **110**, 1445, (1958).
2. E. N. Parker, *Planet. Space Sci.*, **13**, 9, (1965).
3. A. Z. Dolginov and I. Toptygin, *Soviet Phys. JETP*, **24**, 1195, (1967), (Russian edition Dec. 1966).
4. A. Z. Dolginov and I. N. Toptygin, *Icarus*, **8**, 54, (1968).
5. W. I. Axford, E. Leer, and G. Skadron, *Proc. 15th Int. Cosmic Ray Conf., Plovdiv*, **11**, 132, (1977).
6. R. D. Blandford and J. P. Ostriker, *Astrop. J. Lett.*,**221**, L29, (1978).
7. G. F. Krimsky, *Dokl. Akad. Nauk SSSR*, **243**, 1306, (1977).
8. M. A. Forman, *Planet. Space Sci.*, **18**, 25, (1970).
9. L. J. Gleeson, and W. I. Axford, *Ap. J.*, **149**, L115, (1967).
10. F. C. Jones, *Ap. J.*, **361**, 162, (1990).
11. J. Kóta, *Proc. 16th Int. Cosmic Ray Conf., Kyoto*, **3**, 13, (1979).
12. J. R. Jokipii, *Proc. 16th Int. Cosmic Ray Conf., Kyoto*, **14**, 175, (1979).
13. J. R. Jokipii, private communication
14. F. C. Jones and D. C. Ellison, Space Sci. Rev. **58**, 259, (1991).

Quasi-Linear Theory and Transport Theory

Charles W. Smith
Bartol Research Institute, The University of Delaware
Newark, DE 19716, U.S.A.

Abstract

The theory of energetic particle scattering by magnetostatic fluctuations is reviewed in so far as it fails to produce the rigidity-independent mean-free-paths observed. Basic aspects of interplanetary magnetic field fluctuations are reviewed with emphasis placed on the existence of dissipation range spectra at high wavenumbers. These spectra are then incorporated into existing theories for resonant magnetostatic scattering and are shown to yield infinite mean-free-paths. Nonresonant scattering in the form of magnetic mirroring is examined and offered as a partial solution to the magnetostatic problem. In the process, mean-free-paths are obtained in good agreement with observations in the interplanetary medium at 1 AU and upstream of planetary bow shocks.

1 Introduction

The theory of diffusive shock acceleration and particle transport has had remarkable success in explaining the observed distributions of galactic and solar cosmic rays as well as energetic particles upstream of various types of heliospheric shocks. Many papers contained within these proceedings attest to this. While most applications of diffusion theory require careful attention to plasma flow and boundary conditions, the parameterization of diffusion coefficients is generally quasi-empirical. An accurate first-principles derivation of scattering and diffusion coefficients so far has proved elusive.

If there is one result which defines the goal of such efforts, it is the Palmer[1] result. Palmer showed that the mean-free-path for protons and electrons at 1 AU is rigidity-independent over the range 500 kV to 5 GV. The mean-free-path obtained by Palmer varies within the range 0.08 to 0.3 AU. The mean-free-path may still depend on the intensity of the fluctuating magnetic field, and indeed, particle scattering upstream of the Earth's bow shock is enhanced by the presence of upstream waves.

2 Diffusion by Plasma Processes

When the propagation speed of the scattering centers in the rest frame of the plasma is small compared with the particle speed, the diffusive transport equation governing the distribution of energetic test particles, f(x, p, t), takes the form[2,3]:

$$\frac{\partial f}{\partial t} + \mathbf{U} \cdot \nabla f = \nabla \cdot (\boldsymbol{\kappa} \cdot \nabla f) + \frac{1}{3} \nabla \cdot \mathbf{U} p \frac{\partial f}{\partial p} \tag{1}$$

© 1992 American Institute of Physics

where **U** is the plasma flow velocity, **p** is the particle momentum, and κ is the anisotropic spatial diffusion tensor. Determination of κ from first-principles remains the most glaring inadequacy of all diffusion theories of acceleration and transport. Consider the coefficient for spatial diffusion parallel to the ambient magnetic field, $\kappa_\parallel$, and the parallel mean-free-path, $\lambda_\parallel$, which are given by[4]:

$$\kappa_\parallel = \frac{v^2}{2} \int_0^1 \frac{(1-\mu^2)^2}{\Phi(\mu)} d\mu \qquad (2)$$

$$\lambda_\parallel = \frac{3\kappa_\parallel}{v} \qquad (3)$$

where $\mu = \cos(\Theta)$, Θ is the pitch-angle and $\Phi(\mu)$ is the pitch-angle scattering coefficient. Theories for $\Phi(\mu)$ require knowledge of the full spectrum of magnetic fluctuations.

The frequently used magnetostatic approximation to derivations of $\Phi(\mu)$ from quasi-linear theory assumes that the magnetic fluctuations evolve more slowly than the time scale for particle motion through the system, leaving the particles to sample a time-stationary magnetic field. Both resonant processes (particle trajectories matching the spatial variation of the magnetic fluctuation) and nonresonant processes (most notably, mirroring) may contribute to pitch-angle scattering, with the functional dependence of the scattering coefficients depending on the form of the magnetic field spectrum. In the case of the slab geometry, the magnetostatic approximation leads to $\Phi(\mu)$ given by[4]:

$$\Phi_S(\mu) = \frac{2\pi v}{R_L^2 B_0^2} \frac{1-\mu^2}{|\mu|} P_\perp\left(k_r = \frac{1}{\mu R_L}\right) \qquad (4)$$

where B_0 is the magnitude of the mean magnetic field, R_L is the Larmor radius, and $P_\perp(k_r)$ is the power spectrum evaluated at the resonant wavenumber. The form for $\Phi(\mu)$ relevant to isotropic turbulence is more complicated[5], but can be approximated by $\Phi_I(\mu) \approx |\mu| \Phi_S(\mu)$ when $\mu \neq 0$ and the power spectrum is not a steeply falling function of wavenumber. Both of these forms lead to poor agreement with Palmer[1].

3 Nature of IMF Fluctuations

Figure 1 shows a typical spectrum of the fluctuations of the interplanetary magnetic field (IMF) at 0.6 AU from 0.08 to 3 Hz in the spacecraft frame of reference. The computed spectrum resolves the high frequency end of the inertial range, which typically extends from approximately 10^{-4} to 1 Hz with a moderately steep power law form[6,7,8]. Higher frequencies display a dissipation range spectrum that falls significantly more steeply than the inertial range[9,10]. At the highest wavenumbers present (perhaps beyond the measured range) the magnetic field spectrum must fall at least as steeply as k^{-3} [11,12]. There is no evidence of persistent, nonzero polarization (or magnetic helicity) in the inertial range when statistically independent frequencies are averaged together[7,8]. The dissipation range sometimes shows evidence of significant polarization which is suggestive of resonant dissipation processes[13]. Theories that rely on a polarized inertial range spectrum to compute particle scattering coefficients[14] are at odds with virtually all observations, except for limited frequency ranges associated with wave generation by the energetic particle distributions (such as in planetary foreshocks[15,16,17]). Except for

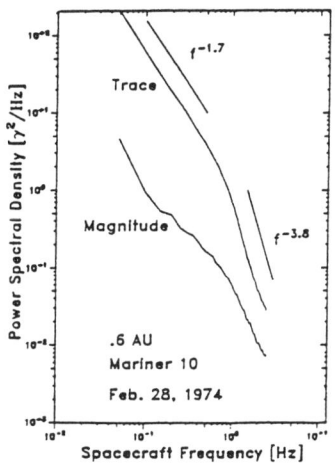

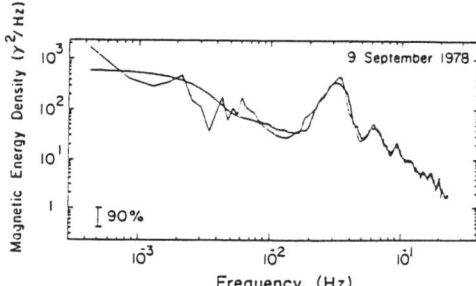

Fig. 2: Power spectrum of upstream waves at Earth's bow shock. Peaks in the interval $2 \times 10^{-2} \leq f_{sc} \leq 10^{-1}$ Hz represent enhancements of the wave power excited by resonant interaction with energetic ions.

Fig. 1: Power spectrum of IMF fluctuations recorded by Mariner 10 at 0.6 AU. Trace of power spectral density matrix and spectrum of the fluctuating magnitude are shown. Note dissipation range spectrum at frequencies greater than 1 Hz.

one analysis of IMF fluctuations[18], and a limited class of upstream waves at the Earth's bow shock[15], the geometry of the IMF fluctuation spectrum remains undetermined.

The presence of a high frequency spectrum that differs from the spectral form of the inertial range fluctuations has generally been overlooked in past attempts to compute scattering coefficients for energetic charged particles, with the inertial range spectrum assumed to continue to arbitrarily small spatial scales[1]. Under this misleading assumption, resonant magnetostatic scattering in slab geometries gives mean-free-paths that are a factor of 10 too small[1], while in the isotropic geometry it leads to infinite mean-free-paths[5]. This seems to suggest that a hybrid geometry (neither completely slab-like nor isotropic) might best describe the IMF fluctuations[18], and if the correct geometry is identified then the predicted mean-free-paths will agree with observations.

Mirroring has been argued to yield rigidity-independent mean-free-paths[18], but there exists the need to suppress resonant scattering in order to bring this result into agreement with Palmer[1]. In the past, resonant scattering processes were suppressed by the adoption of an unphysically steep inertial range spectrum[19,20,21]. Collapse of the magnetic fluctuation spectrum to a 2-D state with wavevectors perpendicular to the mean magnetic field (inferred from the observed collapse of 2-D MHD simulations toward the analogous 1-D state when a DC magnetic field is imposed[22]) has been suggested as one means of eliminating resonant scattering[23]. An extension of this idea has been proposed[24] whereby the solar wind fluctuations evolve toward a quasi-2D state, consisting of 2D fluctuations convecting in a plane orthogonal to the large-scale magnetic field superimposed on a family of Alfven waves propagating along B_0. Likewise, dissipation range spectra at high wavenumbers decrease the available energy for resonant scattering near 90° and increase the mean-free-path due to resonant processes[25].

Nonresonant scattering cannot be the entire solution. Figure 2 shows the power spectrum for an interval of low frequency magnetic waves upstream of the Earth's

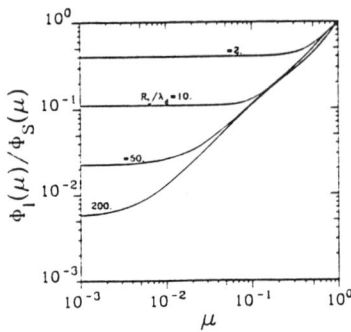 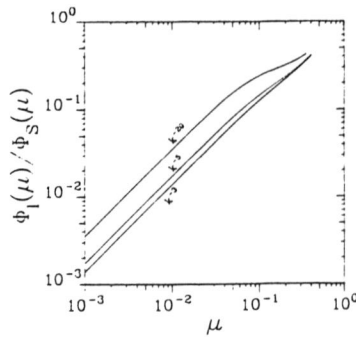

Fig. 3: (left) Ratio of pitch-angle scattering coefficients for isotropic and slab geometries as a function of the cosine of the pitch-angle, μ, when $k^{-5/3}$ inertial range and Gaussian dissipation range are assumed.

Fig. 4: (right) Same as Fig. 2, but with power law dissipation range assumed. Power law indices are given in figure.

bow shock[26]. These waves are generated through resonant interaction with energetic, sunward-streaming protons. When upstream waves are generated by these populations, the mean-free-path of upstream energetic particles is decreased significantly through enhanced resonant scattering[17] to a value of $\sim 7 R_E$ [27]. This decrease in $\lambda_\parallel$ and associated wave energy leads to the formation of shock-accelerated diffuse ion populations upstream of the bow shock. Therefore, resonant scattering does play a role in particle transport, at least under those circumstances when resonant wave energy is enhanced.

4 Laying Resonant Magnetostatic Theory to Rest

When dissipation range spectra are included in resonant magnetostatic theory calculations of pitch-angle scattering coefficients, infinite mean-free-paths are obtained for both the slab and isotropic geometries[28]. Figure 3 shows how $\Phi_I(\mu)$ compares with $\Phi_S(\mu)$ as a function of μ for different Larmor radii when a Gaussian dissipation range form is assumed. The scattering coefficient exhibits a transition from the result of Fisk et al.[5] to a form resembling the Jokipii prediction[4], but with a coefficient that is energy-dependent, so that $\Phi_I(\mu)/\Phi_S(\mu) \to (2/\sqrt{\pi})(\lambda_d/R_L)$ for small μ, where λ_d characterizes the scale size of the dissipation range. The pitch-angle at which this transition occurs is given by the value at which the resonant wavenumber enters the dissipation range. Figure 4 shows the ratio $\Phi_I(\mu)/\Phi_S(\mu)$ when a power law dissipation range is adopted[28]. These results agree more nearly with the Fisk et al.[5] factor of $|\mu|$ when a minimal k^{-3} form is assumed. All forms of the dissipation range spectrum which satisfy this minimum condition lead to departures from the Fisk et al.[5] prediction, with the asymptotic form of $\Phi_I(\mu)$ dependent on the spectral form chosen for the dissipation range.

Since $\Phi_S(\mu) \sim |\mu|^{-1} P_\perp(k_r)$ with $k_r = (\mu R_L)^{-1}$, and since $\Phi_I(\mu) \geq |\mu|\Phi_S(\mu)$, equation (2) possesses a non-integrable singularity at $\mu = 0$ for both isotropic and slab geometries when the presence of a dissipation range spectrum at arbitrarily large wavenumbers is considered and the magnetostatic approximation is employed. Then $\kappa_\parallel$ and $\lambda_\parallel$ are

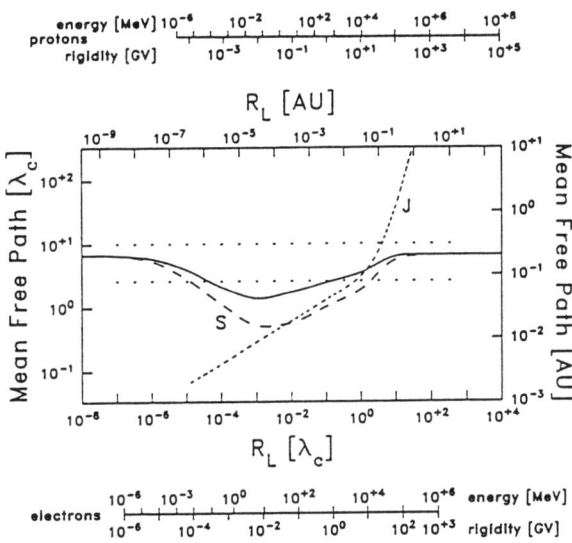

Fig. 5: The mean-free-path for particles propagating under combined influence of resonant and mirrored scattering. Solid line is isotropic scattering while dashed line "S" gives slab result with mirroring added. Dashed line "J" is nondissipative resonant slab without mirroring. Horizontal dotted lines are upper and lower limits of Palmer consensus range.

both infinite. It appears to be a general result of resonant magnetostatic theory that it yields infinite mean-free-paths, so long as the presence of a dissipation range at the smallest spatial scales is acknowledged. Since energetic particles are observed to scatter with known mean-free-paths, we must conclude that resonant magnetostatic theory is inadequate. Indeed, the very ordering of the time scales that defines the magnetostatic approximation is improper when addressing particles with 90° pitch-angles[19,23,25,29].

5 Revisiting Magnetic Mirroring

Past treatments of particle scattering through mirroring suffer from two problems. First, the scattering coefficients for nonresonant processes are derived using the magnetostatic approximation, which again leads to misordering problems. Resonance line broadening calculations help to offset this difficulty. Second, most calculations of resonant and nonresonant scattering have been performed without consideration of the dissipation range. Since resonant scattering will enhance scattering at small and moderate pitch-angles, it is interesting to ask what effect this will have on the rigidity-independent mean-free-path result derived from mirroring effects alone.

Mirroring and resonant scattering theory can be combined in an approximate fashion to lend some insight into the effects of dissipation range spectra on the mean-free-path. To do this, we assume that the magnetic fluctuations can be decomposed into two components: one contributing to resonant scattering while the second is responsible for mirroring. We must adopt a form of mirroring theory that uses resonance line broadening[19,20,30,31] to extend the mirroring contribution to include a finite interval of pitch-angles. Owens[30] performed a calculation based on wave-particle interactions, leading to resonance line broadening that yields a pitch-angle scattering coefficient approximated by $\Phi_M(\mu) = C_1 v (1 - \mu^2)$, where the constant C_1 is related to $\langle \delta B_{rms}^2 \rangle / \langle B_0^2 \rangle$. Likewise, equation (4) becomes $\Phi_S(\mu) = C_2 v R_L^{-2} |\mu|^{-1} (1 - \mu^2) P_\perp(k_r)$. We adopt the spectral forms given by Smith et al.[28] and take $P_\perp(k) \sim (1 + k^2\lambda_c^2)^{-\nu/2}(1 + k^2\lambda_d^2)^{-\delta/2}$. With $\nu = 5/3$ and $\delta = 4/3$ this gives power law inertial and dissipation range spectra with indices $= -5/3$ and -3, respectively. The parameters λ_c and λ_d characterize the correlation and dissipation scales. We take λ_c and λ_d to be 3×10^{-2} and 3×10^{-6} AU, respectively, corresponding to spacecraft frame frequencies of 10^{-4} and 1 Hz. The pitch-angle scattering coefficient for the combined effects of mirroring and a resonant slab geometry then becomes $\Phi_{slab}^{total} = \Phi_M + \Phi_S$. The factor C_1 is determined to give a mean-free-path equal to 0.2 AU when the resonant term is set to zero ($C_2 = 0$). The parameter C_2 is set so that the computed mean-free-path agrees with that given by Palmer for the slab geometry when both λ_d and C_1 are set to zero. Finally, we can repeat this same analysis for the isotropic geometry using the same values for λ_c, λ_d, C_1 and C_2 by adopting the Fisk et al.[5] result $\Phi_I = |\mu|\Phi_S$ which is well supported by Figure 4 for this dissipation range form. The pitch-angle scattering coefficient for isotropic geometry is then given by $\Phi_{isotropic}^{total} = \Phi_M + |\mu|\Phi_S$.

Figure 5 shows the resulting mean-free-path for the slab (dashed line labelled "S") and isotropic (solid line) calculations. The nondissipative slab result ($\lambda_d = 0$) quoted by Palmer is represented by the dashed line labelled "J". The mirroring term acts to supplement scattering near 90° and yield finite mean-free-paths. At high particle rigidities resonant scattering is weak (as was shown by Palmer) and the mean-free-path is dominated by the mirroring term. The mean-free-path for low rigidity particles is also dominated by mirroring due to the depletion of resonant energy. However, when the Larmor radius falls between the correlation and dissipation scales, resonant scattering in both the slab and isotropic calculations augment mirroring to give mean-free-paths below the low end of the Palmer consensus range. Recent analyses of the mean-free-path for protons with rigidities between 3×10^{-2} and 6×10^2 GV [32,33] show an increase with rigidity that is in good agreement with the results shown in Figure 5.

Figure 5 also holds promise for the upstream particle problem. It implies that particles resonant with enhanced wave power experience additional scattering at small and intermediate pitch-angles leading to shorter mean-free-paths. Mirroring provides the means for particles to scatter through 90° that is missing from the resonant paradigm.

While this calculation is intended simply to illustrate the interplay between resonant scattering and mirroring when dissipative spectra are considered, it naturally provides mean-free-paths in approximate agreement with Palmer[1]. In this way, it may not be necessary to invoke spectral collapse of the magnetic fluctuations to a 2-D state (or a partial collapse, the more likely result at 1 AU) in order to account for the role of resonant processes in the scattering and diffusion of energetic particles.

This is only a simple model for the interplay between resonant and nonresonant scattering. It is not a first-principles theory and does not address the 90° problem of magnetostatic theory in a consistent fashion. Two papers in this volume attempt to provide magnetodynamic treatments stemming from distinctly different viewpoints: Schlickeiser offers a wave-propagation theory to provide magnetodynamic scattering at 90° and Bieber offers an approach based on the turbulent decorrelation of the field.

Acknowledgements: The author acknowledges helpful discussions with J. W. Bieber, M. L. Goldstein, J. S. Perko, R. Schlickeiser and G. P. Zank. This work was supported by Jet Propulsion Laboratory contract 959167, NASA grant NAGW-1637, and the NASA

SPTP program under grant NAGW-2076 to the Bartol Research Institute.

References

1. I. D. Palmer, *Rev. Geophys. Space Phys.*, **20**, 335 (1982).
2. E. N. Parker, *Planet. Space Sci.*, **13**, 9 (1965).
3. L. O'C. Drury, *Rep. Prog. Phys.*, **46**, 973 (1983).
4. J. R. Jokipii, *Astrophys. J.*, **146**, 480 (1966).
5. L. A. Fisk et al., *Astrophys. J.*, **190**, 417 (1974).
6. K. Behannon, *Observations of the Interplanetary Magnetic Field Between 0.46 and 1 AU by the Mariner 10 Spacecraft*, Ph.D. Thesis, The Catholic University of America (1976).
7. W. H. Matthaeus and M. L. Goldstein, *J. Geophys. Res.*, **87**, 6011 (1982).
8. W. H. Matthaeus, M. L. Goldstein and C. W. Smith, *Phys. Rev. Lett.*, **48**, 1256 (1982).
9. K. Denskat, H. Beinroth and F. Neubauer, *J. Geophys.*, **54**, 60 (1983).
10. C. W. Smith, W. H. Matthaeus and N. F. Ness, *Proc. 21st Internat. Cosmic Ray Conf.* (Adelaide), **5**, 280 (1990).
11. G. K. Batchelor, *The Theory of Homogeneous Turbulence* (Cambridge University Press, New York, 1953).
12. J. W. Bieber, C. W. Smith and W. H. Matthaeus, *Astrophys. J.*, **334**, 470 (1988).
13. M. L. Goldstein, (private communication, 1992).
14. U. Achatz et al., *Proc. 22nd Internat. Cosmic Ray Conf.* (Dublin), **3**, 240 (1991).
15. M. M. Hoppe and C. T. Russell, *J. Geophys. Res.*, **88**, 2021 (1983).
16. C. W. Smith, M. L. Goldstein and W. H. Matthaeus, *J. Geophys. Res.*, **88**, 5581 (1983).
17. M. A. Lee, *Reviews of Geophysics and Space Physics*, **21**, 324 (1983).
18. M. L. Goldstein, *J. Geophys. Res.*, **85**, 3033 (1980).
19. M. L. Goldstein, *Astrophys. J.*, **204**, 900 (1976).
20. T. B. Kaiser, T. J. Birmingham and F. C. Jones, *Phys. Fluids*, **21**, 361 (1978).
21. R. Schlickeiser, *J. Geophys. Res.*, **93**, 2725 (1988).
22. J. V. Sheblin, W. H. Matthaeus and D. Montgomery, *J. Plasma Physics*, **29**, 525 (1983).
23. W. H. Matthaeus, M. L. Goldstein and D. A. Roberts, *J. Geophys. Res.*, **95**, 20,673 (1990).
24. G. P. Zank and W. H. Matthaeus, *J. Geophys. Res.*, submitted (1992).
25. J. M. Davila and J. S. Scott, *Astrophys. J.*, **285**, 400 (1984).
26. C. W. Smith et al., *J. Geophys. Res.*, **90**, 1429 (1985).
27. F. M. Ipavich et al., *J. Geophys. Res.*, **86**, 4337 (1981).
28. C. W. Smith, J. W. Bieber and W. H. Matthaeus, *Astrophys. J.*, **363**, 283 (1990).
29. F. C. Jones, T. J. Birmingham and T. B. Kaiser, *Phys. Fluids*, **21**, 347 (1978).
30. A. J. Owens, *Astrophys. J.*, **191**, 235 (1974).
31. H. J. Volk, *Astrophys. Space Sci.*, **25**, 471 (1973).
32. W. Wanner and G. Wibberenz, *Proc. 22nd Internat. Cosmic Ray Conf.* (Dublin), **3**, 221 (1991).
33. W. Droge et al., *Proc. 22nd Internat. Cosmic Ray Conf.* (Dublin), **3**, 225 (1991).

PARTICLE TRANSPORT FROM A TURBULENCE PERSPECTIVE

John W. Bieber and William H. Matthaeus
Bartol Research Institute
University of Delaware, Newark, DE 19716, U.S.A.

ABSTRACT

The rate at which energetic charged particles scatter in turbulent magnetic fields is remarkably sensitive to the detailed properties of the turbulence. While the importance of the inertial range power spectrum was recognized early on, more recent studies have emphasized the crucial role played by the geometry of the turbulence, the dissipation range spectrum, and the dynamical character of turbulence. We present the mean free path as a function of energy derived in a turbulence model that incorporates all of these factors, and show that current understanding of turbulence in space permits a wide range of "predicted" behaviors, including some that agree remarkably well with observation.

INTRODUCTION

Understanding the mechanism by which energetic charged particles scatter in turbulent magnetic fields continues to be one of the foremost problems of modern cosmic ray physics. This problem has been put in sharpest focus by observations of cosmic rays in the solar wind. Although we now have a fairly detailed knowledge of the magnitude and energy dependence of the scattering mean free path (or, equivalently, diffusion coefficient) required to explain the cosmic ray observations[1], we are still unable to compute the mean free path from first principles using basic plasma physics and the known properties of turbulence in space.

The problem is not that the predicted mean free path is "too small" or has the "wrong" energy dependence. That common characterization applies only to a particular form of turbulence, the slab model[2], which is but one member of a large class of turbulence geometries that can be considered even when attention is restricted to axially symmetric descriptions[3]. Rather, the problem is that scattering theory is not sufficiently constrained given the existing state of knowledge on turbulence in the solar wind. By making seemingly minor, technical adjustments to the turbulence model employed, one can easily change the predicted mean free path from much too small to much too large. One can also produce many different types of energy variation of the mean free path, including some that agree very well with observation[4].

The remarkable sensitivity of the mean free path to details of the turbulence model can be understood in terms of the relationship between the spatial mean free path and the scattering rate as a function of pitch angle[2,5,6]:

$$\lambda_\| = \frac{3}{4} V \int_{-1}^{+1} \frac{(1-\mu^2)^2}{\Phi(\mu)} d\mu , \qquad (1)$$

where $\lambda_\|$ is the parallel (to the large-scale field) mean free path, V is the particle speed, μ is the cosine of the particle pitch angle, and $\Phi(\mu)$ is the scattering rate, or Fokker-Planck coefficient. (The parallel diffusion coefficient, $K_\parallel$, is related to the mean free path by $K_\parallel = \lambda_\parallel V/3$.) Owing to the presence of $\Phi(\mu)$ in the denominator of the integrand, the behavior of the scattering rate in regimes where it is small has a disproportionate influence upon the mean free path. This characteristic makes sense physically: A charged particle in magnetic turbulence experiences a macroscopic change in direction (spatial scattering) as a cumulative result of many small random changes in its pitch angle (pitch angle scattering). Hence we should expect the spatial mean free path to be regulated by the pitch angle scattering rate at pitch angles where scattering is slowest. Generally this means pitch angles near 90°.

In the "slab" model[2] the wavevectors of the turbulent fluctuations are aligned with the large-scale field. Assuming that a Kolmogoroff ($k^{-5/3}$ where k is the magnitude of the wavevector) energy spectrum extends to arbitrarily high wavenumbers, this model produces a $|\mu|^{2/3}$ variation of the scattering rate as the pitch angle nears 90°. Hence Φ vanishes as $\mu \to 0$, but it does so slowly enough that the integral in equation (1) remains finite and, in fact, yields a mean free path that is much too small at low to intermediate energies[1]. Choosing instead an isotropic turbulence geometry[7,8] produces an extra factor of $|\mu|$ in the scattering rate, so that Φ vanishes as $|\mu|^{5/3}$ (again assuming a Kolmogoroff spectrum to arbitrarily large k), and the mean free path is infinite. Thus, simply changing the assumed turbulence geometry drastically changes the rate of particle scattering and diffusion.

More recent work[9,10] has emphasized the presence of turbulence dissipation processes in the high-k (or small scale) regime which governs the resonant scattering of particles with pitch angles near 90°. Owing to the strong suppression of turbulent energy in the dissipation range, the scattering rate in the magnetostatic approximation vanishes at least as fast as $|\mu|^2$ as $\mu \to 0$, regardless of the turbulence geometry[11,12]. As a result, the predicted mean free path is infinite in resonant magnetostatic theory, even in the venerable slab model.

It is clear that this dilemma is an artificial one caused by the simplifying assumptions of standard quasilinear theory. However, identifying the dominant higher order process that allows the particles to scatter through 90° and yields a mean free path in agreement with observation has proved difficult. Among the proposed mechanisms are "mirroring" by fluctuations of the magnetic field magnitude[13], a variety of nonlinear extensions of the theory[14,15,16,17], wave propagation effects[18], and effects of dynamical turbulence[4]. We will focus upon the latter possibility in the remainder of this report. If turbulent processes are active in the solar wind as evidence strongly suggests[19], then dynamical evolution of the magnetic fluctuations will necessarily occur. Hence, dynamical effects must be considered in any description of particle transport grounded in turbulence theory.

DYNAMICAL MAGNETIC TURBULENCE

Representations of dynamical turbulence are most naturally discussed in terms of a power spectrum matrix that depends upon lag-time t as well as wavevector $\mathbf{k}$ (i.e., the dynamical correlation function is Fourier transformed in its spatial coordinates but not in its time coordinate). We adopt a power spectrum matrix of the general form:

$$P_{ij}(\mathbf{k}, t) = P_{ij}^{(0)}(\mathbf{k}) \, \Gamma(\mathbf{k}, t), \qquad (2)$$

where $P_{ij}^{(0)}(\mathbf{k})$ is a valid magnetostatic representation of the turbulence and $\Gamma(\mathbf{k},t)$ describes the dynamical correlations. For $P_{ij}^{(0)}(\mathbf{k})$ we employ the slab and isotropic models defined by equations (1)–(3) of Bieber[20]. However, we modify these to include a multiplicative factor $\exp(-k\lambda_d)$ representing the dissipation range, with λ_d the dissipation scale. We normalize the spectra to have the same variance, δB_x^2, in one of the perpendicular field components and the same correlation length, λ_c, of the "reduced" perpendicular spectrum, and we choose the spectral index to produce a Kolmogoroff ($k^{-5/3}$) inertial range in the reduced spectrum. We take the dissipation scale to be $\lambda_d = 2.7 \times 10^{-4} \lambda_c$.

Little is known about the dynamical correlation function of turbulence in space. Since $\Gamma(\mathbf{k},t)$ represents a temporal decay process in the two-point correlations, we assume it has the form $\exp\{-t/\tau(\mathbf{k})\}$, where $\tau(\mathbf{k})$ is the decay time. A probable lower limit for the decay time, based on analogy with spectral theories[21], is given by the Alfvén time scale, $\tau(\mathbf{k}) = (kV_A)^{-1}$, with V_A the Alfvén speed. Accordingly, we take the dynamical correlation function to be of the form

$$\Gamma(\mathbf{k},t) = \exp(-\alpha k V_A t), \qquad (3)$$

where the parameter α allows us to adjust the strength of the dynamical effects, ranging from $\alpha = 0$ (magnetostatic limit) to $\alpha = 1$ (strongly dynamical).

Now that the form of the dynamical power spectrum has been specified, the scattering rate (or Fokker-Planck coefficient) and parallel mean free path can be computed through some fairly straightforward extensions of magnetostatic theory. We defer a presentation of this to a future publication and proceed directly to the results.

PITCH ANGLE SCATTERING AND "MIRRORING"

Figure 1 shows the scattering rate as a function of pitch angle in dynamical turbulence with slab and isotropic geometries. To render the results in physical units, we have assumed $B_0 = 5$ nT, where B_0 is the magnitude of the mean magnetic field, $\delta B_x^2/B_0^2 = 0.3$, $\lambda_c = 5 \times 10^9$ m, and $V_A = 40$ km/s, all of which are representative of the solar wind at 1 AU. In each model the scattering rate approaches a finite value as the pitch angle approaches 90° ($\mu \to 0$). As a result the mean free path is finite in both models, specifically 0.32 and 0.19 AU respectively in slab and isotropic geometries.

The dotted curve, marked "Mirror" in Figure 1, isolates the contribution of the Landau resonance to the isotropic scattering rate. This resonance describes scattering attributable

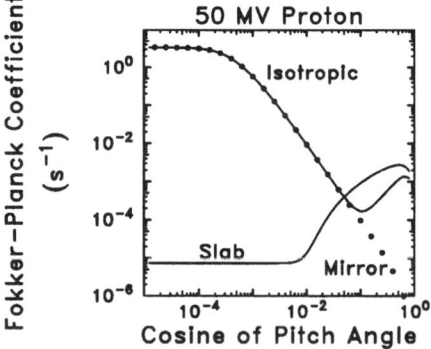

Figure 1. Fokker-Planck coefficient for cosmic ray pitch angle scattering in dynamical magnetic turbulence with $\alpha = 0.1$. Two different turbulence geometries are shown, each of which possesses a dissipation range as well as a realistic inertial range. Dotted line isolates the contribution of mirroring (Landau resonance) in the isotropic model.

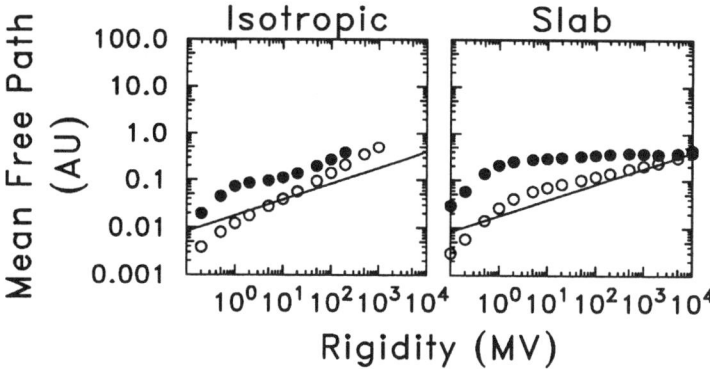

Figure 2. Rigidity dependence of parallel mean free path for scattering in dynamical turbulence. Results are shown for protons in isotropic and slab geometries using $\alpha = 1$ (open circles) and $\alpha = 0.1$ (closed circles) to characterize the strength of the dynamical effects. The line shows the conventional (dissipationless, magnetostatic) slab result for comparison.

to "mirroring" interactions between the particles and fluctuations of the magnetic field magnitude[8,13]. We see that mirroring causes the scattering rate near 90° pitch angle to be much larger in the isotropic model than in the slab model (which in lowest order has no magnitude fluctuations and hence no mirroring). Although the mirroring contribution appears very prominent when plotted logarithmically as in Figure 1, this does not imply that mirroring is the dominant factor determining the spatial mean free path. In fact, the mean free path computed from the Landau resonance alone is 44 AU for the scattering rate displayed in Figure 1, which is much larger than the correct value of 0.19 AU. On the other hand, the mean free path computed when the Landau resonance is ignored while all other resonances are retained is 1.6 AU, which is likewise a poor estimate of the correct value. Evidently mirroring and resonant interactions should in general be considered together to describe properly particle scattering and diffusion in turbulent magnetic fields.

PROTON AND ELECTRON MEAN FREE PATHS

Figure 2 shows the rigidity dependence of the parallel mean free path computed from dynamical turbulence models with two different values of α. Most intriguing is the behavior of the slab model with $\alpha = 0.1$, for which the mean free path remains in the range 0.13–0.38 AU as the rigidity varies from 0.5 to 5000 MV. This matches almost precisely the behavior of the Palmer[1] "consensus" mean free paths derived from cosmic ray observations in the solar wind. The isotropic model with $\alpha = 0.1$ exhibits similar tendencies at low rigidities, but it becomes too large above a few hundred MV. With $\alpha = 1$, both turbulence geometries yield mean free paths that are too small at low rigidities.

A notable feature of scattering in dynamical turbulence is that the mean free path for electrons will generally differ from the mean free path for protons of the same rigidity. This is demonstrated in Figure 3. The effect of dynamical turbulence upon particle scattering depends upon the ratio of the Alfvén speed to the particle speed.

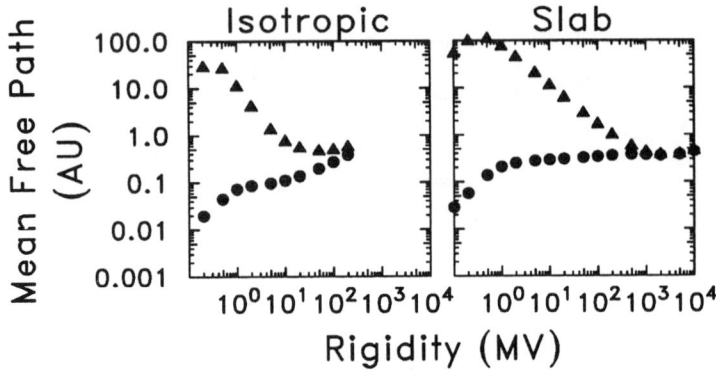

Figure 3. Rigidity dependence of parallel mean free path for protons (circles) and electrons (triangles). Results are shown for dynamical turbulence with $\alpha = 0.1$ in slab and isotropic geometries.

Except in the regime where both protons and electrons are relativistic, this ratio will differ substantially for protons and electrons of the same rigidity. As shown in the figure, the effect is such that electrons are more weakly scattered than protons of the same rigidity. There is in fact evidence for nearly "scatter-free" transport of low rigidity solar electrons[22].

DISCUSSION

Cosmic ray scattering in dynamical magnetic turbulence has the potential to resolve the long-standing discrepancy between mean free paths derived from cosmic ray observations and those derived from quasilinear theory. Earlier magnetostatic models identified the behavior of the inertial range and the geometry of the turbulence as crucial factors governing scattering rates in interplanetary space. These remain important in the dynamical models we have discussed, but the form of the dynamical correlation function and the nature of the dissipation range also play a crucial role. Of all these factors only the properties of the inertial range are well established[19] at present.

To reach a better understanding of the turbulence geometry, we may be able to use the cosmic rays themselves as remote probes of the three-dimensional structure of turbulence in space[20]. The analysis of magnetometer measurements at very high time resolution is providing new clues as to the nature of the dissipation range[23]. Improved descriptions of dynamical turbulence are expected to emerge from MHD theoretical studies[21], and, again, observations of the cosmic rays themselves may provide constraints on acceptable forms of dynamical correlation functions (see Figure 2). In short, gaining a better understanding of the properties of turbulence in space may prove key in finally arriving at a satisfactory description of cosmic ray scattering and diffusion in astrophysical plasmas.

Acknowledgements. We thank Charles W. Smith for useful discussions. Supported by NASA grant NAGW-1637 and the Space Physics Theory Program and by NSF grants ATM-9014806 and ATM-8913627. Computing support provided by the San Diego Supercomputer Center.

REFERENCES

1. I. D. Palmer, *Rev. Geophys. Space Phys.*, **20**, 335 (1982).
2. J. R. Jokipii, *Astrophys. J.*, **146**, 480 (1966).
3. W. H. Matthaeus and C. W. Smith, *Phys. Rev. A*, **24**, 2135 (1981).
4. J. W. Bieber and W. H. Matthaeus, *Proc. 22nd Internat. Cosmic Ray Conf.*, **3**, 248 (1991).
5. K. Hasselmann and G. Wibberenz, *Astrophys. J.*, **162**, 1049 (1970).
6. J. A. Earl, *Astrophys. J.*, **193**, 231 (1974).
7. K. Hasselmann and G. Wibberenz, *Zs. Geophys.*, **34**, 353 (1968).
8. L. A. Fisk, M. L. Goldstein, A. J. Klimas, and G. Sandri, *Astrophys. J.*, **190**, 417 (1974).
9. J. M. Davila and J. S. Scott, *Astrophys. J.*, **285**, 400 (1984).
10. J. W. Bieber, C. W. Smith, and W. H. Matthaeus, *Astrophys. J.*, **334**, 470 (1988).
11. C. W. Smith, J. W. Bieber, and W. H. Matthaeus, *Astrophys. J.*, **363**, 283 (1990).
12. C. W. Smith, this volume (1992).
13. M. L. Goldstein, A. J. Klimas, and G. Sandri, *Astrophys. J.*, **195**, 787 (1975).
14. A. J. Owens, *Astrophys. J.*, **191**, 235 (1974).
15. H. J. Völk, *Rev. Geophys. Space Phys.*, **13**, 547 (1975).
16. M. L. Goldstein, *Astrophys. J.*, **204**, 900 (1976).
17. F. C. Jones, T. J. Birmingham, and T. B. Kaiser, *Phys. Fluids*, **21**, 347 (1978).
18. R. Schlickeiser, R. Dung, and U. Jaekel, *Astron. Astrophys.*, **242**, L5 (1991).
19. W. H. Matthaeus and M. L. Goldstein, *J. Geophys. Res.*, **87**, 6011 (1982).
20. J. W. Bieber, *Proc. 21st Internat. Cosmic Ray Conf.*, **5**, 308 (1990).
21. W. H. Matthaeus and Y. Zhou, *Phys. Fluids B*, **1**, 1929 (1989).
22. R. P. Lin, *J. Geophys. Res.*, **75**, 2583 (1970).
23. C. W. Smith, W. H. Matthaeus, and N. F. Ness, *Proc. 21st Internat. Cosmic Ray Conf.*, **5**, 280 (1990).

RECENT DEVELOPMENTS IN PARTICLE TRANSPORT THEORY: THE WAVE VIEWPOINT

Reinhard Schlickeiser

Max-Planck-Institut für Radioastronomie
Auf dem Hügel 69, D-5300 Bonn, Germany

ABSTRACT

Recent results of weak turbulence quasilinear theory for the acceleration rates and propagation parameters of charged test particles from their resonant interaction with a partially random electromagnetic field are reviewed. The random electromagnetic field is described as the superposition of individual plasma modes, whose dispersion relations reflect the physical state of the wave-carrying background medium. The plasma waves interacting with different cosmic ray particle populations are identified and the corresponding cosmic ray transport and acceleration parameters are discussed. Results are applied to the transport and acceleration of solar cosmic ray particles.

I. HIERARCHY OF COSMIC RAY TRANSPORT EQUATIONS

Theoretical descriptions of the transport and acceleration of cosmic rays in interplanetary and interstellar plasmas are usually based on two transport equations which both are derived from the collisionless Boltzmann-Vlasov equation into which the electromagnetic fields of the interplanetary and interstellar medium enter by the Lorentz force term. The first of the equations, the *Fokker-Planck equation*, results from applying the quasilinear approximation (Kennel and Engelmann 1966; Lerche 1968; Hasselmann and Wibberenz 1968). One considers the behaviour of energetic charged particles in a uniform magnetic field, $\vec{B}_o = B_o \vec{e}_z$, with superposed small-amplitude plasma waves $(\delta \vec{E}, \delta \vec{B})$ in the rest frame of the plasma wave supporting fluid (e.g. the solar wind). The plasma waves are represented as superposition of individual plasma modes of wave vector $\vec{k}$ and frequency $\omega = \omega_j(\vec{k})$, $j = 1, ..., N$, so that the total electromagnetic field is

$$\vec{B}_T = \vec{B}_o + \delta \vec{B} = B_o \vec{e}_z + \sum_{j=1}^{N} \int d^3k \vec{B}^j(\vec{k}) \exp[i(\vec{k} \cdot \vec{x} - \omega_j t)] \quad (1a),$$

$$\vec{E}_T = \delta \vec{E} = \sum_{j=1}^{N} \int d^3k \vec{E}^j(\vec{k}) \exp[i(\vec{k} \cdot \vec{x} - \omega_j t)] \quad (1b).$$

Maxwell's induction law relates $\vec{B}^j(\vec{k}) = \frac{c}{\omega_j}\vec{k} \times \vec{E}^j(\vec{k})$. The effect of the plasma waves on the particles is studied by calculating first-order corrections to the particle's orbit in the uniform magnetic field $\vec{B}_o$, and ensemble-averaging over the statistical properties of the plasma waves (Jokipii 1966). In the *mixed comoving coordinate system*, in which the *space coordinates* are measured in the laboratory system and the *particle's momentum coordinates* are measured in the rest frame of the background plasma, that supports the plasma waves and in which the turbulence is homogenous in space and time, the gyrophase-averaged phase space density $f(z, p, \mu, t)$ evolves according to the Fokker-Planck equation (Kirk et al.1988)

$$\frac{\partial}{\partial \mu}[D_{\mu\mu}\frac{\partial f}{\partial \mu} + D_{\mu p}\frac{\partial f}{\partial p}] + \frac{1}{p^2}\frac{\partial}{\partial p}p^2[D_{\mu p}\frac{\partial f}{\partial \mu} + D_{pp}\frac{\partial f}{\partial p}] =$$

$$-S + \Gamma(1 + \frac{uv\mu}{c^2})[\frac{\partial f}{\partial t} - \frac{1}{c}\frac{\partial u}{\partial t}\Gamma^2 E \frac{\partial f}{\partial p_z}] + \Gamma(u + v\mu)[\frac{\partial f}{\partial z} - \frac{1}{c}\frac{\partial u}{\partial z}\Gamma^2 E \frac{\partial f}{\partial p_z}] \quad (2),$$

where $u(z,t)$ denotes the bulk speed of the background plasma (e.g. the solar wind speed in the case of the interplanetary medium). $\mu = p_z/p$ is the cosine of the particle's pitch angle, S denotes the source term of particles, $\Gamma = (1 - u^2/c^2)^{-1/2}$, and $E = (p^2c^2 + m^2c^4)^{1/2}$ is the total particle energy. Eq.(2) is derived assuming that the flow velocity $\vec{u} = u(z,t)\vec{e}_z$ is parallel to the background magnetic field. For a radial distribution of the solar wind the corresponding transport equation has been derived by Luhmann (1976). The respective Fokker-Planck coeffcients

$$D_{\mu\mu} = <\frac{\Delta\mu\Delta\mu^*}{2t}>, D_{\mu p} = <\frac{\Delta\mu\Delta p^*}{2t}>, D_{pp} = <\frac{\Delta p\Delta p^*}{2t}>, \quad (3)$$

are calculated from the ensemble-averaged particle orbit corrections (Hall and Sturrock 1967; Achatz et al.1991)

In the presence of low-frequency magnetohydrodynamic waves as Alfven waves, whose magnetic field component is much larger than their electric field component ($|\delta\vec{B}| = (c/V_A)|\delta\vec{E}|$, Alfven velocity $V_A \ll c$), the particle's distribution function $f(z,p,\mu,t)$ adjust very rapidly to quasi-equlibrium through pitch-angle diffusion, which is close to the isotropic distribution. In this case a second cosmic ray transport equation can be derived from the Fokker-Planck equation (3) by a well-known approximation scheme (Jokipii 1966, Hasselmann and Wibberenz 1968, Schlickeiser 1989a) which is commonly referred to as the *diffusion-convection equation* for the pitch-angle averaged phase space density $F(z,p,t)$ and which for non-relativistic bulk speed $u \ll c$ reads

$$\frac{\partial F}{\partial t} - S_o = \frac{\partial}{\partial z}[\kappa\frac{\partial F}{\partial z}] - [u + \frac{1}{4p^2}\frac{\partial}{\partial p}(p^2va_1)]\frac{\partial F}{\partial z} + [\frac{p}{3}\frac{\partial u}{\partial z} + \frac{v}{4}\frac{\partial a_1}{\partial z}]\frac{\partial F}{\partial p} + \frac{1}{p^2}\frac{\partial}{\partial p}[p^2a_2\frac{\partial F}{\partial p}] \quad (4),$$

where the spatial diffusion coefficient κ, the rate of adiabatic deceleration a_1 and the momentum diffusion coefficient a_2 are determined by pitch-angle averages of the three Fokker-Planck coefficients (3) as

$$\kappa = \frac{v^2}{8}\int_{-1}^{1}d\mu\frac{(1-\mu^2)^2}{D_{\mu\mu}}, \quad a_1 = \int_{-1}^{1}d\mu(1-\mu^2)\frac{D_{\mu p}}{D_{\mu\mu}}, \quad a_2 = \frac{1}{2}\int_{-1}^{1}d\mu[D_{pp} - \frac{D_{\mu p}^2}{D_{\mu\mu}}] \quad (5).$$

S_o is $(1/2)\int_{-1}^{1}d\mu S$ and v is the cosmic ray particle velocity. Since Eqs.(2) and (4) are derived in the mixed comoving coordinate system the rate of adiabatic deceleration a_1 does not include $\partial u/\partial z$.

Before elaborating on the transport parameters (5) of the diffusion-convection equation we want to emphasize two important points. The first concerns further reduction of the diffusion-convection equation (4) to simpler transport equations:

(a) in the stationary case ($\partial F/\partial t = 0$) the scattering time method allows us to express the distribution function F as an infinite sum of momentum-dependent distribution functions which obey simple "leaky-box" equations (for details see Wang and Schlickeiser 1987; Lerche and Schlickeiser 1988);

(b) one further average of the diffusion-convection equation over turbulent velocity fields replaces the term $(-u\frac{\partial F}{\partial p} + \frac{p}{3}\frac{\partial u}{\partial z}\frac{\partial F}{\partial p})$ by an additional momentum diffusion term of particles controlled by the velocity-correlation function (Dolginov and Silantev 1990);

(c) multiplying the diffusion-convection equation with $4\pi p^3v/3$ and integrating over p one deduces an hydrodynamical equation for the cosmic ray pressure P_c that has been used in treatments of particle acceleration at modified shocks and cosmic ray driven galactic winds (Schlickeiser and Lerche 1985; Fichtner et al.1991). Its hydrodynamical transport parameters are momentum-averages of the transport parameters (5), and so ultimately are determined by the Fokker-Planck coefficients.

Secondly, if appropriate one has to add additional terms to the Fokker-Planck equation and/or diffusion-convection equation representing (i) energy loss processes of particles (Coulomb interactions, ionization, pion production, bremsstrahlung, synchrotron radiation, inverse Compton interactions), (ii) adiabatic focusing if the scale of spatial variation of the background magnetic field $l_B = B_o/|\partial B_o/\partial z|$ is smaller than the scale of interest. Since these additional effects depend on pitch-angle they may very well modify strongly the iteration scheme from the Fokker-Planck to the convection-diffusion equation.

In any case, however, fundamental to all deductions of cosmic ray transport equations and their parameters is the knowledge of the particle's Fokker-Planck coefficients (3). Recently Jaekel and Schlickeiser (1992) have calculated the quasilinear Fokker-Planck coefficients (3) for general electromagnetic fluctuations in the weak turbulence approximation. Their results can be employed once the relevant plasma modes in a given physical system are identified either from in-situ measurements (as in the interplanetary medium) or from calculations of the growth rates of individual modes.

II. RELEVANT PLASMA MODES

To identify the relevant low-frequency plasma modes in a given physical system, that is not accessible by direct in-situ measurements of the electromagnetic fluctuations, the generation and damping of plasma waves has to be considered. A useful start is provided by the plasma wave dispersion relation of a cold magnetized electron-proton plasma.

II.1 COLD MAGNETIZED ELECTRON-PROTON PLASMA

For parallel (to $\vec{B}_o$) propagating waves at frequencies much smaller than the electron gyrofrequency $|\omega| << |\Omega_e|$ the dispersion relation is

$$\frac{k_\parallel^2 c^2}{\omega^2} \simeq \frac{c^2}{V_A^2} \frac{\Omega_p}{\Omega_p - \omega} \tag{6},$$

where we use the convention that solutions of Eq.(6) with positive frequency $\omega > 0$ are physically left-handed polarized while solutions with negative frequency $\omega < 0$ are physically right-handed polarized. The solutions of Eq.(6) are shown in Fig.1a and include several plasma modes with different asymptotic dispersion relation:

1) At small wavenumbers $|k_\parallel| << k_c \equiv 2\Omega_p/V_A$ and frequencies $|\omega| << \Omega_p$ Eq.(8) reduces to $\omega^2 \simeq V_A^2 k_\parallel^2$ which describes non-dispersive ($V_{ph} = \omega/k_\parallel$ =const.) forward ($V_{ph} > 0$) and backward ($V_{ph} < 0$) moving *Alfven waves* which are either right-handed ($\omega < 0$) or left-handed ($\omega > 0$) circularly polarized.

(2) At large wavenumbers $|k_\parallel| >> k_c$ the left-handed Alfven branch develops into the left-handed *ion cyclotron wave* branch, which can propagate both forward and backward and which at all values of $k_\parallel$ have nearly the same frequency $\omega \simeq \Omega_p$.

(3) At large wavenumbers $|k_\parallel| >> k_c$ and frequencies between $\Omega_e < \omega < -\Omega_p$ the right-handed Alfven branch develops into the right-handed *Whistler wave* branch, which is dispersive ($V_{ph} \neq$ const.), because of the quadratic wave number dependence of its dispersion relation $\omega \simeq -2V_A k_\parallel^2/k_c$, and which can propagate both forward and backward.

(4) At very large wavenumbers $|k_\parallel| >> 21 k_c$ the right-handed Whistler wave branch develops into the right-handed *electron cyclotron wave* branch $\omega \simeq \Omega_e$, which again can propagate both forward and backward.

The solution of the cold electron-proton plasma dispersion relation for propagation across the static magnetic field in the high-density limit $\omega_{p,e} >> |\Omega_e|$ is shown in Fig.1b. Each solution comes fourfold as forward, backward, physically right- and left-handed polarized wave. The solutions include several well-known plasma modes with asymptotic dispersion relation. Besides the ordinary and extraordinary electromagnetic wave mode at high frequencies we identify:

(5) At small frequencies $|\omega| \to 0$ the *compressional Alfven waves*. Its dispersion relation $\omega^2 = V_A^2 k_\perp^2$ is formally identical to the one for parallel Alfven waves, but there is one important

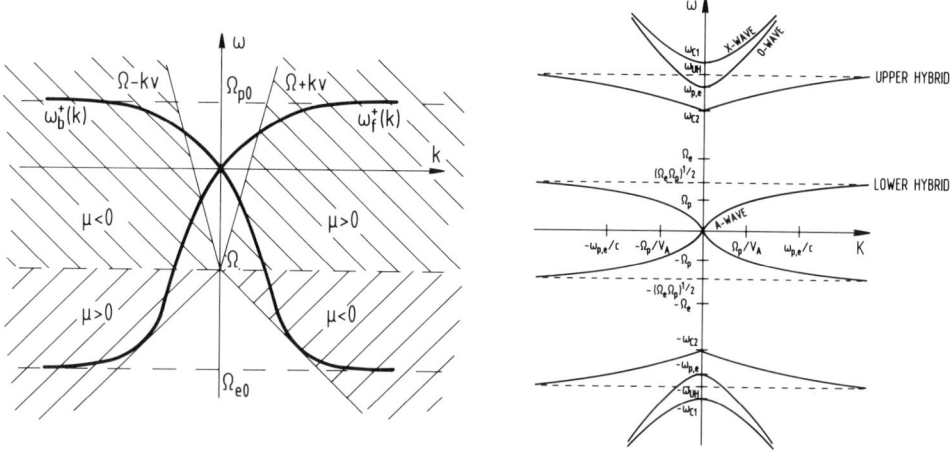

Fig.1: Dispersion relations of transversal plasma waves in a cold electron-proton plasma travelling (a) parallel (from Achatz et al.1992) (left figure) and (b) perpendicular (right figure) to the ordered magnetic field. In the diagram for parallel propagating waves the resonance condition (7) for a relativistic electron with positive and negative pitch angle cosine μ is illustrated.

difference as $|\omega| \to \Omega_p$. As we have discussed, the character of the parallel Alfven mode alters drastically with the left-handed branch degenerating into an ion cyclotron wave (see (2)) and the right-handed branch developing into the Whistler wave (see (3)). Quite differently, the compressional Alfven wave in general undergoes no radical change at the proton gyrofrequency. As can be seen from Fig.1b, in the high-density limit compressional Alfven waves propagate not only below $|\omega| = \Omega_p$ but also through the ion cyclotron resonance. For this mode there is no resonance until the lower hybrid frequency $\omega \to \pm \omega_{LH} = \pm 43 \Omega_p$ is reached.

(6) As the lower hybrid frequency is approached the compressional Alfven waves develop into the *lower hybrid waves* which at all values of $k_\perp$ have nearly the same frequency $\omega = \pm \omega_{LH}$.

(7) In the stop band $\omega_{LH} < |\omega| < \omega_{c2} = \sqrt{\omega_{p,e}(\omega_{p,e} - |\Omega_e|)}$ no waves propagate in the high-density limit, but in the small frequency band $\omega_{c2} \leq |\omega| \leq \omega_{UH}$ the *upper hybrid waves* exist which at all values of $k_\perp$ have nearly the same frequency $\omega = \pm \omega_{UH} = \pm \sqrt{\omega_{pe}^2 + \Omega_e^2}$.

II.2 WARM MAGNETIZED ELECTRON-PROTON PLASMA

The modifications by including finite temperature effects due to a warm Maxwellian background plasma distribution on the cold-plasma results are mainly threefold:

(1) The presence of parallel propagating ion cyclotron waves near Ω_p and of electron cyclotron waves near Ω_e is drastically reduced since these waves are cyclotron damped by the thermal protons and electrons, respectively (Davila and Scott 1984; Achatz et al.1992).

(2) The right-handed polarized Alfven-Whistler branch is unaffected by the background plasma near $-\Omega_p$.

(3) Oblique propagating low-frequency magnetohydrodynamic waves are quickly damped depending on the plasma-beta of the background plasma (Barnes 1969; Foote and Kulsrud 1979; Achterberg 1981). Because of this modification the theory of resonant interactions of cosmic ray particles with plasma waves has concentrated almost exclusively on parallel propagating waves.

As a consequence of all three modifications we expect a drastic change of the plasma wave's net polarization (i.e. magnetic helicity) from frequencies much below the proton gyrofrequency $\Omega_p \sim$

1 Hz in the solar wind, where both right- and left-handed polarized waves should be present, to frequency values above the proton gyrofrequency where only the right-handed polarized Whistler waves remain unaffected.

III. QUASILINEAR THEORY OF RESONANT INTERACTION WITH PARALLEL WAVES

The quasilinear theory of the resonant interaction of cosmic ray particles with parallel low-frequency plasma waves demands the fulfillment of the resonance condition

$$\omega(k_{res}) = v\mu k_{res} + (\Omega/\gamma) \tag{7}.$$

Particles of mass m, charge q and gyrofrequency $\Omega = qB_o/(mc)$, pitch-angle cosine μ and Lorentzfactor γ resonantly interact only with those plasma waves whose frequency ω and wavenumber $k_\| = k_{res}$ obey Eq.(7). This resonance condition has been graphically illustrated in Fig.1 for a relativistic electrons with positive and negative μ, respectively. As an important special case we notice at pitch-angle $\mu = 0$ cosmic ray protons ($\Omega > 0$) can only interact with left-handed waves ($\omega > 0$), while cosmic ray electrons ($\Omega < 0$) interact only with right-handed polarized waves. The situation is more complicated at finite μ (Achatz et al.1991; Steinacker and Miller 1992): most importantly, if waves of both polarization state propagating in both directions are present, there is no resonance gap at any pitch angle both for cosmic ray electrons and protons, i.e. these particles find waves to resonate with at all pitch-angles. While energetic protons resonate with Alfven and ion cyclotron waves, the electrons can resonate with up to three right-handed waves and at least one left-handed waves encompassing Alfven, Whistler, electron and ion cyclotron waves. A real difficulty is encountered only in case of nonrelativistic cosmic ray positrons that cannot be scattered through $\mu = 0$ with parallel waves. For them, the influence of oblique waves will be vital.

However, the actual values of the transport and acceleration parameters (3) and (5) for these resonant interactions will depend on the actual intensities of the different wave modes which, of course, will be determined by the physics of wave generation and wave damping.

IV. STATUS OF COSMIC RAY TRANSPORT PARAMETER CALCULATION

Early calculations of cosmic ray transport parameters based on the "magnetostatic approximation" ($\omega = 0$, $|\delta \vec{E}| = 0$), in which all finite frequency and electric field effects in the rest frame of the background fluid have been neglected, readily implied $D_{\mu p} = D_{pp} = 0$ and thus $a_1 = a_2 = 0$ in the diffusion-convection equation (4); i.e. Eq.(4) then contains no momentum diffusion, and the convection and adiabatic deceleration of particles is only due to the motion $u(z)$ of the wave-carrying background medium. However, this magnetostatic approximation with its legendary resonance gap problems has been identified as the source of singularities and inconsistencies (Schlickeiser et al.1991), and is now surpassed by the complete development of more realistic quasilinear calculations which includes finite frequency effects. There is no compelling reason to use the magnetostatic results anymore.

For parallel propagating Alfven waves (often referred to as "slab model") the calculation of quasilinear Fokker-Planck coefficients and transport parameters of the diffusion-convection equation is fairly complete. Schlickeiser (1989a,b), Dung and Schlickeiser (1990a,b) and Schlickeiser et al.(1991) study in detail the influence of cross and magnetic helicities as well as the dissipation range of the Alfvenic turbulence on the cosmic ray transport. The most noteworthy results are that (i) the famous discrepancy between observed (κ_{fit}) and quasilinear (κ_{ql}) spatial diffusion coefficients for solar flare particles can be resolved phenomenologically, (ii) nonzero values of magnetic and cross helicities enhance the value of κ_{ql}, (iii) dissipation range spectra yield finite values of κ_{ql}, and (iv) momentum diffusion is unavoidable ($a_2 \neq 0$) for nondegenerate ($|H_c| \neq 1$) values of the cross helicity.

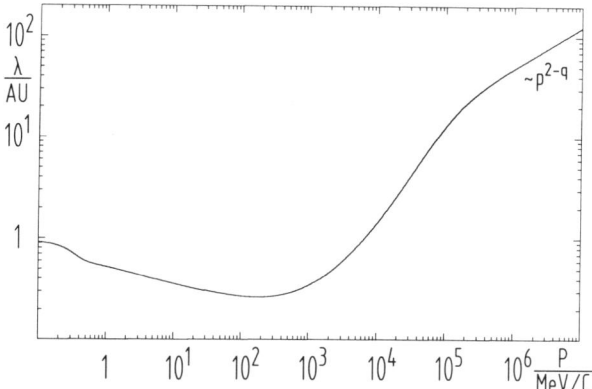

Fig.2: Quasilinear mean free path $\lambda = 3\kappa_{ql}/v$ for cosmic ray electrons in the interplanetary medium calculated with the observed electromagnetic turbulence during the 7 June 1980 solar event (from Achatz et al.1992).

Extension of the theory to oblique Alfven waves is straightforward with the now available Fokker-Planck coefficients of Jaekel and Schlickeiser (1992). Turbulence models are considered where plasma waves can propagate within a cone of opening angle χ with respect to the background magnetic field. Jaekel and Schlickeiser demonstrated analytically for these generalized cone models that all Fokker-Planck coeffients agree with the slab model result at pitch angle cosine values $\mu \geq \tan\chi$ whereas at small values $\mu \leq \tan\chi$ an additional factor of order μ appears, which for isotropic turbulence ($\chi = \pi/2$) has been numerically discovered for $D_{\mu\mu}$ by Fisk et al.(1974). Steinacker et al.(1992) calculate the consequences of this modification for the transport parameters of the diffusion-convection equation: while the parameters a_1 and a_2 remain unchanged, the spatial diffusion coefficient κ changes its absolute value significantly and has a different momentum dependence at nonrelativistic particle energies while at relativistic energies the momentum dependence is the same as in the slab model.

As we have seen in section II, at large wavenumbers dispersive effects come into play and the simple linear Alfvenic frequency-wavenumber-relation $\omega \propto k_{\|}$ no longer holds. Long-wavenumber plasma waves are particular important for the resonant interaction of nonrelativistic ions and mildly relativistic electrons and positrons. Achatz et al.(1991, 1992) and Steinacker and Miller (1992) have developed the complete quasilinear transport theory for parallel propagating waves including all dispersive effects and possible damping of waves in a warm background medium. In Fig.2 the calculated momentum variation of the mean free path of electrons $\lambda = 3\kappa/v$ in the interplanetary medium is shown which has been calculated using the observed electromagnetic turbulence during the 7 June 1980 solar event. It clearly can be seen that the momentum dependence p^{2-q} predicted by resonant interaction with Alfven waves is only attained for ultrarelativistic electron momenta $> 10^5$ MeV/c, but that at lower momenta quite a different energy dependence results. The analogous study for the transport of cosmic ray protons show that modifications to the simple Alfven picture are drastic at nonrelativistic proton energies. The calculated absolute values and the momentum dependence for electrons in Fig.2 agree rather well within factors of 2 with the phenomenologically determined mean free paths from fits to the ISEE-3 particle data in the range 0.2 to 10 MeV/c (see Achatz et al.1992 for details). These results are certainly very encouraging, since they prove the usefulness of quasilinear weak-turbulence theory. The quasilinear mean free paths are systematically larger than the observed ones, so that including additional non-linear scattering effects (as resonance broadening) will certainly improve the agreement with the observations; this is quite contrary to the old discrepancy of the magnetostatic mean free paths which were systematically too small, so that any additional scattering would further decrease the mean free path and thus worsen the discrepancy.

As a summary on the present status of weak turbulence quasilinear transport theory and its confrontation with in-situ observations in the interplanetary medium we want to emphasize that much more work has to be done to come to a complete understanding of cosmic ray transport and acceleration in cosmic plasmas. However, significant process has been achieved in dealing with helicities, dispersion and dissipation of the electromagnetic scattering centers during the last years. Quasilinear theory is certainly in much better shape as compared to ten years ago when singularities and discrepancies with observations severely limited its credibility. Incorporating finite frequency effects has been essential in overcoming these difficulties. It remains to extend the theory to oblique non-Alfvenic waves and to include corrections arising from nonlinear effects as resonance broadening. This may turn quasilinear theory from a qualitatively correct theory into a quantitatively correct theory.

Acknowledgement. I am grateful to my collaborators U. Achatz, W. Dröge, R. Dung, U. Jaekel, J. Steinacker and G. Wibberenz for their cooperation, support and encouragement.

REFERENCES

Achatz,U., Steinacker,J., Schlickeiser,R., A&A 250, 266 (1991)

Achatz,U., Dröge,W., Schlickeiser,R., Wibberenz,G., J. Geophys. Res., submitted (1992)

Achterberg,A., A&A 98, 161 (1981)

Barnes,A., ApJ 155, 311 (1969)

Davila,J.M., Scott,J.S., ApJ 285, 400 (1984)

Dolginov,A.Z., Silantev,N.A., A&A 236, 519 (1990)

Dung,R., Schlickeiser,R., A&A 237, 504 (1990a)

Dung,R., Schlickeiser,R., A&A 240, 537 (1990a)

Fichtner,H., Fahr,H.J., Neutsch,W., Schlickeiser,R., Crusius-Wätzel,A., Lesch,H., Nuovo Cimento 106B, 909

Fisk,L.A., Goldstein,M.A., Klimas,A.J., Sandri,G., ApJ 190, 417 (1974)

Foote,E.A., Kulsrud,R.M., ApJ 233, 302 (1979)

Hall,D.E., Sturrock,P.A., Phys. Fluids 10, 2620 (1977)

Hasselmann,K., Wibberenz,G., Z. Geophys. 34, 353 (1968)

Jaekel,U., Schlickeiser,R., J. Phys. G, in press (1992)

Jokipii,J.R., ApJ 146, 480 (1966)

Kennel,C.F., Engelmann,F., Phys. Fluids 9, 2377 (1966)

Kirk,J.G., Schlickeiser,R., Schneider,P., ApJ 328, 269 (1988)

Lerche,I., Phys. Fluids 11, 1720 (1968)

Luhmann,J.G., J. Geophys. Res. 81, 2089 (1976)

Lerche,I., Schlickeiser,R., Ap. Space Sci. 145, 319 (1988)

Schlickeiser,R., ApJ 336, 243 (1989a)

Schlickeiser,R., ApJ 336, 264 (1989b)

Schlickeiser,R., Dung,R., Jaekel,U., A&A 242, L5 (1991)

Schlickeiser,R., Lerche,I., A&A 151, 151 (1985)

Steinacker,J., Miller,J.A., ApJ, in press (1992)

Steinacker,J., Jaekel,U., Schlickeiser,R., ApJ, submitted (1992)

Wang,Y.-M., Schlickeiser,R., ApJ 313, 200 (1987)

PARTICLE ACCELERATION AND TRANSPORT BY STRONG MHD-TURBULENCE AND SHOCK WAVE ENSEMBLES

A.M. Bykov
A.F.Ioffe Institute of Physics and Technology
194021, St. Petersburg, Russia

1 Introduction

The problems of particle transfer and acceleration by a turbulence in cosmic plasma traditionally attract great attention. This is mainly due to the fact that objects with large energy release produce significant fraction of free energy in the form of large-scale motions and shock waves. A detailed discussion of these problems can be found in the reviews [1 – 5]. In this report we consider two problems of CR kinetic theory. The first one is charged particle acceleration by an ensemble of shocks. We present the temporal evolution of the low energy CR distribution function averaged over a turbulent ensemble. The equation for the particle distribution function has been obtained in [6] (see also the interesting work [7, 8] where analogous problems are discussed). The low energy CR distribution function has an intermittent character with strong perturbations near shocks. Using the nonperturbative connection between average distribution function and its intermittent part an integrodifferential equation is obtained for averaged distribution function. For CR with high enough energy, the particle distribution can be described by the equations from [9], in this case intermittence is not important. The second problem is connected to CR particle spectrum features for the systems with strong smooth fluctuation. We construct a stationary CR spectrum with account taken of the effect of finite energy change over correlation scale. Particularly it is important for the problem of the electrons injection. The formulation of both problems is based on the "renormalization approach" developed in [10].

2 Cosmic Ray Spectra Evolution in Systems with an Ensemble of Strong Shocks

Consider the evolution of nonthermal charged particles of low energy cosmic rays (CR) inside a system with an ensemble of strong shock and violent large scale motions. The CR distribution has an intermittent structure because of the presence of strong shocks. The kinetic equation for the CR distribution function which takes into account intermittency has been obtained by Bykov and Toptygin [6]. The regimes of particle acceleration by an ensemble of shocks are determined by the dimensionless parameter

$$\psi = \frac{ul}{\kappa(p)}, \qquad (1)$$

where $\kappa(p)$ is the longitudinal (with respect to the regular magnetic field) diffusion coefficient produced by small-scale resonant magnetic inhomogeneities.

If $\psi > 1$, the particle distribution function $N(r,p,t)$ averaged over the ensemble of turbulent motions satisfies the transport equation (see [6]):

$$\frac{\partial N}{\partial t} + \frac{N}{\tau_e} = \left(\frac{1}{\tau_{sh}} + B\right)\hat{L}N + \frac{1}{p^2}\frac{\partial}{\partial p}p^4 D\frac{\partial N}{\partial p} + A\hat{L}^2 N + 2B\hat{L}\hat{P}N + F_i(p), \qquad (2)$$

where $\hat{L}$ and $\hat{P}$ are the operators given by

$$\hat{L} = \frac{1}{3p^2}\frac{\partial}{\partial p}p^{3-\alpha}\int_0^p dp'\, p'^\alpha \frac{\partial}{\partial p'}, \quad \hat{P} = \frac{p}{3}\frac{\partial}{\partial p}. \qquad (3)$$

In this case $\alpha = 3\Delta/(\Delta - 1)$ is a power-law exponent of the 'universal' CR spectrum on an individual strong shock front, and Δ is the shock compression ratio. The kinetic coefficients A, B, D, τ_{sh}, and τ_e are expressed in terms of spectral functions which describe the correlations between large scale turbulent motions and shock ensemble cross-correlation properties, for details see [6]. Note here that it is possible to generalize our procedure to account of effect of CR on the shock turbulence. One way is through the "two-fluid" approximation (see [11,12]). Our model of CR generation in the OB - associations (see [13]) suggests that the generation of suprathermal particles (which will be further accelerated by large scale turbulent motions of plasma inside the association) takes place at shock wave fronts. Another source of injected particles for acceleration by shock fronts can be associated with magnetic reconnection. This is likely to be the case for solar flare models and for particle acceleration in the magnetic disk of AGNs ([14]). Energetic particles, as was shown in the work [15], are extracted out from the thermal pool with quite high efficiency by the same scattering process which is responsible for the collisionless shock structure formation. This process is in a good quantitative agreement to the results of the observations of CR acceleration by quasiparallel shocks in the interplanetary medium and near the Earth's bow shock (see for the review [16]). There is no complete quantitative theory now for particle injection phenomena at shocks with the arbitrary strengths and orientations relative to the local magnetic field. So we use two averaged (phenomenological) parameters for the description of the CR injection. One parameter is the quasi-stationary generation rate of suprathermal particles q which can be calculated by the averaging of shock injection rate over the shock ensemble:

$$q = <\eta n u_{sh}\Sigma_{sh}V_{sb}^{-1}>, \qquad (4)$$

where η is a fraction of particles incident on the shock surface (Σ_{sh}) to be injected into the acceleration, V_{sb} -is the system volume. The second parameter is the momentum of injected particles p_i for which the appropriate value is likely to be "several" times more then the downstream thermal particle momentum (see, e.g.,[17,18]). As for the value of η, its estimate is still rather uncertain. The "conventionally" appropriate value is in the range

$$10^{-5} < \eta < 10^{-1} \qquad (5)$$

(see e.g. [18]).The observed injection rate at the Earth's bow shock corresponds to $\eta \sim 10^{-3}$. Actually the value of η (in fact η and p_i are dependent on the angle between the shock normal and the local magnetic field direction) is the local injection rate at a single supernova shock, while in our treatment we use some average value η. So we note that the average injection rate in shock fronts for several early supernovae (which explode before CR get the substantial portion of system's free energy) are likely to accord with Eq.(5). Such a situation takes place for supernova shocks which propagate in the cavity (created by the wind from a massive star) due to the strong modulation of low-energy CR by the wind as well. The shock injection rate should be reduced as a result the of influence of pregenerated CR particles on shock structure for supernova from a lower massive star without a strong wind.

It is important that the injection rate of particles with momentum p_i be substantially suppressed if the energy density of accelerated nonthermal particles reaches the downstream thermal particle pressure value. The exact theory of this process is still not available but the existing models show that it is important for shock acceleration (see the reviews [5, 16]). Thus, actually there is some time dependent process in the system after which the reduced quasistationary injection rate is established. Such a value of η is assumed below. Thus the source function of injected particles has the form:

$$F_i(p) = (\alpha - 3) q \frac{p_i^{\alpha-3}}{p^\alpha} \theta(p - p_i). \tag{6}$$

Some small terms can be neglected during the integration, after which one can obtain:

$$N(p,t) \approx \frac{(\alpha-3)n\eta}{9p_i^3} \begin{cases} \left(\frac{p_i}{p}\right)^3 \left[1 - \frac{\exp(-\sigma(t+t_p))}{1-[3\alpha/(\alpha-3)(2\alpha-3)]}\right], & p < p_*, \\ \left(\frac{p_i}{p}\right)^\alpha \exp\left[\frac{t}{\tau_{sh}}\right], & p_* < p < p_{max}, \end{cases} \tag{7}$$

where $p_* = p_i \exp[(t/\tau_{sh})/(\alpha-3)]$, $t_p = \alpha \tau_{sh} \log(p/p_i).\sigma = 3/(\tau_{sh}(2\alpha-3))$.

Note that the maximum momentum p_{max} of the spectrum Eq. (7) is determined by the condition $\psi(p_{max}) \sim 1$. The equation for the distribution function at $p > p_{max}$ has been obtained by Bykov and Toptygin [9]. The time evolution of this spectrum is discussed elsewhere [13].

3 Non Fokker-Planck Kinetics of Cosmic Rays

The kinetics of charged particles in stochastic electromagnetic fields has been studied in detail in the Fokker-Planck approximation (FPA). The conditions for the validity of FPA reduce, as a rule, to the requirement for the energy or momentum changes of particles to be small along a correlation length (or correlation time). Therefore FPA is restricted to the consideration of rather energetic particles or of "briefly" correlated systems. On the other hand, a solution of many problems, particularly, of the problem how the spectrum of low-energy cosmic rays (CR) is formed, cannot be obtained in FPA. Here we discuss the formation of CR spectrum in systems with a strong long-wavelength turbulence following the work by the author [19]. Consider a plasma system where an MHD, statistically uniform turbulence develops. Electric fields induced by turbulent motions of an ideally conducting medium with a frozen-in magnetic field lead to stochastic changes in energy of charged particles. The magnetic field fluctuations with scales $l \leq r_h$ (r_h - is the particle gyroradius in the mean magnetic field) produce an effective izotropisation of particle momenta, and determine their transport path Λ. Under long-wavelength fluctuations we will mean those whose scales $l > \Lambda$. A distribution function of CR, $G(r, \xi, t)$, averaged over an ensemble of turbulent fluctuations on all scales, satisfies the equation (see [10])

$$\frac{\partial G}{\partial t} - \int_{-\infty}^{\infty} d\xi' \chi_{\alpha\beta}(\xi - \xi') \frac{\partial^2 G(r, \xi', t)}{\partial r_\alpha \partial r_\beta}$$
$$- \left(\frac{\partial^2}{\partial \xi^2} + 3\frac{\partial}{\partial \xi} \right) \int_{-\infty}^{\infty} d\xi' D(\xi - \xi') G(r, \xi', t) = Q\delta(\xi). \qquad (8)$$

Here $\xi = log(p/p_0)$ is a variable determined by a particle momentum p, Q denotes the injection rate of monoenergetic particles (momentum p_0). Eq. (8) is conveniently solved by the Fourier transformation with respect to ξ. The Fourier transforms of the kernels of Eq. (8) are expressed through the turbulence correlation functions. Let us consider a general case of the compressible system with an extended velocity fluctuation spectrum of the form

$$W(k) = \frac{<u>^2}{4} C(\nu) \frac{k_0^{\nu-1}}{(k^2 + k_0^2)^{(\nu/2+1)}}, \quad k \leq k_{max}, \qquad (9)$$

and with the Lorentz-type frequency dependence determined by the dispersion relation $\omega_0 = ku$ and by the resonance width $\gamma/2 = uk_0(k/k_0)^{(3-\nu)/2}$. Here $u \equiv <u^2>^{1/2}$, and $C(\nu) \approx \frac{4\Gamma(\nu/2+1)}{3\pi^{3/2}\Gamma(\nu/2-1/2)}$ -is the normalization constant. In this case, according to [19], one can obtain the transcendental equations for the Fourier transforms of the kernels,

$$D_1(s) = \epsilon + C(\nu) \int_0^b dx \frac{x^2}{(x^2+1)^{\nu/2+1}} \left\{ \frac{\psi}{\psi^2 + \omega_0^2} - \frac{2}{3} D_1 x^2 (\lambda/6 + 1) \frac{\psi^2 - \omega_0^2}{(\psi^2 + \omega_0^2)^2} + \right.$$
$$\left. + \frac{4}{9} \lambda D_1^2 x^4 \frac{\psi^3 - 3\psi\omega_0^2}{(\psi^2 + \omega_0^2)^3} \right\}, \qquad (10)$$

$$D_2(s) = \frac{C(\nu)}{9} \int_0^b dx \frac{x^4}{(x^2+1)^{\nu/2+1}} \frac{\psi}{\psi^2 + \omega_0^2}. \qquad (11)$$

Eqs. (10) and (11) take into account the effects of finite energy change of particles along the correlation scale. In these equations we have introduced the dimensionless variables: $\chi_{\alpha\beta} \equiv uk_0^{-1} D_1(s) \delta_{\alpha\beta}$, $D(s) \equiv uk_0 D_2(s)$, $b = k_{max}/k_0$, $\epsilon = \kappa k_0/u$. Here κ is the small-scale diffusion coefficient,

$$\psi(x, \lambda) = D_1(s) x^2 + \lambda D_2(s) + \frac{\gamma(x)}{2}, \quad \lambda \equiv s(s + 3i) \qquad (12)$$

The above derivation is valid when the particle velocities are larger than the turbulent velocity. The magnitude of ϵ can be arbitrary. If $\epsilon > 1$ (the particle transport is governed by scattering on small-scale fluctuations), the solutions to Eqs. (10) and (11) can be written approximately as $D_1(s) \approx \epsilon$ and $D_2(s) \approx (9\epsilon)^{-1}$. Then we have $\chi_{\alpha\beta}(\xi - \xi') \approx \kappa \delta(\xi - \xi') \delta_{\alpha\beta}$, $D(\xi - \xi') \approx u^2 (9\kappa)^{-1} \delta(\xi - \xi')$. In this case Eq.(8) has the FP-form. If $\epsilon < 1$ (i.e. for rather low-energetic particles) the particle transport is determined mainly by long-wave fluctuations. Then the transport equation is essentially of a non FP-form and the kernels should be renormalized. An analytic analysis allows us to obtain the asymptotic expressions for Fourier-transforms of the kernels:

$$D_1(s) \rightarrow \epsilon/2, \qquad (13)$$

$$D_2(s) \rightarrow b^{(3-\nu/2)/2}(a(\nu)\sqrt{\lambda})^{-1} \qquad (14)$$

Using these expressions one can derive the asymptotics of the steady-state spectrum of particles at $p \rightarrow p_0$:

$$G(p,p_0) \propto -Q\,(uk_0)^{-1}b^{(\nu-3)/2}a(\nu)\ln|\ln(p/p_0)|, \qquad (15)$$

In the vicinity of $p \approx p_0$ the spectrum (15) has an integrable logarithmic singularity. It appears because finite energy changes are allowed along the correlation scale. At $p \gg p_0$ the particle spectrum is determined by the behavior of the kernels in the limit of $s \rightarrow 0$

$$G(p,p_0) \propto Q\,(uk_0 D_2(0))^{-1}(p/p_0)^\mu,\ \mu = -1.5 - \left(2.25 + \zeta D_1(0)D_2^{-1}(0)\right)^{0.5} \qquad (16)$$

This relationship describes also the FP spectrum at all momenta $p > p_0$. In contrast to (15) the FP distribution function (16) remains finite at $p \rightarrow p_0$ and depends only weakly on the b - the fluctuation spectrum extension measure.

It is interesting to note that Eq. (8) describes also the formation of the photon spectrum due to Thomson scattering in an optically thick medium with turbulent macroscopic motions. In the latter case the diffusion coefficient κ is determined by the electron number density. (cf. [20]). **Acknowledgements** The author would like to thank I.N. Toptygin for many hours of joint working and discussions , D.G. Yakovlev and G.D. Fleishman for their help. I am deeply indebted to G.P. Zank for inviting me at the Bartol Research Institute.

References

1. Jokipii, J.R. 1971 *Rev. Geoph. Sp. Sci.* **9**, 27
2. Axford, W.I. 1980. *Proc. IAU symp 94* , 339
3. Ginzburg, V.L., Ptuskin, V.S. 1985 *Sov. Sci. Rev. E*
4. Toptygin, I.N. 1985, Cosmic Rays in Interplanetary Magnetic Fields. Reidel. Amsterdam
5. Berezhko, E.G., Krymsky, G.F. 1988 *Sov. Phys. Uspekhi* **31**, 27
6. Bykov, A.M., Toptygin, I.N. 1990 *Sov.Phys.JETP*,**71**, 702
7. Achterberg, A. 1990 *Astron.Astrophys.* **231**, 251
8. Spruit, H.C. 1988 *Astron.Astrophys.* **194**, 319
9. Bykov, A.M., Toptygin, I.N. 1979 *Proc. 16th ICRC (Kyoto)*,**2**, 66
10. Bykov, A.M., Toptygin, I.N. 1990 *Sov.Phys.JETP*,**70**, 108
11. Zank, G.P., Axford, W.I., McKenzie, J.F. 1990, *Astron.Astrophys.* **233**, 275
12. Webb, G.M. 1983, *Astron.Astrophys.* **127**, 97
13. Bykov, A.M. & Fleishman G.D. 1992 *Mon. Not. Roy. Astron. Soc.*
14. Field, G., Rogers, R. 1991, *Proc. Heidelberg conference on Phys. AGN* II.4

15. Ellison, D.C., Eichler, D. 1984, *Astrophys.J.*, 286, 691
16. Blandford, R., Eichler, E. 1987. *Physics Reports* **154**, 2
17. Bell, A.R. 1987. *Mon. Not. R. Astr. Soc.*, **225**, 615
18. Drury, L.O'C., Markiewicz, W.J., & Vólk, H.J., 1989.*Astron. Astrophys* **225**, 179
19. Bykov, A.M. 1991. *Sov. Phys. JETP Letters* v**54**, (10 Dec.)
20. Blandford, R.D., Payne, D.G. 1981. *Mon. Not. Roy. Astron. Soc.* **194**, 1041

ANOMALOUS DIFFUSION OF COSMIC RAYS ACROSS THE MAGNETIC FIELD

L.G.Chuvilgin, V.S.Ptuskin

Institute of Terrestrial Magnetism, Ionosphere and Radio Wave Propagation, USSR Academy of Sciences, Troitsk, Moscow region 142092, Russia

1. **Introduction.** Diffusing CR particles are extremely magnetized in galactic magnetic field. The scattering on small-scale weak turbulence leads mainly to diffusion along the field. The ratio of gyroradius r_H to mean free path l is $r_H/l \approx 10^{-6}$ for the energy 1GeV ($r_H \approx 10^{12}$cm, $l \approx 10^{18}$cm) /1/. Galactic large-scale random field with the main scale $L \approx 100$pc gives rise to random rotations of the local very anisotropic diffusion tensor and can thereby make the diffusion more isotropic. Thus, long-wave random magnetic field leads to an enhanced (anomalous) diffusion across the mean magnetic field. Similar problem arises in application to investigation of particle transport in tokamaks.

Quantitative consideration of the problem of anomalous diffusion displays some nontrivial results, see reviews /1,6-8/. "Natural" recipe for calculation of the tensor of anomalous diffusion: 'Take the local diffusion tensor and average it over random rotations' turns out to be incorrect in the case of static field. Effect of compound diffusion slows down perpendicular transport of particles.

2. **Compound diffusion.** Compound diffusion is a combination of two independent random walks - diffusion of the particle along the field line and random walk of the field line itself /5/. During the time t the

particle travels along the field line over a distance $s \propto \sqrt{\varkappa_{11} t}$. At the same time the field line is shifted transverse to the mean field H_o by $x^2 \propto A^2 Ls$ ($A = H_1/H_o$ is the relative amplitude of the random field, $A < 1$, $H_1 \perp H_o$). The average displacement of the particle grows as $x \propto t^{1/4}$ (during an ordinary diffusion one has $x \propto t^{1/2}$). Effective perpendicular diffusion coefficient is $D \propto x^2/t \to 0$.

Here we present an equation describing the compound diffusion of CR in static magnetic field with $A \ll 1$, see /3/ for details. The compound diffusion occurs in the plane perpendicular to H_o in this case.

The diffusion of particles along magnetic field lines obeys the equation ($f(t,\mathbf{r})$ is CR number density, $\varkappa_{11} = $ const)

$$\frac{\partial f}{\partial t} - \varkappa_{11} \nabla_s b_s b_j \nabla_j f = 0, \quad b = H/H \qquad (1)$$

Starting with the eq.(1) and using standard technique of the weak turbulence theory /4/, one can obtain the equation for CR number density $N = \langle f \rangle$ averaged over ensemble of magnetic field fluctuations /10/. This equation coarsens on the time intervals $\Delta t > L^2/\varkappa_{11}$ and taking into account terms of the order of A^2 it may be presented in the form ($\nabla_{11}^2 = \partial^2/\partial z^2$, $\nabla_\perp^2 = \nabla^2 - \nabla_{11}^2$, z axis is directed along H_o, $t \gg L^2/\varkappa_{11}$)

$$\frac{\partial N(t,\mathbf{r})}{\partial t} - (1 - \langle A^2 \rangle) \varkappa_{11} \nabla_{11}^2 N(t,\mathbf{r}) - \frac{\langle A^2 \rangle}{2} \varkappa_{11} \nabla_\perp^2 [N(t,\mathbf{r}) - \frac{1}{2} \int_1^\infty dy \, y^{-3/2} N(t - yL^2/\varkappa_{11}, \mathbf{r})] -$$

$$- 2 \langle A^2 \rangle L^2 \varkappa_{11} \nabla_{11}^2 \nabla_\perp^2 \int_1^\infty dy \, y^{-1/2} N(t - yL^2/\varkappa_{11}, \mathbf{r}) =$$

$$= 0 \tag{2}$$

The eq.(2) includes diffusion along the mean field and more complicated propagation in perpendicular directions. The presence of integrals means that the process has a good 'memory'. Unlike the ordinary diffusion, the compound diffusion is not a Markovian process.

One-dimensional CR transport along x axis is described by the eq.(2) with $\nabla_{11}=0$. The solution of this one-dimensional equation for the point instant source $\delta(t)\delta(x)$ (Green function) is

$$G(t,x) = \frac{\vartheta(t)\, x}{2\pi^{3/2} \langle A^2 \rangle æ_{11}^{1/2} L t^{1/2}} \int_0^\infty ds\, s^{-1/2} \exp\left(-\frac{1}{s} - \frac{s^2 x^4}{16 \langle A^2 \rangle^2 æ_{11}^2 L^2 t}\right). \tag{3}$$

here $\vartheta(t)$ is step function. Mean square of particle displacement

$$\langle x^2 \rangle = \int_{-\infty}^{\infty} dx\, x^2 G(t,x) = 2\pi^{-1} \langle A^2 \rangle L æ_{11}^{1/2} t^{1/2} \tag{4}$$

does correspond to the compound diffusion law $x \propto t^{1/2}$.

3. Anomalous perpendicular diffusion of magnetized particles.

The compound diffusion is destroyed if particle runs off the magnetic field line /9,11/. A magnetized particle can get out of the field line due to slow diffusion ($æ_\perp \neq 0$) or drift ($U \neq 0$) across the local magnetic field. Here we consider the problem in the framework of diffusion-convection equation

$$\frac{\partial f}{\partial t} - æ_{11} \nabla_s b_s b_j \nabla_j f + U \nabla f = 0 \quad (\nabla U = 0), \tag{5}$$

where $u(t,r)$ is convection velocity (considered as a small random perturbation).

One can introduce averaging $\langle...\rangle_H$ and $\langle...\rangle_u$ over ensembles of random realizations of H and u. The averaging over joint distribution is $\langle...\rangle_{uH}$. Then, defining $N=\langle f \rangle_{uH}$, $N_H=f-\langle f \rangle_H$, $N_u=f-\langle f \rangle_u$ and neglecting by small terms $\propto uA$, we have $f \approx N + N_H + N_u$. The equation for N may be obtained by the averaging of (5) over the joint distribution. It gives

$$\frac{\partial N}{\partial t} - \nabla_i \varkappa_{11}(1-\langle A^2 \rangle_{uH})h_i h_j \nabla_j N - \nabla_i \varkappa_{11}\langle (1-A^2)A_i A_j \nabla_j(N+$$
$$+ N_H+N_u)\rangle_{uH} - \nabla_i \varkappa_{11}\langle A_i h_j(1-A^2)\nabla_j(N_H+N_u)\rangle_{uH} -$$
$$- \nabla_i \varkappa_{11} h_i \langle A_j(1-A^2)\nabla_j(N_H+N_u)\rangle_{uH} + \langle u_i \nabla_i(N_H+N_u)\rangle_{uH} =$$
$$= 0, \quad (H=H_o/H_o) \tag{6}$$

The problem is now to find N_u, N_H and to average the products of random functions in the eq.(6). The procedure is rather cumbersome /3/. By consequent averaging of the eq.(5) over H and u distributions one can find the equation for $\langle N_H \rangle_u$. It contains non-linear diffusion term $\nabla_n D_{nm}^t \nabla_m \langle N_H \rangle_u$ describing the action of weak turbulent diffusion. This perpendicular term turns out to be important when $|\nabla_{11}\langle N_H \rangle_u| \ll |\nabla_{11}\langle N_H \rangle_u|$. An analysis of the equation shows that this turbulent diffusion gives the effect similar to non-linear resonance broadening in turbulent plasma theory /4/. The final formula for the diffusion coefficient across the mean magnetic field at $u \to 0$ is

$$D_\perp = g \varkappa_{11} \langle A^2 \rangle_H^2, \tag{7}$$

here the coefficient $g = (\pi/15)[(16(2\pi)^{1/2})^{-1}(48 \times$
$\times \Gamma^2(3/4) + 5\Gamma^2(1/4)) - 1] \approx 0.51$ is calculated for

the random field H_1 with Gaussian power spectra $W \propto \exp(-k^2 L^2)$ in k-space and with the polarization corresponding to alfvenic waves. $\Gamma(a)$ is gamma function.

The formula (7) coincides with the expression obtained in the case of small cross-field diffusion ($æ_\perp \neq 0$, u=0) of magnetized particles in the static magnetic field with weak random component A«1 /2,3/. This means that the nature of mechanism which allows particle to get slightly off the magnetic field line is not very important for the final result. The main effect is determined by the divergency of random magnetic field lines themselves /9,11/.

3. Discussion. Perpendicular diffusion of extremely magnetized particles is enhanced in the presence of large-scale random magnetic field. The case of static field has been considered. CR propagation in the direction perpendicular to the mean field may be described as one-dimensional diffusion with the coefficient $\simeq A^2 æ_{11}$ at the time intervals $\Delta t \leqslant L^2/æ_{11}$, as compound diffusion at $L^2/æ_{11} \leqslant \Delta t \leqslant L^2/A^4 æ_{11}$ (described by the eq.(2)), and as anomalous diffusion with $D = gA^4 æ_{11}$ at $\Delta t \geqslant L^2/A^4 æ_{11}$ (here A^2«1).

If we combine the results obtained in the present work with the calculations /2,3/ of perpendicular diffusion in the presence of developed alfvenic turbulence where v_a is not small, we come to the following general expression for the anomalous perpendicular diffusion coefficient

$$D_\perp \simeq æ_\perp + \frac{\pi}{4}\langle A^2 \rangle \sqrt{æ_\perp æ_{11}} + a\sqrt{\pi}\langle A^2 \rangle v_a L + g\langle A^2 \rangle^2 æ_{11} \qquad (8)$$

where A^2«1; $æ_\perp$« $æ_{11}$; a=0.5 for $æ_{11}$« $v_a L$, a=1.5 for $æ_{11}$» » $v_a L$; g is given by (7); Gaussian spectrum of waves is assumed. As compared with the eq.(7), formula (8) additionally takes into account the non-static character of turbulence ($v_a = 0$) and the effect of local perpendicular diffusion ($æ_\perp \ne 0$).

In the interstellar medium one has L = 100 pc, v_a = 3×10^6 cm/s, $æ_{11} = 10^{28}$ cm^2/s (E=1GeV) /1/. Under these conditions, the effective cosmic ray diffusion across the mean magnetic field is

$$D_\perp \simeq \frac{3\sqrt{\pi}}{2}\langle A^2 \rangle v_a L + g\langle A^2 \rangle^2 æ_{11} \qquad (9)$$

The first term in the right-hand part of (9) is energy - independent, while the second term increases as $æ_{11}(E)$. At E=1GeV the two terms prove to be of the same order. It cannot be excluded that the observed flattening of the energy dependence of the mean amount of matter traversed by particles at $E \leqslant 2$ GeV/n /12/ is due to the contribution of the first term of (9). At higher energies the second term of (9) leads to $D_\perp \approx 0.3\, D_{11}$.

It is the perpendicular diffusion coefficient $D_\perp$ which determines the leakage of cosmic rays out of the Galaxy, since toroidal component dominates in the galactic regular magnetic field, see /13/.

Anomalous diffusion is also important for the transport of Jovian electrons in the interplanetary magnetic fields, see /14/. The ratio $r_H/l = 10^{-3}$ is typical for these particles. Anomalous perpendicular diffusion mainly provided by the last term of the eq.(8) gives $D_\perp/D_{11} = 2 \times 10^{-2}$ in a good agreement with observations. The problem will be discussed in details in our forthcoming paper /15/.

References.
1. V.S.Berezinskii, S.V.Bulanov, V.A.Dogiel, V.L.Ginzburg,

V.S.Ptuskin, Astrophysics of Cosmic Rays, North-
-Holland, 1990.
2. L.G.Chuvilgin, L.I.Dorman, V.S.Ptuskin, 21 ICRC 3 (1990) 328.
3. L.G.Chuvilgin, V.S.Ptuskin, Preprint IZMIRAN 49(1990).
4. V.N.Tsytovich, Theory of Turbulent Plasma, Consultants Bureau, N.Y.,1977.
5. G.G.Getmantsev, Sov.Astron. 6(1962)477.
6. W.Horton, in: Basic Plasma Physics, A.A.Galeev and R.Sudan eds.,North-Holland, 1984, p.383.
7. J.A.Krommes, C.Oberman, R.J.Kleva, J.Plasma Phys. 30 (1983) 11.
8. P.C.Liever, Nuclear Fusion 25(1985)543.
9. V.S.Ptuskin, Astrophys.Space Sci. 61(1979)259.
10. V.S.Ptuskin, 19 ICRC 3(1985)75.
11. A.B.Rechester, M.N.Rozenbluth, Phys.Rev.Lett. 40 (1978) 38.
12. J.J.Engelmann, P.Ferrando, A.Soutoul et al., Astron. Astrophys. 233(1990)96.
13. R.J.Rand, S.R.Kulkarni, 1989 Astrophys.J. 343 (1989) 760.
14. T.F.Coulon, J.Geophys.Res. 83(1978)541.
15. L.G.Chuvilgin, P.Ferrando, V.S.Ptuskin, 1991 (in preparation).

Distribution of Particles and Fields in Turbulent Media

A.Z.Dolginov
Bartol Research Institute, The University of Delaware, Newark, DE 19716

1 Statement of the problem

Cosmic ray particles pass different large-scale space regions during their random walk in the Galaxy. The cosmic medium is very inhomogeneous. Parameters of turbulence, magnetic field, particle number density etc. vary by orders of magnitude. The diffusion approximation can be used to describe the CR distribution if the distribution can be considered as almost isotropic. Spatial scales necessary for cosmic rays to reach the isotropic distribution are much smaller than large-scale structures of the Galaxy. Thus the diffusion coefficient as well as other kinetic coefficients must be considered as having both an average and a random part. We show that fluctuations of the kinetic coefficients can lead to the development of aninstability that tends to increase the gradients of distribution. It is also important to note that even for regions ,where the turbulent cosmic media can be cosidered as approximately homogeneous ,the kinetic coefficients must be calculated by taking into account not only the two-point correlation function of the turbulent velocity but also correlation functions of higher ranks. It leads to significant change in the kinetic coefficients values. In particular, it leads to a decrease in the effective diffusivity and to the production of instabilities that may lead to particle bunching.

2 CR diffusion and acceleration in inhomogeneous cosmic media

2.1 Fluctuations of diffusion coefficients

The diffusion equation for CR's averaged over the small-scale magnetic field and turbulent velocity distributions has the well-known form:

$$\left(\frac{\partial}{\partial t} - \nabla D \nabla\right) N(\mathbf{r},p,t) = \frac{1}{p^2}\frac{\partial}{\partial p} p^4 \alpha \frac{\partial}{\partial p} N(\mathbf{r},p,t) \tag{1}$$

where $D = D_0 + \tilde{D}$, $D_0 = D_m + D_T$. D_m is the diffusivity arising from scattering on stationary magnetic inhomogeneities and D_T the part of the diffusivity due to the turbulence of the medium. $\tilde{D}$ determines the diffusivity due to the large scale variations of the medium characteristic along the path of CR. Similary the coefficient of the diffusivity $\alpha = \alpha_0 + \tilde{\alpha}, \alpha_0 = \alpha_m + \alpha_T$ in momentum space. Here $\langle D \rangle = D_0$, $\langle \alpha \rangle = \alpha_0$,

$\langle \tilde{D} \rangle = \langle \tilde{\alpha} \rangle = 0$. The brackets $\langle \cdot \rangle$ denote averaging over large scale variation in the characteristics of the medium. The distribution function $N(\mathbf{r}, p, t) = F(\mathbf{r}, p, t) + f(\mathbf{r}, p, t)$, where $\langle N \rangle = F$ is the average and f the fluctuating part of N, i.e. $\langle f \rangle = 0$. The averaging of (1) gives:

$$\left(\frac{\partial}{\partial t} - \hat{L}_0\right) F = \langle \nabla(\tilde{D}\nabla f)\rangle + \frac{1}{p^2}\langle \frac{\partial}{\partial p}p^4\tilde{\alpha}\frac{\partial f}{\partial p}\rangle; \qquad (2)$$

$$\hat{L}_0 = \nabla D_0 \nabla - p^{-2}\left(\frac{\partial}{\partial p}p^4\alpha_0\frac{\partial}{\partial p}\right); \qquad (3)$$

$$\left(\frac{\partial}{\partial t} - \hat{L}_0\right) f = \nabla(\tilde{D}\nabla F) + \nabla(\tilde{D}\nabla f) - \langle \nabla(\tilde{D}\nabla f)\rangle + \frac{1}{p^2}\frac{\partial}{\partial p}p^4\tilde{\alpha}\frac{\partial f}{\partial p}$$
$$- \langle \frac{1}{p^2}\frac{\partial}{\partial p}p^4\tilde{\alpha}\frac{\partial f}{\partial p}\rangle + \frac{1}{p^2}\frac{\partial}{\partial p}p^4\tilde{\alpha}\frac{\partial F}{\partial p}. \qquad (4)$$

Consider the case of a Gaussian distribution of random quantities $\tilde{D}$ and $\tilde{\alpha}$. The distribution function is a functional of $\tilde{D}$ and $\tilde{\alpha}$ which allows us to use the Furutzu-Novikov formula to calculate $\langle D(\mathbf{r}, p, t)\nabla f(\mathbf{r}, p, t)\rangle$ and $\langle \alpha \frac{\partial f}{\partial p}\rangle$.

$$\langle \tilde{D}(\mathbf{r}_1, p_1, t_1)\nabla f(\mathbf{r}_1, p_1, t_1)\rangle = \int dr_2 \int dp_2 \int dt_2 W(1,2)\nabla_1\langle \frac{\delta f(1)}{\delta \tilde{D}(2)}\rangle;$$

$$\langle \tilde{\alpha}\frac{\partial f(1)}{\partial p}\rangle = \int dr_2 \int dp_2 \int dt_2 Q(1,2)\frac{\partial}{\partial p_1}\langle \frac{\delta f(1)}{\delta \tilde{\alpha}(2)}\rangle.$$

Here and in what follows we use the notation: $f(1)$ for $f(\mathbf{r}_1, p_1, t_1)$, dn for $dr_n dp_n dt_n$ etc., $W(1,2) = W(\mathbf{r}_1, t_1; \mathbf{r}_2, t_2) = \langle D(\mathbf{r}_1, t_1)D(\mathbf{r}_2, t_2)\rangle$, is the correlation function of the random components of diffusivity at two space-time points $\mathbf{r}_1, t_1$ and $\mathbf{r}_2, t_2$ and

$$\langle \frac{\delta f(\mathbf{r}, p, t)}{\delta \tilde{D}(\mathbf{r}_1, p_1, t_1)}\rangle \qquad (5)$$

is the functional derivative of f over $\tilde{D}$. Also, $Q(1,2) \equiv Q(\mathbf{r}_1, p_1, t_1; \mathbf{r}_2, p_2, t_2) = \langle \tilde{\alpha}(1)\tilde{\alpha}(2)\rangle$. Thus (4) can be written in an integral form as follows

$$f(1) = \int d2 G(1,2)\left(\nabla(\tilde{D}(2)\nabla F(2)) + \frac{1}{p_2^2}\frac{\partial}{\partial p_2}p_2^4\tilde{\alpha}(2)\frac{\partial F(2)}{\partial p_2}\right.$$
$$\left. + \nabla(\tilde{D}\nabla f(2)) - \langle \nabla(\tilde{D}\nabla f(2))\rangle + \frac{1}{p_2^2}\frac{\partial}{\partial p_2}p_2^4\tilde{\alpha}\frac{\partial f(2)}{\partial p_2} - \langle \frac{1}{p_2^2}\frac{\partial}{\partial p_2}p_2^4\tilde{\alpha}\frac{\partial f(2)}{\partial p_2}\rangle\right).$$

Here $G(\mathbf{r}_1, p_1, t_1; \mathbf{r}_2, p_2, t_2) = G(1,2)$ is the Greens function of the operator $\hat{L}_0$. For the functional derivative we obtain the expression:

$$\langle \frac{\delta f(1)}{\delta \tilde{D}(3)}\rangle = \int d2 G(1,2)\left[\nabla \delta(2,3)\cdot \nabla F(2) + \delta(2,3)\nabla^2 F(2)\right] = -\nabla G(1,3)\cdot \nabla F(3). \qquad (6)$$

Using $W(1,2)$ and $\langle \frac{\delta f(1)}{\delta \tilde{D}(3)}\rangle$ we can rewrite eq. (2) in the form:

$$\left(\frac{\partial}{\partial t} - \nabla D_0 \nabla\right) F(1) = \frac{1}{p^2}\frac{\partial}{\partial p}\alpha_0\frac{\partial}{\partial p}F(1) - \int d3 W(1,3)\left[(\nabla_1^2\nabla_{3k}G(1,3))(\nabla_{3k}F(3))\right.$$
$$\left. + (\nabla_{1i}W(1,3))(\nabla_{1i}\nabla_{3k}G(1,3))\nabla_{3k}F(3)\right] - \int d3 \left[Q(1,3)\right.$$
$$\left.\left(4p_1 + p_1^2\frac{\partial}{\partial p_1}\right) + p_1^2\frac{\partial Q(1,3)}{\partial p_1}\right]p_3^2\frac{\partial^2 G(1,3)}{\partial p_1 \partial p_3}\frac{\partial F(3)}{\partial p_3}. \qquad (7)$$

The average over step function $H(t_2 - t_3)$ is included in W and Q. The Greens function for a particular case $D_0(p) = D_0 = const$ and $\alpha_0 = const$ has the form

$$G(1,2) = g(R,\tau)\chi(p_1,p_2,\tau);$$
$$g(R,\tau) = H(\tau)[4\pi D_0\tau]^{-3/2}exp[-R^2/4D_0\tau];$$
$$\chi(p_1,p_2,\tau) = [4\pi D_0\tau]^{-1/2}\exp\left[-\frac{(ln(p_1/p_2) + 3\alpha_0\tau)^2}{4\alpha_0\tau}\right].$$

Here $\mathbf{R} = \mathbf{r}_1 - \mathbf{r}_2$, $\tau = t_1 - t_2$, and $H(\tau) = 1$ if $\tau > 0$ and $H(\tau) = 0$ if $\tau < 0$. The condition $D_0 = const$ is valid, for example, if the Larmor radius of the particle is smaller than the spatial scale of magnetic inhomogeneities.

One can approximately transform (7) to a differential equation using the fact that the average distribution function $F(1)$ is a much smoother function of $\mathbf{r}$ and t than $G(1,2)$ and $W(1,2)$. Using the Taylor expansion:

$$F(3) = F(1) + (\mathbf{r}_3 - \mathbf{r}_1)\nabla F(1) + (1/2)(\mathbf{r}_3 - \mathbf{r}_1)_i(\mathbf{r}_3 - \mathbf{r}_1)_k \nabla_i \nabla_k F(1) +$$
$$(p_3 - p_1)\frac{\partial F(1)}{\partial p_1} + (t_3 - t_1)\frac{\partial F(1)}{\partial t_1} + ...$$

we can obtain the differential equation for $F(1)$.

The differential equation is especially simple if the change of the energy is not significant and the correlation time of the random component of the diffusion coefficient is much shorter than the CR free path time, i.e. the correlation function has the form $W(1,2) = W(\mathbf{r}_1, p_1; \mathbf{r}_2, p_2)\delta(t_1 - t_1)$. In this case the Greens function becomes a δ-function and we obtain

$$(\frac{\partial}{\partial t} - \hat{L}_0)F = (\nabla_i \nabla_k W)_0 (\nabla_i \nabla_k)F + W_0 \nabla^4 F. \tag{8}$$

The quantities $(\nabla_i \nabla_k W)_0$ and W_0 are taken at the point $\mathbf{r}_1 = \mathbf{r}_2$.

We can see that fluctuations of the diffusion coefficient lead to a decrease in effective diffusion because $W(1,2)$ is maximum at the point $\mathbf{r}_1 = \mathbf{r}_2$, and the second derivative of $W(1,2)$ is negative. The decrease in diffusion coefficients in media with fluctuating parameters has been demonstrated also in [1]. However the spatial dependence of $\hat{D}$, which is important even if D_0 is constant, was disregarded in [1]. The decreasing of D for the particular case of CR propagation implies that the particles will spend much more time in the Galaxy than they would were the medium homogeneous. The effective diffusion coefficient $D = D_0 + D_1$ may even become negative which implies an increase in gradients of $F(\mathbf{r},p,t)$ and the tendency of the CR density to increase in some random distributed regions in space. Of course, when the gradients become too large, the distribution function can't be considered as smooth and we must return to the integral equation (7).

Besides the case of $D < 0$, there can appear another kind of instability most effective for small-scales. Consider a perturbation of the distribution function in the form: $\delta F \sim exp[\gamma t + i\mathbf{kr}]$. Inserting δF in (8) yields the dispersion relation:

$$\gamma = -k^2 D_0 - k_i k_n [\nabla_i \nabla_n W]_0 + k^4 W. \tag{9}$$

Thuse perturbations increase exponentially if $\gamma > 0$, i.e. for short scale perturbations in which k is large.

2.2 Fluctuations in the medium compressibility

The problem of the CR diffusion in compressible turbulent media was considered detail in [2,3,4] where it was shown that effective diffusion is strongly decreased in such media. In some cases the diffusion coefficient changes sign and an instability with respect to particle cluster formation begins.

Consider, for example, the simplest case of particle propagation described by:

$$\left(\frac{\partial}{\partial t} - D_0 \nabla^2\right) n(\mathbf{r}, t) = -(\mathbf{u}n) - n(\nabla \mathbf{u}). \tag{10}$$

Here $\mathbf{u}$ is the turbulent velocity of the medium and D_0 is the diffusivity due to the scattering in the medium if it were at rest. The distribution function $n = n_0 + \tilde{n}$, where n_0 is the distribution averaged over fluctuations of the medium turbulent velocity. A procedure similar to that described above gives the following equation for the n_0:

$$\left[\frac{\partial}{\partial t} - (D_0 + D_T)\nabla^2\right] n_0 = 0, \tag{11}$$

where

$$D_T = \frac{1}{3}\int d\mathbf{R} \int_{-\infty}^{\infty} d\tau \left[\langle u_i(1)G(1,2)u_i(2)\rangle - R_i\langle u_i(1)G(1,2)\nabla \mathbf{u}(2)\rangle\right]. \tag{12}$$

The Green's function $G(1,2)$ was calculated in [2,3,4] taking into account not only the second rank correlation function of turbulent velocities $\langle u_i(1)u_k(2)\rangle$ but all correlation functions of higher ranks. Self-consistent but approximate values of D_T can be obtained by using the Green's function of eq.(11). The high rank correlation functions become important if the dimensionless parameter (the Strouhal number) $S^2 = \langle u^2 \rangle \tau_c^2 / 3L_c^2$ is larger or of order of 1. Here τ_c is the time scale and L_c the spatial scale of turbulence. In most cases $S \geq 1$ in cosmic media.

We see from (12) that the term with $\nabla \mathbf{u}$ can have a negative sign. It reduces the D_T value and, in some cases, the D_T and even the $D_0 + D_T$ may be negative. It implies the developing of an instability with respect to the formation of clusters of particles.

For example the instability of supersonic turbulent motions (a system of shock waves) can lead to the bunching of the particles. In regions of shock front intersections, the density of the medium increases and the concentration of CR particles increases too. The compression proceeds with the shock fronts supersonic velocity whereas the decompression proceeds only with the sonic velocity. The decompression has insufficient time to be complete before the next series of shock waves enters this region and the compression starts again. Formally this process can be described as negative diffusion which increases, rather than decreases, the gradients of the distribution function instead. Again, the diffusion approximation becomes invalid when the gradients become too sharp. A decrease and possible change of sign of the diffusion coefficient may be important for the problem of the CR propagation and acceleration. If diffusion is suppressed, then the particle spends more time in acceleration regions thereby enhancing its energy gain.

3 Magnetic field in inhomogeneous turbulent media

Fluctuations of the medium parameters may be very important for magnetic field dissipation and generation in cosmic space. Consider the dynamo equation taking the coefficients of magnetic diffusivity and helicity as having the average and random components $D = D_0 + \tilde{D}$, $\alpha = \alpha_0 + \tilde{\alpha}$ and $\mathbf{B}_{tot} = \mathbf{B} + \mathbf{b}$,

$$\frac{\partial \mathbf{B}_{tot}}{\partial t} - \nabla \times D\nabla \times \mathbf{B}_{tot} - \nabla \times \alpha \mathbf{B}_{tot} = 0. \tag{13}$$

We can, using a similar procedure, obtain an integral equation for the average field

$$\begin{aligned}\frac{\partial B_i}{\partial t} &= (\nabla \times D_0 \nabla \times \mathbf{B})_i + (\nabla \times \alpha_0 \mathbf{B})_i - \int d3 W(1,3)(\nabla_k \nabla^2 G_{il}(1,3))(\nabla_l B_k(3) \\ &\quad - \nabla_k B_l(3)) + \int d3(\nabla_k W(1,3))(\nabla_r \nabla_i(G_{kl}(1,3) - \nabla_k \nabla_r G_{il}(1,3))(\nabla_r B_l(3) \\ &\quad - \nabla_l B_r(3)) - \int d3 Q(1,3) \epsilon_{ikl} \epsilon_{mnr}(\nabla_k \nabla_n G_{lm}) B_r(3) \\ &\quad - \int d3(\nabla_k Q(1,3)) \epsilon_{ikl} \epsilon_{mnr}(\nabla_n G_{lm}(1,3)) B_r(3). \end{aligned} \tag{14}$$

Here G_{ik} is the Greens function of equation (13) with average D_0 and α_0 values. In the particular case of constant D_0 and α_0, we obtain

$$\begin{aligned} G_{ik}(R,\tau) &= H(\tau)[\delta_{ik} G_0(R,\tau) + \epsilon_{ikl} \nabla_l G_1(R,\tau)] \\ G_0(R,\tau) &= R^{-1}(4\pi D_0 \tau)^{-3/2} \left[R\cos(\frac{\alpha_0 R}{2 D_0}) + \alpha_0 \tau \sin(\frac{\alpha_0 R}{2 D_0}) \right] exp \left[-\frac{R^2}{4 D_0} + \frac{\alpha_0^2 \tau}{4 D_0} \right] \\ G_1(R,\tau) &= (2\pi R)^{-1}(4\pi D_0 \tau)^{-(1/2)} \sin(\frac{\alpha_0 R}{2 D_0}) exp \left[-\frac{R^2}{4 D_0 \tau} + \frac{\alpha_0^2 \tau}{4 D_0} \right] \end{aligned} \tag{15}$$

Here ϵ_{ikl} is the unit antisymmetric tensor of third rank. We see from (14) that spatial fluctuations of the medium parameters decrease the effective magnetic diffusivity.

4 Nonlinear waves in random inhomogeneous media

There are many interesting and important problems of particles and wave propagation that are described by nonlinear equations. For example, nonlinear waves in cold magnetised plasma can be described by the nonlinear Korteweg-de Vries-Burgers equation:

$$\frac{\partial u}{\partial t} + u\frac{\partial u}{\partial \xi} + \mu \frac{\partial^3 u}{\partial \xi^3} - \chi \frac{\partial^2 u}{\partial \xi^2} = 0. \tag{16}$$

Here $u(\xi, t) = (4 c_A/B_0) b_z(\xi, t) \sin \alpha$, $c_A^2 = B_0^2/4\pi\sigma$, $\omega(k) = c_A k \sin \alpha$ and

$$\xi = x - c_A t; \quad \mu = c_A(c^2/2\omega_{pi}^2)(m_e/m_i - \cot^2 \alpha); \quad \chi = c^2/8\pi\sigma.$$

$b_z(\xi, t)$ is the magnetic field of the wave, and B_0 the average magnetic field in the plasma. The wave is transverse $\mathbf{b} \perp \mathbf{k}$, $\mathbf{b} \parallel \mathbf{k} \parallel OX$, and $\mathbf{B_0}$ is in the XZ plane.

If $\mu = \mu_0 + \tilde{\mu}$, $\chi = \chi_0 + \tilde{\chi}$ and $u = u_0 + \tilde{u}$, where μ_0, χ_0 and u_0 are average quantities and if the correlation time is very short then, as before, we obtain the following equation for u_0:

$$\frac{\partial u_0}{\partial t} + u_0 \frac{\partial u_0}{\partial \xi} + \left[\mu_0 + (\frac{\partial^3 Q}{\partial \xi^3})_0\right] \frac{\partial^3 u_0}{\partial \xi^3} - \left[\chi_0 + (\frac{\partial^2 W}{\partial \xi^2})_0\right] \frac{\partial^2 u_0}{\partial \xi^2}$$
$$= \left[W_0 + 3(\frac{\partial^2 Q}{\partial \xi^2})_0\right] \frac{\partial^4 u_0}{\partial \xi^4} + Q_0 \frac{\partial^6 u_0}{\partial \xi^6}. \quad (17)$$

We see, from equation (17), that fluctuations of μ and χ decrease the effective diffusivity and change radically the character of the differential equation. In particular, the system of waves is unstable with respect to short scale perturbations which may lead to the formation of fractal structures in the wave fronts.

5 Conclusions

1. Spatial fluctuations of the cosmic medium parameters (such as density, magnetic field etc.) along the CR path in the Galaxy change the character of CR propagation and acceleration. The effective diffusivity may be much less than in the case of homogeneous media.
2. High rank correlation functions of the cosmic medium turbulent velocity must be taken into account in most real cases.
3. Turbulence in compressible media can lead to a decrease of the CR diffusion coefficient.
4. In some cases the diffusion coefficient may become negative which implies the development of an instability with respect to the formation of clusters of particles.
5. A similar instability may exist for fields in turbulent media and in media with fluctuating parameters. It may lead to the bunching of magnetic field lines and to the fractal structure of waves propagating in the media.

Acknowledgments. This author accnowledges thanks to the Faculty and Staff of the Bartol Research Institute for their help and support during my visit there and to the following people whose grants this authors visit was supported by:JPL Contract 959167 PI: N.F.Ness;NASA Grants: NAG 5-374 PI:P.Evenson,NAGK 1-1637 PI:J.Bieber, NAG 5-1702 PI:D.Mullan,NAG 5-1573 PI:W.H.Matthaeus. I am thankful to Professors N.F. Ness and G.P. Zank for fruitful discussions.

REFERENCES

1. A.V.Tur, V.V.Yanovsky, Phys.Fluids A, 3(8), p.1969 (1991).
2. A.Z.Dolginov, N.A.Silant'ev, Sov. Phys. JETP, v. 66 ,p. 90 (1987).
3. A.Z.Dolginov, N.A.Silant'ev, Astron. Astrophys., v. 236, p. 516 (1990).
4. A.Z.Dolginov, N.A.Silant'ev, Geophys. Astrophys. Fluid Dyn., v. 54, in press (1991).
5. A.Z.Dolginov, A.V.Klyachkin, Journ. Fluid Mech., in press (1992).

A NEW NONLINEAR DIFFUSION FORMALISM IN A MAGNETIZED PLASMA: APPLICATION TO SPACE PHYSICS AND ASTROPHYSICS

H. KARIMABADI AND D. KRAUSS-VARBAN

Department of Electrical and Computer Engineering and California Space Institute, University of California at San Diego, La Jolla, CA 92093-0407

ABSTRACT

We present a new diffusion formalism that takes into account the finite width of resonances. The diffusion results from the overlap of resonances in phase space. As an application of this new theory, we consider the problem of pitch-angle scattering through 90° which has been a topic of much interest and controversy over the past 30 years.

I. INTRODUCTION

The problem of pitch-angle and velocity diffusion of particles by waves is of fundamental importance in space plasma physics and astrophysics. For instance, it plays an important role in the process of ion thermalization at collisionless shocks,[1] heavy ion acceleration in the magnetosphere,[2] pick-up of cometary ions,[3] electron acceleration in supernova remnants,[4] and the acceleration of cosmic rays in interstellar space.[5] The usual theoretical tool for studying the particle diffusion has been the quasilinear (Q-L) theory. Q-L theory is based on two important assumptions: (1) orbits are described accurately by their unperturbed motion; (2) the wave autocorrelation time is much less than the trapping time. The assumption of unperturbed orbits leads to the well known resonance gap at 90°, i.e. the Q-L diffusion coefficient evaluated at 90° pitch angle is zero. This is in direct conflict with both spacecraft measurements[6] as well as numerical simulations[3] of particle diffusion. Several attempts to overcome this difficulty have been made.[7-9] Such studies are invariably based on some apriori assumed statistical behavior of the orbits.[10] What has been lacking is a theory of particle orbits. Here, we present a theory which includes analytical derivation of trapping width, bounce frequency and the onset conditions for diffusion.

II. NONLINEAR ORBIT THEORY

The physics of the scattering and diffusion depends critically on whether the waves have the same component of phase velocity parallel to the static magnetic field ($N_\| \equiv ck_\|/\omega$) or not.[1] Using the generating function $F_2 = \sum_{i=1}^{M} P_{zi}(z - \frac{\omega}{k_{\|i}}t)$, the Hamiltonian of a charged particle in the presence of a wave spectrum comprised of M-waves becomes a constant of motion and is given by:

$$H = \frac{1}{2}mv^2 + q\sum_{i=1}^{M} \Phi_i \sin\psi_i - \sum_{i=1}^{M} \frac{\omega_i}{k_{\|i}} P_{zi}. \qquad (1)$$

For $N_{\|i} = N_{\|j}$, the last term in eq. (1) is equal to P_z times a constant factor, where $P_z = \sum_{i=1}^{M} P_{zi}$. Thus, the Hamiltonian surface (i.e. the allowed region in $v_\| - v_\perp$ plane that a particle can wander about) becomes a circle of finite but small width. If $N_{\|i} \neq N_{\|j}$, the variables P_{zi}'s remain independent of P_z and the Hamiltonian surface becomes two dimensional (i.e., diffuse occurs both along and perpendicular to the circle) and particles can gain very large energies. This is a generalized concept of second order Fermi acceleration. The acceleration is more efficient at oblique directions and/or for large spread in $N'_{\|i}s$. The center of the diffusion circle is given by the weighted sum of all the parallel phase velocities in the lab frame. For instance, for two waves with equal amplitude and the same phase velocity but propagating in opposite directions, the center of the diffusion circle is at $v_z \equiv v_\| = 0$.

It proves useful to plot the surface defined by eq.(1) in the $v_\| - v_\perp$ plane for various particle energies as shown in Fig. 1a-b. On the same figure, we draw the resonance lines. Finally, we draw the calculated resonance (trapping) width associated with each resonance on the H-surface. Figures 1a-b show the resonance diagram[12] in the presence of an obliquely propagating magnetosonic wave at two different wave amplitudes (see parameters in figure caption). A pitch angle of 90° corresponds to the point where $v_\| = \omega/k_\| = 1.5 V_A$, which also forms the locus of points in resonance with $\ell = 0$. As the wave amplitude increases, the resonance widths get larger and can eventually overlap. The motion of a particle before the resonances overlap is periodic. When the resonances overlap, however, the particle motion becomes diffusive. The trapping width for $\ell = 0$ is clearly nonzero and particles starting at small pitch angles can scatter through 90° if the wave amplitude is above the overlap threshold (Fig.1b). Note that the resonance widths are asymmetric with respect to $\ell = 0$ with the negative ℓ resonances (right side of Fig.1a-b) having larger widths. This would be reversed for a left circularly polarized waves or for a wave propagating in the opposite direction to the static magnetic field.

Figure 1c-d show the orbits of six particles for the same parameters as Fig.1a-b, respectively. The orbits were integrated using a fourth-order Runge-Kutta method. Evidently, the theoretically obtained resonance diagrams (Fig.1a-b) compare well with those obtained from numerical integration of orbits (Fig.1c-d).

Using the resonance diagrams, we have found the following. Scattering is much more efficient when there are two oppositely propagating waves. Second order Fermi acceleration can only occur if $N_{\|i} \neq N_{\|j}$. A linearly polarized wave is a more efficient scatterer than a circularly polarized wave at parallel propagation (for a linearly polarized wave, both resonances $\ell = \pm 1$ are present, for a circularly polarized wave only $\ell = 1$ or $\ell = -1$). So in studying the acceleration process, it is important to determine if both Alfven and magnetosonic modes are present simultaneously. Since diffusion perpendicular to diffusion circle increases with the difference in phase velocities, the case of oppositely directed waves can give larger acceleration than the case where there exists a mixture of the two modes.

III. NEW DIFFUSION EQUATION

One of the main difficulties with Q-L theory is the theory's prediction of a zero diffusion coefficient (i.e. a gap) at 90° pitch-angle. The usual derivations of the zero diffusion at 90° are limited to MHD Alfven waves (i.e. nondispersive and circularly polarized) propagating parallel to the background magnetic field. Then, the center of the diffusion circle is at $v_\parallel = \omega/k$, and a pitch-angle of 90° corresponds to the point $v_\parallel = \omega/k$ or $\ell = 0$ resonance. Since for parallel propagation, $\ell = 0$ resonance does not exist, it follows that the diffusion coefficients are zero for pitch-angle of 90°. Thus, the presence or absence of the gap at 90° pitch angle is intimately related to $\ell = 0$ resonance when $N_{\parallel i} = N_{\parallel j}$.[11] For obliquely propagating waves, however, the $\ell = 0$ resonance is present and the gap no longer exists. The condition for crossing of the 90° pitch angle in parallel propagating waves is for the wave amplitude to be sufficiently large so that the trapping width due to either $\ell = 1$ and/or $\ell = -1$ resonances (depending on wave polarization) reaches $v_\parallel = \omega/k$.

So far we have shown that the finite trapping effects are very important in the nonlinear description of the orbits. A recent theory[13] takes into account trapping effects in the diffusion equation. The idea is quite simple. The decorrelation time or the shift in the resonance condition due to the finite trapping width is $2\omega_b$ in the ℓth resonance. The delta function in the diffusion coefficients is then replaced by a function $f(x)$, where $f(x) = 1/4\omega_b$ if $|x| \leq 2\omega_b$ and $f(x) = 0$ otherwise. The new diffusion equation is applicable to cases where Q-L theory fails, namely when waves are coherent and/or have large amplitudes. The new diffusion coefficient does not in general have a singularity (gap) at 90° pitch angle. Test of this theory against numerically calculated evolution of orbits has yielded excellent results.[13]

We emphasize that this theory differs considerably from the usual resonance broadening theories.[10] In Dupree's theory, the zero order orbits are modified by a term proportional to the diffusion coefficient. The result is an implicit equation for the diffusion coefficient that is difficult to solve. In addition, the finite trapping effects and the difference in the size of trapping widths are not accounted for in Dupree's theory.

IV. SUMMARY

We have developed a nonlinear orbit theory in a magnetized plasma. The importance of finite trapping effects was emphasized. The resonance diagram technique was then shown to reproduce the details of the particle orbits very accurately and can be used to determine the acceleration/scattering in the presence of a given wave spectrum. We also showed how the nonlinear orbits can be incorporated into the diffusion equation. The resulting diffusion equation is an extension of the Q-L theory to cases where the waves have large amplitudes and/or are coherent. This new equation does not have have a gap at 90° in cases where the individual orbits can cross the gap. We also examined the conditions under which the resonance gap at 90° pitch angle exists. The test of the above formalism and its detailed application to cometary ions is currently underway. We hope to reexamine other space physics and astrophysical problems[1,2,4,5] using our new theory.

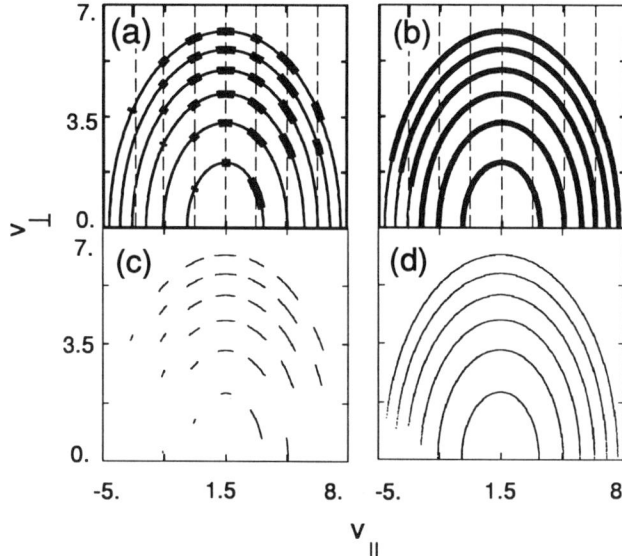

Fig. 1. (a-b) Resonance diagram. The parameters used are: $\omega/\Omega = 0.9$, $\omega_{pi}/\Omega = 4000$, and $\alpha = 30°$. Each circle corresponds to a different value for the Hamiltonian in (1). The H values run from 1 to 18 mV_A^2. (a) $B/B_o = 0.014$, and (b) $B/B_o = 1.4$. The theoretical resonance widths are indicated by heavy lines. Numerically evaluated particle orbits are shown in (c) $B/B_o = 0.014$ and (d) $B/B_o = 1.4$.

ACKNOWLEDGMENTS

Useful conversations with N. Omidi and T. Terasawa are gratefully acknowledged. This research was supported by the NASA Space Physics Theory Program at the UCSD and NASA grant NAGW-1806. Computing was performed on the CRAY Y-MP at the San Diego Supercomputer Center and the CRAY Y-MP at the NASA Center for Computational Sciences.

REFERENCES

1. Sckopke, N., et al., J. Geophys. Res., *95*, 6337, (1990).
2. Omura, Y. et al., J. Geomag. Geoelectr., *40*, 949, (1988).
3. Gary, S. P., R. H. Miller, and D. Winske, Geophys. Res. Lett., *18*, 1067, (1991).
4. Cowsik, R., Astrophys. J., *241*, 1195, (1980).
5. Schlickeiser, R., Astrophys. J., *336*, 264, (1989).
6. Neugebauer, M., et al., J. Geophys. Res., *95*, 18,745, (1990).
7. Goldstein, M. L., J. Geophys. Res., *82*, 1071, (1977).
8. Jones, F. C., T. J. Birmingham, and T. B. Kaiser, Astrophys. J., *180*, L139, (1973).
9. Volk, H. J., Astrophys. Space Sci., *25*, 471, (1973).
10. Dupree, T. H., Phys. Fluids, *9*, 1773, (1966).
11. Karimabadi, H., et al., submitted to J. Geophys. Res., (1992).
12. Karimabadi, H., et al., Phys. Fluids, *B2*, 606, (1990).
13. Karimabadi, H., and C. R. Menyuk, J. Geophys. Res., *96*, 9669, (1991).

Particle Acceleration at Non-Relativistic Shocks

MICROPHYSICS AND STRUCTURE OF QUASI-PARALLEL SHOCKS: OBSERVATIONS, THEORY, AND IMPLICATIONS FOR PARTICLE ACCELERATION

Manfred Scholer
Max-Planck-Institut für extraterrestrische Physik, 8046 Garching, Germany

ABSTRACT

Considerable insight into the physics of collisionless quasi-parallel shocks has recently been gained by hybrid simulations. After briefly reviewing pertinent observations at the quasi-parallel bow shock and the theories developed for quasi-parallel shocks we report on recent hybrid simulations concerning the cyclic behaviour of shocks, the specular reflection of part of the incident plasma as a relatively cold beam, and the acceleration of part of the incoming plasma to superthermal energies. The latter point is of particular importance in cosmic ray physics, since these particles constitute a seed particle population for a first order Fermi acceleration process.

INTRODUCTION

The structure of collisionless quasi-perpendicular shocks has been studied both observationally and theoretically in detail. Until recently, this has not been the case for quasi-parallel shocks, although these shocks are known to be of great importance for the diffusive or first order Fermi acceleration process. Quasi-parallel shocks observed in interplanetary space, in particular the Earth's bow shock, exhibit a rather complicated structure in the magnetic field profiles. At the Earth's bow shock it was not clear to what extent the complications are due to the quasi-perpendicular foreshock being swept into the quasi-parallel foreshock region by the solar wind or are inherent to the shock structure itself. After briefly reviewing pertinent observations at the quasi-parallel bow shock and the theories developed for quasi-parallel shocks we report on recent hybrid simulations concerning the cyclic behaviour of shocks, the specular reflection of part of the incident plasma as a relatively cold beam, and the acceleration of part of the incoming plasma to superthermal energies.

OBSERVATIONS

Quasi-perpendicular collisionless shocks have been extensively studied in the past both observationally [1] as well as theoretically [2,3]. It has been shown that dissipation and heating at the quasi-perpendicular shock is mainly achieved by the specular reflection of part of the incident ions. These ions are accelerated during their gyration in the upstream magnetic field by the motional $\mathbf{v} \times \mathbf{B}$ electric field parallel to the shock surface. The gyration brings them then downstream, where they constitute a hot distribution around the transmitted and decelerated ions. The combination of these two distributions is unstable to the Alfvén ion cyclotron instability and the two downstream distributions are eventually thermalized. Observations at the quasi-perpendicular Earth's bow shock have shown the existence of a second type of reflected ions. These ions are not specularly reflected, but are field-aligned and have energies considerably exceeding the shock ram energy [4]. One possible explanation for these ions is reflection of part of the incident solar wind under conservation of the magnetic moment. Since ions leaving the quasi-perpendicular portion of the Earth's bow shock in the upstream direction along the magnetic field are convected into the quasi-parallel foreshock region, it was assumed for a long time that these field-aligned beams could constitute a seed particle population for a first order Fermi acceleration process at the quasi-parallel bow shock. This idea has been challenged by recent composition measurements in field-aligned beams: whereas the more energetic diffuse ions upstream of the quasi-parallel bow shock exhibit about the same proton to alpha particle ratio as in the solar wind [5], alpha particles in the field-aligned beams are considerably underabundant [6]. This suggests that the quasi-perpendicular shock itself accelerates particles out of the thermal particles to a superthermal population, which then can be further accelerated by a first order Fermi mechanism.

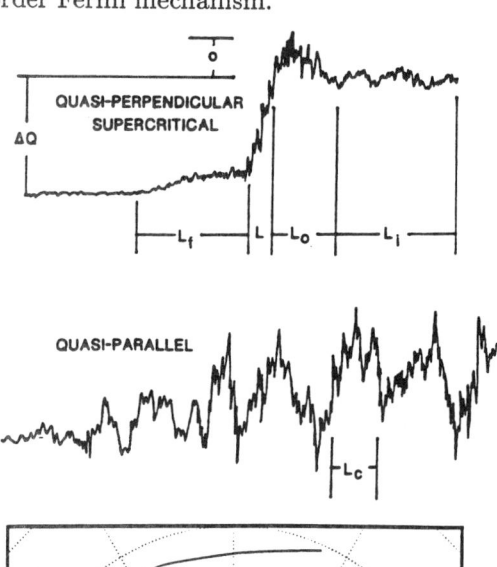

Fig. 1. Schematic of the magnetic field measured at the quasi-perpendicular (top) and at the quasi-parallel (bottom) bow shock (after Greenstadt et al.[7]).

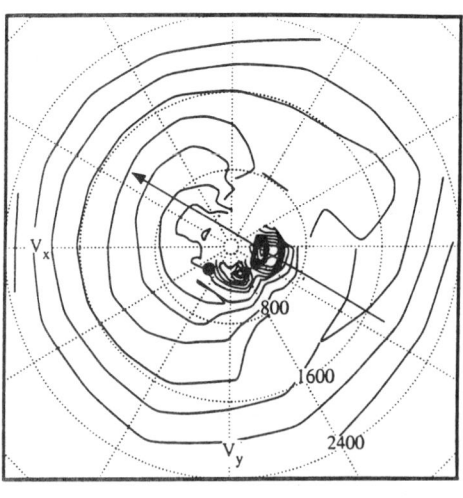

Fig. 2. Iso-intensity contours of protons measured upstream of the Earth's bow shock in the $v_x - v_y$ plane. The sun is to the left, the center is the spacecraft frame (after Onsager et al.[10]).

The early magnetic field measurements at the quasi-parallel bow shock have shown that the magnetic field profile does not show a smooth transition from upstream to downstream, but exhibits nonlinear waveforms known as pulsations. Figure 1 shows schematically in the upper part the magnetic field profile at a quasi-perpendicular shock and in the lower part a profile at a quasi-parallel shock (from [7]). Studies using dual spacecraft measurements have shown two types of pulsation signatures: convecting signatures, which are pulsations carried along by the solar wind, and nested signatures, in which the pulsations are nearly at rest in the spacecraft frame or propagating sunward [8]. The scale length of short spiky pulsations is typically 300 km and the relative amplitude $\delta B/B$ is larger than 1 and can be as large as 5. It was rather a surprise when it was found that the quasi-parallel bow shock also specularly reflects part of the incident solar wind ions. Gosling et al.[9] reported observations of relatively cold beams of ions in the shock ramp, whose position in velocity space are consistent with the assumption of specular reflection. An example of such an observed ion distribution is shown in Figure 2 (from [10]). This Figure shows iso-intensity contours of protons in $v_y - v_x$ phase space (GSE coordinates). The contour plot center corresponds to zero speed in the spacecraft frame. In addition to the solar wind distribution on the v_x axis one can see a second beam type distribution. The location in phase space is close to what is predicted when the ions are specularly reflected (dot). The projection of the magnetic field into the ecliptic is indicated by the large arrow. The beams of specularly reflected ions are seen over a wide range of upstream parameters but only fairly close to the shock. The latter fact suggests that the beam ions are quickly scattered as they propagate upstream. Since alpha particles were also seen to be specularly reflected at the quasi-parallel bow shock [11], it was speculated that these beams could be the seed particles for a Fermi type process. This is most probably not so, as we have learned from numerical simulations of quasi-parallel shocks, and as will be discussed later. In the region downstream of the quasi-parallel bow shock Thomsen et al.[12] found two ion distributions alternating between each other: a cool and denser distribution and a hotter, less dense, more Maxwellian type distribution This is evidence that the dissipative dynamics at quasi-parallel shocks is non-stationary and cyclic in nature.

THEORY

A number of theoretical concepts has been developed over the past three decades in order to describe parallel shock formation (see Quest[13] for a recent update). Parker[14] suggested that the overlapping upstream and downstream plasmas generate a fluid type instability similar to the firehose instability. Parker showed that an additional parallel pressure is created by two plasma beams, and the marginal firehose condition $p_\parallel - p_\perp > B^2/4\pi$ is exceeded. In the Parker model a long ion precursor exists upstream by ions evaporating from the downstream region which generates low-frequency electromagnetic waves by the firehose, or more properly termed the ion-ion nonresonant instability as two ion streams interact. Later Kennel and Sagdeev[15] considered the firehose instability driven by a temperature anisotropy in the downstream region which is due to steepening of ion acoustic waves at the shock front. Golden et al.[16] proposed the excitation of the resonant electromagnetic ion beam instability to be responsible for shock formation. The instability is driven by the interaction of the upstream and the downstream ion distribution. A third mechanism for quasi-parallel shock formation involves the excitation of short wavelength whistler mode waves. Jackson and Golden[17] proposed that these waves are excited by ions reflected off the shock ramp.

Quest[13] has argued that at a steady parallel shock the firehose instability is not excited, as the density of the ions backstreaming from the shock is too low. Instead, the electromagnetic resonant ion/ion beam instability generates waves upstream of the shock which are then convected into the shock and amplified, thus providing the necessary downstream heating. It was found by numerical simulations that the backstreaming ions are generated by incoherent scattering of some ions of the incoming solar wind in the compressed wave field. This seems to be inconsistent with the observations of cold specularly reflected beams reported above. Furthermore, the simulations of Quest[13] resulted in sinusoidal upstream waveforms rather than in the observed nonlinear pulsations. Since the classic paper by Quest on the structure of exactly parallel shocks interest has shifted somewhat and the numerical work has concentrated on quasi-parallel shocks.

NUMERICAL SIMULATIONS ON SHOCK STRUCTURE

A great deal of insight has recently been gained by numerical simulations of quasi-parallel shocks. Most of these simulations have been carried out by using the so-called hybrid code. The hybrid code treats the ions as macro-particles and the electrons as an inertialess, charge neutralizing fluid. Thus, this method does not correctly describe electron heating and frequencies considerably larger than the iongyrofrequency. Two different methods have been used in order to generate a collisonless shock: in one method, the ions are distributed in the simulation system in such a way that their moments (bulk velocity $\mathbf{V}$, density n, pressure p) in the upstream and downstream region fulfill the Rankine-Hugeniot conditions. The upstream and downstream quantities are connected by some arbitrary function and the system is followed in time, with values at the upstream and downstream boundaries fixed according to the Rankine-Hugoniot jump conditions. This method allows the evolution of the shock to be followed for long periods of time. The second, more popular method involves reflection of an incident beam off a wall positioned at one end (say, right hand end) of the simulation system. The reflected and incident beam couple with each other via an electromagnetic ion/ion beam instability and a shock is launched which propagates to the left in the simulation frame. Ions are continously injected from the left hand side. This method has the advantage that the shock is not predetermined and boundary conditions are easier to handle, but it has the disadvantage that the upstream regions becomes smaller as the shock develops and propagates toward the left hand boundary. In these simulations dimensionless units are used; time is expressed in units of the inverse of the ion gyrofrequency, $\Omega_{ci} = eB_o/mc$, where c is the speed of light, e is the magnitude of the electron charge, B_o is the upstream magnetic field strength and m is the ion mass. Distances are expressed in units of the ion inertial length, $\lambda = c/\omega_{pi}$, where ω_{pi} is the ion plasma frequency. For typical solar wind conditions upstream of the bow shock one ion inertial length corresponds to $\sim 100\,\mathrm{km}$. The unit velocity is the upstream Alfvén velocity v_A. The shock is specified by the Alfvén Mach number M_A, the angle between the upstream magnetic field and the shock normal, $\Theta_{Bno} = \sin^{-1}(B_{zo}/|Bo|)$, and $\beta_i(\beta_e)$, the ratio of ion (electron) thermal to magnetic pressure.

By carrying out the hybrid simulations of a supercritical collisionless shock ($M_A = 6, \Theta_{Bno} = 30°$) to long times Burgess[18] found that the shock structure is not steady but constantly re-forms itself: foreshock perturbations impinging on the shock caused the front of the shock to disrupt and then to re-form periodically. Figure 3, taken from [18], shows a time sequence of the magnetic field versus the simulation coordinate x. Time runs from bottom to top. One can clearly see the shock running to the left and the convection of upstream waves which steepen up as they are convected into the shock and subsequently lead to shock re-formation. Scholer and Terasawa[19] pointed out that the re-formation cycle starts with the specular reflection of a large part of the incident ions. Figure 4 shows the temporal development of the distribution function versus v_x immediately upstream of the shock (top panel) and further upstream (bottom panel) during a re-formation event. Positive v_x corresponds to the solar wind, backstreaming ions can be found at negative v_x. The shock re-forms at a position upstream of the old location so that the incident and reflected beam finds itself downstream of the new shock position. The two beams are eventually thermalized and constitute a hot downstream distribution. Although ions are reflected specularly at the shock and can in principle propagate upstream, shock re-formation upstream of the old position will keep them confined to the (new) downstream region. Thus, it is rather unlikely that specularly reflected ions can be the seed particles for a first order Fermi acceleration process.

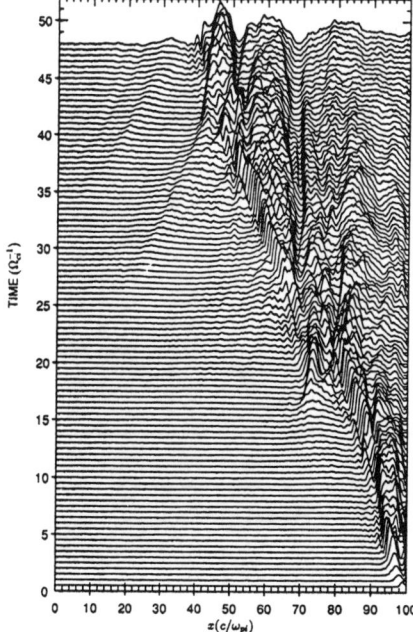

Fig. 3. Stackplot of the magnetic field from a quasi-parallel shock simulation. The solar wind is reflected at the right hand side, which produces a shock runing to the left (from Burgess[18]).

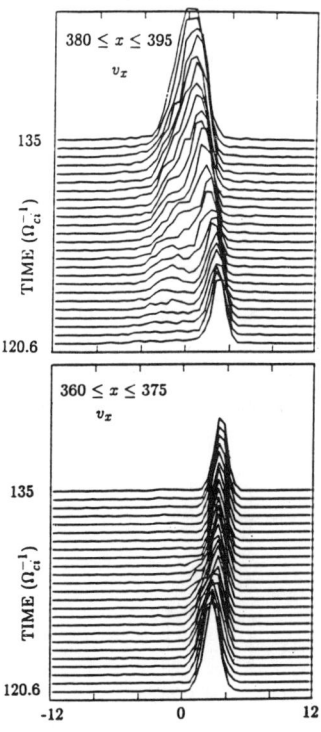

Fig. 4. Stackplot of the distribution function at two positions during a re-formation event (from Scholer and Terasawa[19]).

Various processes can couple the reflected ions with the incident solar wind and can thus lead to a re-formation process. Burgess[18] and Scholer and Terasawa[19] found that the ultra low-frequency (ULF) wave generated upstream and convected into the shock steepen up as they interact with the specularly reflected ions and actually become the new re-formed shock. The ULF waves are produced by the more energetic diffuse upstream ions as will be discussed later. However, when eliminating ULF waves by taking the diffuse backstreaming ions out of the simulation system the shock still re-forms. Therefore, Winske et al.[20] proposed that an interface instability, involving an ion/ion instability located where the incident solar wind overlaps the the downstream heated population, is responsible for shock re-formation. The linear analysis of the interface instability results in wave lengths at maximum growth rate which are close to the average re-formation distance observed in the simulation, whereas the linear analysis of the ion/ion beam instability resulted in wave lengths which are much too large. Onsager et al.[21] investigated the nonlinear coupling of high density finite length beams with a uniform background plasma (where the beam was streaming parallel to the magnetic field) in detail. They found that the linear analysis is considerably modified by the beam having a finite length. This analysis was extended for arbitrary angle Θ_{Bn} between beam and magnetic field and it was found that the coupling beween beam and solar wind occurs at about

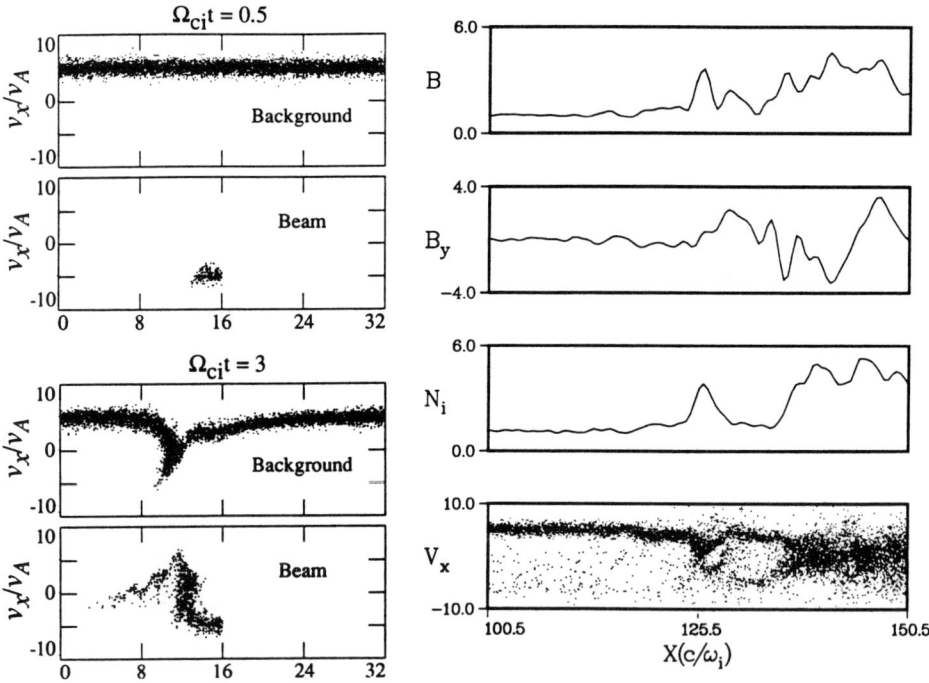

Fig. 5. Simulation results of finite beam - background plasma interaction at two times (after Onsager et al.[22]).

Fig. 6. Magnetic field B, B_y component, density, and $v_x - x$ phase space plot during a re-formation cycle (after Thomas et al.[23]).

the time and location where the reflected ions become deflected transverse to the shock normal direction [22]. Figure 5 shows the interaction of a finite lenght beam with 40% of the solar wind density placed initially at $x = 16c/\omega_{pi}$ at two time steps. The angle Θ_{Bno} was 30°. Shown are $v_x - x$ phase space plots of both the background and of the beam ions. At $\Omega_{ci}t = 3$ a shock like structure has formed and a fraction of the incident ions have been reflected. This is due to the fact that specularly reflected ions in the $\Theta_{Bno} = 30°$ case will reach $v_x = 0$ at some point in their gyration and will accumulate. This is very similar to what happens in "real" shock simulations: Figure 6 shows magnetic field B and the B_{By} component, density n an $v_x - x$ phase space during a re-formation cycle after Thomas et al.[23] for a $\Theta_{Bno} = 30°$ shock. One can clearly see the reflected ions, their deflection upstream of the shock, and the magnetic field and density increase at that position. Even in the case of a small upstream value of Θ_{Bno} the occurence of ULF waves with large amplitudes will lead to large local values of Θ_{Bn}. Using initial upstream conditions derived from self-consistent simulations of quasi-parallel shocks Scholer and Burgess[24] have shown that a finite length beam will get deflected and a large magnetic pulse appears at the position where the beam ions reach an approaching wave crest and attain zero velocity in the shock normal direction. Finally, it should be pointed out that in addition to ion/ion beam instability, interface instability and deflection of reflected ions, in some simulations large-amplitude whistler waves near the shock have led to shock re-formation [26] It is, however, not clear to what extent these whistler waves are a numerical artifact [20].

DIFFUSE SUPERTHERMAL IONS

The diffusive shock or first order Fermi acceleration theory is not concerned with the question how a certain part of the ambient particles are injected into the acceleration process, but starts with a source of seed particles, which can either be in the upstream flow or be injected at the shock. However, the important question is how particles that are originally part of the thermal plasma begin to be accelerated. It is this injection or "seed particle" problem which will be addressed in the following. In the past, three possible injection models for diffuse ions have been proposed in the literature. These models are either based on reflection of part of the incident ions or on thermal leakage. Reflection at the quasi-perpendicular bow shock has already been dismissed on grounds of different proton to alpha particle ratio in field-aligned beams and in diffuse ions. The second reflection model assumes time dependent reflection at a shock which locally alternates between quasi-parallel and quasi-perpendicular states. This model has the same problem in explaining heavy ion abundances as the beam disruption model. Furthermore, when in the simulations reflected ions occur the shock re-forms upstream of the old shock position and the reflected and the incident ions constitute the thermalized population downstream of the newly re-formed shock. In other words, reflected ions do not escape in any appreciable quantities upstream and seem to be different from the diffuse ions. The thermal leakage injection model assumes that part of the shock heated solar wind ions can freely scatter back across the shock into the upstream region. This model has first been advocated by Ellison[25] and has been investigated in consideable detail. From self-consistent hybrid simulations of quasi-parallel shocks Lyu and Kan[26] have recently also concluded that backstreaming ions are predominantly leakage ions. The leakage model is also discussed in the paper by Kucharek and

Scholer (these Proceedings).

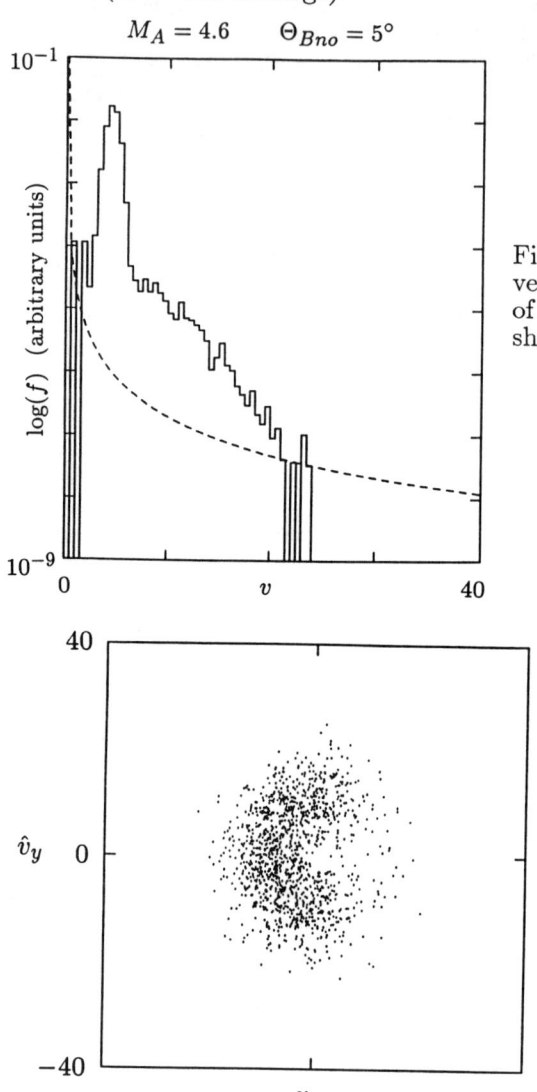

Fig. 7. The distribution function versus velocity v in the shock frame of reference upstream of a simulated shock (after Kucharek and Scholer[28]).

Fig. 8. $v_y - v_x$ phase space plot of all backstreaming ions upstream of a simulated shock (after Kucharek and Scholer[28]).

A different mode of injection of ions at quasi-parallel shocks into a diffusive acceleration mechanism was found from hybrid simulations first by Quest[13] for a parallel shock and more recently by Scholer[27] for a quasi-parallel shock. The self-consistent simulations of quasi-parallel shocks resulted in a large percentage (∼ 4%) of diffuse upstream ions. The majority of these ions was not due to leakage of thermalized downstream ions, nor due to specularly reflected ions: the diffuse ions were accelerated from the incident thermal plasma to high energies while riding for a long period of time (about 15 ion gyroperiods) close to the subshock. They eventually escape upstream and were scattered again by upstream waves of their own making. Figure 7 (from [28]) shows the distribution function versus velocity in the shock frame of reference of all particles within $100 c/\omega_{pi}$ upstream of a $\Theta_{Bn} = 5°$, $M_A = 4.6$ shock in a log-linear representation. Also shown (dashed line) is the one-particle limit of the simulation. One can clearly see a high energy tail in addition to the thermal distribution which peaks at $v = 4.6$. That the particles are rather diffuse can be seen from Figure 8: this figure shows $v_y - v_x$ phase space plots of the backstreaming ions for the same shock run. The distributions are diffuse although not isotropic: the majority of the upstream particles have a negative velocity v_x and escape the system.

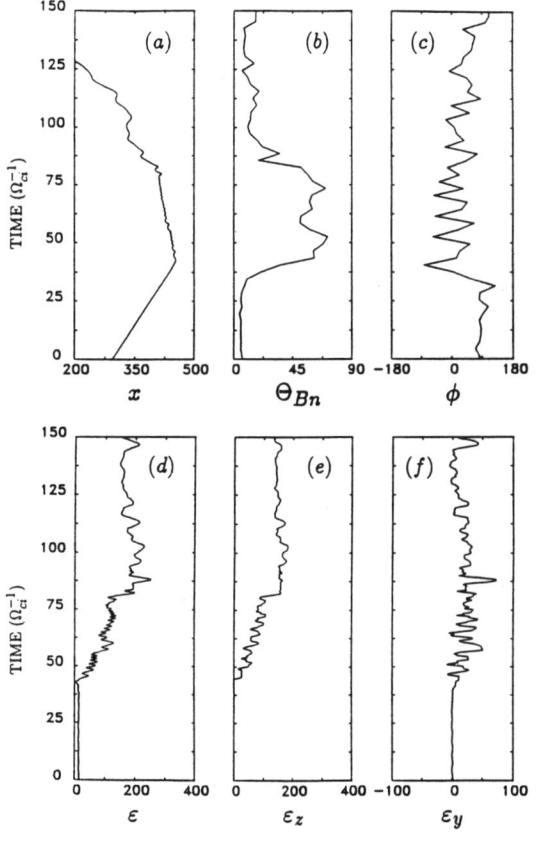

Fig. 9. For an upsteam diffuse particle: (a) time versus position x; (b) time versus local Θ_{Bn} at the particle's position; (c) time versus phase of the magnetic field at the particle's position; (d) time versus particle energy ε in the shock frame; (e) time versus the contribution $\int E_z v_z \, dt$ to the particle's energy gain/loss $\int \mathbf{E} \cdot \mathbf{v} \, dt$; (f) time versus the contribution $\int E_y v_y \, dt$ (after Kucharek and Scholer[28]).

As an example, Figure 9 shows for one particular upstream particle in the top part of (a) time versus position, (b) the instantaneous value of Θ_{Bn} at the particles position, and (c) the phase ϕ of the magnetic field at the particles position. The phase ϕ is defined by $\phi = \tan^{-1}(B_y/B_z)$. Θ_{Bn} as well as ϕ are highly fluctuating quantities and have been averaged over one gyration period. The bottom part of Figure 9 shows (d) time versus total energy in the shock frame, and the the individual contributions $\varepsilon_z = \int E_z v_z \, dt$ (c) and $\varepsilon_y = \int E_y v_y \, dt$ (d) to the total energy ε, where E_y and E_z are the y and z components of the electric field, $\mathbf{E}$. The contribution of $\varepsilon_x = \int E_x v_x \, dt$ in the shock frame is, on the average, zero. Two surprising facts can be seen from Figure 9b and 9c. Firstly, although this is a $\Theta_{Bno} = 5°$ shock, the local value of Θ_{Bn} is larger than

$\sim 40°$ during the particle's trajectory near the shock when it gains a large part of the final energy. Secondly, during this time period the phase is, on average, closer to $0°$ than to $90°$, i.e., near the shock the noncoplanarity magnetic field component B_y is larger than the B_z component. This is also the position of the largest magnetic field gradient, so that the $\mathbf{B} \times \nabla B$-drift is in the z direction and leads to a corresponding energy gain ε_z. The particle is boosted up again in energy when it leaves the shock in the upstream direction, what happens mainly during re-formation cycles.

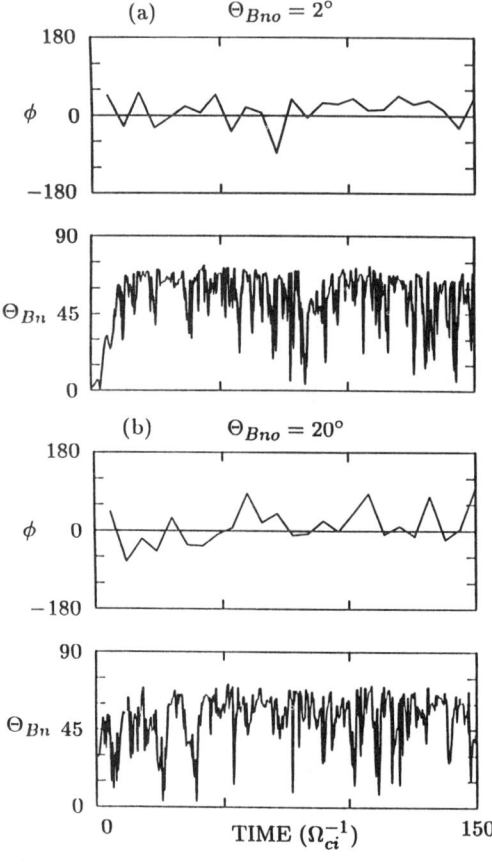

Fig. 10. The phase ϕ of the magnetic field and the local value of Θ_{Bn} at the shock versus time for two different shock runs: $\Theta_{Bno} = 2°$ and $\Theta_{Bno} = 20°$. For both cases $M_A = 4.6$. The phase has been averaged over $6\,\Omega_{ci}^{-1}$ (after Kucharek and Scholer[28]).

Surprisingly, the drift is for small initial Θ_{Bno} mainly in the z direction. This is due to a large noncoplanarity magnetic field component B_y. Figure 10 shows the phase ϕ and the local value of Θ_{Bn} at the shock versus time for two shock runs: $\Theta_{Bno} = 2°$ and $\Theta_{Bno} = 20°$ with $M_A = 4.6$. The shock position has been defined, as the position where the density gradient is largest and the density exceeds the upstream value by a factor 3. The Θ_{Bno} curves have not been averaged; the phases have been averaged over $6\,\Omega_{ci}^{-1}$, which corresponds to about one ion gyroperiod. From Figure 10 we see the following: the local value of Θ_{Bn} near the shock for quasi-parallel shocks with Θ_{Bno} between $2°$ and $20°$ is rather large and reaches in both cases values exceeding $60°$. Near shocks with $\Theta_{Bno} \leq 20°$ the gyroperiod averaged phase is close to $0°$, i.e., the magnetic field at the shock is largely perpendicular to the coplanarity plane. The consequences for the particles are that firstly, there will be drift, and secondly, the drift is mainly in the z direction, i.e. within the coplanarity plane.

One of the crucial questions is whether alpha particles, like protons, are also directly accelerated out of the incident thermal particle distribution to suprathermal energies in any appreciable quantity, so that they can constitute a seed particle population for a first order Fermi process. By including alpha particles self-consistently into the hybrid code it has been indeed shown that the quasi-parallel shock produces diffuse upstream protons as well as alpha particles with a density ratio bias factor f of the order of 1, i.e., the ratio of diffuse He^{2+} to diffuse H$^+$ is about equal to the same ratio in the solar wind [29]. Figure 11 shows energy spectra in the shock frame of protons and alpha particles for a shock with $\Theta_{Bn} = 20°$, $M_A = 4.6$, and a solar wind alpha to proton ratio of 5%. The spectra are shown in a log-log representation versus energy per charge. Energy is expressed in units of $m_p v_A^2/2$ (v_A is Alfvén speed). The exact value of f depends on the solar wind He^{++} to H$^+$ density ratio and on the shock Mach number. The density ratio bias factor for a low Mach number shock ($M_A = 4.6$) is $f \sim 0.5$ when the alpha particles are treated as test particles and increases to 1.0 when the solar wind He^{++} to H$^+$ ratio is 0.2. It was found in the simulations that the density ratio bias factor considerably increases with increasing Mach number, a fact which is at present not understood.

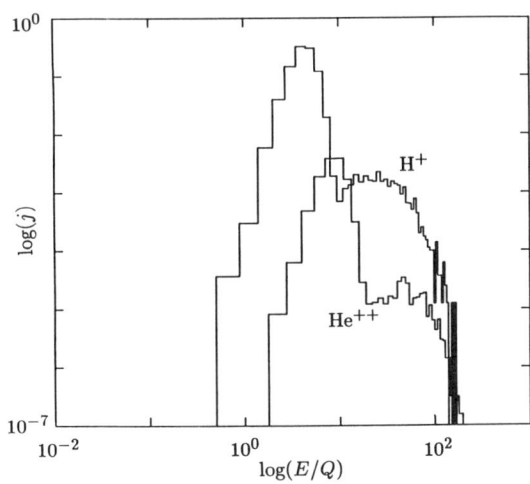

Fig. 11. Energy spectra (omnidirectional differential flux) in the shock frame of protons and alpha particles upstream of a simulated shock. The spectra are shown in a log-log representation versus energy per charge. Energy is expressed in units of $m_p v_A^2/2$ (after Trattner and Scholer[29]).

SUMMARY

Considerable insight into the physics of collisionless quasi-parallel shocks has recently been gained by hybrid simulations. These simulations have led to a better understanding of the shock re-formation process, of the dissipation mechanism, and of acceleration of part of the incident plasma to a superthermal population. It has been demonstrated that energetic particle acceleration, upstream wave production, occurrence of specularly reflected ions, and shock re-formation at quasi-parallel shocks cannot be considered separately, but are closely interconnected. The work reported here and in other papers given at this conference has demonstrated that large-scale hybrid simulations have closed the gap between the diffusive shock acceleration theory and the micro-physics of the shock itself.

REFERENCES

1. N. Sckopke, G. Paschmann, S. J. Bame, J. Gosling, and C. T. Russell, J. Geophys. Res. **88**, 6121, (1983).
2. M. M. Leroy, D. Winske, C. C. Goodrich, C. S. Wu, and K. Papadopoulos, J. Geophys. Res. **87**, 5081, (1982).
3. D. Burgess, W. P. Wilkinson, and S. J. Schwartz, J. Geophys. Res. **94**, 8783, (1989).
4. G. Paschmann, N. Sckopke, J. R. Asbridge, S. J. Bame, and J. T. Gosling, J. Geophys. Res. **85**, 4689, (1980).
5. F. M. Ipavich, J. T. Gosling, and M. Scholer, J. Geophys. Res. **89**, 1501, (1984).
6. F. M. Ipavich et al., Geophys. Res. Lett. **15**, 1153, (1988).
7. E. Greenstadt et al., in Solar Terrestrial Physics: Present and Future, NASA Ref. Publ. 1120, (1984).
8. M. F. Thomsen, J. T. Gosling, S. J. Bame, and C. T. Russell, J. Geophys. Res. **95**, 957, (1990).
9. J. T. Gosling, M. F. Thomsen, S. J. Bame, and C. T. Russell, J. Geophys. Res. **94**, 10027, (1989).
10. T. G. Onsager, M. F. Thomsen, J. T. Gosling, S. J. Bame, and C. T. Russell, J. Geophys. Res. **95**, 2261, (1990).
11. S. A. Fuselier, O. W. Lennartson, M. F. Thomsen, and C. T. Russell, J. Geophys. Res. **95**, 4319, (1990).
12 M. F. Thomsen, J. T. Gosling, S. J. Bame, and T. G. Onsager, J. Geophys. Res. **95**, 6363, (1990).
13. K. B. Quest, J. Geophys. Res. **93**, 9649, (1988).
14. E. N. Parker, J. Nucl. Energy **C2**, 146, (1961).
15. C. F. Kennel and R. Z. Sagdeev, J. Geophys. Res. **72**, 3303, (1967).
16. K. I. Golden, L. M. Linson, and S. A. Mani, Phys. Fluids **16**, 2319, (1973).
17. R. W. Jackson and K. I. Golden, J. Plasma Phys. **22**, 491, (1979).
18. D. Burgess, Geophys. Res. Lett. **16**, 345, (1989b).
19. M. Scholer and T. Terasawa, Geophys. Res. Lett. **17**, 119, (1990).
20. D. Winske, N. Omidi, K. B. Quest, and V. A. Thomas, J. Geophys. Res. **95**, 18821, (1990).
21. T. G. Onsager, D. Winske, and M. F. Thomsen, J. Geophys. Res. **96**, 1775, (1991).
22. T. G. Onsager, D. Winske, and M. F. Thomsen, J. Geophys. Res. **96** 21,183, (1991).
23. V. A. Thomas, D. Winske, and N. Omidi, J. Geophys. Res. **95**, 18809, (1990).
24. M. Scholer and D. Burgess, J. Geophys. Res., in press, (1992).
25. D. C. Ellison, Geophys. Res. Lett.**8**, 991, (1981).
26. L. H. Lyu and J. R. Kan, Geophys. Res. Lett. **17**, 1041, (1990).
27. M. Scholer, Geophys. Res. Lett. **17**, 1821, (1990).
28. H. Kucharek and Scholer, J. Geophys. Res. **96**, 21,195, (1991).
29. K. J. Trattner and M. Scholer, Geophys. Res. Lett. **18**, 1817, (1991).

DIFFUSIVE SHOCK ACCELERATION:
ACCELERATION RATE, MAGNETIC-FIELD DIRECTION
and the
DIFFUSION LIMIT [†]

J.R. Jokipii
University of Arizona

ABSTRACT

This paper reviews the concept of diffusive shock acceleration, showing that the acceleration of charged particles at a collisionless shock is a straightforward consequence of the standard cosmic-ray transport equation, provided that one treats the discontinuity at the shock correctly. This is true for arbitrary direction of the upstream magnetic field. Within this framework, it is shown that acceleration at perpendicular or quasi-perpendicular shocks is generally much faster than for parallel shocks. Paradoxically, it follows also that, for a simple scattering law, the acceleration is faster for less scattering or larger mean free path. Obviously, the mean free path can not become too large or the diffusion limit becomes inapplicable. This is shown to occur when the dimensionless parameter $\Upsilon = r_g U_{shock}/\kappa_\perp \approx (\lambda_\parallel/r_g)(U_{shock}/w) \approx 1$. Recent Monte-Carlo simulations which show what happens near and beyond this limit are briefly discussed. Gradient and curvature drifts caused by the magnetic-field change at the shock play a major role in the acceleration process in most cases. Recent observations of the charge state of the anomalous component are shown to require the faster acceleration at the *quasi-perpendicular* solar-wind termination shock.

INTRODUCTION

In figure 1 is shown the observed cosmic-ray energy spectrum, over the energy from 10^5 eV to 10^{20} eV. The spectrum below 10^8 eV is depressed because of solar modulation. Above 10^9 eV, the continuity and smoothness of the spectrum over more than 10 decades in energy, with only a small steepening at some $10^{15}-10^{16}$ eV (the "knee"), suggests strongly that one or perhaps two closely-related mechanisms are accelerating the particles. As emphasized by Jokipii and Morfill (1991), the steepening of the spectrum above 10^{16} eV cannot be produced, short of a "cosmic conspiracy", if the higher-energy particles were accelerated by a process unrelated to that producing the lower-energy particles. Thus, if the bulk of the cosmic rays are generated by supernova blast waves in our galaxy, the UHE cosmic rays must also be produced by a mechanism associated with supernova and/or the interstellar medium.

Diffusive shock acceleration of charged particles at collisionless astrophysical shocks has emerged over the past decade as the most viable acceleration mech-

---

[†] Contribution # 92-07 of the University of Arizona Theoretical Astrophysics Program

anism, primarily because it tends to produce a universal spectrum close to that observed, and because it is efficient.

The diffusive transport of charged particles in astrophysical plasmas is governed by the transport equation for the distribution function f as a function of momentum magnitude p, position x_i and time t, first derived by Parker (1965). It may be written in the form

$$\begin{aligned}\frac{\partial f}{\partial t} &= \frac{\partial}{\partial x_i}\left[\kappa_{ij}\frac{\partial f}{\partial x_j}\right] & (diffusion) \\ &- U_i\frac{\partial f}{\partial x_i} & (convection) \\ &- V_{di}\frac{\partial f}{\partial x_i} & (guiding-center\ drift) \\ &+ \frac{1}{3}\frac{\partial U_i}{\partial x_i}\left[\frac{\partial f}{\partial \ell n p}\right] & (energy\ change) \\ &+ Q(x_i, t, p) & (source)\end{aligned} \quad (1)$$

where the various terms are labelled with the associated effect, and where κ_{ij} is the symmetric part of the diffusion tensor, U is the fluid flow velocity, $V_d = (pcw/3q)\ \nabla \times [B/B^2]$ is the particle drift velocity, c is the speed of light and Q the local source.

This equation is extremely general and appears to require only that the scattering be rapid enough that the distribution function is nearly isotropic in the plasma frame. Note that *all* of diffusive shock acceleration is contained in this equation, if the shock is properly treated. In the late 1970's it was realized by Blandford and Ostriker (1977); Krimsky (1976); Axford, Leer, and Scadron (1977); and Bell (1977a, 1977b) that diffusive shock acceleration occurring in shock waves obeying the previous equation has many desirable features.

We apply this equation to a collisionless shock propagating in a fluid with a uniform, ambient magnetic field **B**, at an angle θ_1 relative to the upstream flow velocity U_1. B is taken to be small enough that it has a negligible effect on the fluid dynamics. The shock thickness is generally small enough that the shock may be taken to be locally plane. We work in the shock frame in which the shock is stationary and the upstream flow $U_1\hat{e}_x$ is normal to the shock plane, taken to be the $y-z$ plane. Since the field is weak, the downstream flow is also in the x direction, and is less by a factor $1/r$, where $r \leq 4$ is the shock ratio. The angle θ_2 between U_2 and B_2 is readily shown to be $\tan^{-1}[r\ \tan(\theta_1)]$.

The solution to equation (1) near a shock may be obtained by solving the differential equation upstream and downstream of the shock and matching the solutions by using an appropriate matching condition. This condition can be obtained by simply integrating the transport equation across the shock from a very

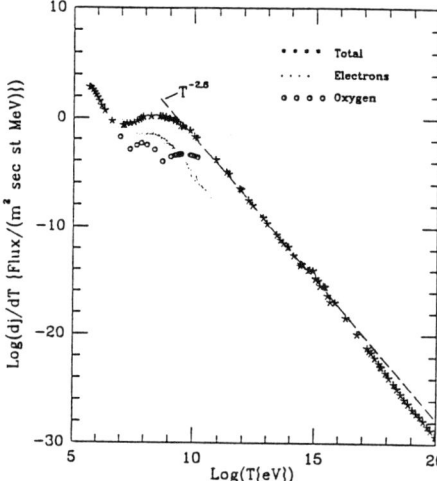

Figure 1. The energy spectrum of cosmic rays.

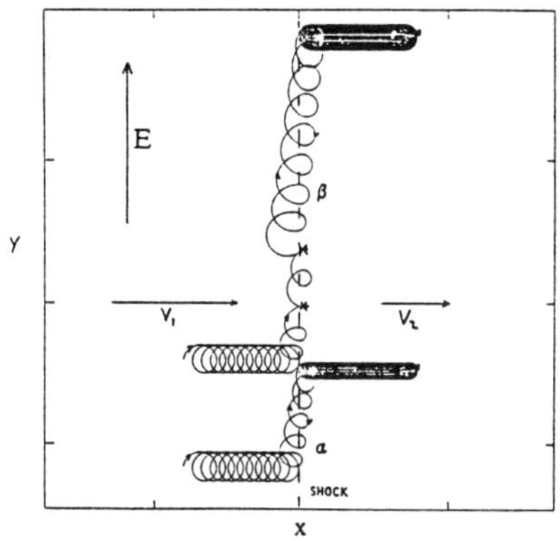

Figure 2. Trajectories of two particles interacting with a shock. Particle α is not scattered. Particle β is scattered and can gain more energy.

small distance upstream of the shock to a very small distance downstream. In the process one obtains the general matching condition for the solution at the shock (Jokipii, 1982, 1987).

$$\left[\kappa_{xx}\frac{\partial f}{\partial x} + \frac{V_x}{3}\frac{\partial f}{\partial \ell n p} - \frac{pcw}{3q}\frac{B_z}{B^2}\frac{\partial f}{\partial y}\right]_1^2 = Q_*. \qquad (2)$$

where Q_* is that part of the source which is singular at the shock,

The third term on the right in equation (2) reflects the change in the magnetic field at the shock, and contains the effect of gradient and curvature drifts. Note that the drift term vanishes if there is no derivative in the distribution function along the shock in the direction of drift motion. Hence, if we have a purely one dimensional system, as is often discussed, the standard jump condition applies and the drifts at the shock do not affect the solution. This is because new particles are always drifting in to replace those drifting out of a given volume element. It is important, however, to realize that the drifts are still present and contribute to the physics of the problem, in spite of the fact that they do not formally appear in the solution.

140 Diffusion Shock Acceleration

All such one-dimensional systems give exactly the same time-asymptotic solution, which may be written

$$f(x,p) = Ap^{-q} \qquad\qquad x > x_{shock}$$

$$ = Ap^{-q} \exp[U_1(x - x_{shock})/\kappa_{xx1}] \quad x \leq x_{xshock},$$

where $q = 3r/(r-1)$ and κ_{xx1} and κ_{xx2} are, respectively, the upstream and downstream diffusion coefficients in the flow direction. In terms of the parallel and perpendicular diffusion coefficients $\kappa_\parallel$ and $\kappa_\perp$ we have $\kappa_{xx1} = \kappa_\parallel \cos^2(\theta_1) + \kappa_\perp \sin^2(\theta_1)$, and a similar expression for κ_{xx2}. Note that the only parameter to enter into the index of the power law is the shock ratio r. Since $r = 4$ for strong shocks, which should produce most of the cosmic rays, we have what is essentially a universal cosmic ray spectrum, which is close to that observed.

Of particular importance is the fact that, since the particles must cross the shock many times to gain energy, acceleration takes time. For example, if one injects energetic particles which have a injection momentum p_0, then the spectrum above p_0 will be approximately the power law p^{-q} out to a cut-off momentum p_c which increases with time according to the approximate relation (Forman and Morfill, 1979, Jokipii, 1987 a,b)

$$\frac{1}{p_c}\frac{dp_c}{dt} \approx \frac{U_1^2(r-1)}{3r[\kappa_{xx1} + r\kappa_{xx2}]} \qquad (3)$$

Near p_c, the spectrum will begin to decrease more-rapidly than the power law and asymptotically approach zero. The location of this break will occur at a characteristic energy which depends on the boundary conditions and time available for acceleration.

In general, therefore, to accelerate particles most effectively in any given situation, we wish to make the diffusion coefficient normal to the shock as small as possible. This particular consequence of shock acceleration was explored in detail for supernova shocks by Lagage and Cesarsky (1983). They considered the maximum energy which might be obtained by this mechanism in a supernova blast wave. They suggested that the minimum value of the diffusion coefficient for a particle of speed w and gyroradius r_g is the Bohm value $\kappa_{Bohm} = wr_g/3$. This simply states that the smallest value of the diffusion mean free path in such a system should be approximately the particle gyroradius. They then integrated equation (3) for a supernova blast wave using the modified Sedov solution for U_1. In this procedure they found a "firm upper limit" of a few times $10^{14}Z$ eV on the maximum energy obtainable in a typical supernova blast wave. This clearly would be a very severe constraint on the allowed Z, as conventional wisdom states that the particles below the knee at $\approx 10^{16}$ eV, where the slope of the spectrum changes, comes from supernova shock waves. This picture clearly has difficulties in the above scenario, particularly since the mean free path is likely to exceed the gyro-radius by a considerable amount.

However, for quasi-perpendicular shocks, $\kappa_\perp$ plays a more-important role than $\kappa_\parallel$, and it can be much *less* than the Bohm limit, alleviating the above problem.

THE ROLE OF THE MAGNETIC FIELD

If, as is usual, the particle parallel mean free path $\lambda_\parallel >$ several $\times\, r_g$, it may be shown that the acceleration of cosmic rays in equation (1) is due in general to two quite distinct physical processes: compression along field lines and particle drift motion across the field (Jokipii, 1979, Kota, 1979). If the magnetic field is time-independent in the observer's frame, this is simply equivalent to

$$\frac{dT}{dt} = \frac{T}{3}\frac{d\ln T}{d\ln p}\left[\nabla\cdot\mathbf{U}_\parallel + \nabla\cdot\mathbf{U}_\perp\right]$$
$$= -q\mathbf{V_d}\cdot\frac{\mathbf{U}\times\mathbf{B}}{c} + \frac{T}{3}\frac{d\ln T}{d\ln p}\nabla\cdot\mathbf{U}_\parallel, \tag{4}$$

where $\mathbf{E}$ is the electric field. Essentially the same result applies for a time-dependent field, except for a term involving $\partial B/\partial t$. Within the framework suggested by equation (4), then, acceleration at *parallel* shocks is then due to the *compression* of the fluid at the shock, whereas at a *perpendicular* shock the acceleration may be viewed as being due to *particle drift* along the shock face. Of course, as revealed in equation (4), the latter is also equal to the divergence of the flow at the shock.

The acceleration of a particle at a perpendicular shock is illustrated in figure 2. The particle $\mathbf{E}\times\mathbf{B}$ drifts toward the shock at the fluid flow velocity $\mathbf{U_1}$. As its orbit intersects the shock, it drifts along the shock face in the direction of the electric field and gains energy. If there were no scattering (and hence no diffusion) the particle (α) would also continue to drift through the shock and never encounter the shock again. Under these circumstances it may be shown that the particle (averaged over pitch angle) will gain an energy roughly equal to a factor of the shock ratio times the original energy in its interaction with the shock. However, if the particle (β) is scattered while it is interacting with the shock it can clearly be scattered back upstream and pick up the same amount of energy many times, as illustrated in the figure 2. This is very analogous to ordinary diffusive shock acceleration discussed above, except that in this case the particle drift along the shock is the dominant source of energy gain, and the particle can gain much more energy between scatterings. Equation (3) still applies, but since the value of $\kappa_\perp$ can be much smaller than the Bohm limit, the acceleration is in general much faster. The relevant diffusion coefficient becomes $\kappa_\perp$ which, for simple scattering, may be related to $\kappa_\parallel$ by

$$\kappa_\perp/\kappa_\parallel = 1/\left\{1 + (\lambda_\parallel/r_g)^2\right\}. \tag{5}$$

The Bohm limit applies only to $\lambda_\parallel$ which must be greater than the particle gyroradius r_c. Note that since $\kappa_\perp$ decreases as we increase $\lambda_\parallel$ it can be much smaller than the Bohm limit.

LIMITS

From the preceding analysis it is evident that for perpendicular (or highly oblique) shocks, as $\lambda_\parallel$ becomes larger the acceleration can become faster. However, if $\lambda_\parallel$ becomes too large (or $\kappa_\perp$ becomes too small), the diffusion approximation becomes invalid. Hence, in contrast to quasi-parallel shocks where the particles must be scattered to return to the shock, acceleration at perpendicular shocks is not so clearly related to scattering, and one must impose an additional constraint to ensure that the diffusion approximation is valid. The associated maximum value of $\lambda_\parallel$ can be obtained from a number of different considerations (Jokipii, 1987 a,b), all of which lead to the same conclusion. One may require that the particle scatter often enough to keep the distribution function isotropic at the shock or that the particle can diffuse upstream fast enough to stay ahead of the shock. Each of these leads to the condition

$$\Upsilon = U_{shock} \frac{r_g}{\kappa_\perp} \ll 1. \tag{6a}$$

or, for our simple scattering model,

$$\Upsilon = \frac{\lambda_\parallel}{r_c} \frac{U_{shock}}{w} \ll 1. \tag{6b}$$

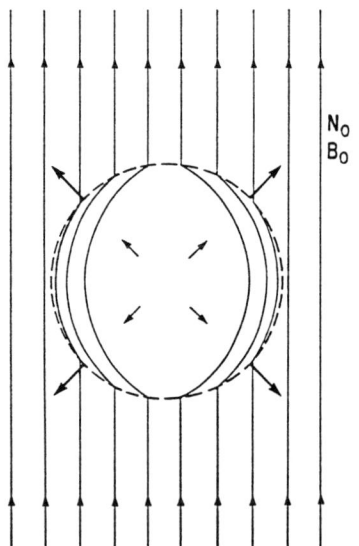

Figure 3. A supernova blast wave in a uniform magnetic field B_0

If the particle speed is significant larger than the shock speed $\lambda_\parallel/r_c$ can be very large, and we can have a significant enhancement above the Bohm limit. As the particle speed approaches the shock speed, which is true for nearly thermal particles, the possible enhancement will be very small.

Hence, the maximum energy at a supernova shock wave can be significantly larger than the $10^{14} Z$ discussed by Lagage and Cesarsky, since the shock will be quasiperpendicular over much of its area. Figure 3 diagrams schematically the expansion of a supernova blast wave into a homogenous uniform magnetic field. Clearly, the shock is perpendicular over a great deal of its face and we expect a significant effect.

These considerations then lead to a revised maximum energy for diffusive shock acceleration which is obtain by substituting the maximum value of the mean free path in the expression for dp/dt, and then using this in the expression for the shock acceleration rate. This yields

$$T_{max} \approx 10^{17} ZeV, \tag{7}$$

which is adequate to accommodate the interpretation of the cosmic- ray spectrum in which the change in slope occurs when the supernova acceleration mechanism ceases. The actual energy at which this occurs will be determined by the scattering, and will in general be significantly *less* than T_{max}.

MONTE-CARLO SIMULATIONS

Now consider what happens as the diffusive limit becomes invalid (ie. as Υ in equation (6) becomes of the order of unity or greater). This can not be studied analytically because the appropriate Boltzmann equation is too complex. Monte Carlo simulations of shock acceleration have been carried out previously by a number of authors (Passes, Decker, and Armstrong, 1982, Chiueh, 1988, Takahara and Teresawa, 1991), but the diffusion limit was not investigated. Jokipii and Terasawa (1992) have used this approximation to study in detail the transition from diffusive to scatter-free behavior.

For simplicity, I discuss here preliminary results for the special case of a perpendicular shock. Assume that the scattering of the particles is given by a phenomenological scattering rate which make the particles isotropic in the local frame of the plasma. This scattering rate is taken to scale with particle velocity v (relative to the local plasma frame) as $\tau \propto v$ so that the value of Υ is a constant in each simulation. In figure 4 are illustrated the energy spectra for a two of values of Υ, .25 and 5. The energy spectra for are clearly quite close to the predicted spectrum with the value at $\Upsilon = .5$ being slightly softer. Only as Υ approaches passes unity does the spectrum become significantly softer, verifying that Υ is the relevant parameter in determining when diffusion becomes invalid. Also, for $\Upsilon = .5$ the *rate* of acceleration is clearly larger than for $\Upsilon = .25$, verifying the prediction that an increase in $\lambda_\parallel$ leads to an enhanced acceleration rate. The correlation between the change of the particle and the potential energy in the $\mathbf{U} \times \mathbf{B}/c$ electric field was also investigated. In each case the energy gain of a particle was very closely correlated with the displacement in the y - direction, and equal to the electrostatic potential change from the point of injection. This verifies that even in the diffusion limit, the energy gain comes almost entirely from drift in the electric field. Jokipii and Terasawa (1982) have extended this much further and investigated shocks with magnetic field angles differing from 90° and confirmed that similar behavior occurs, although the breakdown of diffusion was slightly different. A complete discussion of this problem, including oblique shocks is presented in Jokipii and Terasawa (1992).

ACCELERATION OF ANOMALOUS COSMIC RAYS

In view of the somewhat counter-intuitive nature of the above conclusions for quasi-perpendicular shocks, where *less* scattering can result in *more- rapid* acceleration, an observational example is desirable. Recent studies of the anomalous

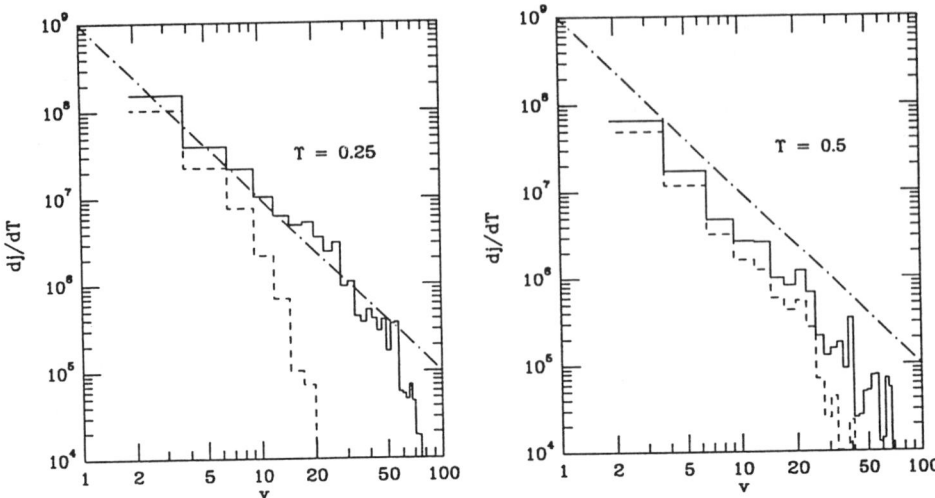

Figure 4. The velocity spectra of accelerated particles for two values of Υ. In each case the dashed line gives the spectrum at the same time, which is earlier than that given by the solid line. The dot-dash line in each case gives the valu predicted from diffusion theory.

component of cosmic rays are relevant in this regard (Jokipii, 1992). The anomalous component of cosmic rays was observed first in the early 1970's, and now is thought to be the result of diffusive shock acceleration at the termination shock of the solar wind (Jokipii, 1968, Pesses et al, 1981, Jokipii, 1986).

The anomalous component was discovered in the early 1970's (Garcia-Munoz, et al, 1973, Hovestadt, et al, 1973, and McDonald, et al, 1973). It consists of enhanced fluxes of helium, nitrogen, oxygen, neon and perhaps Hydrogen (Christian et al., 1988), in a region of the energy spectrum ranging from a kinetic energy of 20 MeV to perhaps 300 MeV. The observed radial intensity gradient of this component is positive out to the maximum distance reached by spacecraft, indicating that it is not of solar origin, but probably originates in the outer solar system.

Fisk, et al (1974) suggested that the anomalous component was the result of heliospheric acceleration (by some unspecified mechanism) of freshly- ionized interstellar neutral particles which stream into the solar system. Once ionized, the then singly-charged particles are then swept out of the inner solar system by the solar wind and subsequently accelerated. This accounts naturally for the presence of oxygen and near-absence of carbon. The charge state of anomalous oxygen at Earth, at an energy of $\approx$ 10 MeV/Nucleon, at 1 A.U., has been observationally determined (Adams, et al, 1991) to be 0.9(+0.3,- 0.2), consistent with the interstellar neutral origin hypothesis.

Recently, Adams and Leising (1991) showed that if 10 MeV/nuc singly-charged oxygen propagates a distance greater than 0.2 pc in the local interstellar medium, electrons will be stripped to make the charge state greater than that allowed by the observations. They also pointed out that this upper limit on the distance to the source, together with the fact that the intensity is still increasing with distance at the furthest spacecraft effectively limits the site of the acceleration to the region of interaction of the solar wind with the interstellar medium.

This stripping limit refers to the total path length, and it is more appropriate here to regard it as an upper limit on the *time* since the particles were accelerated. Adams and Leising based their estimate on the local interstellar neutral-atom density. Because of their charge neutrality, it is expected that these atoms will penetrate into the outer heliosphere, with little attenuation, so the upper limit on the lifetime applies also to anomalous cosmic-ray propagation in the outer heliosphere. Hence the time limit is also an upper limit on the time since the particles were accelerated to energies of the order of 10 MeV per nucleon, or the characteristic acceleration time at this energy. At the observed energy of 10 MeV/nucleon, this implies that acceleration must occur in less than about 4.6 years.

In addition, the adiabatic cooling in the expanding upstream solar wind, at a rate $2V_w/3r$, acts to decelerate the particles. This occurs in a time $\approx .5(r/100(A.U)$ year and would be a much more stringent constraint than the observational limit based on the charge state, if it acted continuously. However, the actual effect will be somewhat less, since the particles will spend some time downstream, where the adiabatic cooling rate is considerably less. Nonetheless, this additional effect will make more stringent the limit imposed by the observed charge state.

Jokipii (1992) showed that this constraint can be used to rule out most proposed acceleration mechanisms since they take much longer to accelerate the particles.

The maximum rate of 2nd-order Fermi acceleration can be estimated from the expression

$$\tau_{F2} > \frac{4\kappa_{Bohm}}{V_A^2}, \tag{8}$$

where κ_{Bohm} is defined above. Using observed solar-wind parameters leads to a time of more than 100 years to accelerate singly-charged oxygen particles to ≈ 200 MeV in the outer heliosphere. Transit-time acceleration gives a similar result.

This leaves only diffusive shock acceleration, either at propagating interplanetary shocks or at the termination shock of the solar wind. From equation (3) it is clear that the acceleration rate varies as the square of the shock velocity (relative to the upstream flow speed), which is relatively small for propagating shocks. In addition, they have only a small time to accelerate the particles.

The dominant shock in the outer heliosphere is the solar-wind termination shock, which is strong and has the highest upstream flow speed. However, even for this strong shock, the standard Bohm limit in equation (3) yields a time scale of more than 12 years for acceleration to ≈ 200 MeV. An This is probably ruled out by the observed charge state, particularly since $\lambda_\parallel$ is likely to be several times r_g.

On the other hand, if we realize that the termination shock is indeed very nearly *perpendicular* over most of its area, we find that the acceleration can proceed comfortably in a time period which is significantly less than one year. The maximum rate, obtained by choosing $\lambda_\|$ such that $\Upsilon \approx 1$, gives a minimum acceleration time of less than one month for singly- charged oxygen at 200 MeV. Using typical transport coefficients typically used in studies of solar modulation of galactic cosmic rays results in times of the order of one year. These times are small enough to comfortably satisfy the constraints.

One may draw from this a strong conclusion that the only viable mechanism for acceleration of the anomalous component, which is consistent with the observations, is diffusive acceleration at the perpendicular solar-wind termination shock, at a rate significantly higher than that obtained using standard expressions with the Bohm limit.

SUMMARY AND CONCLUSIONS

Diffusive shock acceleration at collisionless astrophysical shocks remains an attractive mechanism which may well be the source of most cosmic rays. The process requires the charged particle to cross the shock many times to gain significant energy, and hence the *rate* of acceleration must be considered. The generally-accepted expression for its maximum acceleration rate, based on the Bohm limit for the diffusion coefficient, was shown to be applicable only to parallel shocks. Diffusive acceleration at quasiperpendicular shocks can be much faster than the standard view would permit. A more-general upper limit on the acceleration rate was derived and a Monte-Carlo analysis verified these conclusions.

Recent observations of the charge state of anomalous oxygen place a stringent upper limit on the acceleration time which rules out most acceleration mechanisms. If these particles are to be accelerated at the solar-wind termination shock, the maximum acceleration rate based on the Bohm limit is insufficient by more than a factor of two. Only acceleration at the *quasiperpendicular* termination shock is sufficiently rapid to satisfy the constraint imposed by the charge state.

ACKNOWLEDGEMENTS

The contribution of Prof. Toshio Terasawa was essential to the Monte Carlo analysis reported herein. This research was supported, in part by the National Science Foundation under Grant ATM-8922151 and by the National Aeronautics and Space Administration under Grant NAGW 1931.

REFERENCES

Adams, J.H., M. Garcia-Munoz, N.L. Grigorov, B. Klecker, M.A. Kondratyeva, G.M. Mason, R.E. McGuire, R.A. Mewaldt, M.I. Panasyuk, Ch.A. Tratyakova, A.J. Tylka, and D.A. Zhuravlev, 1991, *Ap. J. Lett.*, 375, L45.

Adams, J.H., and M.D. Leising, 1991, Proceedings 22nd Int. Cosmic Ray Conf, paper SH 4.2.8, Dublin.

Axford, W. I., Leer, E., and Skadron, G., 1978, Proc.15th Int. Cosmic Ray Conf., Plovdiv. **11**, 132.
Bell, A. R., 1978a Mon. Not. Roy. Astr. Soc. **182**, 147.
———, 1978b Mon. Not. Roy. Astr. Soc. **182**, 443.
Blandford, R. D. and Ostriker, J. P., 1978, Ap. J. **221**, L29.
Chiueh, T., 1988, Ap.J.,, bf 333, 366.
Christian, S., A.C. Cummings, and E.C. Stone, 1988, Ap. J., 334, L77.
Fisk, L.A., B. Kozlovsky, and R. Ramaty, 1974, Ap. J. Lett., 190, L35.
Forman, M. A. and Morfill, G., 1979, in Proc. 16th Internat. Cosmic Ray Conf. (Kyoto), **5**, 328.
Garcia-Munoz, M., G.M. Mason, and J.A. Simpson, 1973, Ap. J. (Letters), 182, L81.
D. Hovestadt, O. Vollmer, G. Gloeckler, and C.Y. Fan, 1973, Phys. Rev. Letters, 31, 650.
Jokipii, J. R., 1968, Ap.J., bf 152, 799.
———, 1979, Proc. 16th Internat. Cosmic Ray Conf. (Kyoto), **14**, 175.
———, 1982, Ap. J., **255**, 716.
———, 1986 , J. Geophys. Res., 91, 2929.
———, 1987a, Ap. J.,313, 842.
———, 1987b, Proceedings of the 6th Intl. Solar Wind Conf., Volume II, pg 481, Estes Park, Colorado.
———, 1992, Ap. J. Lett. in press.
Jokipii, J.R. and G. Morfill, 1992, Astrophysical Aspects of the Most Energetic Cosmic Rays, pg 261, World Scientific, Singapore.
Jokipii, J.R. and Terasawa, T., 1992, Ap. J., in preparation.
Kota, J., 1979, Proc. 16th Internat. Cosmic Ray Conf. (Kyoto), **3**,13.
Krimsky, G. F., 1977, Doklady Akad. Nauk SSSR, **234**, 1306.
Lagage, P.O., and Cesarsky, C.J., 1983, Astron. Astrophys, **118**, 223.
McDonald, F.B., B.J. Teegarden, J.H. Trainor, and W.R. Webber, 1974, Ap. J. (Letters), 187, L105.
Parker, E. N., 1965, Planet. Space Sci., **13**, 9.
M.E. Pesses, J.R. Jokipii, and D. Eichler, 1981 Ap. J. (Letters), 246, L85.
Pesses, M. E., Decker, R. B., and Armstrong, 1982, T. A., Space Sci. Rev., **32**, 185.
Takahara, F. and T. Terasawa, Astrophysical Aspects of the Most Energetic Cosmic Rays, pg 291, World Scientific, Singapore, (1991).

NUMERICAL SIMULATIONS OF TIME-DEPENDENT COSMIC RAY MEDIATED SHOCKS

T. W. Jones
University of Minnesota, Minneapolis, MN 55455

Hyesung Kang
Princeton University Observatory, Princeton, NJ 08544

ABSTRACT

Considerable progress has been made in establishing the character of steady, CR mediated plane shocks. Much less is clear about the behavior of unsteady shocks that are influenced by nonlinear aspects of diffusive particle acceleration and that involve multidimensional flows. We discuss some of what has been learned for the latter classes of shocks, and outline some of the difficulties in making further progress.

INTRODUCTION

From its discovery, workers recognized that the diffusive shock acceleration process could be very efficient at transferring kinetic energy from gas motion into random energy of high energy particles (Cosmic Rays, or CR, for short). Indeed it was clear that in strong shocks the pressure from the CR could even dominate. With that came the realization that any serious treatment should consider nonlinear behavior. Clearly this included dynamical feedback or backreaction as well as feedback between the CR and the Alfvén waves that were responsible for scattering the particles. In addition, other important details such as those processes that lead to injection of fresh CR at low energy from the thermal plasma are likely to depend on nonlinear aspects of the problem. All of these features have been widely discussed in the literature and their roles established to varying degrees in the limit of one dimensional, steady state flows. Their character and impact are described from different viewpoints in a number of excellent reviews[1,2,3,4].

However, in many astrophysical situations it seems likely that assumptions of steady state conditions and/or one dimensional behavior are inappropriate. For example, the timescale to accelerate particles from low energy may be comparable to the dynamical timescales in the problem. Likewise, the length scales associated with diffusive behavior may not be negligibly small compared to the other physical scales in the problem. In addition, it has been recognized that when CR pressures are large, flows can become unstable due both to the diffusive nature of the CR and their "lightness"[5,6,7]. Although such issues have by no means been totally ignored, they are certainly not well resolved.

In order to try to improve understanding of some of these rather complex problems we have begun to explore nonlinear, time dependent flows incorporating diffusive shock acceleration by using techniques of numerical hydrodynamics coupled to the standard CR convection-diffusion (a.k.a. "transport") equation (as derived, for example, by Skilling[8]) or its energy moments. Our work follows the early examples of Dorfi[9], Drury and Falle[10], Falle and Giddings[11] and Bell[12] in this regard.

The basic system of equations in the work described below is

$$\frac{d\rho}{dt} = \rho \vec{\nabla} \cdot \vec{u}, \tag{1}$$

$$\frac{d\vec{u}}{dt} = -\frac{1}{\rho}\vec{\nabla}(P_g + P_c), \tag{2}$$

$$\frac{de}{dt} = -\frac{1}{\rho}\vec{\nabla} \cdot \{(P_g + P_c)\vec{u}\} + \frac{1}{\rho}P_c\vec{\nabla} \cdot \vec{u} - \frac{S}{\rho}. \tag{3}$$

and

$$\frac{df}{dt} = \frac{1}{3}(\vec{\nabla} \cdot \vec{u})p\frac{\partial f}{\partial p} + \vec{\nabla} \cdot (\kappa \vec{\nabla} f) + Q, \tag{4}$$

or the energy moment equation from eqn [4]

$$\frac{dE_c}{dt} = -\gamma_c E_c (\vec{\nabla} \cdot \vec{u}) + \vec{\nabla} \cdot (\langle \kappa \rangle \vec{\nabla} E_c) + S, \tag{5}$$

where ρ and P_g, are the gas density and pressure, $f(x,p,t)$ is the isotropic part (assumed dominant) of the CR momentum distribution function, E_c is the CR energy density, d/dt is the total time derivative, κ is the diffusion coefficient (most generally a tensor, but considered a scalar here), and $\vec{u}$ is the background fluid velocity. In eqn [5] $\langle \kappa \rangle$ represents an energy weighted average of $\kappa(p)$ over the distribution function. $e = (1/2)u^2 + (1/(\gamma_g - 1))P_g/\rho$ and γ_g and γ_c are the ratios of specific heats for the gas and CR, respectively. The use of eqn [5] instead of eqn [4] effectively treats the CR as a massless fluid, so that this treatment is often called a "two-fluid" model.

The source terms S and Q represent particle injection at the gas subshock. We generally use a simple model for injection of new CR suggested by Falle and Giddings[11]. It assumes that a fixed fraction, ϵ of the thermal gas particle flux is transformed into low energy CR, but only at the discontinuous gas subshock [13].

We solve the modified ideal fluid equations [1-3] using the 2nd order explicit Piecewise Parabolic Method[14] while the two versions of the transport equation are solved using the 2nd order implicit Crank-Nicholson scheme. Details are spelled out in our previous work[13,14]. Eqn [1-4] neglect such things as magnetic stresses, radiative cooling and work done on the gas by the Alfvén waves generated through CR particle streaming. These are certainly important[16,17], and we are currently incorporating the latter two into our methods.

We do not yet attempt to determine the CR diffusion coefficient fully self consistently. Rather we generally assume it takes a simple form

$$\kappa = k \frac{p^\alpha}{\rho^\beta}, \qquad (6)$$

where k is a constant, and α and β are parameters. In two-fluid calculations we either assume the mean diffusion coefficient is time independent for simplicity or we model its time dependence by integrating eqn [6], assuming a power law distribution function, $f(p) \sim p^{-4}$ extending to a maximum momentum given by eqn [7], below. Typical values of α are 0 and 1; the latter representing Bohm diffusion for relativistic particles. In most cases we have used $\beta = 0$. The density dependence with $\beta = 1$ in eqn [6] represents a crude attempt to include the effects of magnetic field compression. Similar methods have been employed by others carrying out these kinds of simulations.

NONSTEADY BEHAVIORS

The only characteristic time that can be identified from eqn [1-5] is the diffusion time, $t_d = \kappa/u^2$. If κ depends upon particle momentum, t_d is not unique. Since it more closely relates to dynamical questions, we prefer to associate the diffusion time with the mean diffusion coefficient, so that $t_d = \langle \kappa \rangle / u^2$. Not surprisingly, the timescale for individual test particles to be accelerated to high energies, and the timescale for the development of significant CR pressure through a compressed flow scale directly with t_d. The mean time for a test particle to be accelerated through a shock whose compression ratio is r can be written for a spatially constant diffusion coefficient as[18]

$$t_a = \frac{3(r+1)r}{r-1} \frac{\int_{p_o}^{p_1} \frac{\kappa(p)dp}{p}}{u_s^2} \sim 20 \frac{\kappa(p_1)}{\alpha u_s^2}, \qquad (7)$$

where u_s is the shock speed, and the last form assumes $r = 4$. For Bohm diffusion ($\kappa \sim pc^2/(eB)$) this would give $t_a(100 GeV) \sim 10^{10}$ sec in a 1 μG magnetic field. Although there may be some circumstances under which κ becomes smaller than the Bohm limit (see the paper in this volume by Jokipii, for example) or larger, it does provide a practical starting point for estimating relevant acceleration timescales. If $\kappa(p)$ increases with momentum ($\alpha > 0$) then t_a can be much longer for high energy particles. As we will see below that can make what one calls a steady state delicate to define.

Plane Parallel Flows

The simplest situations to try to understand are plane shocks. But even here there are a number of time dependent issues related to nonlinear behavior. We examined them in our own work in some detail using the transport eqn [4] [13,19], (see also work by others[11,12]) and the two-fluid model eqn [5] [15,20] (see also [9,10]).

The numerical simulations show particle acceleration times that are reasonably consistent with what is expected from test particle theory. We found[13] that a simply defined steepening in the distribution function evolving from particles injected in plane shocks at low momenta increased in a manner consistent with eqn [7] even as the shock structures began to be modified by finite P_c. As one would expect, the CR pressure around a shock increases while the particle momenta increase, so that the timescale for buildup in P_c is generally proportional to t_a or t_d. The relative rate of pressure increase depends, of course, upon the particle distribution function. The pressure tends to multiply more rapidly for a given κ when the particles are primarily nonrelativistic (so that $\gamma_c \approx 5/3$), than when they are largely relativistic (so that $\gamma_c \approx 4/3$). That difference is seen both in transport calculations and in two-fluid calculations (see, in addition to our own work the paper in this volume by Drury). It is a consequence of the softer equation of state for relativistic fluids. Since shocks that begin by processing a preexisting flux of relatively low energy CR or an equivalent momentum flux of freshly injected particles will multiply the associated CR pressure fairly rapidly when those particles are nonrelativistic, significant mediation of the shock can develop even before highly relativistic CR momenta are achieved in some cases.

The time required for plane, piston driven shocks to evolve a small preexisting P_c (or injected low momentum particles) to a steady P_c around the shocks depends upon the strength of the shock and the relative value of the preexisting P_c or the injection parameter, ϵ. However, based on two-fluid simulations of moderately strong shocks we have found it a convenient rule of thumb that one requires $\gtrsim 100 t_d$ for this development when $\gamma_c = 5/3$, but much longer times when $\gamma_c = 4/3$. Since the acceleration process should ultimately produce relativistic particles, γ_c should begin to decrease towards $4/3$ at some point (confirmed by transport calculations). As we will point out for one particular application below, the timing of that development compared to dynamical events can have important consequences for the ultimate conversion of kinetic energy to CR energy.

Mediation tends to weaken the gas subshock, and as was recognized early on can even eliminate it for strong shocks given sufficient time[21]. An interesting feature forms in shocks that are in the process of being strongly modified by CR. Compression of gas in the CR precursor to the shock can lead to total compressions through the shock that are considerably in excess of the steady state values. For strong shocks a density "hump" several times more dense than surroundings develops and then separates downstream from the shock just as the subshock becomes smoothed out. This was first noticed by Dorfi[5] and later studied in more detail by us[23]. It is also seen in nonplanar shock evolution, such as the supernova remnant simulations described below.

The case $\gamma_c = 4/3$ is special with regard to the development of steady state conditions, for, as it has often been pointed out, P_c would appear to diverge eventually as particles evolve to extremely high energies. This means, of course, that one should be cautious in applying steady state calculations which assume $\gamma_c = 4/3$[22,23]. A related issue involves two-fluid $\gamma_c = 4/3$ solutions generating finite downstream P_c from zero upstream pressures. To deal with such questions some have emphasized the importance of particle escape at high energies being a modifying influence[4]. That may well be important in some cases. But we would also point out that because of the long timescales needed to approach steady states in those same cases, it may often be inappropriate to assume steady state solutions. One must then ask the age and history of a particular shock. On the other hand our transport calculations suggest that shock structures dynamically resembling those predicted by steady state calculations can develop eventually even though the particle distribution functions continue to evolve. A particularly interesting example involves what happens to flows that evolve with a CR population derived purely from particles injected at low energy in the discontinuous gas subshock. For at least moderately strong shocks, the two-fluid steady state calculations predict such high efficiencies for even modest injection rates, ϵ, (say $\epsilon \gtrsim 0.001$) that the gas subshock should be smeared out. Since that outcome violates asumptions for the model, presumably such steady shocks cannot exist. But our transport calculations beginning with such injection conditions and no preexisting CR show that even after P_c builds up to the point that the subshock disappears and the injection process is suppressed, the continued acceleration of CR *already present* can maintain a strongly modified shock with conditions that qualitatively resemble what one expects from steady state calculations[19,20]. This comes about through repeated CR diffusion *upstream* through the shock followed by downstream advection. The adiabatic index of the CR in this model clearly begins near 5/3, but evolves asymptotically towards 4/3.

Another nonsteady aspect of nonlinear CR shocks is that once shocks are strongly CR modified they become subject to instabilities directly related to the properties of the CR "fluid". Some of these explicitly depend upon interactions with the magnetic field[7], so our current methods do not allow us to examine their nonlinear behaviors. However, the CR acoustic instability is accessible to us and perhaps serves as a representative of what may be expected. This instability results from the diffusive nature of the CR pressure and was first explained by Drury 1984. Because CR can diffuse rapidly for small scale density perturbations P_c will not respond adiabatically. Consequently density fluctuations (sound waves) propagating from upstream towards a strong CR pressure gradient are subjected to an approximately constant volume force, leading to nonadiabatic growth of sound waves in precursors to CR modified shocks. Drury and Falle[10] suggested and we later confirmed[19] that this can result in formation of "minishocks" within the precursor. Our transport calculations show that the formation of these minishocks has only minor influence on the acceleration of CR within the shocks, although it can result in significant gas entropy production and so alter expectations about downstream temperatures. The minishocks might provide additional sites for the injection of fresh CR, as well. Growth of the instability is limited by the fact that it only works within the precursor, however.

More recently we have explored this instability further by simulating it in two dimensions (see the paper by Kang, Ryu and Jones in this volume). The formation of the minishocks has one particularly significant consequence for multidimensional flows. The downstream sides of the minishocks can be unstable to the Rayleigh-Taylor fluid instability. Thus we find that small scale density and velocity structure can become rather chaotic in the downstream region as a result. This might be important for second order Fermi acceleration processes in that space, but does not seem likely to lead to disruption of the shock front itself.

Spherical Flows

Several aspects of spherical flows distinguish them from plane ones. For instance the existence of an interior region means that one cannot generally assume particles to be simply advected away from the shock. Expansion in the post shock flow leads to adiabatic cooling. Curvature of the shock front could have direct consequences, especially at large momenta where the mean free paths could become large compared to the radius. In supernova remnant blast waves (SNR) an especially important feature of the spherical nature of the problem is the fact that the dynamical time scale of the remnant is comparable to its age. We have studied the evolution of SNR during the adiabatic, Sedov phase assuming spherical symmetry and using transport methods [13] or two-fluid models[15,20]. Other, comparable two-fluid calculations have been carried out by Dorfi[17,24], and simplified two-fluid models by Drury, Markiewicz, and Völk [21,25] (see the paper in this volume by Drury).

The two most generally discussed aspects of SNR with regard to CR are the expected CR momentum distribution and the overall efficiency; *i.e.*, what fraction of the blast energy is converted into CR energy. With regard to the momentum distribution our own work suggests that it is difficult to make definitive predictions, especially from test particle theory, be it based on plane or spherical shocks. Even though the predictions from eqn [7] for the location of a steepening in the CR spectrum as a consequence of the finite lifetime of the SNR are crudely followed (but not as well as for plane shocks), we have not been able to confirm any simple predictions about the shape of the distribution function below this momentum. Unless the diffusion coefficient is so small that acceleration is quite rapid compared to the dynamical age of the SNR we do not find distributions that really resemble power laws, for example. Due to computational resource limits we have been able to carry out only one transport simulation that was based on a diffusion coefficient small enough that a power law momentum distribution began to form out of injected low energy particles. After about 4 dynamical times from the start of the Sedov phase this simulation showed $f(p) \propto p^{-4.3}$ for $p \lesssim mc$. Because this simulation showed signs of having the shock be near equilibrium conditions, we suggested that this result should not depend upon our detailed choices of diffusion coefficient and speculated that by the end of the Sedov phase (about 60 of the same time units) such a distribution function might reasonably extend up to $\sim 10^{14}$eV if one adopted a Bohm diffusion model. However, it is not quite clear why this particular slope developed. It is not the value that one would simply predict from test particle theory for the shocks that evolved it. On the other hand, P_c was sufficient to modify the structure of the shock, so that the acceleration was certainly in a nonlinear stage.

The efficiency question seems similarly complicated. All of the recent nonlinear SNR computations mentioned above agree that it may be possible to expect net efficiencies in the acceleration process of 10% or more as is seemingly required to replenish galactic CR. But, it has also become apparent that the efficiency of a given model is sensitive to some rather uncertain details. Most important of these may be whether significant CR pressure develops during the early stages of the Sedov phase. In the calculations we report most recently[20] and those described at this meeting by Drury such dependence manifested itself through high sensitivity to the evolution of γ_c during this time. That is understood through the comment made earlier about the more rapid way in which P_c develops for γ_c closer to 5/3 than 4/3. It is apparent that if a significant seed pressure develops at these early times it can continue to absorb considerable kinetic energy from the blast wave and more than compensate for adiabatic losses in the interior of the SNR. Other details that, likewise, may be important at later times are related to the timing of the onset of radiative cooling. If that begins too soon, then it will limit the net amount of energy available to CR and presumably reduce the maximum energies obtainable. Generally one expects this to begin when the post shock gas temperature drops below $\sim 10^6$ K. Large CR pressures in the shock structure will tend to reduce the gas temperature (basically on account of total energy conservation), allowing cooling to set in earlier. On the other hand the energy in the Alfvén waves generated by streaming CR is presumably dissipated by the gas and may provide heating in the precursor that counteracts this. Dorfi[17] has now carried out some initial calculations including these two effects together. Similarly, the acoustic instability described earlier could also heat the gas and influence this behavior.

More Complex Flows

Most real flows, probably including the ones we have already modeled, are not truly one dimensional. We may expect this to introduce additional complications into the picture. Some hints about that may come from the acoustic instability described above. As a further initial attempt to learn about modeling diffusive acceleration in flows that are inherently multidimensional we have carried out simulations of a plane shock colliding with a small dense gas cloud. These were computed in the two-fluid model in two dimensions. On the face of it, one might expect the impact of the shock on the cloud to augment the acceleration of CR by the shock. For example, the shock that penetrates into the cloud as a result of the impact is stronger than the incident shock and can generate several times the total pressure of the incident shock. In addition the impact produces other shocks, such as a bow shock that could be expected to augment CR acceleration. But these flows are complex and there are several scaling parameters to be considered. The diffusion time t_d needs to be compared to at least three hydrodynamical time scales. One is the so-called "cloud crushing time" during which the shock penetrates the cloud. A second is the time for the shock to wrap itself externally around the cloud. The third is the time for the Kelvin-Helmholtz instabilities to disrupt the cloud. These three times are all related, of course, but depend in different ways on the density contrast between the cloud and the ambient medium. Even more than for the SNR calculations these simulations show quite a bit of sensitivity to the detailed assumptions. That is especially true when one asks specific questions about when and where CR particles are being concentrated. We find, however, that the clouds generate at most a modest acceleration augmentation. On time scales longer than the cloud crushing time the net effect in most cases we examined seems to be a small increase in the total CR energy, primarily through acceleration in the tail shock of the cloud[26].

CONCLUSIONS

From time dependent simulations that we and others have carried out it is possible to draw the following general conclusions:

- In many situations it is probably not appropriate to assume steady states for astrophysical CR shocks. This results both from the need to be concerned about the finite rate at which CR pressure builds and modifies a shock and from the existence of instabilities.

- Even so it seems likely that CR modified shocks can develop to a state with substantial CR pressure.

- The overall acceleration efficiency and the CR particle momentum distribution produced in time dependent flows typically depend on several details such as the ratio of various time scales and the initial proportion of relativistic to nonrelativistic particles in the CR population. Before models can be expected to provide real quantitative predictions about CR acceleration in astrophysical settings we will need to improve our understanding of the related microphysics.

Acknowledgments

This work was supported in part by the NSF through grants AST-8720285 and AST-9100486, by NASA through grant NAGW-2548 and by the University of Minnesota Supercomputer Institute. We are grateful to Paul Woodward for making his PPM hydrodynamics code available to us.

REFERENCES

1. Drury, L. O'C. *Rep. Prog. Phys.*, **46**, 973 (1983).

2. Blandford, R. D., and Eichler, D. *Phys. Rept.*, **154**, 1 (1987).

3. Berezhko, E. G. and Krymskii, G. F. *Soviet Phys. Usp.*, **31**, 27 (1988).

4. Jones, F. C, and Ellison, D. C. *Space Science Reviews*, **58**, 259 (1991).

5. Drury, L. O'C. *Adv. Space Res.*, **4**, (2-3), 185 (1984).

6. Chalov, S. V. *Sov. Astron. Lett.*, **14**, 114 (1988).

7. Zank, G. P., Axford, W. I., and McKenzie, J. K. *A&A*, **233**, 275 (1990).

8. Skilling, J. *MNRAS*, **172**, 557 (1975).

9. Dorfi, E. A. *Adv. Space Res.*, **4**, (2-3), 205 (1984).

10. Drury, L. O'C., and Falle S. A. E. G. *MNRAS*, **223**, 353 (1986).

11. Falle, S. A. E. G., and Giddings, J. R. *MNRAS*, **225**, 399 (1987).

12. Bell, A. R. *MNRAS*, **225**, 615 (1987).

13. Kang, H. and Jones, T. W. *MNRAS*, **249**, 439 (1991).

14. Colella, P. and Woodward, P. R. *J. Comp. Phys.*, **54**, 174 (1984).

15. Jones, T. W. and Kang, H. *ApJ*, **263**, 499 (1990).

16. McKenzie, J. K. and Völk, H. J. *A&A*, **116**, 191 (1982).

17. Dorfi, E. A. *A&A*, , in press (1991).

18. Lagage, P. O. and Cesarsky, C. J. *A&A*, **125**, 249 (1983).

19. Kang. H., Jones, T. W. and Ryu, D. *ApJ*, **385**, 193 (1992).

20. Jones, T. W. and Kang, H. ApJ, submitted (1992).

21. Drury, L. O'C., Markiewicz, W. J., & Völk, H. J. *A&A*, **225**, 179 (1989).

22. Kang, H. and Jones, T. W. *ApJ*, **253**, 149 (1990).

23. Drury, L. O'C., and Völk, H. J. *ApJ*, **248**, 344 (1981).

24. Dorfi, E. A. *A&A*, **234**, 419 (1990).

25. Markiewicz, W. J., Drury, L. O'C., Völk, H. J. *A&A*, **236**, 487 (1990).

26. Jones, T. W. and Kang, H. , in preparation (1992).

PARTICLE INJECTION AND THE STRUCTURE OF ENERGETIC PARTICLE MODIFIED SHOCKS

G.P. Zank[†], G.M. Webb[‡] and D.J. Donohue[†]
[†]Bartol Research Institute, The University of Delaware, Newark, DE 19716
[‡]Department of Planetary Sciences, University of Arizona, Tucson, AZ 85721

INTRODUCTION

Particles can be energized at shocks, taking up momentum from the large scale flow field and reacting non-linearly back onto the background fluid. An increasingly important issue in the general theory of particle acceleration at shocks concerns the "injection" of particles from the thermal background gas into an energetic particle gas. To properly explore this problem requires a detailed understanding of the microphysics of quasi-parallel shocks, something that currently eludes us. However, some dynamical models have been advanced in which some form of "heating" mechanism is used to simulate the injection process[1-4]. In this paper, we consider the problem of injection anew and extend some ideas of Eichler, Ellison and Drury[1,2,5] in developing a new "self-consistent" model of cosmic ray hydrodynamics within the context of a two-fluid description. Our motivation is two-fold: (i) can the nature and model of particle injection directly affect the structure and dynamics of the shock, and hence the efficiency of particle acceleration? and (ii) can the shock dynamically regulate particle injection? As noted we do not (and cannot) address these issues at a microphysical level but introduce instead a simple but general model which we hope captures some of the essential physics behind this complex problem.

THE SHOCK STRUCTURE EQUATION

In the following, we utilize the approach of Axford *et al.*[6] (ALM). As usual, in the two-fluid formulation, energetic particles are described in terms of a pressure p_c and energy E_c, both related via a specific heat ratio γ_c. The energetic particles are assumed to have negligible mass density and to satisfy the transport equation

$$\frac{\partial f}{\partial t} + u\frac{\partial f}{\partial x} - \frac{\partial}{\partial x}\left(\kappa_\parallel \frac{\partial f}{\partial x}\right) = \frac{p}{3}\frac{\partial f}{\partial p}\frac{\partial u}{\partial x}, \quad (1)$$

where f denotes the isotropic part of the energetic particle distribution function, u the bulk fluid velocity, p the particle momentum and $\kappa_\parallel$ the diffusion coefficient of the scattered particles. In (1), we have neglected momentum diffusion and a "source" term—the source term being typically included in studies which involve particle injection. Our reason for excluding a source term in (1) is that we assume in principle that no particular distinction exists between the thermal and non-thermal particle populations, and that all particles are described by a single distribution function, however complicated. In spirit, therefore, our calculations are similar to the Monte Carlo calculations of Ellison[7]. To render our calculations more tractable, we do not consider the full distribution function but instead separate the particles into a high energy bin (having momenta $p > p_0$ for some p_0) and a thermal bin ($p < p_0$). Particles can migrate from one population to the other and, for simplicity, we will suppose that the high energy particles propagate according to (1). Clearly, in this model no distinct source term is allowed in (1) unless particles are created spontaneously (such as at comets). Instead, a "source term" will enter through an integrated form of (1) and this term will represent a measure of the particle flux across the momentum boundary p_0. On taking moments of (1) between $[p_0, \infty]$, we obtain a modified form of the diffusive cosmic ray energy equation

$$\frac{\partial E_c}{\partial t} + \frac{\partial}{\partial x}(uE_c) + p_c\frac{\partial u}{\partial x} - \frac{\partial}{\partial x}\left(\bar{\kappa}\frac{\partial E_c}{\partial x}\right) = -\alpha\frac{\partial u}{\partial x}, \qquad (2)$$

where $\alpha = 4\pi/3 \cdot mc^2 p_0^3 f(p_0)\sqrt{1 + p_0^2/m^2c^2}$ is the injected "energetic particle pressure" crossing the lower momentum boundary and $\bar{\kappa}$ is the momentum averaged spatial diffusion coefficient $\kappa_\parallel$. Equation (2) differs from the usual two-fluid formulation in the RHS. The term $\alpha\partial_x u$ has a nice interpretation in terms of particles crossing the lower momentum boundary—as the fluid decelerates, the fluid pressure increases, so leading to an increased particle flux across p_0. Thus particle injection must occur whenever the flow is decelerating and not just at a subshock. Obviously, the RHS becomes increasingly important as the flow gradient steepens with the most important effects occurring at a subshock. In this sense, particle injection at shocks modified by energetic particles is regulated dynamically.

For smooth flows, the conservation of total energy can be reduced to an equation for the pressure of the thermal gas p_g,

$$\frac{\partial p_g}{\partial t} + u\frac{\partial p_g}{\partial x} + \gamma_g p_g\frac{\partial u}{\partial x} = (\gamma_g - 1)\alpha\frac{\partial u}{\partial x}, \qquad (3)$$

γ_g the thermal gas adiabatic index. The RHS of (3) illustrates the exchange in particles (via a "pressure" exchange) between the thermal and energetic particle populations. On integrating (3) in conjunction with the continuity equation, we obtain an equation of state for the thermal gas

$$p_g = A\rho^{\gamma_g} + \alpha\frac{\gamma_g - 1}{\gamma_g}, \qquad (4)$$

where ρ denotes the fluid density and A is a constant of integration. Equation (4) is rather different from the usual adiabatic equation of state and reveals that injection into the background flow modifies the sound speed of the system since now

$$C_s^2 = \frac{\gamma_g p_g}{\rho}\left(1 - \frac{(\gamma_g - 1)}{\gamma_g}\cdot\frac{\alpha}{p_g}\right),$$

a point noted first by Drury[5].

Like ALM, the steady-state two-fluid equations with injection can be combined as a first order shock structure equation which describes the smooth flow. It is particularly notable that the injection flow model experiences a self-reversal at a location different from that of the non-injection two-fluid model, now at a point where the local fluid speed matches the modified sound speed C_s. This result can modify both the location and strength of the subshock. We also find that injection tends to strengthen and broaden the smoothed shock slightly.

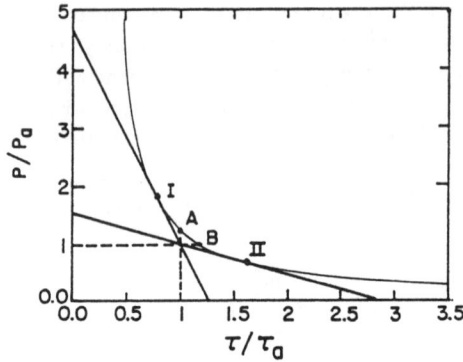

Figure 1: Schematic of the particle injection Hugoniot.

THE SUBSHOCK

In those parameter regimes for which a smooth monotonic transition is impossible, a subshock is necessary to take the flow to its final downstream state. Following ALM, we take moments of the energetic particle streaming flux (between the same limits as used in deriving [2]) to obtain the subshock conditions

$$[p_c] = 0;$$
$$\left[\frac{\gamma_c}{\gamma_c - 1} u p_c - \frac{\bar{\kappa}}{\gamma_c - 1}\frac{dp_c}{dx}\right] = -[\alpha u]. \qquad (5)$$

Since we expect considerable viscous heating of the thermal fluid within the subshock, we can anticipate an enhanced energetic particle flux across the lower momentum boundary. Thus, if the subscripts a/b denote ahead/behind the (sub)shock, we must have $\alpha_b > \alpha_a$ and so $[\alpha u] \equiv \alpha_a u_a - \alpha_b u_b$. The Rankine-Hugoniot (R-H) conditions appropriate to the gas subshock then take the form

$$[\rho u] = 0; \qquad (6)$$
$$[\rho u^2 + p_g] = 0; \qquad (7)$$
$$\left[\rho u \left(\frac{1}{2}u^2 + \frac{\gamma_g}{\gamma_g - 1}\frac{p_g}{\rho}\right)\right] = [\alpha u], \qquad (8)$$

a form quite different from those advanced in previous investigations[3,4], and again the energy exchange between the two fluids is evident throught the RHS's of (5) and (8). The analysis of (6)-(8) is closely related to the physics of combustion shocks, and we can summarize our results in the form of the "particle injection" Hugoniot (Fig. 1). The region between A and B in Fig. 1 does not correspond to physically realizable solutions of the R-H conditions since here we have $m^2 \equiv (\rho u)^2 < 0$. The lines through the upstream state (τ_a, p_{ga}) $(\tau \equiv 1/\rho)$ tangent to the Hugoniot at points I and II are the "Rayleigh" lines, and the points I and II correspond to the "Chapman-Jouguet" points of combustion theory. The *strong compression* branch above I corresponds to subsonic flows downstream of the supersonic incident state. A *weak compression* regime, which takes a supersonic state to a decelerated but still supersonic state downstream of the

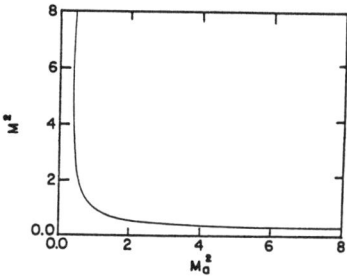
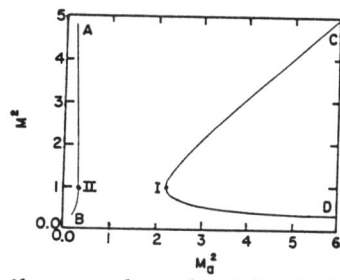

Figure 2: On the left, the classical gas dynamical case and on the right the injection case.

injection front, is located between I and A. The sub- and supersonic expansion regimes are located below B. It can be shown quite rigourously[8] that all compressive solutions to the R-H relations (6)-(8) are physically admissible. Furthermore, the structure of the weak compressive transition is such that the incident flow experiences a deceleration to the local sound speed ($u = C_s$) followed by an isentropic acceleration to the final supersonic ($u > C_s$ locally) downstream state. Such compound shocks have no classical gas dynamical analogue and are a consequence of particle injection at the subshock providing an "exothermic" reaction.

Further insight into the properties of particle injection fronts is to be gained by considering the solutions of (6)-(8) as a function of the upstream Mach number M_a ($= u_a/C_{sa}$) where it should be noted that, like ALM, we use M_1 to denote far upstream of the shock (at $-\infty$) and M_a to denote immediately ahead of the subshock. The absence of a subscript, e.g., M, indicates a downstream state. In Fig. 2, we plot $M^2 = u^2/C_s^2$ as a function of M_a^2. In the left panel, we plot the classical gas dynamical hyperbola and, on the right, the particle injection relation. As in Fig. 1, I and II locate the "Chapman-Jouguet" points, ID identifies the "strong" gas shocks and IC the compound shock solutions. Clearly, flows possessing injection fronts can be very different from non-reacting flows—compound shocks are possible and there even exists a parameter regime for which no stable simple or compound shocks can exist.

Having clarified the general theory of gas dynamic shocks with particle injection, it is now necessary to "fit" the subshock in order to complete the stationary shock structure problem. By balancing the momentum and energy fluxes across the (sub)shock and matching to the shock structure equation, one obtains an algebraic relation governing the location and strength of the injection front required to achieve a final downstream state. For the purposes of illustration, we compare directly the non-injection results of ALM (Fig. 3) with an injection plus viscous heating set of results (Fig. 4). Careful inspection of Fig.'s 3 and 4 reveals the following differences between reacting and non-reacting fronts: (i) for small M_1^2 values, the inverse compression ratio $y_a \equiv u_a/u_1$ is larger for the injection (Fig. 4a) than the non-injection model (3a); (ii) the injection subshock (4d) is somewhat stronger than the non-injection subshock (3b); (iii) it was found that for solutions to exist in the weak shock regime, it is necessary to bound α_b from above (i.e., α_b is not completely arbitrary and depends on the strength of the subshock), and (iv) for injection fronts, there exists an upstream parameter regime for which $M^2 > 1$! Thus, one can indeed have a cosmic ray modified shock with a post-

162 Particle Injection

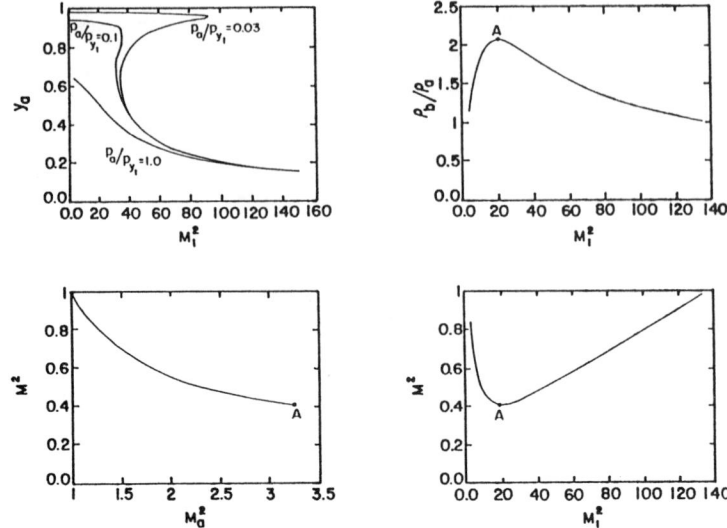

Figure 3: The non-injection case.

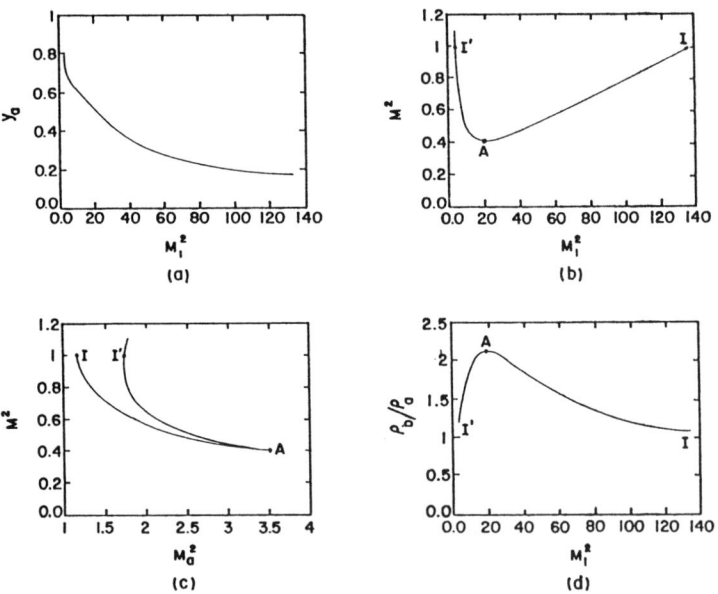

Figure 4: The injection case.

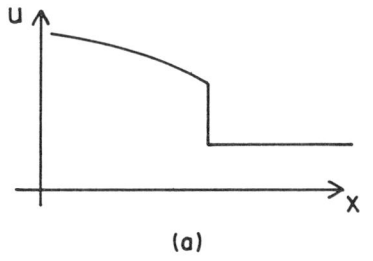 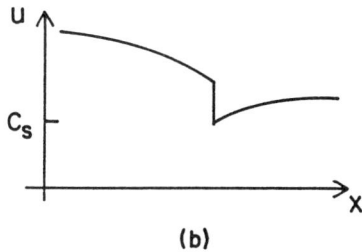

Figure 5: Energetic particle modified shock structure with and without a post-cursor.

cursor (in Kirk's[9] terminology) and so it is possible to have two different structures for stationary energetic particle modified shocks, both of which are illustrated in Fig. 5.

Conclusions

In summary, we have developed a macroscopic, "self-consistent", non-linear, two-fluid model for energetic particle modified shocks which incorporates particle exchange between a thermal gas and an energetic particle population via a "thermal leakage" mechanism. A precise form for the source term has been identified for both the smoothed flow ($\alpha \partial_x u$) and for the subshock jump ($[\alpha u]$). We have demonstrated that the theory of "injection fronts" can be quite different from shocks in non-reacting gas dynamics, and this has important implications for the Riemann problem and for numerical schemes using Riemann solvers. It was found that particle injection is controlled dynamically by the shock, both through the flow gradient and in determining an upper limit on the injection efficiency at the subshock. Finally, we find a richer, more complex class of shock transitions than in the non-injection models, all of which demonstrates that indeed the structure and dynamics (and hence particle acceleration efficiency) can all be affected by the nature of the particle injection process. Preliminary results for non-stationary injection fronts are presented in Donohue & Zank[10].

Acknowledgements: This work has been supported at the BRI by the NASA SPTP program under grant NAGW-2076 and GMW by NASA grant NSG-7101 and NSF grant 8317701.

References

1. D. Eichler, *Ap. J.*, **229**, 419, 1979; *Ap. J.*, **277**, 429, 1984.
2. D.C. Ellison & D. Eichler, *Ap. J.*, **286**, 691, 1984.
3. S.A.E.G. Falle & J..R. Giddings, *M.N.R.A.S.*, **225**, 399, 1987.
4. H. Kang & T.W. Jones, *Ap. J.*, **353**, 149, 1990.
5. L.O'C. Drury, *Rep. Prog. Phys.*, **46**, 973, 1983.
6. W.I. Axford, E. Leer & J.F. McKenzie, *Astron. Astrophys.*, **111**, 317, 1982.
7. see e.g., F.C. Jones & D.C. Ellison, *Space Sci. Rev.*, **58**, 259, 1991 for a summary.
8. G.P. Zank, G.M. Webb & D.J. Donohue, *Ap. J.*, submitted 1992.
9. J.G. Kirk, *Astron. Astrophys.*, **239**, 404, 1990.
10. D.J. Donohue & G.P. Zank, these proceedings.

Time-Dependent Cosmic Ray Shocks with Injection: A Progress Report

D. J. Donohue and G. P. Zank
Bartol Research Institute, The University of Delaware
Newark, DE 19716, U.S.A.

Abstract

Preliminary results are shown of time-dependent solutions to the two-fluid equations with injection as derived by Zank et.al.[1]. The emphasis is on the structure of cosmic-ray-modified shocks.

1 Introduction

We present prelimary results of a two-fluid hydrodynamical simulation of cosmic-ray-modified shock structure by extending the steady-state calculations of Zank et.al.[1] to the time-dependent case. The two-fluid formalism consists of a thermal background gas and an energetic 'cosmic ray' fluid which interact non-linearly through momentum exchange. The system is self-consistent in including the backreaction of the cosmic ray fluid on the thermal gas.

These simulations model diffusive shock acceleration (1st-order Fermi) of the energetic particle component at planar shocks. We present a new model for the injection of cosmic ray particles. Particles with momentum $p > p_0$ comprise the cosmic ray fluid, whose transport is described by a cosmic ray convection-diffusion equation. The background thermal gas consists of the cooler particles in the core ($p < p_0$) of the distribution. Injection consists of momentum transport across p_0 in a region of decelerating flow (near the subshock) where the thermal fluid is compressed and heated.

2 Two-fluid Equations

The dynamics of the thermal fluid are described by the usual mass, momentum and energy conservation equations with additional terms representing the interaction with the cosmic ray fluid. The equations are

$$\frac{\partial \rho}{\partial t} + \frac{\partial}{\partial x}(\rho u) = 0 ; \tag{1}$$

$$\frac{\partial}{\partial t}(\rho u) + \frac{\partial}{\partial x}(\rho u^2 + p_g + p_c) = 0 ; \tag{2}$$

$$\frac{\partial}{\partial t}\left(\frac{1}{2}\rho u^2 + \frac{p_g}{\gamma_g - 1} + \frac{p_c}{\gamma_c - 1}\right) +$$

$$\frac{\partial}{\partial x}\left(u\left[\frac{\gamma_g}{\gamma_g-1}p_g+\frac{1}{2}\rho u^2\right]+\frac{\gamma_c}{\gamma_c-1}up_c-\frac{\kappa}{\gamma_c-1}\frac{\partial p_c}{\partial x}\right)=0. \tag{3}$$

Here ρ, u, p_g are the thermal gas density, velocity, and pressure and p_c is the cosmic ray particle pressure. The energy density of each fluid is related to the pressure via an adiabatic index (γ_c,γ_g) such that $E_i = p_i/(\gamma_i-1)$.

The dynamics of the cosmic ray fluid are described seperately by a cosmic ray transport equation

$$\frac{\partial p_c}{\partial t}+u\frac{\partial p_c}{\partial x}+\gamma_c p_c\frac{\partial u}{\partial x}-\frac{\partial}{\partial x}\left(\kappa\frac{\partial p_c}{\partial x}\right)=-(\gamma_c-1)\alpha\frac{\partial u}{\partial x}, \tag{4}$$

where κ is a momentum averaged diffusion coefficient for the cosmic ray fluid and includes all details of particle scattering. Two source-like terms appear in the transport equation. The particle injection term described in § 1 appears on the right-hand side. The parameter α determines the strength of the injection. The third term on the left-hand side represents acceleration of the existing cosmic ray particles through conversion of the thermal gas flow energy.

3 Numerical Scheme

The modified gas dynamic equations (1-3), written above in the flux-conserved form

$$\frac{\partial \mathbf{q}}{\partial t}+\frac{\partial \mathbf{F}}{\partial x}=0, \tag{5}$$

are finite-differenced and advanced in time with a two-step Lax-Wendroff[2] scheme. The method first takes the intermediate step

$$\mathbf{q}^*_{j+1/2}=\frac{1}{2}\left(\mathbf{q}^n_j+\mathbf{q}^n_{j+1}\right)-\frac{1}{2}\frac{\Delta t}{\Delta x}\left(\mathbf{F}^n_{j+1}-\mathbf{F}^n_j\right), \tag{6}$$

where $\mathbf{q}^*_{j+1/2}$ is the intermediate solution at the half time step $t_{n+1/2}$ and the half mesh point $x_{j+1/2}$. After updating the fluxes $\mathbf{F}^*_{j+1/2}$, the system is advanced to timestep t_{n+1} by

$$\mathbf{q}^{n+1}_j=\mathbf{q}^n_j-\frac{\Delta t}{\Delta x}\left(\mathbf{F}^*_{j+1/2}-\mathbf{F}^*_{j-1/2}\right). \tag{7}$$

A quadratic artificial viscosity is added to the system after the solution at time step t_{n+1} is obtained.

The cosmic ray transport equation (4) is advanced seperately in time using an implicit Crank-Nicholson[2] finite-difference scheme with mass operator. The thermal gas velocity u is taken from the Lax-Wendroff solution. The source terms described above are calculated at each time step and added into the cosmic ray pressure.

4 Initial Conditions

In the simulation results, all dependent variables are non-dimensionalized with respect to the undisturbed upstream state

$$\rho = \frac{\rho}{\rho_u}; \quad u = \frac{u}{c_{su}}; \quad p_g = \frac{p_g}{p_{g_u}}; \quad p_c = \frac{p_c}{p_{c_u}}; \quad \alpha = \frac{\alpha}{p_{g_u}}; \quad \kappa = \frac{\kappa}{c_{su}L}, \quad (8)$$

where ρ_u, c_{su} and p_{g_u} are the gas density, sound speed and pressure far upstream of the shock and L is a normalization scale for the spatial dimension. The non-dimensionalized fluid variables are fixed to the following boundary values far upstream

$$\rho = 1, \quad p_g = 1, \quad u = 0, \quad p_c = \begin{cases} 0, & \text{with injection } (\alpha \neq 0); \\ 1, & \text{no injection } (\alpha = 0). \end{cases} \quad (9)$$

The gas subshock is located initially at $x = 1.5$. The strength of the subshock is specified by the gas pressure ratio across the shock. The Mach number, compression ratio and downstream fluid velocity are then calculated according to the ordinary gas-dynamic Rankine-Hugoniot relations. As the simulation advances in time, the downstream state is held fixed while the shock propagates into the upstream gas ($u_u=0$). Thus, the system solves the Riemann problem, connecting the fixed upstream and downstream states. This model has application to the expansion of a wind into the interstellar medium or even SNR blast waves.

5 Results

In this section we show two different simulation results to demonstrate the types of shock structure we have studied to date. Figure 1 is a simulation of a strong shock with gas Mach number 8.5 and adiabatic indexes $\gamma_g=5/3$, $\gamma_c=5/3$. Particle injection is turned off, so that the shock accelerates only a small, pre-existing upstream cosmic ray population ($p_{c_u} = p_{g_u}$). The initial conditions resemble those used by Kang and Jones[3] (their figure 1a) who solve the same equations (without injection) using the piecewise parabolic method (PPM). The results compare favorably, but it should be noted that Kang and Jones consider piston-driven shocks while our simulations insert the subshock directly by fixing the upstream and downstream states. We set the normalized diffusion coefficient to a value of 0.1, and this diffusion scale can be seen in the subshock precursor. Profiles of each of the fluid variables are shown after 200, 800, and 1200 time steps. After 1200 steps, the system has reached a steady-state in the shock reference frame. We have found that the quadratic artificial viscosity algorithm is numerically unstable and therefore unusable for strong cosmic ray shocks. There is therefore visible ringing and overshoot in the various profiles. Note that a large enhancement in the fluid density appears downstream of the subshock. This 'over-compression' results from the upstream fluid being compressed in the precursor, then traversing a strong shock with compression ratio near the adiabatic limit of four. As the shock reaches a steady-state and is further smoothed by the cosmic ray pressure gradient, the enhancement convects downstream. Note that there is no corresponding effect in either the fluid velocity or gas pressure.

Without showing the result here, we have also found the strong shock structure to be similar when the energetic 'seed' particles are injected rather than prexisting upstream. Specifically, we find the acceleration term in (4) is mostly responsible for the growth in cosmic ray pressure, regardless of the pressure source. This behavior does not seem to hold for weak shocks, however. Figure 2, for example, is a simulation with gas Mach number 1.2. Here the 'seed' particles are injected. Unlike the strong shock result,

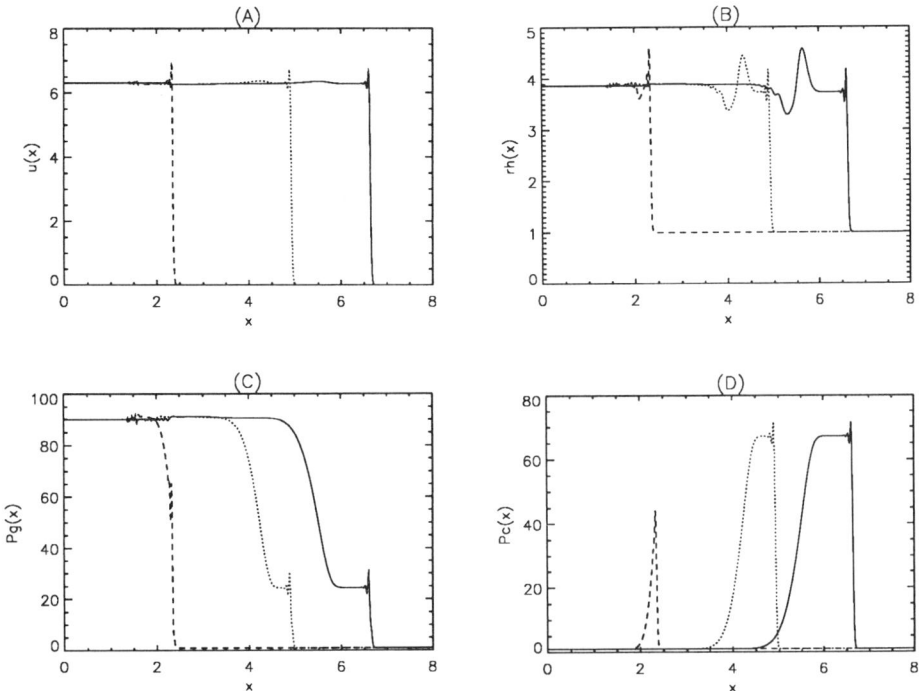

Figure 1: Strong shock (M_s=8.5), γ_g=5/3, γ_c=5/3, κ=0.1, no injection (α=0, p_{c_u}=1), no artificial viscosity. Results are shown after 200, 800, 1200 time steps. The shock propagates from left to right. Shown from (a) to (d) are u, ρ, P_g, and P_c, respectively.

the steady-state cosmic ray pressure downstream is far less than the corresponding gas pressure. The small cosmic ray pressure is nonetheless sufficient to completely smooth the gas subshock. The result shown is after 7500 time steps. Comparing to figure 1, the time taken to reach the steady-state is about one order of magnitude larger for the weak shock. This scaling of the equilibrium time with shock strength is not as strong as the κ/u_s^2 referred to in Kang and Jones[3] and elsewhere. The artificial viscosity algorithm is used in figure 2 and is found to be quite effective for weak shocks, eliminating most of the ringing and overshoot seen in figure 1. The small scale ripples in figure 2 are due to an instability of the Lax-Wendroff scheme caused by the diffusive term in the total energy equation (3). We have found that the instability is largely removed by using the MacCormack predictor-corrector scheme instead of Lax-Wendroff.

As a final comment, we note that the injection parameter α in these calculations is simply taken to be a function of the fluid density, $\alpha = \alpha_0 \rho/\rho_0$. Such a parameterization leads to ordinary gas-dynamic type shock transitions in the calculations of Zank et.al.[1] The compound shock structures discussed in that work arise only when $[\alpha u] \neq 0$. Future papers will consider such transitions and apply the time-dependent model to other examples of cosmic-ray-modified shocks such as the solar wind termination shock.

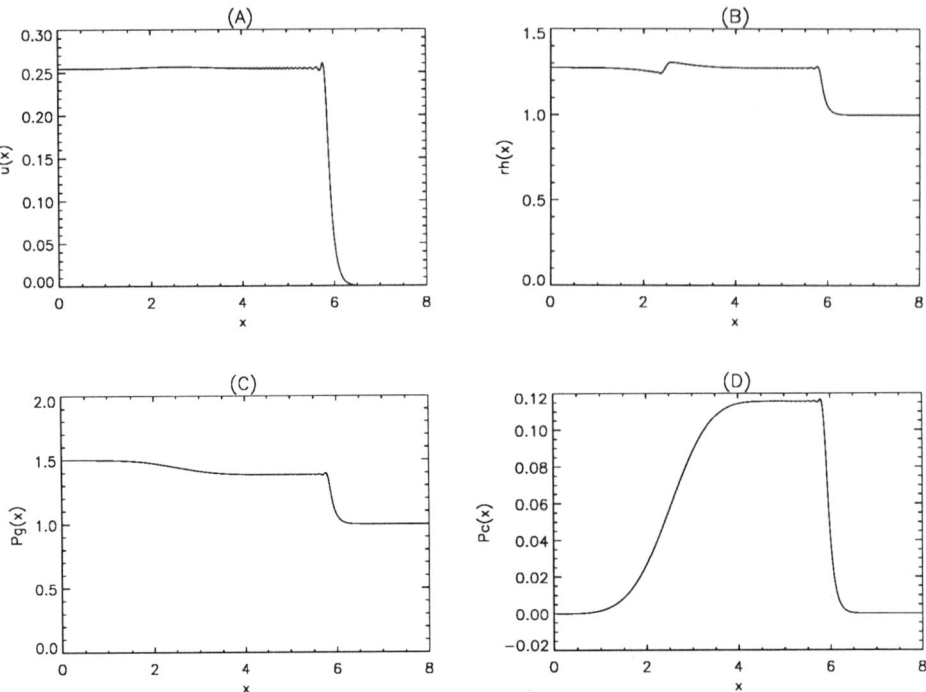

Figure 2: Weak shock ($M_s=1.2$), $\gamma_g=5/3$, $\gamma_c=5/3$, $\kappa=0.08$, with artificial viscosity, with injection ($\alpha \neq 0$, $p_{cu}=0$). Results are shown after 7500 time steps.

Acknowledgements: This work was supported by the NASA SPTP program under grant NAGW-2076 to the Bartol Research Institute.

References

1. G. P. Zank, G. M. Webb and D. J. Donohue, these proceedings.
2. C. A. J. Fletcher, Computational Techniques for Fluid Dynamics, V1 and V2, (Springer-Verlag, Berlin, Heidelberg, 1991).
3. T. W. Jones and H. Kang, *Ap. J.*, **363**, 499 (1990).

ON THE STABILITY OF COSMIC RAY DOMINATED SHOCKS

Hyesung Kang and Dongsu Ryu
Princeton University, Princeton, NJ 08544

T.W. Jones
University of Minnesota, Minneapolis, MN 55455

ABSTRACT

We report that a secondary, Rayleigh-Taylor type instability can exist in cosmic-ray dominated media which are perturbed by the 1D acoustic instability. Using the local WKB approximation, the growth rate of the secondary instability is shown to be comparable to that of the 1D acoustic instability itself in the cases we have considered. We show that flows in the precursor and postshock regions can become highly turbulent due to the secondary instability. As in the 1D acoustic instability, however, the cosmic-ray pressure is not significantly affected by the presence of the secondary instability, because the cosmic ray diffusion timescale from the perturbations is much smaller than the growth timescale of the instability.

1D ACOUSTIC INSTABILITY

It has been shown previously that small compressional disturbances propagating through a cosmic-ray (CR, hereafter) dominated medium can be amplified by the 1D acoustic instability[1,2,3,4]. If the wavelengths of the disturbances are shorter than the scale height of the CR pressure, and if the scale height of the CR pressure is shorter than the CR diffusion length associated with sonic flow, disturbances traveling in the direction of *increasing* CR pressure can grow, while those traveling in the direction of *decreasing* CR pressure can be damped out.

This instability occurs because the rapid CR diffusion greatly reduces otherwise adiabatic CR pressure perturbations, leaving an almost constant CR pressure gradient force that is inversely proportional to the gas density. As a result, an excess force, $\vec{\nabla} p_c \delta\rho/\rho^2$, is applied to the gas in disturbances. On the other hand, the damping of the CR pressure due to diffusive *bulk* viscosity (Ptuskin damping) provides a friction against traveling disturbances[5]. The instability exists when the force due to the CR pressure gradient dominates over the Ptuskin damping force.

We showed in a previous study that the CR precursor in a high Mach number, CR dominated shock is always unstable to the growth of sound waves with wavelengths shorter than the scale height of the precursor and that these waves can steepen into small scale shocks[4]. However, the instability does not significantly change the overall CR energy density nor the CR acceleration efficiency. On the other hand, it may still produce interesting effects in strong astrophysical shocks, because it can generate large amplitude density oscillations and significant heating of gas in the precursor as a consequence of multiple small scale shocks.

SECONDARY INSTABILITY

In strong, CR dominated shocks, downstream facing sound waves are compressed and grow nonlinearly, as they are advected through the CR shock precursor. A Lagrangian fluid element comoving with the flow will experience the repeated passage of these nonlinear traveling waves. The gradients of the gas density and the gas pressure felt by a local fluid element will change sign at each wave passage, whereas the sign of the CR pressure gradient remains the same. We recognize that regions where gradients of the gas density and CR pressure have opposite signs ($\vec{\nabla}\rho \cdot \vec{\nabla}p_c < 0$) can become unstable against the Rayleigh-Taylor type instability.

In order to see the characteristics of the instability, we considered an exponential background defined as $\rho_0 \propto p_{g0} \propto \exp(\frac{x}{L_1})$, $p_{c0} \propto \exp(\frac{x}{L_2})$, and $u_{x0} = u_{y0} = 0$, with the mean CR diffusion coefficient, κ_0, assumed to be constant. Unless $L_1 = L_2$, no true eigenmode exists with the above background, and self-consistent eigenmode analysis is not possible. Nevertheless, some insights into the physics of the instability can be gained by considering the local WKB approximation in the short wavelength limit, where the perturbations are assumed to be proportional to $\exp(\Gamma t + ik_y y)$ and $k_y \to \infty$. Here, Γ should be a function of x and changes over the scale of L_1 or L_2. But, in the limit $(1/k_y) \ll |L_1|$ and $|L_2|$, the spatial dependence of Γ can be neglected.

From the hydrodynamical conservation equations and the CR energy equation using the two-fluid CR transport model, we find the following dispersion relation in the short wavelength limit,

$$\Gamma^2 \approx -\frac{(\gamma-1)}{\gamma}\frac{1}{L_1 L_2}\frac{p_{c0}}{\rho_0} - \frac{(\gamma-1)}{\gamma}\frac{1}{L_1^2}\frac{p_{g0}}{\rho_0}, \qquad (1)$$

where γ is the adiabatic index of the gas. The instability exists if $\Gamma^2 > 0$, or if $\vec{\nabla}\rho_0 \cdot (\vec{\nabla}p_{c0} + \vec{\nabla}p_{g0}) < 0$. Note that the condition for the Rayleigh-Taylor instability of an initially static, incompressible fluid[6] is $\vec{\nabla}\rho_0 \cdot \vec{\nabla}p_0 < 0$.

The perturbations in the CR pressure much shorter than the scale height of the precursor diffuse away with the time scale $\tau_{diff} \sim \lambda^2/\kappa_0$ (pure diffusion limit), where λ is the perturbation wavelength. Since this time scale is generally much smaller than the growth time scale of the instability, $\tau_{inst} \sim 1/\Gamma$, the perturbations in the CR pressure are negligible. On the other hand, the perturbation in the gas pressure in the short wavelength limit is related to the perturbation in the gas density as

$$\frac{\delta p_{g0}}{p_{g0}} \approx \left(\frac{1}{\gamma L_1 L_2 k_y^2}\frac{p_{c0}}{p_{g0}} - \frac{1}{\gamma L_1^2 k_y^2}\right)\frac{\delta \rho}{\rho_0}. \qquad (2)$$

$\delta p_g/p_{g0} \ll \delta\rho/\rho_0$ in the cases we have considered. Hence the perturbations in the gas pressure are much less noticeable than those in the gas density, as expected from the nature of the Rayleigh-Taylor instability.

NUMERICAL RESULTS

In order to study the nonlinear behavior of the secondary instability we have used a time dependent 2D numerical code based on the two-fluid model[7]. We have calculated the evolution of CR dominated plane shocks with $u_s = 10$, propagating into a uniform medium of $\rho = 1$, $p_g = 0.1$ with a preexisting CR pressure $p_c = 0.1$ far upstream. The gas Mach number of the shock is about 24.5. The assumed values of the specific heats and the mean diffusion coefficient for the CRs are $\gamma_c = 4/3$ and $\langle \kappa \rangle = 1.5$, respectively. The calculations were done in the frame where the initial, unperturbed shock is at rest. The plane shock normal was pointed towards positive x, with a spatial domain for the calculation defined by $x = [0, 3]$ and $y = [0., 0.75]$.

We have chosen the analytic steady-state structure as initial conditions. For the above parameters, the initial shock is completely smoothed out and dominated by the CR pressure. The flow incoming through the right ($x = 3$) boundary is perturbed so that: 1) sound waves are generated, propagating to the left with the normalized x-velocity amplitude A_x, and 2) those sound waves carry random y-velocity fluctuations with normalized amplitude A_y. Thus the flow velocity has x and y components: $u_x = A_x c_{s0} \sin\left[(c_{s0} + u_s)k_x t\right] - u_s$, and $u_y = A_y c_{s0} R_n$, where c_{s0} is the unperturbed, upstream sound speed and R_n is a random number ranging from -0.5 to 0.5. The flow at the left boundary is assumed to be continuous.

The growth of perturbations due to the secondary instability has been tracked by computing the amplitude of the density perturbations given by

$$A_k(x) = \left| \frac{1}{\Delta y} \int_{y_{min}}^{y_{max}} \rho(x,y) \, e^{iky} dy \right|, \tag{3}$$

where $\Delta y \equiv y_{max} - y_{min}$ and is equal to 3/4 in our calculations. The instability growth rate has been estimated from the growth of $A_k(x)$ in the precursor ($1.2 < x < 1.7$) as follows: Using $dx/dt = v_0$ where v_0 is the unperturbed flow velocity, $A_k(x)$ was converted into $A_k(t)$. Then using chi-square fitting, the growth rate, Γ, was estimated from $\ln A_k(t) = \Gamma t + C$ where C is a constant. Finally, 10 statistically independent data sets spanning $t = 0.5$ to 0.7 were averaged for a single estimate of Γ. The value of the maximum Γ calculated in this way is ~ 30 for $\lambda = 0.5$. On the other hand, the linear growth rates estimated with the dispersion relation (1), using values of the unperturbed background fluid quantities, is ~ 25 for $\lambda = 0.5$. Considering the uncertainties in the growth rate estimates and the probable effects of nonlinearity, the agreement between the linear growth rate and the growth rate from numerical calculations seems to be good.

Fig. 1 contains a grey-scale 2D image of gas density, ρ, for the shock described above at $t = 0.6$, using $\lambda = 0.5$, $A_x = 0.3$, and $A_y = 0.3$. Within the precursor the density shows a strong sinusoidal structure on the background, due to the 1D acoustic instability. On the downstream end of the precursor and into the post shock flow the density has developed a highly turbulent pattern due to the secondary instability. However, beyond those due to the 1D instability, perturbations in the gas pressure remain relatively small compare to those in the gas density, as expected from the local analysis described in the previous section. The CR pressure shows almost no perturbations at all.

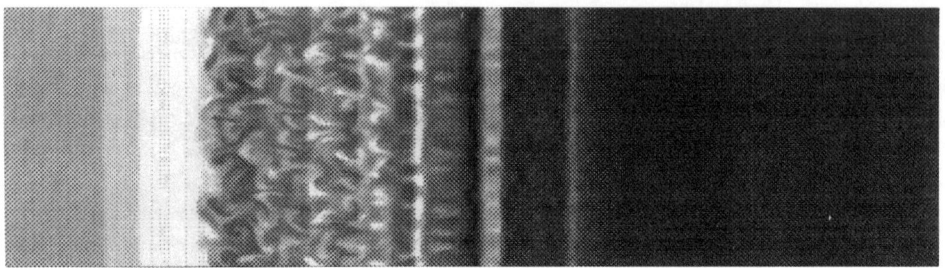

Fig. 1

ACKNOWLEDGMENTS

The work by TWJ and HK was supported in part by the NSF through grant AST-8720285, by NASA through grant NAGW-2548 and by the University of Minnesota Supercomputer Institute. The work by DR was in part supported by David and Lucille Packard Foundation Fellowship through Jeremy Goodman at Princeton University.

REFERENCES

1. Drury, L. O'C., and Falle S. A. E. G., *Mon. Notices R. Astr. Soc.*, **223**, 353, (1986).
2. Chalov, S. V., *Sov. Astron. Lett.*, **14**, 114 (1988).
3. Zank, G. P., Axford, W. I., and McKenzie, J. F. , *Astron. Astrophys.*, **233**, 275, (1990).
4. Kang, H., Jones, T. W., and Ryu, D., *Astrophys. J.*, **385**, 193 (1992).
5. Ptuskin, V. S., *Ap. Space Sci.*, **76**, 265 (1981).
6. Chandrasekhar, S., *Hydrodynamic and Hydromagnetic Stability* (Clarendon Press, Oxford, 1961).
7. Jones, T. W., and Kang, H., *Astrophys. J.*, **363** , 499 (1990).

TWO-FLUID MODELLING OF PARTICLE ACCELERATION IN MODIFIED NON-RELATIVISTIC SHOCKS

Matthew G. Baring
Department of Physics, Box 8202,
North Carolina State University,
Raleigh, NC 27695

ABSTRACT

In two-fluid analytical approaches to the modelling of particle acceleration in cosmic-ray modified shocks, the shock structure and the acceleration efficiency can be found as a function of the shock Mach number for a given shock velocity, from purely hydrodynamic considerations. Heretofore, two-fluid models have usually included no information about the possibility of particle escape from the shock region. Such escape from the environs of a non-relativistic shock creates small but significant anisotropies in the particle distribution. This paper addresses this deficiency in part by including particle anisotropies in a hydrodynamic determination of the shock structure. Modified shocks are examined here, and anisotropies consistent with the diffusion approximation are found to substantially alter the acceleration efficiency predictions of two-fluid models.

INTRODUCTION

Standard two-fluid models predict the particle acceleration efficiency in non-relativistic shocks from simple hydrodynamic considerations, without any consideration of the details of the spectrum of the accelerated cosmic-rays (e.g. Drury and Völk, 1981; Axford, et al. 1982; Heavens, 1984; Achterberg, et al. 1984). This technique has recently been generalized to relativistic shocks by Baring and Kirk (1991, hereafter BK91). This two-fluid approach has the advantage that it is analytically simple and elegant, but lacks the detailed spectral information that is generated in Monte-Carlo simulations of particle acceleration at shocks (e.g. see Ellison, et al. 1990). The solution of the hydrodynamic conservation equations for the momentum and energy fluxes in two-fluid models has heretofore assumed that non-relativistic shocks produce quasi-isotropic cosmic-rays (CRs), and that there is no postcursor to the subshock in the downstream portion of the shock velocity profile (borne out in Monte-Carlo simulations: e.g. see Ellison, et al. 1990). These assumptions simplify the hydrodynamic analysis. However, here it is concluded that even slight anisotropies have a significant role in determining the shock profile and the particle acceleration efficiency.

Cosmic-ray anisotropies have to arise in many, if not all, realistic shocks: the spatial finiteness of the shock region leads to the escape of cosmic-rays from the boundaries of the region. Associated with these boundaries are escaping energy fluxes and a naturally-occurring anisotropy in the highest energy cosmic-rays, which are the most likely to escape because they generally have longer scattering mean-free-paths (e.g. see Giacalone, et al., these proceedings). Positive curvature (e.g. Ellison and Eichler, 1984) of the cosmic-ray spectrum enhances the anisotropy. The escaping energy flux is found to be significant in Monte-Carlo simulations (for a review, see Jones and Ellison, 1991), and indeed particle escape must be included in a self-consistent shock solution in order that the cosmic-ray energy flux be finite in shocks of large Mach number (Ellison and Eichler, 1984). Here the focus will be on how the inclusion of anisotropy alters the two-fluid solutions, leaving the consideration of the influence of the escaping energy flux far from the subshock to later work. Hydrodynamic solutions for the shock profile structure are obtained, relating the Mach number of the shock and the shock compression ratio, and they are found to depend sensitively on the degree of CR anisotropy, even within the confines of the diffusion approximation.

In a two-fluid shock model there are two components to the plasma. The first is an ideal gas, for which we specify the equation of state $P_g = (\gamma_g - 1)(e_g - \rho_g)$ where P_g is the gas pressure, e_g the energy density (including the rest-mass), ρ_g the rest-mass density ($c = 1$ units are used throughout) and γ_g is the ratio of specific heats for the gas. These and all other thermodynamic quantities are measured in a reference frame in which the fluid is locally at rest. One then has $\rho_g u(z) = \Phi_g$ for the conservation of rest-mass of the gas, where Φ_g is a constant mass flux, and $u(z)$ is the flow velocity (as a function of the distance z from the plane of the subshock) in the frame in which the shock is at rest. Writing the specific enthalpy as $w_g = e_g + P_g$, we can define the energy and momentum fluxes of the gas in the observer's frame to be $\alpha_g[u(z)]$ and $\beta_g[u(z)]$, respectively, scaled by the mass flux Φ_g: $\Phi_g \alpha_g \equiv w_g u$, $\Phi_g \beta_g = w_g u^2 + P_g$. All definitions and conventions in this paper follow BK91.

The second component is the cosmic-ray or accelerated population, which may be anisotropic, and has an equation of state $P_c = (\gamma_c - 1)e_c$, with $\rho_c \ll e_c$. Analogously, we can write down energy and momentum fluxes for the cosmic rays (see BK91 for the case of relativistic flows): $\Phi_g \alpha_c \equiv w_c u + t_c$, $\Phi_g \beta_c = w_c u^2 + P_c + 2t_c u$ for $w_c = e_c + P_c$. The relationships between the macroscopic hydrodynamic quantities e_c, P_c and t_c and the cosmic-ray particle distribution function can be found in BK91. Here it is sufficient to note that $|t_c| < e_c$. The principal hydrodynamical premise of the two-fluid model is that the total (gas plus cosmic-ray) energy and momentum fluxes, $\alpha_g + \alpha_c$ and $\beta_g + \beta_c$ respectively, are conserved throughout the shock profile. These are used to define the set of shock solutions below that include the modifying effects of the CR pressure.

SHOCK STRUCTURE SOLUTIONS

Obtaining the hydrodynamic solutions to the flux conservation relations for the shock structure amounts to solving to find the flow velocities just upstream (u_1) and just downstream (u_+) of the subshock given the upstream flow velocity u_- and the Mach number $\mathcal{M}$ of the flow far upstream. Hereafter, the subscripts $-$, 1 and $+$ will respectively denote quantities far upstream, just upstream, and downstream of the subshock (there is no downstream postcursor). For non-relativistic shocks, these solutions can be expressed as compression ratios $r = u_-/u_+$ for the shock and $r_{pr} = u_-/u_1$ for the shock precursor. In addition, the Mach numbers $\mathcal{M}_1$ just upstream and $\mathcal{M}_+$ just downstream of the subshock can be determined. They are related, as are u_1 and $u_+ = u_2$ by the familiar Rankine-Hugoniot jump conditions (e.g. see Blandford and Eichler, 1987; BK91): for example $u_+/u_1 = [(\gamma_g - 1)\mathcal{M}_1^2 + 2]/(\gamma_g + 1)\mathcal{M}_1^2$. The energy and momentum flux conservation relations can be evaluated at u_- and u_+ to obtain hydrodynamic solutions. Details of the method are outlined in BK91. For a truly isotropic CR population the resulting equation takes the form of eq. (A3) of BK91. Here, we specify the CR anisotropy by the parameter a_{c+}:

$$t_{c+} = a_{c+} u_+ e_{c+} \quad , \quad |a_{c+} u_+| \ll 1 \quad . \tag{1}$$

Then combining a_{c+} and γ_c into the single parameter $\lambda = (\gamma_c + a_{c+})/(\gamma_c - 1)$, the shock solutions are defined by a form that is similar to eq. (A3) of BK91:

$$\begin{aligned}\frac{\gamma_g + 1}{\gamma_g - 1} \mathcal{M}^2 &\left\{ r_{\mathrm{pr}}^2 \left(\mathcal{M}^2 + \frac{2}{\gamma_g - 1} \right) - \left(\mathcal{M}^2 + \frac{2 r_{\mathrm{pr}}^{\gamma_g+1}}{\gamma_g - 1} \right) \right\} \\ &= 2\lambda \left(\mathcal{M}^2 + \frac{2 r_{\mathrm{pr}}^{\gamma_g+1}}{\gamma_g - 1} \right) \left\{ r_{\mathrm{pr}} \left(\mathcal{M}^2 + \frac{1}{\gamma_g} \right) - \left(\mathcal{M}^2 + \frac{r_{\mathrm{pr}}^{\gamma_g+1}}{\gamma_g} \right) \right\} \quad .\end{aligned} \tag{2}$$

Specifying the precursor compression ratio r_{pr}, the solution for the Mach number $\mathcal{M}$ can be obtained, and then r found. Solutions to eq. (2) are displayed in Fig. 1 for different anisotropies. Also depicted there is the CR acceleration efficiency $\eta_c = [\alpha_g(u_-) - \alpha_g(u_+)]/(\alpha_g(u_-) - 1)$ as defined in BK91: for non-relativistic shocks this assumes the form (see BK91)

$$\eta_c = 1 - \frac{1}{r_{\mathrm{pr}}^2} \frac{2 r_{\mathrm{pr}}^{\gamma_g+1} + (\gamma_g - 1)\mathcal{M}^2}{2 + (\gamma_g - 1)\mathcal{M}^2} \quad . \tag{3}$$

The solutions in Fig. 1 are displayed for different anisotropies, though because of the inclusion of γ_c in the definition of λ used in eq. (2), they could equally well represent different γ_c. The illustrative choice of anisotropies in Fig. 1 is possibly extreme. Note that the overall compression ratio r deviates dramatically from the usual range of values $4 \lesssim r \lesssim 7$. For non-relativistic shocks $|a_{c+}| \gg 1$ is possible in principle, though no solutions to eq. (2) are obtained when $a_{c+} \lesssim -1$. The major point to draw from the figure is that when the anisotropy t_{c+} is a significant fraction of $u_+ e_{c+}$, the solution space deviates from the usual two-fluid predictions for truly isotropic cosmic-rays (case (a)). The general trends are that when the anisotropy increases in a positive sense, the compression ratios increase and so does the acceleration efficiency.

The diffusion approximation predicts an anisotropy $t_c \sim -\kappa(de_c/dz)$, where κ is effectively a spatial diffusion coefficient (e.g. see Drury and Völk, 1981; BK91). In the shock profile, dz must be of the order of the mean free path for thermal gas particles, which is proportional to κ/v_{th}, where v_{th} is a typical thermal particle speed and de_c must be of the order of e_{c+}. So it follows that $t_{c+} \sim -v_{th} e_{c+}$ would be expected and if v_{th} is not much less than u_+, the anisotropies expected from the diffusion approximation would be enough to cause significant deviations in η_c from the truly isotropic case. Clearly the degree of anisotropy depends on the microphysical properties of scattering.

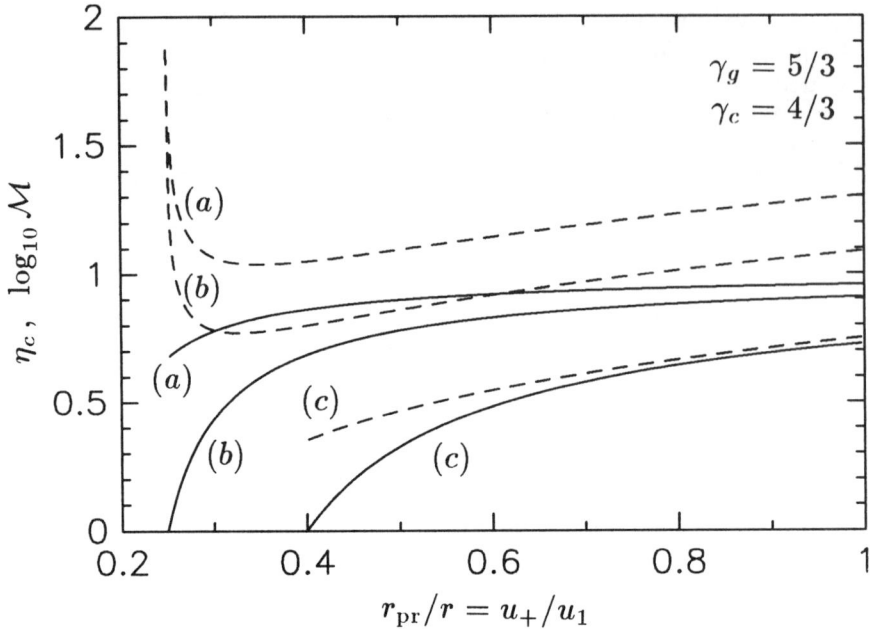

Fig. 1: The solutions of eq. (2), which describes the energy and momentum flux conservation of the gas plus CRs, depicting the shock mach number $\mathcal{M}$ (dashed curves) and the CR acceleration efficiency (solid lines) η_c as a function of $r_{\rm pr}/r$ for three different CR anisotropies downstream: (a) $a_{c+} = -0.5$, (b) pure CR isotropy $a_{c+} = 0$, and (c) $a_{c+} = 0.5$. When the $a_{c+} \geq 0$, the highest efficiency solutions are for $r_{\rm pr} = r$, i.e. CR-dominated shocks.

In conclusion, this paper finds that in non-relativistic shocks, anisotropies caused by particle escape that are consistent with the diffusion approximation can dramatically alter the shock structure and the particle acceleration efficiency for modified shocks. This provides ample motivation for extending this analysis to relativistic shocks, and further including the possibility of escaping energy flux in the conservation equations, in order to fully describe particle escape from shocks using the two-fluid model. Work on these aspects is under progress.

REFERENCES

Achterberg, A., Blandford, R.D. and Periwal, V.: 1984 *Astron. Astr.* **132**, 97
Axford, W.I., Leer, E. and McKenzie, J.F.: 1982 *Astron. Astr.* **111**, 317
Baring, M. G. and Kirk, J.G.: 1991 *Astron. Astr.* **241**, 329
Blandford, R.D. and Eichler, D.: 1987 *Phys. Reports* **154**, 1
Drury, L.O'C. and Völk H.-J.: 1981 *Astrophys. J.* **248**, 344
Ellison, D.C. and Eichler, D.: 1984 *Astrophys. J.* **286**, 691
Ellison, D. C., Möbius, E. and Paschmann, G.: 1990 *Astrophys. J.* **352**, 376
Heavens, A.F.: 1984 *Mon. Not. R. astr. Soc.* **210**, 813
Jones, F. C. and Ellison, D. C.: 1991 *Space Sci. Rev.* **58**, 259

PARTICLE ACCELERATION IN MODIFIED OBLIQUE NON-RELATIVISTIC SHOCKS

Matthew G. Baring and Donald C. Ellison
Department of Physics, Box 8202,
North Carolina State University,
Raleigh, NC 27695

and

Frank C. Jones
Laboratory for High Energy Astrophysics,
NASA/Goddard Space Flight Center,
Greenbelt, MD 20771

ABSTRACT

Nonlinear Monte-Carlo simulations of particle acceleration at astrophysical shocks have been restricted to the specialized case of plane-parallel shocks, while Monte-Carlo work with oblique geometries has to date been confined to test-particle models. The Monte-Carlo approach has significant advantages over hybrid plasma simulations in that it can determine particle spectra and acceleration efficiencies over extremely wide energy ranges with modest computing resources. Previous applications of the Monte-Carlo technique at the quasi-parallel earth bow shock led to very successful modelling of proton and heavy ion spectra, as well as other observed quantities (e.g., Ellison, et al. 1990). This has motivated the extension of this technique to oblique shock geometries typical of those found in most astrophysical shock environments. In addition, such a generalization will permit the thorough examination of theoretical predictions of rapid acceleration times at quasi-perpendicular shocks. Therefore, we have embarked on the modification of our existing Monte-Carlo code, and in this paper outline the major technical aspects involved in developing a simulation of cosmic-ray acceleration at modified, oblique, non-relativistic shocks.

INTRODUCTION

First-order Fermi acceleration at shocks in space plasmas has been widely invoked as a mechanism to accelerate particles to high energies in a number of different astrophysical environments (for a review see Jones and Ellison 1991, hereafter JE91). The understanding of this acceleration process has been enhanced by many studies, most of which specialize to the case of plane-parallel (i.e., the shock normal is parallel to the magnetic field) shocks in non-relativistic flows. However, limiting models to the plane-parallel case has the drawback that probably very few astrophysical shocks fit this description. Shocks in supernova remnants, lobes of radio galaxies, and also the earth bow shock show obliquity of the field to the shock normal. It has been suggested by Jokipii (1987) and Ostrowski (1988) that quasi-perpendicular shocks can accelerate ions much more quickly than parallel ones and, by implication, be more efficient particle accelerators. However, none of these calculations have included the effects of the accelerated ions on the shock structure. Most studies of oblique shocks have been confined to the test particle case, which is unrealistic since the modifications to the shock structure caused by the accelerated population can dramatically affect the shock acceleration efficiency. A major advantage of our Monte-Carlo technique is that it can readily be applied to cosmic-ray modified shocks and therefore provide meaningful astrophysical predictions: predictions that differ significantly from those of test particle models (e.g., Ellison and Eichler 1984; Ellison and Reynolds 1991). Further, Monte-Carlo codes can easily describe mildly-relativistic particles, a feature absent in analytic models (e.g. JE91).

The method is to treat all particles as a single population and follow their time histories in the environs of a shock, allowing them to diffuse, as if scattering off magnetic irregularities anchored in the fluid, with a phenomenological mean free path in the *fluid* frame of $\lambda = \lambda_0 p^\alpha / \rho$ (e.g. see JE91), where p is the particle momentum, ρ is the plasma density, λ_0 is an input parameter as is the gyroradius r_{g0} of a thermal particle; the ratio r_{g0}/λ_0 is important for quasi-perpendicular shocks (Jokipii, 1987). The choice of the index α is somewhat flexible, however, it should be noted that comparisons with bow shock observations (Ellison, et al. 1990) predict $\alpha = 1$, and the form of λ can be determined from plasma simulations (Giacalone, et al., these proceedings). Clearly a variety of microphysics can be incorporated in specifying λ, and both pitch angle and large angle scattering can be described, with λ defining the particle turn-around length scale. Since no distinction is made between thermal and superthermal ions, and all shock heated ions can, as determined by statistics, recross the shock, gain energy and be "injected" into the Fermi process, we have a self-consistent description of injection. This is an important feature that is not addressed in standard two-fluid model approaches to shock acceleration.

While particles are scattered in three dimensions, the one-dimensional code only requires a particle's momentum normal to the shock and its pitch angle. This contrasts with the Monte-Carlo method of Ostrowski (1991) and plasma simulation approaches, where the exact helix of the particle's gyration in the magnetic field is specified at all times, an extremely time-consuming feature in numerical codes. For parallel shocks, knowledge of the gyroradius and gyrophase of a particle is irrelevant to the scattering problem, so the calculation of a helix of gyration is unnecessarily cumbersome. This renders a guiding center approach distinctly advantageous, particularly when acceleration to extremely high energies is to be modelled, and we propose to apply it to oblique geometries. However, in oblique shocks some phase information about the particle must be known in order to determine various fluxes normal to the plane of the shock. These can be specified when needed by randomly choosing the phase without biassing the results of the code, since the gyrophase will be randomized by scattering and the phase change between scatterings will also vary randomly.

MONTE-CARLO SIMULATIONS IN OBLIQUE GEOMETRIES

Our current plane-parallel code follows a particle's history until either (i) its energy exceeds a specified maximum energy, E_{max}, or (ii) it passes beyond boundaries far upstream or downstream of the subshock *and* is determined to escape from the system. In the downstream region, this determination is made by calculating the probability of return with a simple analytic formula (e.g., Bell 1978; see also JE91). These particle escape features are to be included in similar fashion in the oblique Monte-Carlo coding.

Besides a self-consistent description of injection, the most important feature of our code is that we include the nonlinear effects of the pressure of the accelerated particles on the hydrodynamic structure of the shocked flow. To do this, we calculate the energy, momentum, and number fluxes at different spatial locations, which define a grid of planes parallel to the shock front. This accounting is performed and the Rankine-Hugoniot (R-H) conservation conditions are used to determine the spatial shape of the flow velocity profile for the next iteration (see Decker 1988 for the R-H conditions at oblique shocks). The grid spacing is adjustable, and is normally shorter than the thermal mean free path near the subshock. At each grid point, the magnitudes and angles of both the speed of the flow and the magnetic field are tabulated, and between grid points the flow speed is found by linear interpolation, while the other quantities are binned in a histogram fashion. This is depicted in the normal incidence (shock) frame (NIF) in Fig. 1; in this frame the flow velocity far upstream is normal to a stationary shock front, while the magnetic field has some far upstream value, θ_{B1}. It is planned to complete all of the flux accounting in the NIF, following the convention used in plane-parallel shocks.

Practically, the flux balancing is an iterative procedure, starting with a discontinuous shock profile with the compression ratio determined from the R-H relations. If particles are accelerated, however, a modified shock profile with a precursor will result after one iteration. If particles escape from the system, as is inevitable in steady-state shocks of Mach number $\gtrsim 4$ (see JE91), the compression ratio must be increased above that predicted by the R-H relations without escape. The new compression ratio is estimated using the calculated fluxes for the total shock, and the next iteration is performed with the new compression ratio and the smoothed shock profile (and angles in the oblique case). This method convergences rapidly and results in curved profiles, as depicted in Fig. 1, that are simultaneously hydrodynamical solutions for the converging flow, and also include the details of Fermi acceleration self-consistently.

The major differences between the plane-parallel and oblique cases appear in the treatment of the particle transport in the converging flow. Each particle is permitted to travel between scatterings for a collision time that is calculated in the local fluid frame, i.e., $t_c = \lambda/v_F$, where v_F is the particle speed in the fluid frame, and t_c is exponentially spread. If large-angle scattering is assumed, the pitch angle cosine, μ, is randomized in each collision. Pitch angle diffusion can also be treated. The particle transport between scatterings is best described in the well-known de Hoffmann-Teller (HT) frame (1950), where not only is the shock front stationary, but the fluid flow (with speed u) is everywhere parallel to the magnetic field ($\mathbf{u} \times \mathbf{B} = \mathbf{0}$). In this force-free frame of reference, which is obtained from the NIF by a boost of $u_1/\cos\theta_{B1}$ in the plane of the shock, the electric field is everywhere zero and there is no $\mathbf{E} \times \mathbf{B}$ particle drift in the shock layer. When the HT frame exists (i.e., most of phase space for non-relativistic flows), it is uniquely defined *throughout the flow* since $\mathbf{u} \times \mathbf{B}$ conservation is

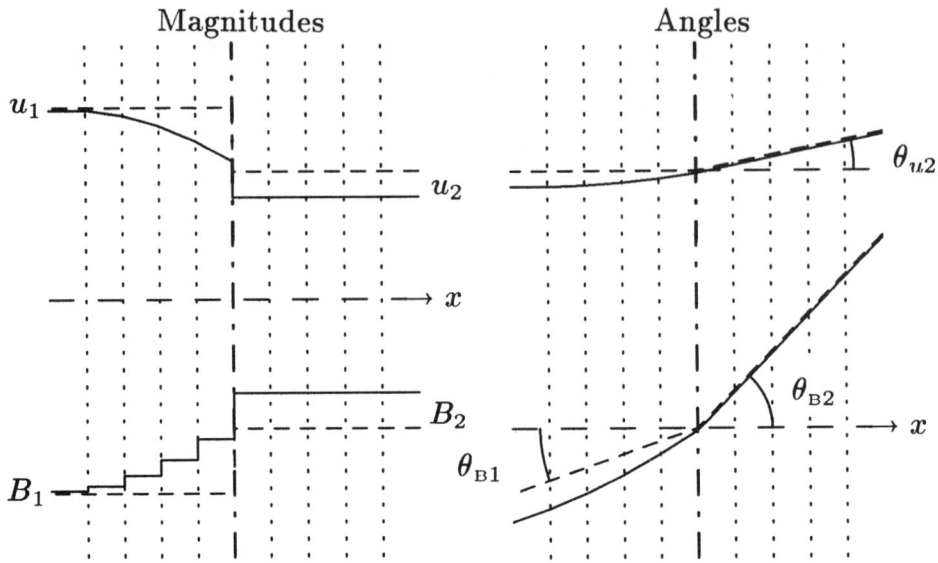

Fig. 1: Here the angles and magnitudes of the velocity (u, θ_u) and magnetic field (B, $\theta_{\rm B}$) vectors are depicted in the normal incidence frame for flow through an oblique shock, with the short dashed lines representing the initial iteration and the solid lines the final iteration and solution to the problem. The vertical dash-dot line is the stationary plane of the subshock and the dotted lines are the grid points, while the long dashes define the direction normal to the shock. Subscripts 1 and 2 denote far upstream and downstream values, respectively.

demanded (e.g. see Decker, 1988). In due course, the coding will be adapted for extremely perpendicular and superluminal relativistic shocks, where the HT frame does not exist.

The absence of electric fields in the HT frame simplifies the description of particle convection and shock or grid-point crossings immensely. As in the plane-parallel case, the details of the particles' gyropaths are largely irrelevant during propagation, and a particle is convected along the field lines, keeping track of its total momentum and its pitch angle cosine, $\mu_{\rm HT}$ in the HT frame. Energy and momentum Lorentz transformations between the local fluid and HT frames are simple, preserving the particle's gyroradius and gyrophase. Particle convection between scatterings then maintains $\mu_{\rm HT}$ and the gyrophase of the particle is unaccounted for. What happens to the particle upon the crossing of a grid-point or the subshock is now, however, intrinsically different from the parallel shock case.

In oblique geometries, particle transmission at a flow discontinuity or in the upstream precursor is not automatically guaranteed, and reflection of particles flowing downstream can occur because of the increasing magnetic field. We liken grid-points to small subshocks, and treat grid-point crossings by particles in the same manner as crossings of the subshock, thereby including the possibility of particle reflection in the upstream precursor region. Shock or grid-point crossings are dramatically simplified in the HT frame, where we assume adiabatic passage across the grid-point. In the HT frame this amounts to conservation of the magnitude of the particle's momentum and also conservation of the particle's magnetic moment (e.g. Toptyghin, 1980),

$$(1 - \mu_u^2)/B_u = (1 - \mu_d^2)/B_d , \quad (1)$$

where the subscripts u and d denote the upstream and downstream sides of the grid-point. The effects of particle drifts, and therefore shock drift acceleration, in the NIF are included in this description, being just manifestations of Lorentz transformations between the NIF and HT shock frames. Eq. (1) gives the condition that reflection of particles convecting downstream occurs when

$$\mu_u < \sqrt{1 - B_u/B_d} , \quad (2)$$

for which $\mu \to -\mu$. The assumption of adiabatic passage, which amounts to a random choice of gyrophase (Drury, 1983), will be made initially and relaxed in later versions of the code. Note, however, that when fluxes in the NIF are computed, Lorentz transformations of particle energies and momenta from the HT frame require the gyrophase of the particle to be included randomly.

One major difference between parallel and oblique shocks is the importance of cross-field diffusion. When a particle is scattered, it changes its gyroradius and gyrophase, and as a result its gyrocentre "jumps" to another field line. Associated with this is diffusion normal to the shock, which can be seen by an inspection of Fig. 2. In the HT frame, if the particle has gyroradii r_g and r_g' (also the fluid frame gyroradii) and gyrophases ϕ and ϕ', before and after collision, then the translation of the particle's gyrocentre amounts to a change

$$\delta x_{\mathrm{NIF}} = \delta x_{\mathrm{HT}} = \left[r_g \cos\phi - r_g' \cos\phi' \right] \sin\theta_{\mathrm{B}} , \quad (3)$$

in distance normal to the shock (here θ_{B} is the angle the field makes in the HT frame to the shock normal) in both the HT and NIF frames. The phases are chosen randomly and while δx_{HT} averages to zero, it has a diffusive contribution. Near a grid-point, particle scattering overrides any adiabatic conditions so that cross-field diffusion can then transport particle gyrocentres across grid-points producing the energy gains associated with shock drift acceleration.

In conclusion, we have outlined the major technical features of the extension of a steady-state, plane-parallel, Monte-Carlo particle acceleration code to oblique shocks. By following the particle transport in the de Hoffmann-Teller frame, we have kept all of the advantages of the plane-parallel code and minimized the computational complexity. The Fermi mechanism is expected to be an efficient accelerator both from theoretical considerations and from direct observations at heliospheric shocks (JE91). If this is the case, accelerated particles will modify the shock structure and test-particle results may be unrealistic.

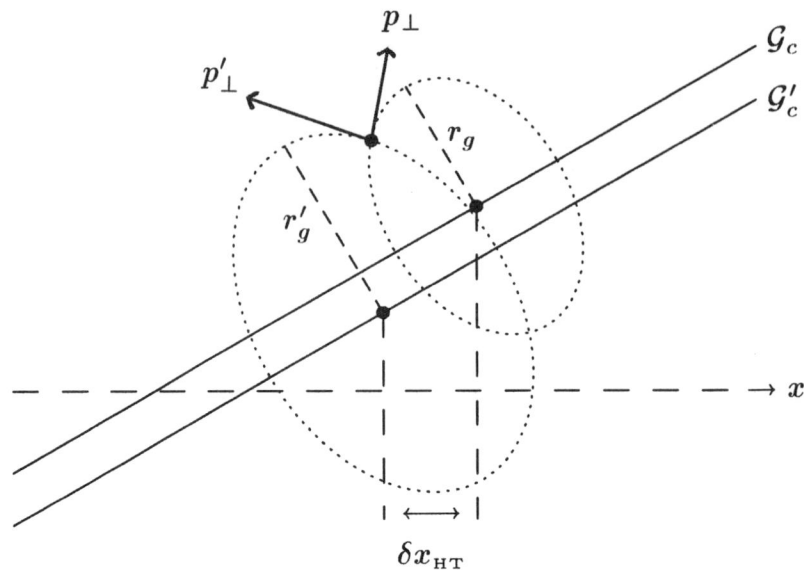

Fig. 2: A depiction of the change in a particle's gyrocentre and position x in the HT frame during a scattering. Before collision, the particle has a guiding center (or field line) $\mathcal{G}_c$, gyroradius r_g and component of momentum $p_\perp$ normal to the field, with primes denoting these quantities after collision. The change $\delta x_{\rm NIF} = \delta x_{\rm HT}$ in x is given by eq. (3), and the small dashed lines denote zero phase angle. The angle between the guiding centers and the x-axis is $\theta_{\rm B}$.

This effect may be more pronounced in oblique shocks than parallel ones if the high efficiencies claimed for oblique shocks are realized. We expect to be able to model *modified* oblique shocks and produce particle spectra and acceleration efficiencies to energies of astrophysical significance.

REFERENCES

Bell, A. R.: 1978 *Mon. Not. R. astr. Soc.* **182**, 147
Decker, R. B.: 1988 *Space Sci. Rev.* **48**, 195
de Hoffmann, F. and Teller, E.: 1950 *Phys. Rev.* **80**, 692
Drury, L. O'C: 1983 *Rep. Prog. Phys.* **46**, 973
Ellison, D. and Eichler, D.: 1984 *Astrophys. J.* **286**, 691
Ellison, D. and Reynolds, S. P.: 1991 *Astrophys. J.* **382**, 242
Ellison, D. C., Möbius, E. and Paschmann, G.: 1990 *Astrophys. J.* **352**, 376
Jokipii, J. R.: 1987 *Astrophys. J.* **313**, 842
Jones, F. C. and Ellison, D. C.: 1991 *Space Sci. Rev.* **58**, 259 (JE91)
Ostrowski, M.: 1988 *Mon. Not. R. astr. Soc.* **233**, 257
Ostrowski, M.: 1991 *Mon. Not. R. astr. Soc.* **249**, 551
Toptyghin, I. N.: 1980 *Space Sci. Rev.* **26**, 157

SHOCK DRIFT ACCELERATION

R. B. Decker
Johns Hopkins University Applied Physics Laboratory
Laurel, MD 20723-6099

ABSTRACT

We review basic aspects of shock drift acceleration at fast-mode collisionless shocks, and describe recent modeling efforts that incorporate such effects as shock structure, shock curvature, magnetic loops, rippled shocks, and magnetic turbulence.

INTRODUCTION

Shock drift acceleration, SDA, or drift mechanism all refer to the same basic mechanism of charged particle acceleration at collisionless, fast-mode shock fronts. This topic has been covered in previous reviews, from both the observational[1] and theoretical[2] viewpoints. The present brief review is concerned with developments in theory and computer modeling within the past five years or so.

Acceleration during a drift interaction, which typically involves several shock crossings, results from a net displacement, due to a grad-B drift, of an ion (electron) guiding center parallel (anti-parallel) to the convection electric field. The energy gain is proportional to this drift distance, which depends upon the plasma and shock parameters, the particle species and velocity, and the intensity of electromagnetic fluctuations in the vicinity of the shock as well as within the shock transition itself.

We refer to the shock frame as that where the shock is stationary and the upstream plasma velocity, **U**, is incident normal to the shock. Angle θ is that between the upstream magnetic field, **B**, and the shock normal, **n**. The convection electric field, $\mathbf{E} = -\mathbf{U} \times \mathbf{B}/c$, is continuous across the shock. For a planar shock and uniform **B**, the intersection point between **B** and the shock surface moves along the surface at speed $W = U\tan\theta$. For W subluminal, i.e., W<c, by moving with this point, one transforms to the deHoffmann-Teller (dHT) frame, where plasma flow is along the magnetic field, and the electric field vanishes, on both sides of the shock. For W superluminal, one can always transform to a frame where the shock is perpendicular (i.e., $\theta=90°$).[3]

At naturally occurring shocks, **B** and **n**, and thus θ, exhibit spatial and temporal variations over a wide range of amplitudes and scales. Therefore, one retains the greatest generality in modeling shock acceleration by working in the shock frame. Drift acceleration is largest in the quasi-perpendicular regime, $45° < \theta \leq 90°$, where **E** and b≡ (ratio of downstream to upstream magnetic field strength) are largest.

SINGLE SHOCK INTERACTION

The basic aspects of drift acceleration emerge from the simple model of a uniform upstream magnetic field convecting into a planar shock. If one assumes (particle gyroradius, r_g) >> (shock thickness, L_s) and neglects any shock-associated turbulence, one can replace the shock by a simple discontinuity and approximate particle motion as scatter-free on both sides of the shock. One can either integrate many particle orbits numerically and perform ensemble averages, or else use the approximate invariance of particle magnetic moments to derive analytic expressions for reflection coefficients, energy and angular distributions, etc.[4,5,6]

Figure 1 shows sample trajectories for particles reflected (left), transmitted downstream (middle), and transmitted upstream (right) at the shock (x=0). The upstream magnetic field lies in the x-z plane, pointing nearly out of the figure, with $\theta=80°$, b=4. The initial particle speed $v_0=10U$. The ordinate is the particle energy E (units if initial energy E_0), which is also proportional to the y-coordinate, because the electric field is along the y-axis, i.e., upward.

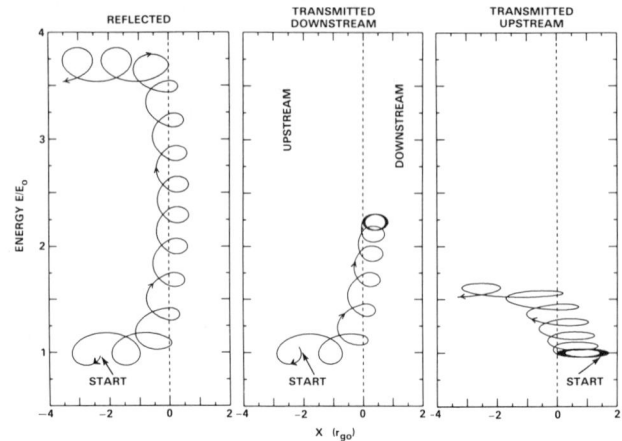

Figure 1. Sample orbits for scatter-free shock drift acceleration at a quasi-perpendicular shock. The upstream magnetic field is nearly normal to the plane of the figure.

For nonrelativistic speeds, i.e., $v^2, W^2 \ll c^2$, a few general results for single, scatter-free shock interactions are as follows.[2] (Note that the magnetic field jump b increases monotonically from 1 to r, the shock compression ratio, as θ increases from 0° to 90°, and that for a ratio of specific heats of 5/3, $1 \leq r \leq 4$.) (i) Energy gains and pitch angle changes are largest when the speed of the incident particle is comparable to the dHT frame speed, i.e., $v_0 \sim W$. (ii) Particles can reflect upstream only if their post-acceleration speed v exceeds W. (iii) Peak fractional energy gains, $\Delta E/E_0$, are about 4(b-1) for reflected particles, 2(b-1) for particles transmitted downstream, and somewhat less than this for those transmitted upstream. At perpendicular shocks or quasi-perpendicular shocks for which $v_0 < \sqrt{b}W$, all particles are transmitted downstream and the peak fractional energy gain is about b-1. (iv) Upstream, accelerated distributions exhibit large field-aligned, anti-shockward anisotropies. Downstream, peak intensities shift from shockward along the field line for $v \sim W$, to nearly perpendicular to the field for $v \gg W$.

The addition of a spectrum of transverse hydromagnetic waves to a quasi-perpendicular shock like that in Fig. 1 perturbs particle drift orbits, producing a wide range of drift displacements and energy gains. As compared to the scatter-free case, for a single drift interaction, the magnetic fluctuations (i) increase the downstream transmission of particles incident from upstream, (ii) produce broader energy distributions with steep power-law tails extending to several times the peak energy gained in the scatter-free case, (iii) reduce anisotropies near the shock, and (iv) destroy the magnetic moment invariance on a single-particle and ensemble-averaged level.[7]

The effects of shock structure on drift acceleration of mildly superthermal ions, ~1-10 keV, at nearly perpendicular shocks has been studied via test particle simulations using a hybrid code.[8] Signatures of drift acceleration were clearly evident in the hybrid runs at the lower ion (proton) energies for which $r_g \sim L_s$. It was also noted that because θ will fluctuate over a wide range at real shocks, the drift process can not only accelerate incident superthermal ions, but can also accelerate ions out of the thermal core when $\theta \sim 30°-60°$, and reflect them upstream where they will be available for further acceleration. A simpler model of internal shock structure was used to study the effects of magnetic overshoot on mildly superthermal ions.[9] It was shown that the overshoot enhances ion reflection when $r_g \sim L_s$, and also that reflection occurred at lower initial ion energies than those in the absence of an overshoot.

The role of shock curvature on drift acceleration of superthermal protons has been modeled using two and three dimensional bow shock surfaces.[10] Relative to the planar case, curvature (i) increases the fraction of reflected particles, (ii) lowers the energy threshold above which reflection can occur, and (iii) decreases reflected energy gains at increasing initial energies because shock curvature sets a limit on drift distance.

At higher energies, it has been suggested that Forbush decreases in the intensity of cosmic ray ions >30 MeV or so can arise from variable drifts along planar, but nonuniform, shocks.[11] The basic idea is that a variation in the shock strength along a shock front in the direction of ion drift paths will produce variable drift distances and energy gains. Behind nearly perpendicular shocks, this would produce local intensity voids with signatures similar to those observed at Forbush decreases.

It well known that electrons are accelerated to superthermal energies at the Earth's bow shock and at interplanetary shocks. These electrons are in the $r_g \ll L_s$ regime, and it has been reasonable to assume that electron interactions with quasi-perpendicular shocks are adiabatic, i.e., the electrons conserve their first adiabatic invariant or magnetic moment.[12] This approach has been adopted in theoretical analyses of electron energization and reflection at the Earth's bow shock by drift acceleration, sometimes referred to as fast Fermi or gradient drift acceleration in the context of electrons.[13,14] Test particle studies using time-dependent fields from a 1-D hybrid code have confirmed that the adiabatic assumption is indeed a good one.[15] This technique has been extended to investigate the effects of shock curvature on electron acceleration and reflection at planetary bow shocks.[16]

On the astrophysical scale, the scatter-free, but nonadiabatic, acceleration of relativistic electrons at relativistic superluminal shocks was exploited in a model of hot spots in extragalactic radio sources.[17] The increase, above the adiabatic value, of the mean energy gain due to single, nonadiabatic drift interactions of transmitted electrons, plus the increased compression across a relativistic shock (i.e., large proper compression ratio relative to the fluid frames) yields enhancements in the downstream electron synchrotron emissivity that are sufficient to explain the observed contrasts in surface brightness between the hot spots and the upstream flow.

MULTIPLE SHOCK INTERACTIONS

Single scatter-free shock drift interactions at quasi-perpendicular shocks can accelerate particles to at most a few times the shock compression ratio. Weak scattering during single drift interactions can increase this upper limit for a small fraction of an incident particle distribution, but the energy spectra are still rather steep. One anticipates large energy gains and flatter energy spectra that extend to high energies if some particles can return to the shock for many drift interactions.

In the diffusive limit where the particle distribution is maintained at near-isotropy by sufficiently intense magnetic fluctuations, drifts are included in the transport equation and, therefore, in the formalism of diffusive shock acceleration theory.[18] However, the region between the scatter-free and diffusive limits remains largely unexplored in any systematic fashion. We consider models other than diffusive shock acceleration (which is reviewed elsewhere[19,20] and addressed by other papers in this volume) that invoke multiple drift interactions with quasi-perpendicular shocks.

The clearest signatures of drift acceleration are found in shock spike events, which are spiky enhancements of superthermal ion and electron fluxes observed at interplanetary shocks. These events are narrow (sometimes only~few r_g), often multi-peaked intensity increases usually characterized by a large impulsive peak at the shock passage, rapid fluctuations in energy spectra, large anisotropies, and wide variations in structure from event to event.[1]

Single drift interactions of an incident ambient particle population under the ideal conditions depicted in Figure 1 cannot reproduce the impulsive structure of the observed spike events, and often underestimate the intensity increases. Recent models have emphasized effects of large-scale inhomogeneities and downstream turbulence.

It has been noted that some intense ion shock spike events display bidirectional, field-aligned anisotropies with loss-cone signatures in the region upstream of the shock.[21] This is suggestive of the classic case of a collapsing magnetic bottle, and is the basis of a model of proton trapping and acceleration due to multiple drift interactions along magnetic loops that convect through a planar quasi-perpendicular shock.[22] Upstream particles bounce back and forth along a loop and gain parallel energy at each reflection until they fall within the loss cone and transmit downstream. Proton orbits were integrated numerically in a steady-state model with an incident upstream proton population that was isotropic and distributed in energy as a power-law.

The left panel of Figure 2 shows the time (or space) evolution of the pitch angle distribution for 91-147 keV protons observed at an actual shock ($\theta=84°$), and the right panel shows the model predictions ($\theta=85°$) for 100 keV protons. The model nicely reproduces the qualitative behavior of the upstream bidirectional anisotropies, and also predicts a pitch-averaged peak intensity at the shock, as observed. Downstream, the observations and model differ, but this can easily be explained by an asymmetrical loop, or by a spacecraft path that was offset from the axis of a symmetrical loop.[22]

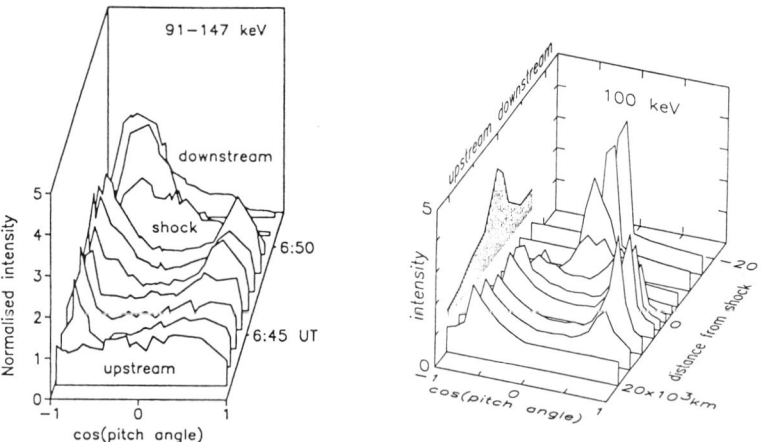

Figure 2. Left: Pitch angle distribution of 91-147 keV ions observed at a quasi-perpendicular interplanetary shock. Right: Predicted results for magnetic loop model.[22]

In a topologically similar model, drift acceleration takes place within shallow, large-scale ripples of sinusoidal form assumed to exist on the surface of an otherwise perpendicular shock.[23] As an incident (uniform) upstream magnetic field line convects through the ripple, it intersects the shock at two points, forming a temporary magnetic trap. A few particles injected into this trap undergo many drift reflections at the shock and are accelerated nonadiabatically before being convected downstream. In some cases the particles form a high-energy power-law tail on the accelerated energy spectrum. Large-amplitude, spike-like flux enhancements of width ~r_g are coincident with the shock passage. Pitch angle distributions upstream vary from unidirectional to bidirectional along the field, and those downstream are generally peaked nearly transverse to the field. The model reproduces many features of observed shock spike events, including their singly or multiply spiked intensity structures, the variability of these structures with particle energy in a given event, and the large variety of impulsive intensity structures observed from event to event.

The heating and acceleration of an incident thermal electron distribution by multiple shock interactions within collapsing magnetic traps, formed either by magnetic loops or rippled shocks, has been modeled assuming adiabatic shock interactions.[24] In addition to bulk heating, some electrons are accelerated to several times their initial energy. Predicted angular distributions are similar to those of the ions, with field-aligned beams upstream, and pancake distributions peaked transverse to the field downstream.

A multiple shock interaction model has been advanced to account for the radio flare seen two days after the explosion at SN1987A.[25] The emission is assumed to be synchrotron radiation from electrons accelerated to relativistic energies by multiple reflections from the blast wave. As the blast wave propagates outward through the progenitor star's spiral magnetic field, some electrons remain ahead of the shock, being continually overtaken and re-reflected as the spiral angle tightens and the shock approaches perpendicularity. Ultimately the shock becomes superluminal, and all electrons are deposited downstream, where they emit synchrotron radiation.

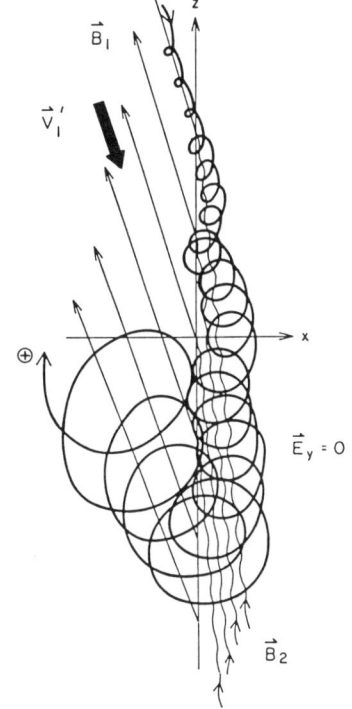

Models of multiple shock interactions by pitch angle scattering and spatial diffusion at quasi-perpendicular shocks require both a relatively high threshold speed ($v_0 \geq W$) of injected particles and also sufficiently intense magnetic fluctuations on both sides of the shock.[2,18] Chiueh has questioned these assumptions and has proposed a model wherein ions can be injected from only mildly superthermal energies and accelerated to high energies at quasi-perpendicular shocks devoid of upstream magnetic turbulence.[26,27] Ions undergo multiple shock drift interactions by cross-field diffusion driven by scattering off compressional mode hydromagnetic turbulence downstream of the shock.

Figure 3 depicts the basic physical process viewed from the dHT frame, where energy is conserved. Initially, $v_0 \ll W$ in the shock frame, so the ion speed is ~W in the dHT frame. The effect of the downstream turbulence is to return the ion to the shock several times, where parallel energy is converted to perpendicular energy, until finally the ion can return upstream, with a large energy gain in the shock frame. Broad energy spectra with power-law forms are possible, although the details depend upon a number of model parameters.

Figure 3. Ion orbit depicted in dHT frame.[26]

CONCLUSIONS

We close by posing a few questions prompted by the models described above. On the observational side: (i) How common are bidirectional ion and electron anisotropies at quasi-perpendicular shocks? (ii) What can systematic analyses of spatial structures and anisotropies as a function of particle species and energy tell us about the medium (e.g., distance to and conditions at the shock, level of turbulence, etc.)? (iii) Is there observational evidence for Chieuh's[26,27] proposed injection and acceleration scenario at interplanetary shocks? On the theoretical side: (i) What are the implications for plasma wave growth driven by counterstreaming, bidirectional ion or electron beams upstream of quasi-perpendicular shocks? (ii) What are the implications for local shock stability at localized "hot spots" of enhanced particle pressure at points where transient magnetic traps collapse? (iii) How are electron shock spikes at energies ~tens of keV to several MeV produced at interplanetary shocks?

This work was supported in part by NASA Grant NAGW-2619 and in part by NASA under Task I of Navy contract N00039-89-C-0001.

REFERENCES

1. T. P. Armstrong, M. E. Pesses, and R. B. Decker, in Collisionless Shocks in the Heliosphere (Geophys. Monogr. Ser. **35**, 1985), p. 271.
2. R. B. Decker, Space Sci. Rev., **48**, 195 (1988).
3. P. D. Hudson, Mon. Not. R. Astr. Soc., **131**, 23 (1965).
4. M. E. Pesses, R. B. Decker, and T, P. Armstrong, Space Sci. Rev., **32**, 185 (1982).
5. I. N. Toptyghin, Space Sci. Rev., **26**, 157 (1980).
6. G. M. Webb, W. I. Axford, and T. Terasawa, Astrophys. J., **270**, 537 (1983).
7. R. B. Decker and L. Vlahos, J. Geophys. Res., **90**, 47 (1985).
8. D. Burgess, J. Geophys. Res., **92**, 1119 (1987).
9. J. Giacalone, R. B. Decker, and T. P. Armstrong, J. Geophys. Res., **96**, 3521 (1991).
10. J. Giacalone, J. Geophys. Res., in press (1992).
11. A. F. Cheng, E. T. Sarris, and C. Dodopoulos, Astrophys. J., **350**, 413 (1990).
12. C. S. Wu, J. Geophys. Res., **89**, 8857 (1984).
13. M. Vandas, Bull. Astron. Inst. Czechosl., **40**, 175 (1989).
14. D. Krauss-Varban and C. S. Wu, J. Geophys. Res., 94, 15367 (1989).
15. D. Krauss-Varban, D. Burgess, and C. S. Wu, J. Geophys. Res., **94**, 15089 (1989).
16. D. Krauss-Varban and D. Burgess, J. Geophys. Res., **96**, 143 (1991).
17. M. C. Begelman and J. G. Kirk, Astrophys. J., **353**, 66 (1990).
18. J. R. Jokipii, Astrophys. J., **255**, 716 (1982).
19. R. Blandford and D. Eichler, Phys. Reports, **154**, 1 (1987).
20. F. C. Jones and D. C. Ellison, Space Sci. Rev., **58**, 259 (1991).
21. G. Erdos and A. Balogh, Planetary Space Sci., **38**, 343 (1990).
22. A. Balogh and G. Erdos, J. Geophys. Res., **96**, 11583 (1991).
23. R. B. Decker, J. Geophys. Res., **95**, 11993 (1990).
24. G. Gisler and D. Lemons, J. Geophys. Res., **95**, 14925 (1990).
25. J. G. Kirk and M. Wassmann, Astron. and Astrophys., in press (1992).
26. T. Chiueh, Astrophys. J., **333**, 366 (1988).
27. T. Chiueh, Astrophys. J., **341**, 497 (1989).

PARTICLE ACCELERATION IN SUPERNOVA REMNANTS

L. O'C. Drury
Dublin Institute for Advanced Studies
5 Merrion Square, Dublin 2, Ireland

ABSTRACT

The evidence for particle acceleration in supernova remnants is briefly surveyed and theoretical models discussed.

OBSERVATIONAL EVIDENCE

It is interesting to ask what direct observational evidence for particle acceleration in supernova remnants (SNRs) there is, or might be, before turning to theoretical models. The oldest, and in some ways the clearest, is the observation of radio synchrotron emission, indicating the presence of relativistic electrons in SNRs. However it must be said that the observations, although they do indicate that electrons were accelerated at some time during the evolution of most SNRs, do not unambiguously demonstrate continuing acceleration and the spectra are not obviously consistent with the simplest version of shock acceleration theory. Nevertheless, the very sharp limb-brightened edges of remnants such as Tycho (in radio emission) if interpreted, as discussed at this meeting by R. Blandford, in terms of a very small diffusion coefficient for the radio-emitting electrons also implies production and acceleration of these electrons at the outer shock; a population of electrons produced, or left over, in the interior would not be able to diffuse against the down-stream flow to the shock front and produce the sharp rim observed. It is of course possible that the sharp change in synchrotron emissivity reflects a sharp change in the magnetic field and not a change in the electron density, in which case this argument fails.

The radio synchrotron emission tells us nothing about accelerated ions; in principle these could be detected by the gamma rays produced (mostly through pion decay) when the accelerated ions suffer nuclear collisions with the material of the remnant. There have been claims for an enhanced gamma-ray emission from the region of the north polar spur on the basis of the COS-B data[1,2], but though tantalising the evidence is, in my view, very weak. Hopefully GRO will be able to produce some stronger results in this area. It is perhaps worth noting that in conventional propagation models the cosmic ray spectra in the sources should be harder power-laws (by about 0.6 in the exponent) than in the general ISM; thus in looking for evidence of acceleration in discrete objects (such as SNRs) we should look for two signatures; an enhanced emission and a harder spectrum.

In theory strong particle acceleration could be detected through the resulting modifications of the remnant dynamics; if one could get accurate estimates of the kinetic and thermal energy content of the remnant, its expansion speed and the density and pressure of the ambient medium, then an energy balance would reveal whether significant quantities of mechanical energy were being used for acceleration (or magnetic field generation!). However as these quantities can usually only be estimated to within factors of about two an energy budget accurate to 10% (which would be required) is currently impossible. A more promising approach is to look for the enhanced MHD turbulence predicted from resonant wave excitation just outside the bounding shock. S. Spangler has looked for such wave activity using interstellar scintillation studies[3], but without success.

A final possibility, very different in character from these astronomical methods, is to recognise that the solar system is an interstellar probe which must have passed through many SNRs and to look for evidence of this in the record of cosmogenic nuclei. This is discussed in more detail in the talk by G. Kocharov.

In summary, the evidence is not overwhelming. However observations of shocks in the Heliosphere show convincingly that at plasma parameters quite comparable to those of the ISM strong shocks are associated with copious particle acceleration. In addition there are good theoretical reasons, discussed in many presentations at this meeting, for believing that particle acceleration is intrinsic to strong collisionless shocks and should therefore occur in SNRs. And finally, the argument made thirty years ago by Ginzburg and Syrovatskii, that on energetic grounds the SNRs are the most plausible sources for the bulk of the galactic cosmic rays, retains its force; there is not much else capable of doing the job!

Thus we need to study acceleration in SNRs to see whether this can in fact explain all, or most, of the Galactic cosmic ray production and also to see what consequences there are for the dynamical evolution and structure of SNRs.

Before leaving the topic of observational evidence, there is a very interesting negative observation to which T. A. Lozinskaya has recently drawn attention[4]; although wind blown bubbles around WR stars have shock speeds, sizes and densities comparable to those of SNRs the radio emission is thermal rather than synchrotron. Does this indicate that such shocks are not significant electron accelerators or that the magnetic fields in the wind are very low?

THEORETICAL MODELS

Particle acceleration in SNRs is a very broad topic, and a full review would have to cover acceleration of electrons in the first few hundred days (the radio supernovae) and acceleration by pulsars in the plerions or filled-centre remnants. Nor, even excluding the plerions and composite remnants, are SNRs a uniform population; there are clear morphological differences between the O-rich remnants, exemplified by Cas A, and the sharp-rimmed almost spherical remnants exemplified by Tycho. The best hope for theoretical modelling would seem to be to start with the SNRs which look closest to a theoretician's ideal, the Tycho like remnants which are thought to originate in the thermonuclear detonation of a white dwarf as a SN of type Ia. This should be reasonably approximated by a point explosion releasing of order 10^{51} erg and a few solar masses of ejecta into a relatively undisturbed interstellar medium.

The explosion energy is initially contained almost entirely in the kinetic energy of expansion of the ejecta (adiabatic losses during the expansion rapidly convert the thermal energy of the ejecta into kinetic energy). There is no method of directly tapping the energy of this uniform large-scale expansion for particle acceleration (except perhaps at very high particle energies, cf the talks by Ip and Axford at this meeting); however there are two ways in which this energy can be made available on small scales more suitable for particle acceleration. First, when the swept up mass of ambient matter is roughly equal to the ejecta mass the expansion decelerates and a Rayleigh-Taylor instability can drive turbulence in the remnant; the total energy put into the turbulence is of order the explosion energy and may produce second-order Fermi acceleration[5,6]. Secondly, and currently more popular, the system of shocks associated with the SNR process a total amount of energy several times that of the explosion* and this is available for diffusive acceleration, a first-order Fermi process[7]. The main worry, based on experience of steady planar shock models, is that the mechanism may be too efficient and radically alter the appearance of SNRs.

In an ideal world we would simply run a very large plasma simulation (preferably treating the electrons as particles) of a SNR. While this is obviously impractical it does rather graphically illustrate the fundamental hope, that could such a calculation be performed, the cosmic rays would emerge naturally

---

* There is of course no contradiction with energy conservation; the expansion converts some of the downstream pressure back into kinetic energy which is then 'recycled' through the shock. To put it in quantitative terms, in the Sedov phase

$$\frac{1}{2}\dot{R}^2 \frac{4\pi}{3}\rho_0 R^3 = \theta E$$

where ρ_0 is the external density, R the shock radius, $\dot{R}$ the shock speed, E the explosion energy and θ a constant of order 0.7. The total flux of energy through the outer shock integrated over the duration of the Sedov phase is

$$\int 4\pi R^2 \frac{1}{2}\rho_0 \dot{R}^3\, dt = \int 3\theta E \frac{dR}{R} \approx 2E \ln(R_m/R_0)$$

where R_0 is the radius at the start of the Sedov phase and R_m that at the end. To this should be added the energy flux through the forward and reverse shocks during the preceeding sweep-up phase. Clearly the total energy available for driving diffusive acceleration can be several times the explosion energy, so that the constraints on the acceleration efficiency are somewhat milder than usually thought.

as the non-thermal high-energy tails of the distribution functions of the shock heated ions and electrons. With current resources we must reduce and simplify the problem to the point where it becomes possible to make some progress. Depending on where the emphasis is put this can be done in various ways.

One approach, which represents perhaps the extreme limit of this policy of reduction and simplification, is the 'simplified models' programme[8,9,10]. In the short space available to me I would now like to briefly discuss these models and what, in my opinion, we can learn from them.

The motivation for this work was that as important parts of the physics can only be handled in a rather crude and *ad hoc* manner it is sensible to begin studying the problem with a 'cheap and cheerful' model which incorporates all the important physical processes, but in a very crude and basic way. To do this we divide the SNR into four spherical shells; the undisturbed exterior, the cosmic ray precursor, the immediate postshock region and the interior. We neglect the thickness of the post-shock region and the inertia of the material in the interior. By using the conservation equations for mass, energy and (approximately, the spherical geometry causes some problems) momentum we derive a system of ordinary differential equations describing the evolution of the mass, thermal energy, particle energy and 'mean' radial velocity of each region.

There are at least two points relating to the acceleration on which this model can be criticised. First, the distinction between thermal energy and accelerated particle energy is artificial; there is no sharp distinction between thermal particles and the high energy part of the distribution function. Thus the injection has to be inserted 'by hand' using some parametrization of the injection rate. As a SNR description at the individual particle level is likely to remain hopelessly impractical for the conceivable future this problem cannot be avoided. However collisionless shock simulations should soon be able to suggest reasonable parametrizations and I do not personally regard this as a serious problem.

The second, and more serious, point which can be criticised is the assumption that the cosmic ray precursor can be characterised by a single spatial scale. This is certainly an extreme over-simplification. In reality particles of different energies should have different diffusion coefficients and different associated length scales. The same criticism applies to all 'two-fluid' models where the accelerated partiocles are treated as a second (mass-less) fluid coupled to the background fluid through diffusion and pressure.

The simplified models, however, also have some very positive aspects. They are dynamic, in that the discretized conservation equations are integrated forward in time to determine the evolution; we do not assume that the shock is in some quasi-static state (although it often is). Further the closure parameters, the mean diffusion coefficient and the effective adiabatic exponent for the accelerated particles, are determined dynamically during the evolution using our 'informed guess' as to the probable spectral shape and the information contained in the model.

But for exploratory work the great virtue of the simplified models is that they are trivial to compute; it is only necessary to integrate a system of eleven slightly stiff ordinary differential equations and the complete evolution of a SNR can be calculated on a workstation in a few seconds. Because of their simple structure it is also easy to incorporate additional physical effects, or experiment with different formulations. We have included in this way the effects of wave heating in the precursor region and diffusion between the post-shock region and the interior of the remnant, both of which turn out to be important in moderating the evolution.

The uncertainty of course is whether, in view of the extreme simplifications made in setting up the models, the results are of any use. Certainly we would not wish to defend them as exact quantitative models, and they were never intended to be such; our hope was that they would reproduce the correct qualitative behaviour. Apart from stating that the results look physically reasonable the only way to test this is to compare with more detailed calculations. Jones and Kang [11] did this for one case and found a worrying discrepancy; however I find that on rerunning the simplified model with the same prescription for the closure parameters as used by Jones and Kang the agreement is reasonable. Thus I conclude that in the one case where they have been tested the simplified models are in reasonable agreement with detailed hydrodynamic two-fluid models (which in turn are in reasonable agreement with calculations incorporating the full spectrum). Unfortunately the main thing we learn from the models is that they are very sensitive to the initial conditions, the choice of closure parameters and the parametrization of the injection! However the production of the bulk of the Galactic cosmic rays at energies below about 10^{14} eV in SNRs still appears quite possible.

REFERENCES

1. F. Lebrun and J. Paul *Proc. 19th Int. Cosmic Ray Conf. (La Jolla)* **1** 309 (1985).

2. C. L. Bhat, M. R. Issa, C. J. Mayer and A. W. Wolfendale *Nature* **314** 515 (1985).

3. S. R. Spangler *Proc. Varenna-Abastumani Int. School and Workshop on Plasma Astrophysics*, ESA SP-285 **1** 153 (1988).

4. T. A. Lozinskaya *Proc. 22nd Int. Cosmic Ray Conf. (Dublin)* **5** 123 (1991).

5. T. R. Gull *Mon. Notices R. Astr. Soc.* **161** 47 (1973).

6. D. Ryu and E. T. Vishniac *Astrophys. J.* **368** 411 (1991).

7. R. Blandford and D. Eichler *Phys. Rep.* **154** 1 (1987).

8. H. J. Völk, L. O'C. Drury and E. Dorfi *Cosmical gas dynamics: proceedings of the Manchester conference* ed. F. D. Kahn (VNU SCience Press, Utrecht) 101 (1985).

9. L. O'C. Drury, H. J. Völk and W. J. Markiewicz *Astron. Astrophys.* **225** 179 (1989).

10. W. J. Markiewicz, L. O'C. Drury and H. J. Völk *Astron. Astrophys.* **236** 487 (1990).

11. T. W. Jones and H. Kang *Astrophys. J.* **363** 499 (1990).

Observational Tests of Particle Acceleration Theory
— Shell Supernova Remnants

Martha C. Anderson, Lawrence Rudnick
University of Minnesota, Minneapolis, MN 55455

Abstract

We discuss ways in which radio continuum observations of shell supernova remnants can be applied to the study of astrophysical particle acceleration. As an example, radio observations of SNRs Cas A, G39.2-0.3, and G41.1-0.3 are presented.

Introduction

Galactic supernova remnants (SNRs), generally considered to be the birthplace of galactic cosmic rays, provide an observational laboratory in which we can test various models of astrophysical particle acceleration. Although in general we have no means of detecting the nuclear component of the cosmic rays accelerated in SNRs *in situ*, we *can* trace the effects of the acceleration process on the electronic component. Different acceleration mechanisms will have distinct signatures in the synchrotron radiation emitted by relativistic electrons.

We outline here some possible signatures of particle acceleration which should be manifested in the non-thermal radio emission from shell SNRs. We will concentrate on the shell class of SNR (as opposed to the plerionic or composite type remnants) because it presents the simplest acceleration scenario — there is no additional input of high energy particles from a central pulsar. Of primary importance to this study is the spatial distribution of the synchrotron spectral index in these remnants, as it is intrinsically related to variations in energy distribution amongst the accelerated electron populations. To this end, we discuss observations of three shell SNRs which exhibit spatially varying spectral indices.

Radio Synchrotron Emission from Shell SNRs: Signatures of Particle Acceleration

Radio Brightness Distribution:

Two different mechanisms have been suggested to explain the radio emission observed in shells. In most older remnants, where the shocks are radiative and highly compressive, compression of the interstellar magnetic field and cosmic ray electrons behind the SNR shock wave can account for observed synchrotron emissivities. For younger remnants, where the compression factors are unlikely to be greater than 4, a local acceleration of existing cosmic ray electrons may be needed. In this case, surface brightness variations may reflect variations in shock strength, acceleration efficiency, or perhaps weather in the pre-existing cosmic ray distribution.

Spatial Spectral Index Distribution:

Shell remnants have typically been found to have a fairly uniform radio spectral index[1,2] (α, where $S_\nu \sim \nu^\alpha$). However, a few have been shown to exhibit spatial spectral variations. There are several possible explanations for these variations. We might be viewing a single radiating electron population with a curved energy spectrum in regions with differing magnetic field strengths. We could be observing a mixture or superposition of thermal and non-thermal particle populations. Alternatively, the differing spectral indices may be highlighting regions of locallized particle acceleration. In this case, particles are accelerated and assume an energy distribution characteristic of the local shock and magnetic field parameters. Estimates of magnetic field and shock strength, in addition to measurements of the spectral index distribution, then yield tests for various acceleration models.

Polarization:

Fractional polarization intensity in synchrotron sources represents the degree to which the magnetic field is ordered within the effective beam area (in the absence of an irregular Faraday screen). It is therefore an important key to understanding the efficiency of acceleration in the presence of ordered and disordered fields. Maps of linear polarization orientation can be used to identify regions of dynamical turbulence, where second-order mechanisms may be most important, and to investigate the relative efficiencies of quasi-parallel *vs.* quasi-perpendicular shock acceleration.

Radio Observations of Three Shell Remants

Cassiopeia A:

Cas A, at 300 years of age, is currently undergoing the rapid dynamical changes associated with the transition between the free-expansion and Sedov phases of SNR evolution. Gull predicted[3] that this transition is accompanied by a rapid amplification of magnetic field, brought on by the onset of large scale deceleration of the ejecta shell. In such a turbulent scenario, one might expect a rampant acceleration of particles.

In Reference 4, we presented a study of the spatial distribution of spectral index in Cas A. We found that the "bow shock" structures (see ref. 5) are significantly steeper than other regions in the remnant. There is no evidence for curvature in the integrated spectrum of Cas A, nor is there evidence for significant amounts of intervening thermal material, thus these spectral variations may very well be tied to local particle acceleration conditions. The synchrotron emissivity in these compact features ranges up to 10^9 times a typical interstellar value — such an enormous increase cannot be acheived by simple compression. Some form of particle acceleration seems necessary.

Simple first-order Fermi models, however, are unable to explain, concurrently, the steep spectral indices and high emissivities associated with the bow shocks. The average spectral indices and the opening angles of the bow shocks are consistent with Mach numbers between 2 and 6.[5] However, Mach numbers three orders of magnitude larger are needed to increase the pressures in relativistic particles from typical cosmic ray values up to the minimum particle pressures measured in the bow shocks (assuming first-order acceleration in the steady state test particle limit, with k=100). Such high Mach numbers preclude the interpretation of these features as being "bow shocks", and would lead to spectral indices of -0.5 — much flatter than observed.

The conclusion to be drawn from this study is that simple first-order models do not work for the bright compact features in Cas A. Other types of mechanisms, including turbulent models and cosmic ray-mediated shock models, need to be explored.

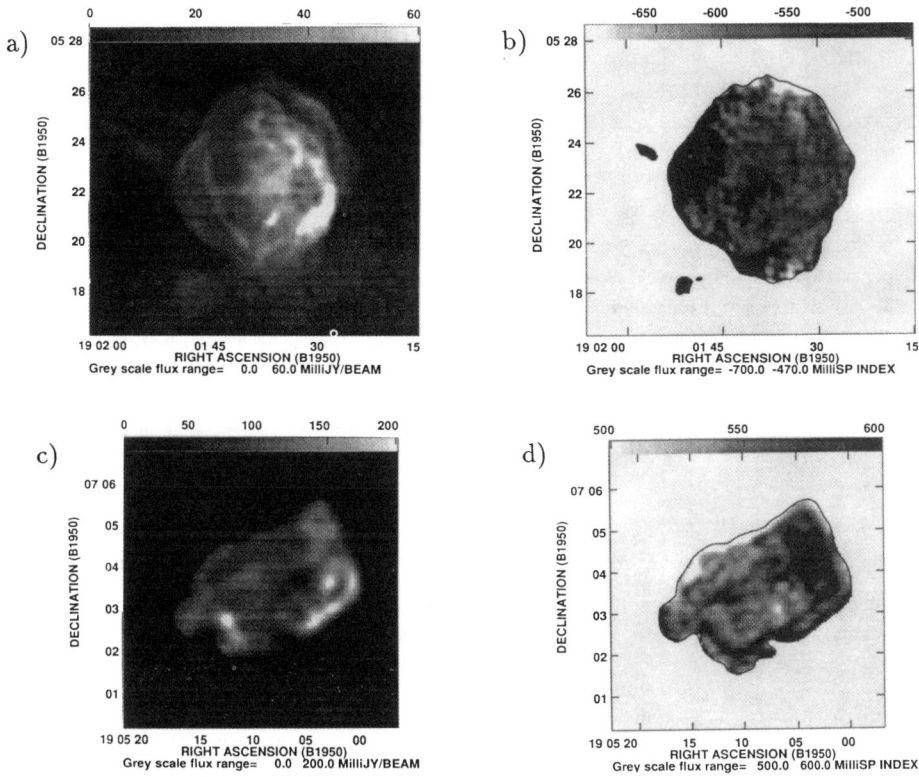

FIG. 1 Greyscale images (λ20cm) and spectral index distributions (measured between λ6 and 20cm) of G39.2-0.3 (a,b, respectively) and G41.1-0.3 (c,d).

G39.2-0.3:

There has been an ongoing debate concerning the classification of SNR G39.2-0.3 (Fig. 1a). Morphologically speaking, it has characteristics of both the shell class (a well-defined shell) and the plerionic class (a centrally enhanced brightness distribution), and thus might be typed as a composite remnant. The consensus recently, however, has been to classify it as a pure shell. Becker, et al.[6] found no apparent spectral difference between the central emission and the shell. Moreover, the integrated spectral index[7] of G39.2 ($\alpha_{int} = -0.4$) is more in line with that of the shell class than that of the plerions. Thus it has been suggested that the central component is a fragment of an uneven shell, seen in chance projection.[6,8]

New observations[9] at λ6 and 20cm indicate a slightly flatter spectral index associated with the central emission in G39.2-0.3, as compared with the shell (see Fig. 1b). In a plot of spectral index versus flux density at 1.4 GHz the central emission appears to be flatter than other regions of similar flux density by ∼0.04 in α. This suggests that the central enhancement may in fact be physically distinct from the shell emission. Whether particles are being preferentially accelerated in this region, *e.g.*, we are seeing a plerion in a very early stage, has yet to be determined (see Ref. 9 for further discussion).

G41.1-0.3:

G41.1-0.3, along with Cas A, is among the six smallest, brightest, and apparently youngest known shell remnants in our galaxy[10]. Fig. 1c is a total intensity image of G41.1 at λ20cm. It is predominantly filamentary in structure, with brightening in the direction of the galactic plane. A map of spectral index in G41.1 is shown in Fig. 1d. It is apparent that the western region of the remnant (roughly coincident with the region of brightening) is uniformly somewhat steeper than other parts of the remnant. The average spectral index in the west is -0.60, compared to an average in the central region of -0.55, with typical errors < 0.01 in α (see Ref. 9 for further discussion).

One possible explanation for the spatial coincidence of steeper and brighter regions in the west might be that enhanced deceleration is occurring in this direction, the lower Mach numbers thus yielding steeper spectra (see, *e.g.*, the discussion of Dubner, *et al.* [11] concerning Puppis A). The opposite result (bright regions being flatter) has also been found in some shells (*e.g.*, the Cygnus Loop[12] and IC 443[13], although it can be argued that the compression of electrons with a curved spectra might be important in these cases.

Polarized intensity and total intensity appear to be fairly well correlated in this remnant[14]. This would seem to indicate that a well-ordered field is important, in this case, to the efficient acceleration of relativistic electrons.

SUMMARY

Multi-wavelength radio observations of shell SNRs can be used to study the particle acceleration mechanisms considered responsible for the generation of galactic cosmic rays. Spatial variations in radio spectral index, observed in a handful of shell remnants, in conjunction with measurements of radio brightness and polarization can provide constraints for various acceleration models.

This work was supported, in part, by NSF grants AST-8720285 and AST-9100486 at the University of Minnesota, and by a grant of computing time from the Minnesota Supercomputer Institute. The observations presented here were obtained with the Very Large Array, operated by the NRAO through AUI's contract with the NSF.

REFERENCES

1. P.A.G. Scheuer, *Adv. Space Res.*, **4**, 337 (1984).
2. D.A. Green, *Genesis and Propagation of Cosmic Rays*, ed. M.M Shapiro & J.P. Wefel (Reidel, Dordrecht, 1988), p. 205.
3. S.F. Gull, *M.N.R.A.S.*, **161**, 47 (1973).
4. M. Anderson, L. Rudnick, P. Leppik, R. Perley & R. Braun, *Ap. J.*, **373**, 146 (1991).
5. R. Braun, S.F. Gull and R.A. Perley, *Nature*, **327**, 395 (1987).
6. R.H. Becker and D.J. Helfand, *A. J.*, **94**, 1629 (1987).
7. N.E. Kassim, *Ap. J.*, **347**, 915 (1989).
8. A.R. Patnaik, G.C. Hunt, C.J. Salter, P.A. Shaver and T. Velusamy, *Astron. Astrophys.*, **232**, 467 (1990).
9. M.C. Anderson, L. Rudnick (1992). In preparation.
10. J.L. Caswell, R.F. Haynes, D.K. Milne and K.J. Wellington, *M.N.R.A.S.*, **200**, 1143 (1982).
11. G.M. Dubner, R. Braun, P.F. Winkler, and W.M. Goss, *A. J.*, **101**, 1466 (1991).
12. D.A. Green, *A. J.*, **100**, 1927 (1990).
13. D.A. Green, *M.N.R.A.S*, **221**, 473 (1986).
14. R.H. Becker, T. Markert and M. Donahue, *Ap. J.*, **296**, 461 (1985).

COSMIC RAY ORIGIN FROM SUPERNOVA REMNANTS - A COMPARISON WITH EXTERNAL GALAXIES

H.J. Völk
Max-Planck-Institut für Kernphysik
D-6900 Heidelberg, Germany

ABSTRACT

The origin of Cosmic Ray nucleons from within individual Supernovae Remnants for energies below about 10^{15} eV is quite plausible theoretically, but a direct observational verification has been difficult up to now. An important question is therefore whether the properties of other spiral galaxies are consistent with this picture. The problem is considered through the study of observed statistical correlations for disk galaxies and a detailed analysis of the nearby starburst galaxy M82. Both types of evidence impressively confirm the supernova origin if pulsars are indeed not born with millisecond periods.

INTRODUCTION

Ever since Baade's and Zwicky's original suggestion [1] Supernova Remnants (SNR's) have been favoured as the sources of the Cosmic Rays (CR's) in the Galaxy on energetic grounds. A simple estimate shows this. Assuming SNR's to make the dominant contribution to the hydrodynamical energy input into the interstellar medium, as discussed below, this input rate is $\simeq 10^{42}$ erg/sec if we take $\simeq 10^{51}$ erg total energy per SNR and a Galactic supernova (SN) rate of one in thirty years. If we further assume that the grammage of interstellar matter seen by a CR particle during its lifetime depends on its rigidity proportionally to $R^{-\delta}$, where $\delta \simeq 0.5$, then the energy input into CR's is $\simeq 10^{41}$ erg/sec or even somewhat higher [2]. Thus, the CR power is a sizeable part of the total energy input and cannot be easily provided by any other type of sources.

TEST PARTICLE MODELS

On the theoretical side a number of models have been calculated, based on diffusive shock acceleration theory [3]. They approximate the time-dependent evolution of a single SNR and its CR production in various ways. If, on the one hand, their back-reaction on the SNR evolution is neglected, the CR's are considered as test particles. And since the acceleration scale of even the highest energy particles is $\simeq 1/30$ of the SNR radius in the adiabatic phase, one can at any epoch of evolution assume that a quasisteady particle distribution is generated behind the outer shock and is subsequently cooled adiabatically in the expanding interior of the SNR. For a SNR expanding into a uniform interstellar medium and assuming the shock to be quasiparallel, the CR momentum distribution may be calculated together with the selfexcited (and gradually dissipated) scattering wave field as they evolve in time. An example of such a kind of model is given in Fig. 1 [4]. It shows the volume integrated CR momentum spectrum F(p) that is released into the interstellar medium towards the end of the SNR evolution. Since different spectra are generated at different epochs, the resulting F(p) is not a simple power law but still approximately so at momenta p > mc with a spectral index $\simeq$ 4.2 to 4.3. The index varies somewhat depending on whether nonlinear wave damping is included or not. Such a source spectrum is indeed consistent with the observed spectral index $\simeq$ 4.75 of CR nucleons if $\delta \simeq 0.5$. Whether this result also carries over to the case of a SN Type II exploding into the wind of the (massive) precursor star remains to be seen although this appears quite likely. In addition, the assumption of a quasiparallel shock is not fulfilled over large parts of the SNR surface. Not only must diffusion and injection across the magnetic field then be taken into account, but especially the calculation of the scattering wave field poses a difficult theoretical question. Such studies are still in their infancy [5].

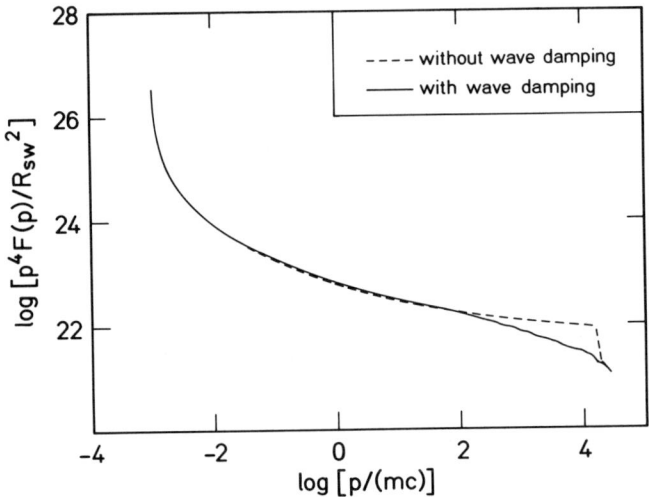

Fig. 1. The source distribution F(p) from a SNR in a $7 \cdot 10^5$ K hot interstellar medium, plotted as $p^4 \cdot F(p)$.

NONLINEAR MODELS

Since the efficiency of acceleration must, even on average, be at least 10 percent as we have seen, the backreaction of the CR's on the dynamics of the SNR cannot really be neglected. No full numerical solutions for this nonlinear problem exist for SNR's although serious attempts in this direction have been started [6]. Approximate calculations for the total CR energy density in SNR's have, however, been made extensively [2,7,8,9,10] and we show a particular result in Fig. 2 [7]. As a function of remnant age, in terms of the sweep-up time t_{sw}, it gives the fractional values of the internal energies E_g and E_c of the thermal plasma and the CR component, respectively. Although the solutions depend on estimates of the (time dependent) mean diffusion coefficient and CR adiabatic index, and especially on the chosen injection rate at suprathermal energies, they can be made consistent with the expected acceleration efficiency for reasonable estimates of these quantities [11]. Especially it is noteworthy [7] that, everything else being equal, the acceleration efficiency is essentially independent of the Galactic interstellar gas density n_o within rather broad limits $10^{-3} < n_o < 1$ H-atoms/cm³. This may justify the astronomically extremely convenient assumption that within limits SNR's are standard candles for the production of CR nucleons. Below we will make extensive use of this assumption.

Thus, it is probably fair to conclude that despite a number of shortcomings and even principal problems, existing theoretical models are consistent with a CR origin from SNR shocks.

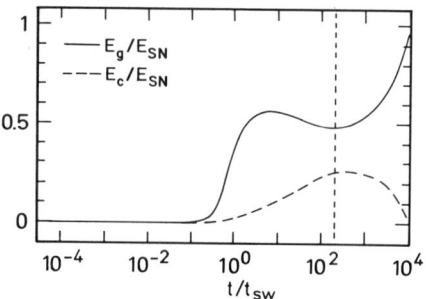

Fig. 2. *Temporal evolution of an adiabatic SNR: total internal energy in the gas (E_g) and CR's (E_c) in terms of energy released ($E_{SN}=10^{51}$ erg). The dashed vertical line indicates the onset of radiative cooling.*

Gamma ray observations with the SAS-2 and COS-B satellites on the other hand have at best resulted in circumstantial evidence [12] that SNR's are associated with CR nucleons. The reason is partly the insufficient instrumental sensitivity since the gas mass involved which can act as a target for π^o-generating collisions is quite limited. Another factor is the limited angular resolution of these γ-ray telescopes. The great hope is that the EGRET instrument on GRO will decisively improve on this situation.

THE RADIO FAR-INFRARED CORRELATION FOR DISK TYPE GALAXIES

In the face of inconclusive direct evidence for the physical nature of the CR sources in our Galaxy, it is important to study external galaxies which are governed by the same physical process but are quantitatively different not only in extensive parameters like size but also in intensive parameters as, for example, in their star formation rate per unit mass. Taking the latter to be proportional to the SN rate ν_{SN}, we can then ask whether it tracks the CR production rate and what types of SN precursor stars are associated with this CR origin.

Perhaps the best statistical result is the tight correlation between the spatially integrated radio emission per unit bandwidth for frequencies between about 0.15 and 10 GHz for disk galaxies, and their spatially integrated far infrared (FIR) luminosity over the wavelength range from about 40 to 120 μm as observed by the IRAS satellite [13,14]. This global correlation has been confirmed by many others. We show here in Fig. 3 our "own" correlation [15], using an optically selected sample of late type galaxies [16] which are qualitatively similar to our Galaxy. The correlation is somewhat steeper than linear and extends over 3 1/2 orders of magnitude.

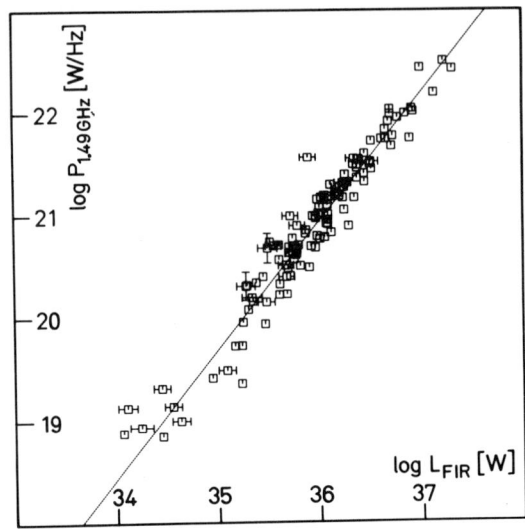

Fig. 3. Radio power at 1.49 GHz for a sample of 121 late-type galaxies vs. FIR luminosity.

Normalizing the correlation through division by total galactic mass shows that this dynamic range is not only a size effect. The radio emission is mostly synchrotron emission from nonthermal electrons whereas the FIR luminosity is mostly due to radiation from dust grains heated by stellar photons. The usual qualitative interpretation has been that the dust is heated by the radiation from massive stars which later explode as SN and then produce the CR electrons (and nucleons) responsible for the local synchrotron emission. Apart from the fact that the local synchrotron emission depends not only on the CR production rate but also on the magnetic field and the electron lifetime in the disk-halo system of the galaxies, also the interstellar photon field contains a large contribution from older stars which are more or less unrelated to ν_{SN} in gas-rich galaxies. In fact, the diffuse "Cirrus"-type dust emission in our Galaxy has been traditionally associated with the general Galactic interstellar radiation field outside HII-regions. However, it has been argued [17] that the synchrotron emission integrated over the halo depends linearly on ν_{SN} but only weakly on other galactic properties. More detailed models confirm the original approximate argument in ref. 17. In addition, by separating the galactic FIR luminosity into a "diffusive, cold" and a "warm, HII-region associated" component one can show that not only the latter component, clearly due to the radiation field from massive stars with HII regions, but also the "diffuse, cold" component correlates very well with the synchrotron emission. This is consistent with the argument [18] that also the "cold" component is predominantly heated by non-ionizing UV-radiation with wavelengths $91.2 < \lambda < 300$ nm from stars with masses $M > 5 M_\odot$; in fact at $\lambda = 200$ nm, about 60 percent of the stellar UV-radiation is absorbed within a galaxy like ours. It is very likely that stars with $5 < M/M_\odot < 8$ are born in equal proportion to core collapse SN progenitor stars with $M > 8 M_\odot$ over galactic epochs as short and recent as 10^8 yrs, the approximate lifetime of $5 M_\odot$ stars. Under these circumstances essentially all galactic

FIR emission is dominated by or at least proportional to the number of massive SN precursor stars, i.e. to the contemporary value of ν_{SN}.

AN ANALYSIS OF M82

If the previous analysis showed statistically that massive stars ultimately dominate the CR production, it is possible to investigate the nearby starburst galaxy M82 (also obeying the radio/FIR correlation) in detail. Its starburst nucleus has $\nu_{SN} \simeq 1/3$ yr. Except in the nucleus CR's are propagating convectively in an extremely fast galactic wind of $\simeq 4000$ km/sec speed, driven by the hot (T $\simeq 10^8$ K) SNR-heated gas. From the observed synchrotron emission of electrons one can derive a minimum energy density of CR's of $\simeq 80$ eV/cm^3 - two orders of magnitude larger than in our Galaxy - with a proton-to-electron ratio of 100, consistent with all the above parameters [19,20]. The total CR energy density for the observed ν_{SN} is $E_c \simeq 10^{50}$ erg per SN. It strongly suggests that the physical processes of CR production are the same both in our Galaxy and this unusually active galaxy whose starburst phase may not last much more than 10^7 yrs.

SNR's vs. STELLAR WINDS AND PULSARS

The strong observational evidence that CR production in normal galaxies is associated with massive stars, does not by itself exclude massive star stellar wind terminal shocks or even pulsars to be the CR sources. Energetically, mainly stellar winds from SN progenitor stars are important. However, they typically lose < 10 M$_\odot$, on average probably more like 5 M$_\odot$, with a speed of $\simeq 2000$ km/sec corresponding to a total energy loss $< 0.4 \cdot 10^{51}$ erg. Thus stellar winds would have to convert their energy almost completely into CR's which is obviously not the case. The only choice left besides SNR's are then pulsars. Assuming most core collapse SN to leave a pulsar behind, we could estimate their initial rotational energy $E_{rot} = (1/2) \cdot I \cdot (2\pi/P_o)^2 = 2 \cdot 10^{48} (P_o/100 \text{ msec})^{-2}$ ergs, if we knew their period P_o at birth, assuming $I = 10^{45}$ g cm^2 as a typical moment of inertia [21]. Most pulsars have rather low observed periods somewhat lower than a second, and it has been argued [22] that most pulsars should also be born as relatively slow rotators. Thus, unless most pulsars are born with periods $P_o < 10$ msec, their rotational energy is negligible compared to the canonical SNR energy of 10^{51} ergs. Up to now only 3 millisecond pulsars have been found. If such high rotation rates are indeed due to a secondary spin-up in a binary system as presently believed, then pulsars are energetically excluded as the dominant CR sources. From a physicist's point of view it would be important if we were able to strengthen these arguments beyond any reasonable empirical doubt.

Acknowledgements

I would like to thank Drs. J. Arons and S. Reynolds for a discussion concerning the energetics of stellar winds and pulsars. Especially I would like to thank my colleagues E. Wunderlich and U. Lisenfeld for the permission to use some yet unpublished empirical results regarding the radio/FIR correlation.

REFERENCES

1. W. Baade, F. Zwicky, *Phys.Rev. 46*, 76 (1934).
2. L.O'C. Drury, W.J. Markiewicz, H.J. Völk, *Astron.Astrophys., 225*, 179 (1989).
3. R.D. Blandford, D. Eichler, *Phys.Rep. 154*, 1 (1987).
4. H.J. Völk, L.A. Zank, G.P. Zank, *Astron.Astrophys. 198*, 274 (1988).
5. J.R. Jokipii, *these Proceedings*.
6. T.W. Jones, *these Proceedings*.
7. W.J. Markiewicz, L.O'C. Drury, H.J. Völk, *Astron.Astrophys. 236*, 487 (1990).
8. H. Kang, T.W. Jones, *Astrophys.J. 353*, 149 (1990).
9. T.W. Jones, H. Kang, *Astrophys.J. 363*, 499 (1990).
10. E.A. Dorfi, *Astron.Astrophys. 234*, 419 (1990).
11. L.O'C. Drury, *these Proceedings*.
12. C.L. Bhat, C.J. Mayer, A.W. Wolfendale, *Astron.Astrophys. 140*, 284 (1984).
13. T. de Jong, U. Klein, R. Wielebinski, E. Wunderlich, *Astron.Astrophys. 147*, L6 (1985).
14. G. Helou, B.T. Soifer, M. Rowan-Robinson, *Astrophys.J. 298*, L7 (1985).
15. E. Wunderlich, U. Lisenfeld, H.J. Völk, in preparation.
16. M. Aaronson et al., *Astrophys.J.Suppl. 50*, 241 (1982).
17. H.J. Völk, *Astron.Astrophys. 218*, 67 (1989).
18. C. Xu, *Astrophys.J. 365*, L47 (1990).
19. H.J. Völk, U. Klein, R. Wielebinski, *Astron.Astrophys. 213*, L12 (1989).
20. H.J. Völk, U. Klein, R. Wielebinski, *Astron.Astrophys. 237*, 21 (1990).
21. J.H. Taylor, D.R. Stinebring, *Ann.Rev.Astron.Astrophys. 24*, 285 (1986).
22. R.A. Chevalier, R.T. Emmering, *Astrophys.J. 304*, 140 (1986).

ON THE FIRST EXPERIMENTAL CONFIRMATION OF HYPOTHESIS OF COSMIC RAY GENERATION IN SUPERNOVA EXPLOSION

Grant E. Kocharov
A. F. Ioffe Physico-Technical Institute, St. Petersburg, Russia 194021

ABSTRACT

Very recent experimental data on cosmogenic ^{10}Be and ^{14}C isotope abundance in natural archives have been considered. It is shown that new data confirm the conclusion on cosmic ray intensity increase in the solar system due to SNE about 35,000 y ago.

INTRODUCTION

At the present time there is only one possibility of obtaining year by year time variations of cosmic ray intensity in the past. By measuring cosmogenic radioisotope concentration as a function of time in natural archives such as tree rings, polar ice cores, corals, stalagmites, and oceanic and sea bottom sediments, we can determine the properties of primary cosmic rays generated by such sources as supernova explosions and solar flares. The most detailed data available presently on the variations of the cosmic ray flux in the past 150,000 y have been obtained by studying the ^{10}Be and ^{14}C isotopes produced in the Earth's atmosphere. The analysis of the experimental data shows a synchronous increase of cosmic rays produced ^{10}Be and ^{14}C radionuclides abundances in the solar system 30-40 ky B.P. This phenomena was explained to be the result of the nearby supernova explosion.[1,2,3]

RECENT EXPERIMENTAL DATA

At the Tucson International Conference (May 1991) J. Beer and G. Raisbeck presented new experimental results on ^{10}Be abundance in ice cores spread to ^{10}Be concentration peak (around 35 ky B.P.) observed in the ice core from Vostok and Dome C Stations, Antarctica.[4] Based on a search in the ^{10}Be ice core records of Bird Station, Antarctica, and Camp Century, Greenland, Beer et al.[5] confirm the ^{10}Be abundance peak at ~ 35 ky B.P. established by Raisbeck et al.[4] According to Beer et al.[5], despite the coarse time resolution and missing data there is a good indication that the ^{10}Be peak at 35 ky B.P. is also present in the Camp Century record. Comparison of the ^{10}Be records in ice cores for Vostok and Byrd Stations shows the small time shift between the peaks which is due to uncertainties ($\pm$ 5 ky) of the time scales.[5]

An important question is whether the peak is caused by an increased production rate of ^{10}Be in the Earth's atmosphere or by special climatic conditions (precipitation rate change) or exchange between stratosphere and troposphere. Analysis of δ^{18}O and δD data shows that the observed peak cannot be due to precipitation rate change.[4] To investigate the exchange effect between the stratosphere and the troposphere Beer et al.[5] have measured sulfate concentration in the Camp Century core. SO_4 is a non-cosmogenic atmospheric tracer, which is transported by aerosols. Its main sources are marine and biotic activity and volcanic emissions. Based on SO_4 and δ^{18}O data, Beer et al.[5] have concluded that there is no indication that changes in atmospheric transport processes caused the observed peak. Therefore the most likely explanation for the considered peak is an increase of the ^{10}Be production rate by cosmic rays.

Grant Raisbeck has presented at the Tucson conference a new high resolution series of measurements in ice at Vostok Station, Antarctica. For the time interval covered ^{10}Be peak, Raisbeck et al.[6] even reached the highest time resolution about 50 y. Uncertainty in the ^{10}Be records is estimated as ~ 6%. It is shown that the ^{10}Be flux appears to have been relatively constant (± 20%) during the last 50 ky, except for ~ 1-2 ky interval 35,000 years ago, when it increased by a factor of 2. Based on the analysis of new data, Raisbeck et al.[6] concluded that this peak is "most likely a reflection of increased production and not a simple climatic effect." These new results completely confirm the existence of the clear ^{10}Be peak at 35 ky B.P. which was discovered by Raisbeck et al. in 1987.[6] Moreover, the very high time resolution allowed them to obtain the fine temporal structure of the considered peak which needs deep theoretical consideration. I hope the obtained fine structure can play a key role in a theory of this astrophysical phenomenon.

The results of radiocarbon and thermoluminescence dating of baked clay from aboriginal fireplaces at Lake Mungo, Australia, were published recently.[7] Radiocarbon and TL ages for the fireplaces F6, F7, F8, and F9 (in 1000 yr) are as follows: ^{14}C, 27.0 ± 0.5; 31.7 ± 0.5; 29.2 ± 0.4; 28.4 ± 0.4; TL, 31.4 ± 2.1; 36.4 ± 2.5; 32.7 ± 2.2; 33.5 ± 2.3. The author of these data, W. T. Bell[7], concludes "that radiocarbon ages from around 30,000 years ago are likely to be approximately 4000 years too young. This age correction can be explained by assuming that the radiocarbon production rate was higher 30,000 years ago than today due to a reduced geomagnetic dipole moment." Obtained results are in qualitative agreement with both ^{10}Be data[4,5,6] and radiocarbon records in stalagmites.[8]

Therefore the synchronization of southern and northern hemisphere ^{10}Be ice core records[4,5,6] and enhanced abundance of ^{14}C in the Earth's atmosphere[7,8] during 30-40 ky B.P. testifies for the global character of the cosmogenic isotope production rate increase at that time.

Bard et al.[9,10] have obtained very interesting results on U-Th and ^{14}C age measurements of corals from Barbados island. The comparison between the U-Th and ^{14}C ages for the Holocene samples demonstrates that the U-Th ages are accurate because they are in agreement with the dendiochronological calibration. It is shown that beyond 9 ky B.P. the ^{14}C ages are systematically younger than the U-Th ages with a maximum difference of about 3000 yr. reached at about 20 ky B.P. Calculated[6,10] atmospheric concentration of radiocarbon is presented in Fig. 1. It is seen that ^{14}C concentration in the past 30 ky was higher than the present day (pre-industrial) value. Excess reaches 40% at 20-30 ky B.P.

SUMMARY AND DISCUSSION

1. There is no doubt of the existence of a ^{10}Be peak at ~35,000 y B.P. in ice cores.

2. There is synchronization of southern and northern hemisphere ice core records for the considered peak.

3. Available radiocarbon records in stalagmites, baked clay, and corals, obtained independently using different experimental methods, show that atmospheric ^{14}C abundance was continuously increased in time to the past with maximum around 35,000 y B.P. by a factor of 2 in comparison with present day value.

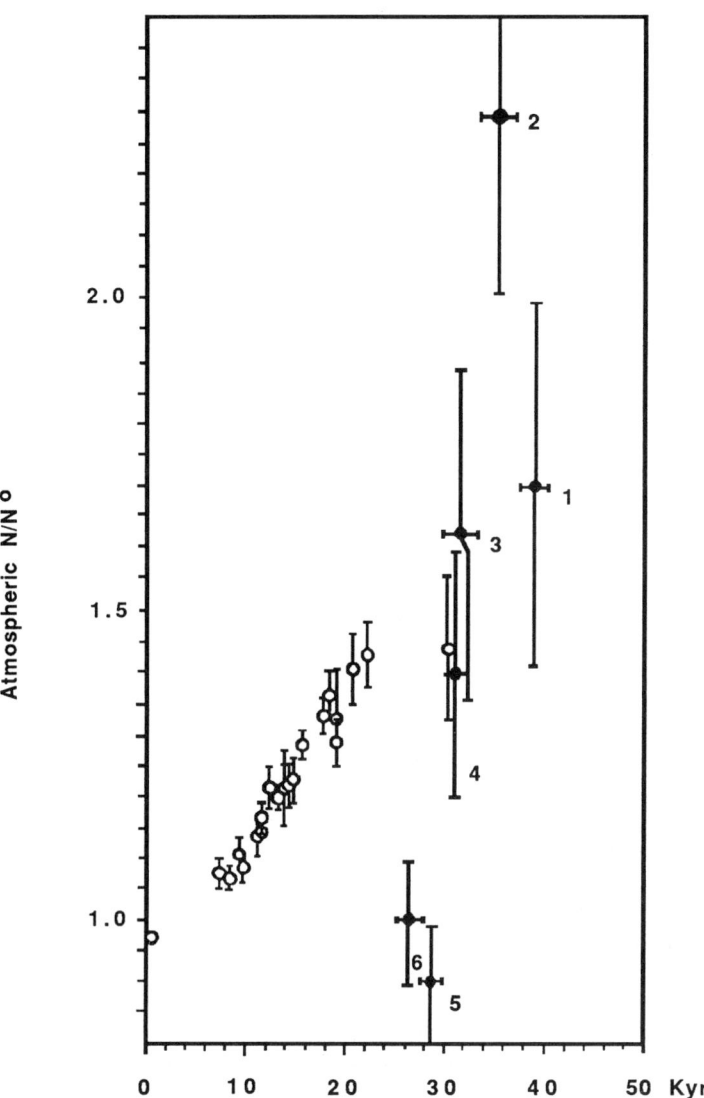

Fig. 1. Calculated atmospheric concentration of radiocarbon. Open circles are from coral data[10]. Solid circles are from stalagmite data[8].

4. Cosmogenic isotopes ^{10}Be and ^{14}C differed strongly in their geochemical and geophysical behaviour. Common for these isotopes is the fact that both have the same parents: cosmic rays.

Thus we can conclude that cosmic ray intensity at the top of Earth's atmosphere was considerably higher than now during the time interval of ^{10}Be and ^{14}C peaks.

Now first of all let us consider ^{10}Be and ^{14}C data for the time interval 30-40 ky B.P. Obtained time profiles of ^{10}Be peaks for both southern[6] and northern[5] hemispheres show the same rise and decrease time, ~ 2 ky. Maximum ^{10}Be abundance occurred for both cases at ~ 35 ky B.P. Unfortunately stalagmite radio-carbon data are not as detailed and reliable as ^{10}Be records. Nevertheless, they contain very important information. Time of maximum production rate is ~ 34-36 ky B.P. Rise and decrease times are practically the same, ~ 2-4 ky. Therefore we can say that both time profiles and amplitudes for ^{10}Be and ^{14}C peaks are practically the same. Detailed theoretical considerations[1,2,3,11,12] show that ^{10}Be and ^{14}C data for the time interval 30-40 ky B.P. can be considered as the first experimental confirmation of the longstanding fundamental hypotheses of W. Baade and F. Zwicky[13] on cosmic ray generation during supernova explosions.

An important question is the nature of the monotonic variation of radiocarbon atmospheric abundance during the last 30,000 y. Is it possible to explain it in the frame of SNE cosmic rays? In principle, yes, it is. The point is the following. Due to peculiarities of the cosmic ray energy spectrum and of nuclear reactions in which ^{14}C and ^{10}Be are generated, there is a difference between effective energy intervals for considered isotopes. ^{10}Be is generated mainly by cosmic rays with energies (2-5) GeV. The low energy region is more important for ^{14}C, and depending on the cosmic ray spectrum it may be from (0.5-0.8) to (2-3) GeV. As the diffusion coefficient of cosmic rays in the interstellar medium increases with energy the monotonous tail in ^{14}C time profile may be due to low energy particles which don't generate ^{10}Be isotopes. The discussed time profile of ^{14}C generated in the Earth's atmosphere by cosmic rays from SNE was predicted more than 20 years ago.[14,15] In the frame of theoretical considerations[14,15] it is impossible to explain the deep minimum (points 5 and 6 on Fig. 1). But if we combine two theoretical possibilities it may be possible to explain all of the time profiles: (i) a point-like instantaneous source of cosmic rays and subsequent diffusive propagation in interstellar media; (ii) cosmic ray generation in the vicinity of the shock wave from a supernova explosion. It is clear that we need new experimental data for the time interval from 20 to 50 ky B.P. with good time resolution and precision. Discussions with R. G. Fairbanks (Lamont-Doherty Geological Observatory of Columbia University, Palisades) show that there is a real possibility of prolonging coral ^{14}C data to 50 ky B.P. It is extremely important to have simultaneous measurements of ^{10}Be and ^{14}C in the same ice cores for the considered interval. A. Wilson (Arizona University) has recently developed a technique to measure the ^{14}C levels in the atmospheric gas trapped in polar ice. Thus the task of simultaneous measurements of ^{10}Be and ^{14}C abundances in ice cores may be solved in the near future.

ACKNOWLEDGMENTS

I have had helpful discussions with G. M. Raisbeck and F. Yiou on various aspects of this paper. A. N. Konstantinov and V. A. Levchenko have worked with me directly on this project[1,2,3] and now we are preparing a joint review paper.

REFERENCES

1. G. E. Kocharov, A. N. Konstantinov, and V. A. Levchenko, Proc. 21 ICRC, Adelaida, 7, 120 (1990).
2. A. N. Konstantinov, G. E. Kocharov, and V. A. Levchenko, Sov. Astron. Lett. 16, 799 (1990).
3. G. E. Kocharov, Nuclear Instruments and Methods in Physics Research B52, 583 (1990).
4. G. M. Raisbeck, F. Yiou, D. Bourles, C. Lorius, J. Jouzel, and N. I. Barkov, Nature 326, 273 (1987).
5. J. Beer, S. J. Johnsen, G. Bonani, R. C. Finkel, C. C. Langway, H. Oeschger, B. Stauffer, M. Suter, and W. Woelfli, to be published (1991).
6. G. M. Raisbeck, F. Yiou, J. Jouzel, J. R. Petit, N. I. Barkov, and E. Bard, In the last deglaciation: absolute and radiocarbon chronologies (E. Bard and W. S. Brocker, eds., NATO ASI series, Springer-Verlag, 1991, in press).
7. W. T. Bell, Archaeometry 33, 43 (1991).
8. J. C. Vogel, Radiocarbon 25, 213 (1983).
9. E. Bard, B. Hamelin, R. G. Fairbanks, and A. Zindler, Nature 345, 405 (1990).
10. E. Bard, B. Hamelin, R. G. Fairbanks, A. Zindler, G. Mathieu, and M. Arnold, Nuclear Instruments and Methods in Physics Research B52, 461 (1990).
11. G. Kocharov, Nuclear Physics B (Proc. Suppl) 22B, 153 (1991).
12. A. E. Ammosov, E. G. Berezhko, A. N. Konstantinov, G. E. Kocharov, G. F. Krymsky, V. A. Levchenko, Izv. Akad. Nauk SSSR, Serya Fiz (in Russian) 55, 2037 (1991).
13. W. Baade and F. Zwicky, Phys. Rev. 46, 76 (1934) and Proc. Natl. Acad. Sci. USA 20, 259 (1934).
14. G. E. Kocharov, V. A. Dergachev, and S. A. Rumyantsev, Proc. of Workshop on Astrophysical Phenomena and Radiocarbon, Tbilisi University, USSR, 1970, p. 11.
15. R. E. Lingenfelter and R. Ramaty, in Radiocarbon Variations and Absolute Chronology, ed. I. U. Ulsson, John Wiley and Sons Inc., New York, 1970, p. 513.

Stochastic Particle Acceleration

PARTICLE ACCELERATION IN SOLAR FLARES: OBSERVATIONS

Donald V. Reames
Laboratory for High Energy Astrophysics, Code 661
NASA Goddard Space Flight Center, Greenbelt, MD 20771

ABSTRACT

Contrary to our historical understanding, the energetic particles in most major solar proton events do not come from the flare itself. The particle abundances, ionization states, time evolution and longitude distributions all indicate that the particles are accelerated from the ambient plasma by a shock wave driven by a coronal mass ejection in these events. In contrast, the particles that *do* come from impulsive solar flares are unique in character. These particles are electron rich, have ^{3}He/^{4}He enhancements of up to 10^4, and enhancements in heavy elements such as Fe/C by factors of 10. The high ionization state of Fe, +20, indicates that the material has been heated to temperatures of $\sim 2\times10^7$K. It is generally believed that preferential heating by selective absorption of plasma waves is combined with stochastic acceleration in these events. Recent studies of the broad gamma-ray lines emitted by *energetic* particles within the flare loops indicate that they are also Fe rich, ^{3}He rich and proton poor like the particles seen at 1 AU. In large impulsive events, particles from the impulsive phase may be re-accelerated by a coronal blast-wave shock.

INTRODUCTION

To understand particle acceleration in solar flares, there are clear advantages in observing the accelerated particles directly, rather than trying to infer particle properties from the secondary photons that they may emit. Historically, however, our exploitation of solar particle observations at 1 AU for this purpose has been hampered by difficulty in disentangling the effects of multiple acceleration and transport mechanisms that we believed to occur in the large complex events that we saw. Observing with instruments of limited sensitivity, our early measurements were largely confined to events that we now regard as major proton events (MPEs). These events evolve gradually on a time scale of days and attempts to associate them with flares having durations of only hours led to models with extremely inefficient particle transport. A process of "coronal diffusion" was proposed[1] *ad hoc* to allow particles to diffuse uniformly around the corona from a point source of acceleration, followed by slow leakage into interplanetary (IP) space.

Abundances of the elements C through Fe were observed[2,3] in MPEs and were found to be comparable with abundances in the solar atmosphere insofar as the latter were known. Subsequently, it was found (see Meyer[4]) that underlying element abundances in the corona, in the solar wind, and in MPEs all agree. These abundances differ, however, from those in the photosphere because of a fractionation process that distinguishes neutral and singly ionized atoms, depending upon their first ionization potential (FIP), as they are transported up into the corona.

With improvements in instruments, a new type of particle event, the ^{3}He-rich event, was observed [5,6,7]. These events are distinguished by values of ^{3}He/^{4}He in the range 0.01 to 10; corresponding values in the solar atmosphere or solar wind are $\sim 5\times 10^{-4}$. Enhancements in heavy nuclei such as Fe by factors of 10 are also observed in ^{3}He-rich events (see Reames [8] and references therein). The absence of the isotopes ^{2}H and ^{3}H in events with ^{3}He/^{4}He ~ 1 eventually excluded the possibility that the ^{3}He might result from nuclear fragmentation, and resonant heating processes in the flare plasma (e.g. Fisk [9]) were suggested as the cause of the abundance enhancements.

As instrument sensitivity increased, many ^{3}He-rich events were observed and we found that they were correlated with the numerous non-relativistic electron events [10] that are observed at a rate of ~ 300 events/year. The 10-100 keV electrons in these events are the same electrons that produce type III radio bursts as they stream outward, nearly scatter free, from impulsive events at the Sun. The ^{3}He ions were observed to stream outward along the IP magnetic field with only slightly more scattering than the electrons. The particles arrive in a time that is defined by the distance along the field line divided by their velocity and events last only a few hours. Both the time profiles and the angular distributions suggest minimal scattering. The hard and soft X-ray flares associated with the ^{3}He-rich events [11] are impulsive events of duration < 20 min, but they are not distinguished in any way, e.g. by their size or complexity, from impulsive flares generally.

Until a few years ago, it was generally believed that ^{3}He-rich events implied some unusual conditions in the flare plasma where the enhancements originated [9] and, furthermore, that unusual conditions in the IP medium allowed the ions to propagate with minimal scattering. As we observe more and more events, however, these beliefs become less and less attractive. Figure 1 shows a recent collection of ^{3}He-rich events from the *ISEE-3/ICE* spacecraft, the lower panel shows the event rate corrected for the poor coverage of the spacecraft after 1983.

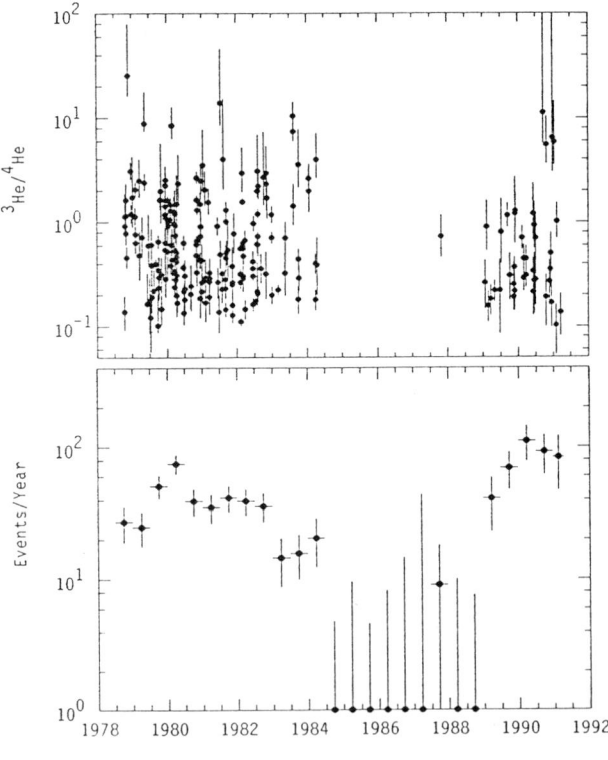

Fig. 1. Individual ^{3}He-rich events and their rate of occurrence during a solar cycle.

WHICH PARTICLES COME FROM FLARES?

Over the last few years there has been growing evidence that there are two well-resolved populations of energetic particles derived from at least two acceleration processes on or near the Sun. The first attempt to distinguish these two processes was made when Cane, McGuire and von Rosenvinge [12] studied the soft X-ray durations of flares producing relativistic electron events at Earth. The events with impulsive X-ray flares were found to be electron rich, or proton poor, relative to those with gradual flares. Pallavicini et al. [13] had previously observed that impulsive (< 1 hr.) soft X-ray events came from compact flares, low in the corona, while gradual events came from extended sources that occurred high in the corona. Moses et al. [14] found that electron spectra from the impulsive flares had spectral flattening above about 1 MeV that was absent in gradual events. Meanwhile Reames, von Rosenvinge and Lin [10] linked ^{3}He-rich events with the electrons that produce type III radio bursts and it became clear that these ^{3}He-rich, Fe-rich, electron-rich events arose from highly impulsive events on the Sun [15,11].

Energetic particle abundances were clearly distinguishing impulsive-flare events with enhanced ^{3}He, electrons and heavy elements from most MPEs that had heavy element abundances near or below those in the corona or solar wind. In fact, the Fe/C abundance ratio seemed to provide surprisingly clear discrimination between the two populations. Figure 2 shows the distribution of Fe/C for a sample of 90 electron events where both impulsive and gradual events are observed [16]. Events that are clearly ^{3}He-rich are blackened in the figure; other events have ^{3}He/^{4}He<0.1.

A key to the origin of the two particle populations was the measurement of the charge state of Fe. Luhn et al. [17] found that Fe ions in ^{3}He-rich events had a mean charge of 20.5±1.2, indicating a temperature of ~2×10^7K. In contrast a charge state for Fe of 14.1±0.2 is found by averaging over MPEs, indicating a temperature of ~2×10^6K as might be found in the ambient corona or solar wind. These observations led to the idea that the ^{3}He-rich, Fe-rich material has been heated in the flare, while the material observed in MPEs has been accelerated by a shock wave, for example, well away from the flare itself.

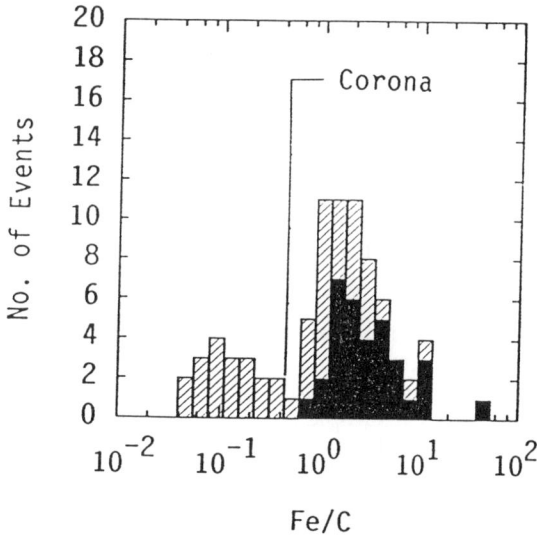

Fig. 2. Distribution of Fe/C for electron events. ^{3}He-rich events are blackened.

Understanding MPEs in terms of shock acceleration also provided an explanation for two other distinguishing features, namely, the long time duration of the events and their wide extent in solar longitude. Typical temporal behavior of the two classes of events is shown in Figure 3. Why do most MPE events last for several days, much longer than optical or X-ray events at the Sun, while the ^{3}He-rich events often have nearly scatter-free profiles? If the profiles respond to pre-existing fluctuations in the IP field, why do the MPEs never appear to encounter the same quiet field regions as the ^{3}He-rich events? If the MPEs come from shocks rather than impulsive flares, then the extended profiles result from continued acceleration rather than inefficient transport. Recently Mason, Reames and Ng [18] have refit the time profiles and anisotropies of protons observed at three radial distances in the classic MPEs of 1977 Nov. 22 and Dec. 27. The fits assumed extended injection and a scattering mean free path, $\lambda = 0.8$ AU. These events are quite consistent with a shock acceleration model.

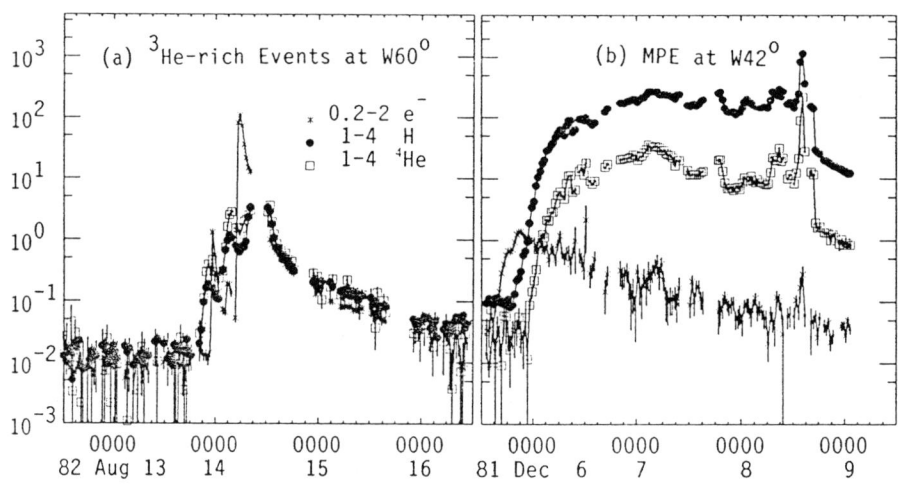

Fig. 3. Time-intensity profiles of electrons, H and ^{4}He are shown at the same scale for (a) several ^{3}He-rich events and (b) a major proton event.

Solar longitude distributions of particle events found using three different selection criteria are shown in Figure 4. The top panel shows the longitude distribution of a sample of ^{3}He-rich events, the bottom panel shows Fe/C vs. longitude for a sample of 36 MPEs used in an element-abundance study [19], and the middle panel shows Fe/C vs. longitude for the same sample of electron events studied in Figure 2 [20]. In the middle panel, the Fe-rich events are clustered near W50°, near the base of the field line to Earth while the Fe-poor events are distributed more uniformly across the Sun. In MPEs, shock waves cross magnetic field lines much more easily than particles and can accelerate particles near the base of the observers field line over a longitude range as large as ~180°. The maximum intensity for protons up to ~40 MeV from MPEs actually occurs near central meridian [21], not W50°.

Many MPE time profiles last until the IP shock passes the observer. These interplanetary shocks are driven by coronal mass ejections (CMEs) (see Kahler [22]). Furthermore, it was found [23] that 96% of MPEs had associated CMEs. Recently, Kahler, Reames and Sheeley [24] plotted intensities of MeV electrons and 175 MeV protons as a function of distance of the leading edge of the CME from the Sun, rather than as a function of time. They found that these intensities often peak when the CME is already 6-10 solar radii from the Sun. In fact, it is not surprising that CME-driven shocks play a major role in particle acceleration since the kinetic energy in a CME can be an order of magnitude larger than the radiative energy in the associated flare [25]. As noted by Kahler [22], the role of CMEs was not appreciated in early flare theories because CMEs were first observed 20 years after radio and X-ray bursts from flares. Similarly, CMEs were not included in early particle-acceleration theories since CMEs were first observed 10 years after the first energetic particles from MPEs.

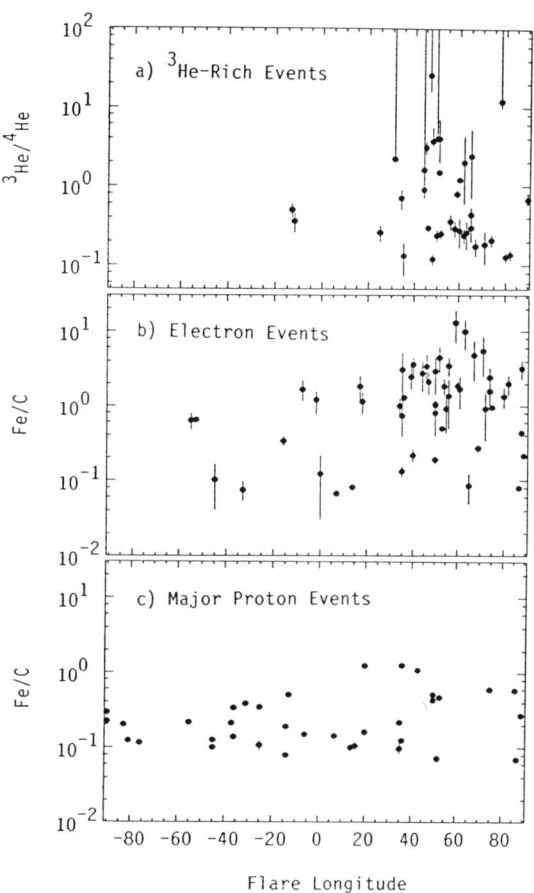

Fig. 4. Solar longitude distributions for three event selection criteria.

Thus, if we wish to study energetic particles from solar flares, it seems that we must study the electron-rich, ^{3}He-rich, Fe-rich, proton-poor particle population that is accelerated in impulsive flares. MPEs are not appropriate for such studies since the particles observed in MPEs are shock-accelerated from the ambient plasma far from the flare. In the process, we must leave behind such historical contrivances as "coronal diffusion"; in nearly 30 years we have been unable to identify any special mechanism to slowly transport particles across magnetic field lines to distant longitudes while holding them within the corona.

Before leaving the subject of flare classes, it is important to add two final comments. First, there *are* "two-phase" events. In magnetically well-connected two-phase events, we see Fe-rich material from the impulsive phase early in the event followed by Fe-poor material from the shock[26]. Second, the distinction between flare classes is *not* solely based on particle intensity. While most impulsive events are small, there are also extremely large impulsive-flare events with ³He-rich, Fe-rich, electron-rich particles [27]. Often these events are gamma-ray-line flares, and an analysis of broad lines produced by the interacting beam particles in one of these events shows the same abundance enhancements that we see in the particles at 1 AU [28].

ABUNDANCES OF PARTICLES FROM IMPULSIVE FLARES

Particles from impulsive flares are characterized by enhancements in electrons, ³He and heavy ions. They are proton poor in comparison with MPEs since MPEs are proton rich in comparison with the corona. Thus there are four abundance ratios, ³He/⁴He, Fe/C, e/p, and H/⁴He that distinguish impulsive events and MPEs. Surprisingly, however, among the impulsive events as a group, none of the four abundance ratios is correlated with any other. It was originally noted by Mason et al. [29] that Fe/C was not correlated with ³He/⁴He. It is possible that Fe ions reach different charge states because of event-to-event variations in temperature. Since the ion gyrofrequency depends upon Q/A, changes in the resonance conditions are one possible explanation for uncorrelated variations. Perhaps, also, the e/p and H/He ratios are affected by spectral variations.

Abundances of individual elements between C and Fe *do* show interesting event-to-event correlations, however. Figure 5 shows correlation plots of O/C, Ne/C, Mg/C, and Si/C as a function of Fe/C for ³He-rich events and MPEs [30]. Least-squares fit lines are shown for the two event types in each panel. The slope of the fitted line for the MPEs increases smoothly with atomic number. This is a consequence of the relatively smooth variation of abundances with Q/A in major events [31,19]. For the ³He-rich events, a sudden increase in slope occurs in going from O to Ne but the slope then changes little in progressing to Mg or Ne.

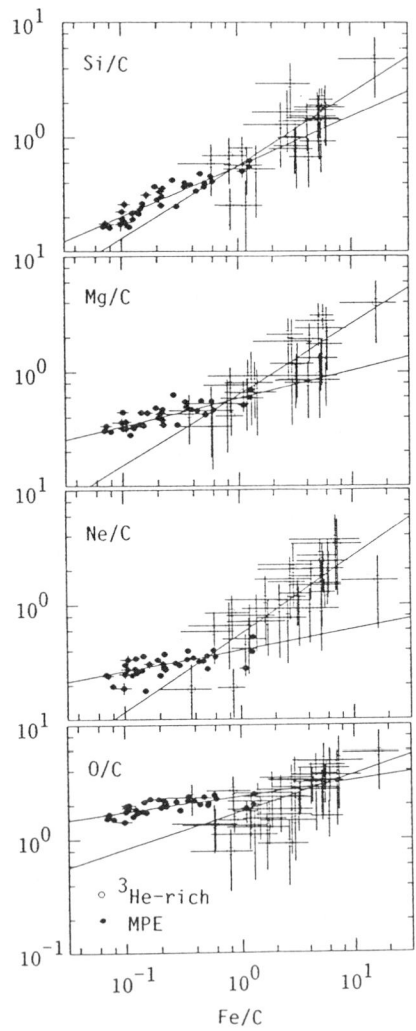

Fig. 5. X/C *vs.* Fe/C and fit lines for ³He-rich events and MPEs.

The variation in slope of the event-to-event abundance variations is shown in Figure 6 as a function of atomic number. We can discuss the behavior of the ^{3}He-rich events in the figure in terms of three element groups:
1) He - O
2) Ne - S
3) Fe

Abundances within a group vary little from event to event. For example, Ne/Mg remains completely constant, within statistics, as Fe/C varies [30]. Since Ne/Mg is a good indicator of the "FIP effect" [4], this observed, constant value of Ne/Mg identifies the material in ^{3}He-rich events as being of coronal, not photospheric, origin.

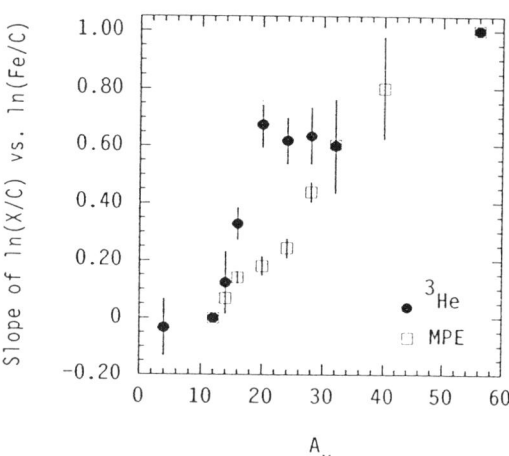

Fig. 6. Slope of the fitted lines of Fig. 5 vs. mass A_x for both event types.

The behavior of the abundances in these impulsive events is not understood quantitatively. However, an interaction of particles of different gyrofrequency with wave turbulence of unknown spectral form must take place. The elements C through S have the same Q/A as ^{4}He when they are fully ionized, hence they resonate at the same wave frequency and cannot be selectively enhanced. Thus, ^{4}He, C and N may be fully ionized throughout the process and O ionizes rapidly. The elements Ne through S are partially ionized in the ambient corona but fully ionized when observed at MeV energies [17]. Energetic Fe has changed its ionization state from ~14 to ~20 during the flare, but its gyrofrequency always remains less than that of C. Beyond these general considerations, however, there is currently no acceleration model that will produce the observed abundances.

ENERGY SPECTRA OF PARTICLES FROM IMPULSIVE FLARES

It is easiest to measure particle energy spectra in large events where sufficient numbers of particles are available over a wide energy interval. For this reason most spectral measurements come from MPEs or from large impulsive events [27] where secondary acceleration by a shock is possible. To study the "pure" impulsive-phase acceleration mechanism it is preferable to look at relatively small events where shock formation is less likely and there is no evidence of shocks in the form of type II radio bursts.

Ion spectra in ^{3}He-rich events were studied by Möbius et al. [32] and the spectra were found to be in agreement with a model of stochastic acceleration. The He spectra, observed in the 0.5 to 5 MeV/amu region, were not power law in form but steepened with energy. Unfortunately, some of the time periods in this early study were also long enough to contain contributions from multiple events. Recently, we have re-examined spectra in impulsive events [33] and extended the spectra over a wider energy interval. ^{4}He spectra from four events in this study are shown in Figure 7. These spectra can be described by two power-law regions with a break at ~2.5 MeV/amu. The power changes from ~2.0 to ~3.5 across the break. Heavier-element spectra seem to follow the ^{4}He spectra, at least above 2.5 MeV/amu, but proton spectra are more rounded in shape. At present there is no quantitative understanding of these spectra.

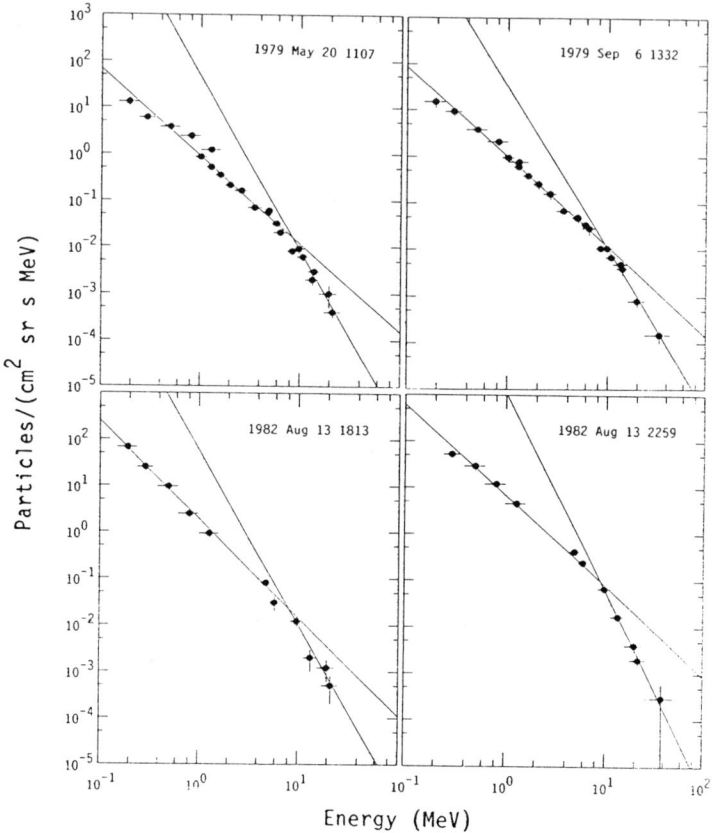

Fig. 7. Fitted ^{4}He spectra in impulsive events show sharp breaks near 10 MeV.

SUMMARY AND CONCLUSIONS

We have seen that the early attempts to understand the relationship between solar flares and energetic particles was biased by observations that were effectively limited to MPEs. Lacking information on the existence of CMEs, on one hand, and impulsive-flare particle observations, on the other, observers turned to models with slow diffusion through the corona and through IP space. Modern observations allow us to understand the properties of MPEs in terms of a CME driven shock that provides an acceleration source that evolves more gradually in space and time. This shock accelerates particles with abundances and ionization states similar to those of the local plasma on the magnetic field line of the observer. In fact, time variations in the particle abundances during an MPE may be attributed to the sampling of different field lines by the observer [34]. In any case, these particles do *not* come from flares. The properties of gradual and impulsive-flare events are shown in the adjacent table.

	IMPULSIVE	GRADUAL
PARTICLES	ELECTRON-RICH ^{3}He/^{4}He ~ 1 Fe/O ~ 1.0 H/He ~ 10	PROTON-RICH ~ 0.0005 ~ 0.1 ~ 100
	Q_{Fe} ~+20	~+13
DURATION	HOURS	DAYS
LONGITUDE CONE	<30 DEG	~180 DEG
RADIO TYPE	III, V (II)	II, IV
X-RAYS	IMPULSIVE	GRADUAL
CORONAGRAPH	-	CME (96%)
SOLAR WIND	-	IP SHOCK
FLARES/YEAR	~1000?	~10

The particles that *do* come from impulsive flares have unique abundances and spectra. They are electron, ^{3}He and Fe rich and, relative to MPEs, they are proton poor; yet none of these properties is correlated on a flare to flare basis. The Fe is in an ionization state that suggests heating to ~2×10^7 K. The abundances of the elements from C to Fe show correlated enhancements much different from MPEs, with evidence of clumping of the elements in groups. The same general behavior of the particle abundances is deduced from the spectra of broad gamma-ray lines from energetic particles accelerated in the magnetic loops during a gamma-ray flare [28].

The spectra of He and heavier elements in impulsive events break to a steep power-law spectrum with a slope of ~3.5 above ~2.5 MeV/amu. With these spectra it is not at all surprising that early observers, sensitive to ions above ~10 MeV/amu, saw relatively few of these events. In very large impulsive-flare events ^{3}He-rich, Fe-rich material is seen [27] with flatter spectra in the 10-50 MeV/amu region. It is possible that this material is re-accelerated by the blast-wave coronal shock that accompanies these large events.

Large ^{3}He-rich events can emit up to ~10^{31} ^{3}He ions above 1 MeV/amu, assuming emission over ~1 steradian. This can deplete nearly all of the ^{3}He in typical flare volume at a density of 10^9 protons/cm^3. The observation of smaller ^{3}He/^{4}He ratios in very intense events [11,27] may be a simple consequence of depletion of all of the ^{3}He that is available in the flare region.

Observations throughout the heliosphere have provided many opportunities to study particle populations that have been accelerated by shocks. However, the particles from impulsive solar flares are rather unique. While it is clear that plasma resonances near the ion gyrofrequencies are involved, no existing theory comes close to explaining the full range of observations. Resonant plasma processes involving electrons are common in radio observations of the Sun [35], but such processes involving ions may only be visible in ion abundances at 1 AU.

I would like to thank S. W. Kahler, C. K. Ng, and T. T. von Rosenvinge for helpful comments on this paper.

REFERENCES

1. Reid,G.C. 1964, *J. Geophys. Res.*, **69**, 2659.
2. Fichtel,C.E. and Guss,D.E. 1961, *Phys. Rev. Letters*, **6**, 495.
3. Bertsch,D.L., Fichtel,C.E. and Reames,D.V. 1969, *Ap. J. (Letters)*, **157**, L53
4. Meyer,J.P. 1985, *Ap. J. Suppl.*, **57**, 173.
5. Hsieh,K.C., and Simpson,J.A. 1970, *Ap. J. (Letters)*, **162**, L191.
6. Anglin,J.D., Dietrich,W.F. and Simpson,J.A. 1973, in *Proc. Symposium on High Energy Phenomena on the Sun*, eds. R. Ramaty and R.G. Stone (NASA SP-342), 315.
7. Garrard,T.L., Stone,E.C., and Vogt,R.E. 1973, in *Proc. Symposium on High Energy Phenomena on the Sun*, eds. R. Ramaty and R.G. Stone (NASA SP-342), 341.
8. Reames,D.V. 1990, *Ap. J. Suppl.* **73**, 235.
9. Fisk,L.A. 1978, *Ap. J.*, **224**, 1048.
10. Reames,D.V., von Rosenvinge,T.T. and Lin,R.P. 1985, *Ap. J.*, **292**, 716.
11. Reames,D.V., Dennis,B.R., Stone,R.G., and Lin,R.P. 1988, *Ap. J.* **327**, 998.
12. Cane,H.V., McGuire,R.E., and von Rosenvinge,T.T. 1986, *Ap. J.*, **301**, 448.
13. Pallavicini,R., Serio,S., and Vaiana,G.S. 1977, *Ap. J.*, **216**, 108.
14. Moses,D., Dröge,W., Meyer,P. and Evenson,P. 1989, *Ap. J.* **346**, 523.
15. Reames,D.V. and Stone,R.G. 1986, *Ap. J.*, **308**, 902.
16. Reames,D.V., Cane,H.V., and von Rosenvinge,T.T. 1990, *Ap. J.* **357**, 259.
17. Luhn,A., Klecker,B., Hovestadt,D., and Möbius,E. 1987, *Ap. J.*, **317**, 951.
18. Mason,G.M., Reames,D.V. and Ng,C.K. 1991, *Proc. 22st Int. Cosmic Ray Conf.* (Dublin), SH 4.1.2 (in press).
19. Cane,H.V., Reames,D.V. and von Rosenvinge,T.T. 1991, *Ap. J.*, **373**, 675.
20. Reames,D.V., Cane,H.V., and von Rosenvinge,T.T. 1990, *Ap. J.* **357**, 259.
21. Cane,H.V., Reames,D.V. and von Rosenvinge,T.T. 1988, *J. Geophys. Res.*, **93**, 9555.
22. Kahler,S.W 1992, *Ann. Rev. Astron. Astrophys.* (in press).
23. Kahler,S.W., Sheeley,N.R.,Jr., Howard,R.A., Koomen,M.J., Michels,D.J., McGuire,R.E., von Rosenvinge,T.T., and Reames,D.V. 1984, *J. Geophys. Res.*, **12**, 209.
24. Kahler,S.W., Reames,D.V. and Sheeley,N.R.,Jr. 1990, *Proc. 21st Internat. Cosmic Ray Conf.* (Adelaide), **5**, 183.
25. Webb,D.F. *et al.* 1980. *Solar Flares* (ed. P.A. Sturrock) Colo. Assoc. Univ. Press, Boulder 471.
26. Reames,D.V. 1990, *Ap. J. (Letters).* **358**, L63.
27. Van Hollebeke,M.A.I., McDonald,F.B., and Meyer,J.P. 1990, ApJ Suppl. 73, 285.
28. Murphy,R.J., Ramaty,R., Kozlovsky,B., and Reames, D.V. 1991, ApJ, 371, 793.
29. Mason,G.M., Reames,D.V., Klecker,B., Hovestadt,D., and von Rosenvinge,T.T. 1986, *Ap. J.*, **303**, 849.
30. Reames,D.V., Cane, H.V., von Rosenvinge, T.T. and Meyer,J.-P. 1991, *Proc. 22st Int. Cosmic Ray Conf.* (Dublin), SH 5.1.2 (in press)
31. Breneman,H. and Stone,E.C. 1985, *Ap. J. (Letters)*, **299**, L57.
32. Möbius,E., Scholer,M., Hovestadt,D., Klecker,B., and Gloeckler,G. 1982, *Ap. J.*, **259**, 397.
33. Reames,D.V., Richardson,I.G. and Wenzel,K.P. 1991, *Proc. 22st Int. Cosmic Ray Conf.* (Dublin), SH 2.2.1 (in press); also 1992, *Ap. J.* (in press).
34. Mason,G.M., Gloeckler,G. and Hovestadt,D. 1984, *Ap. J.*, **280**, 902.
35. Benz,A.O. 1982, *Solar Physics*, **104**, 99.

STOCHASTIC ACCELERATION IN IMPULSIVE SOLAR FLARES

James A. Miller* and Reuven Ramaty
Laboratory for High Energy Astrophysics, Code 665
NASA/Goddard Space Flight Center, Greenbelt, MD 20771

ABSTRACT

We consider ion energization by Alfvén waves, and suggest that a combination of nonlinear Landau damping and gyroresonant acceleration could account for proton acceleration in impulsive solar flares. Based on simulations which include both processes in a H plasma, we find that 10^{-5} of the ambient protons can be accelerated above 1 MeV and that these particles contain about 10% of the total available wave energy. From a simulation in a H–He plasma, we find a large ratio of accelerated alpha particles to protons. Furthermore, arguments based upon the nonlinear Landau damping rate indicate heavy element enhancements. These results are consistent with accelerated particle observations from impulsive flares.

INTRODUCTION

Particles accelerated at the Sun have been observed in the Earth's atmosphere and in interplanetary space for several decades, but only recently has the existence of two distinct classes of solar energetic particle events been clearly established[1]. Particles in first class or impulsive events are probably accelerated from hot flare plasma, a conclusion based on the temperature (10^7K) implied by the charge state of Fe ions observed[2] in these events. Particles in second class or gradual events are probably accelerated from ambient coronal gas by shocks, which we do not consider in the present paper. However, at the sites of impulsive energy release, where the magnetic field is high, the Alfvén speed is expected to exceed the velocity of mass motions. In such an environment shocks will probably not develop. Impulsive flares, therefore, offer a promising environment for studying stochastic acceleration.

Gamma ray production in solar flares is probably due to particles accelerated by the same mechanism as that which operates in impulsive events, a conclusion based on the impulsive nature of the emission[3] and enhanced heavy element abundances[4]. Observations of gamma ray line emission indicate[3] that acceleration to about 30 MeV occurs in about one to a few seconds, while pion radiation shows[5] that protons can be energized to a GeV in less than about 10 s. Likewise, time profiles of hard X-ray[6] and gamma ray[7] bremsstrahlung are diagnostics of electron acceleration, and indicate that acceleration to about 100 keV (10 MeV) can occur in as little as 10^{-2} s (2 s).

These time scales have immediate consequences for the level of plasma turbulence. Assuming that the spectral density of the turbulence is a power law in wavenumber with index $\approx 5/3$, the electron acceleration time[8] to 100 keV and 10 MeV is $\approx 5 \times 10^{-3}/Y$ and $1.5/Y$ s, respectively, while the proton acceleration time[9] to 30 MeV and 1 GeV is $0.2/Y$ and $1.5/Y$ s, respectively. Here,

---

* Universities Space Research Association

$Y = 4(W_t/U_B)(k_{\min}/10^{-6} \text{ cm}^{-1})$, where W_t is the total energy density in the turbulence, U_B is the magnetic field energy density, and $k_{\min}$ is the minimum wavenumber in the spectrum. Taking $k_{\min} = 10^{-5} \text{ cm}^{-1}$ (the minimum resonant wavenumber for a 1 GeV proton) and $U_B = 400 \text{ ergs cm}^{-3}$ requires $W_t \approx 10 \text{ ergs cm}^{-3}$ to account for the above times. This value of W_t is close to the thermal particle energy density, which should not be exceeded. The observed acceleration time scales thus require intense turbulence at relatively short wavelengths.

The number of accelerated particles and their energy content place further constraints. Using gamma ray observations, the total number of accelerated protons[10] can be as high as $\simeq 10^{35}$, and contain $U_p = 10^{30}$ ergs. From bremsstrahlung gamma rays above 300 keV, we find that the number of > 300 keV electrons in large flares[11] is about 5×10^{34}. The electron number and energy content above 100 keV are $\approx 10^{36}$ and $U_e = 10^{29}$ ergs, respectively. Hence, for an ambient hydrogen density $n_H = 10^{10} \text{ cm}^{-3}$ and a flare volume V of 10^{27} cm^3, about 1% of the ambient protons and at least 1–10% of the ambient electrons are energized. Further, $U_p = WVf_p$ and $U_e \approx 0.1WVf_e$, where W is the total amount of turbulent wave energy available, the f's are efficiencies, and the 0.1 factor takes into account the fact that electrons interact with waves which do not contain the bulk of the total energy. A 100% energetic efficiency implies that $W \approx 10^3 \text{ ergs cm}^{-3}$, which could be supplied if 150 G of magnetic field were annihilated throughout the acceleration region. Comparing with the thermal particle energy density limit mentioned above, we see that all this energy cannot be deposited at once, implying an extended injection of turbulence.

In early stochastic acceleration models[12,13], particles were energized by elastic scatterings with hard spheres, for which the diffusion coefficient is the same as that resulting from transit time damping of magnetosonic waves[14]. These waves, however, are strongly Landau damped[15], and thus not very useful for acceleration. Alfvén waves are not affected by Landau damping (except in second order, see below), and thus are a more attractive candidate for stochastic acceleration. Stochastic acceleration due to gyroresonance with Alfvén waves has been considered previously[16,17,18]. Unlike these steady state calculations, we treat here time dependent models, including a preacceleration mechanism.

ION ACCELERATION BY ALFVÉN WAVES

We consider stochastic acceleration due to Alfvén waves propagating either parallel or antiparallel the magnetic field B_0, with either left-handed (L) or right-handed (R) polarization. While the origin of the waves in flares is not well understood, there is evidence from our work in progress that electron beams are able to excite oblique Alfvén waves under flare conditions. For an effective interaction between an ion of Lorentz factor γ and parallel speed $v_\parallel$ and a wave of frequency ω and wavenumber k, the gyroresonance condition $\omega - kv_\parallel = -s\Omega_i/\gamma$ must be satisfied. Here $\Omega_i = Z_i eB_0/(A_i m_p c)$ is the ion gyrofrequency and $s = +1$ (-1) for R (L) waves. Since for Alfvén waves, $\omega \ll \Omega_p$, the resonant wavenumber $k_r = s\Omega_i/(v_\parallel \gamma)$. Using the dispersion relation $\omega = |k|v_A$, interaction with Alfvén waves requires that $|v_\parallel| \gg (Z_i/A_i)v_A$. The threshold kinetic energy per nucleon E_{th} is then $(1/2)m_p c^2(Z_i^2/A_i^2)\beta_A^2$, where $\beta_A = v_A/c$ is the Alfvén speed in units of c. For $B_0 = 100$ G and $n_H = 10^{10} \text{ cm}^{-3}$, the proton threshold energy is ≈ 25 keV, which is much greater than the thermal energy of ≃ 1 keV at 10^7 K, so that a preacceleration mechanism is needed.

Preacceleration by Nonlinear Landau Damping

It was suggested[15,19] that nonlinear Landau damping[20] could serve as such a mechanism. In this process, two parallel Alfvén waves combine to form a beat wave which drives a longitudinal electric field, and is subject to Landau damping. The maximum damping rate due to a particular plasma species is proportional to $n(m/T_\parallel)$, where n is the particle density, m the mass, and $T_\parallel$ the parallel temperature, and occurs when the beat wave phase speed v_φ equals the species thermal speed. If the two waves propagate in opposite directions and have the same polarization, $|v_\varphi| \leq v_A$; if they propagate in opposite directions and have opposite polarizations, $|v_\varphi| \geq v_A$; and if they propagate in the same direction, $|v_\varphi| = v_A$. For a typical flare case, the proton thermal speed $\ll v_A \ll$ the electron thermal speed. Therefore, interactions between antiparallel waves of the same (opposite) polarization yield beat waves which damp on the ions (electrons). However, since the damping rate is proportional to m, ion damping will be the most important.

We have derived a general expression for the nonlinear Landau damping rate of a wave due to another wave and a particular particle species. We assume an equipartition of energy among waves propagating both parallel and antiparallel to $\boldsymbol{B}_0$, a further equipartition of energy among waves of both polarizations in a given direction, and the same power law $|k|^{-n}$ ($n = 5/3$) for the spectral density of all four types of waves. The rate of decrease of the total Alfvén wave energy density W_A due to species j is given by

$$\left(\frac{dW_A}{dt}\right)^j \approx -\frac{1}{32}\frac{v_A}{U_B}X^j k_{\min} W_A^2 = -\gamma_{\mathrm{nll}} W_A. \tag{1}$$

The quantity X^j must be evaluated numerically. For isothermal plasma temperatures between about 10^5 and 10^8 K, $X^j/R_j \approx 8.7, 26, 74, 96, 105$, and 535, for ^{1}H, ^{4}He, ^{12}C, ^{14}N, ^{16}O, ^{56}Fe, respectively, where R_j is the elemental abundance relative to H. These numbers are not strongly dependent upon the charge states of the ions. It can readily be seen that the ion heating rate is approximately proportional to its mass, and may lead to abundance enhancements consistent with impulsive flare observations[1].

Combined Preacceleration–Acceleration Simulation

Here we present a combined treatment of heating and acceleration in a H or H–He plasma. The same distribution of Alfvén waves heat the plasma by nonlinear Landau damping and accelerate the particles by gyroresonant interactions. In the calculation of the nonlinear Landau damping rate γ_{nll}, we assume that the H and He distributions remain Maxwellian, justified by the short collision time scales. In the calculation of the acceleration damping rate[15] $\gamma_{\mathrm{acc}}(k)$, we assume that all ions heated above threshold remain at this energy, even though the ensemble as a whole continues to be energized. This approximation is fairly good because the resultant particle spectra are quite steep, but means that we will not obtain a detailed particle spectrum. We ensure conservation

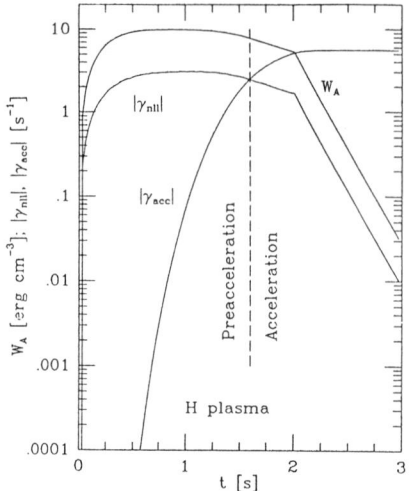
Fig. 1. H plasma damping.

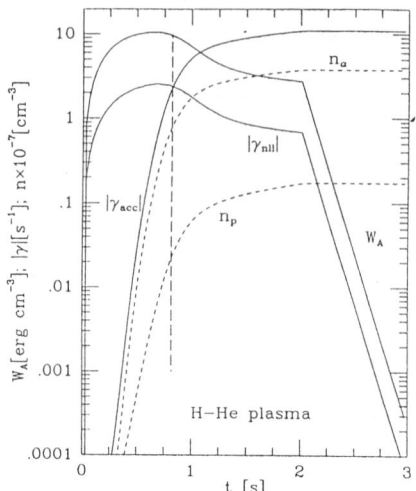
Fig. 2. H–He plasma damping.

of energy by allowing the normalization of the spectral density $W_A(k)$ to decrease, but assume that its shape remains constant. This assumption is valid if the cascading time scale of the turbulence is shorter than the damping time scale. Unless otherwise stated, we take $B_0 = 100\,\text{G}$, $n_H = 10^{10}\,\text{cm}^{-3}$, $n = 5/3$, $k_{\min} = 10^{-6}\,\text{cm}^{-1}$ (results for $k_{\min} = 10^{-5}\,\text{cm}^{-1}$ do not differ substantially), and assume that $30\,\text{ergs}\,\text{cm}^{-3}\,\text{s}^{-1}$ of turbulence is injected over $2\,\text{s}$.

The results for a H plasma are shown in Figure 1. The regime where $|\gamma_{\text{acc}}| < |\gamma_{\text{nll}}|$ is the time where proton heating dominates, whereas at later times, acceleration dominates and protons are energized to relativistic energies. After $2\,\text{s}$, W_A declines exponentially with an e-folding time of $1/|\gamma_{\text{acc}}|$. Most of the $60\,\text{ergs}\,\text{cm}^{-3}$ of available energy go to heating ($\approx 42\,\text{ergs}\,\text{cm}^{-3}$), with the rest being expended upon acceleration. The total number of accelerated protons is about $3\times 10^7\,\text{cm}^{-3}$, with an average energy $\langle E \rangle$ of about $0.47\,\text{MeV}$, obtained by dividing the energy spent on acceleration by the number of accelerated protons. The fraction of protons which are accelerated is therefore 3×10^{-3}.

The time dependencies of the damping rates and W_A are shown in Figure 2 (solid curves) for a H–He plasma with $R_{He} = 0.1$, along with the number of accelerated protons and alphas (dashed curves). The most striking result is that the number of accelerated alphas exceeds that of the protons. While the alpha particle spectrum is softer ($\langle E \rangle = 0.29$ and $0.17\,\text{MeV}$ for protons and alphas, respectively), the alphas outnumber the protons by more than an order of magnitude. The thermal protons and alphas gain about 15 and $1\,\text{ergs}\,\text{cm}^{-3}$, respectively, while the accelerated protons and alphas absorb 5 and $39\,\text{ergs}\,\text{cm}^{-3}$. The relatively large number of accelerated alphas is a consequence of their low threshold energy and higher acceleration efficiency.

Proton Acceleration and Escape from a Magnetic Trap

The previous simulation provided information only on the number and energy content of the accelerated particles but not on their spectrum. Modeling the

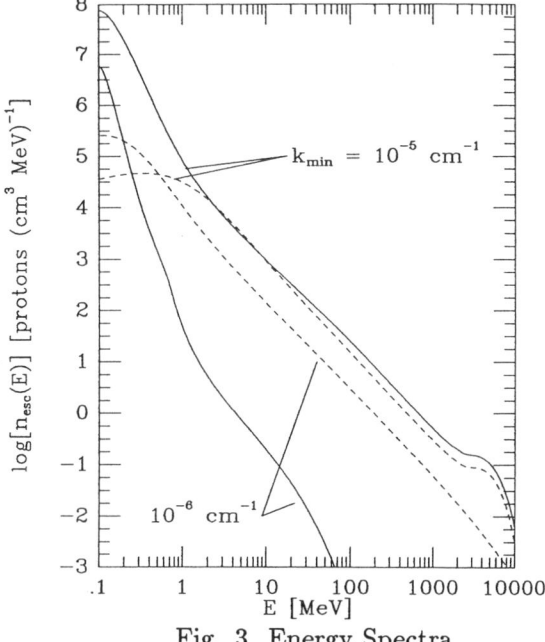

Fig. 3. Energy Spectra.

accelerated particle spectra by a K_2 Fermi spectrum[13], in which case $\langle E \rangle \approx 1.8 m_p c^2 (\alpha T_{esc})^2$, we find that $\alpha T_{esc} \approx 0.017$ for the H plasma, and 0.013 and 0.01 in the H–He plasma for protons and alphas, respectively.

One of the most severe limitations of previous stochastic acceleration models has been the lack of a consistent treatment of particle escape from the acceleration region. In particular, since the turbulence required to accelerate particles also traps them, the particles will escape only after the waves are damped. The modeling of the acceleration, therefore, must be time dependent. Such modeling has been carried out[21] for a H plasma under the same assumption as in the previous section, namely that the shape of the Alfvén spectral density does not change. However, while in the case of heating this assumption has probably no drastic consequences, in the case of acceleration the redistribution of the turbulence results in very soft spectra because most of the wave energy flows into the low energy particles. Here we present new results for a H plasma, taking into account the wavenumber dependent damping, without allowing any cascading.

We assume a magnetic trap as in ref. 21, and follow the evolution of the proton spectrum due to the damping of the waves and acceleration of the particles, but neglecting nonlinear Landau damping. Such a situation occurs after about 1.6 s in Figure 1. In Figure 3 we show the spectrum of protons which have diffusively escaped from a magnetic trap. The solid curves are for the impulsive injection of 10^7 protons cm^{-3} and 8 ergs cm^{-3} of turbulence. The dashed curves, for 10^5 protons cm^{-3} and 5 ergs cm^{-3} of waves, illustrate the effect of lowering the number of accelerated particles. We also take k_{min} to be either 10^{-5} or 10^{-6} cm^{-1}.

With the exception of the lowermost curve, the spectra are power laws above a few MeV of index $s \approx 5/3$, in contrast with the soft spectra obtained when cascading and turbulence redistribution are assumed[21]. While such a hard spectrum is probably not consistent with observations, an intermediate situation where the turbulence is partially redistributed could produce spectra which are more consistent with the data. The lowermost curve represents a case of very inefficient acceleration. Since k_{min} has been extended to 10^{-6}, more of the available wave energy is concentrated at lower wavenumbers. Since these resonate with very high energy protons, there is not enough wave energy to accelerate the protons which are initially near threshold to energies much in excess of a few tens of MeV before escaping from the trap. As a result, much of the wave energy at lower k's remains undamped and few particles are accelerated to high energies.

CONCLUSIONS

The simulations carried out above in the H plasma show that about 10^{-5} of the ambient protons can be accelerated to energies > 1 MeV, and that the energy contained in these protons could be as high as 10% of the total available Alfvén wave energy. We obtain these values by combining the results of Figures 1 and 3. For the ambient density of 10^{10} cm^{-3} assumed in the calculations, this particle efficiency is insufficient for a large flare, but then it is possible that either or both the flare volume or the ambient density are larger. In addition, the calculated particle efficiency could be increased if more of the preaccelerated protons above threshold would be accelerated to energies > 1 MeV, which could happen if some wave cascading was incorporated. On the other hand, the calculated energetic efficiency for proton acceleration seems to be more adequate, even though we have not yet been able to produce particle spectra consistent with observations.

REFERENCES

1. Reames, D. V. 1990, ApJS, 73, 235.
2. Luhn, A., Klecker, B., Hovestadt, D., & Möbius, M. 1987, ApJ, 317, 951.
3. Chupp, E.L. 1984, Ann. Rev. Astron. Astrophys., 22, 359.
4. Murphy, R. J., Ramaty, R., Kozlovsky, B., & Reames, D. V. 1991, ApJ, 371, 793.
5. Forrest, D.J., etal. 1986, Adv. Space Res., 6, No. 6, 115.
6. Kiplinger, A.L., etal. 1984, ApJ, 287, L105.
7. Rieger, E. 1989, Solar Phys., 121, 323.
8. Steinacker, J., & Miller, J.A. 1992, ApJ, 393, in press (20 July).
9. Steinacker, J. & Miller, J.A. 1992, this volume.
10. Ramaty, R., Dennis, B.R., & Emslie, A.G. 1988, Solar Phys., 118, 17.
11. Vestrand, W.T., etal. 1987, ApJ, 332, 1010.
12. Parker, E.N., & Tidman, D.A. 1958, Phys. Rev., 111, 1206.
13. Ramaty, R. 1979, in Particle Acceleration Mechanisms in Astrophysics, ed. J. Arons, C. Max, & C. McKee (New York: AIP), 135.
14. Achterberg, A. 1981, A&A, 97, 259.
15. Miller, J.A. 1991, ApJ, 376, 342.
16. Barbosa, D.D. 1979, ApJ, 233, 383.
17. Dröge, W., & Schlickeiser, R. 1986, ApJ, 305, 909.
18. Miller, J.A., Guessoum, N., & Ramaty, R. 1990, ApJ, 361, 701.
19. Smith, D.F., & Brecht, S.H. 1992, ApJ, submitted.
20. Lee, M.A., & Völk, H.J. 1973, Astro. Space Sci., 24, 31.
21. Miller, J.A. 1990, PhD Thesis, University of Maryland.

Simulations of Second-Order Fermi Acceleration of Electrons: Solving the Injection Problem

GALEN GISLER

Space Plasma Physics, *Los Alamos National Laboratory*,
Los Alamos, NM 87545.

ABSTRACT

The boosting of electrons from a Maxwellian distribution into a suprathermal power-law tail has long been recognized as an important bottleneck governing the subsequent acceleration of some of these electrons to relativistic energies. This is the seed or injection problem. I study this boosting process using a test-particle simulation code, following the full equations of motion of tens of thousands of electrons chosen from a thermal population as they move through general time-dependent magnetic fields. Inhomogeneities in the magnetic field are provided by finite swarms of moving current loops with Maxwellian velocity distributions and power-law distributions of loop size and dipole moment strength.

Whether bulk heating or boosting occurs is found to depend on the size of the swarm thermal speed compared to the electron thermal speed. When the swarm thermal speed is comparable to the electron thermal speed the entire electron population is heated by encounters with the rapidly moving current loops, approximately preserving the Maxwellian character of the electron distribution. On the other hand, at very low swarm thermal speeds there is no bulk heating; instead one percent or fewer of the electrons are boosted into a power-law suprathermal tail with a differential energy spectral index between −1 and −2. Individual boosts of 2000 and more have been observed in samples of 50,000 electrons. Most of the strongly boosted electrons have initial energies that are well below the peak of the initial Maxwellian.

INTRODUCTION

Stochastic acceleration, or second-order Fermi acceleration, is usually treated using a diffusion or Fokker-Planck approach[1,2]. The production of a power-law output energy spectrum from a monoenergetic input is readily accomplished, and there are well-known expressions for the acceleration efficiency and for the diffusion coefficients.

In this paper I am interested in the more general problem of what stochastic acceleration does to a Maxwellian input energy spectrum. This is related to the famous "seed" problem: how does the process of particle energization get started? Arguably, most of the diffuse matter in the universe has an energy spectrum that is roughly Maxwellian, and the extraction of energetic particles from that thermal matter is a core part of the problem of the production of cosmic rays. A diffusion approximation is not well suited to this extraction problem. The second law of thermodynamics demands that the production of a suprathermal tail be an inefficient process, selecting a very few particles for special treatment. Any process that acts on the majority of particles can only transform one Maxwellian into another, that is, by bulk heating or cooling.

Related to this consideration is the fact that a thermal population of particles in contact with a population of scatterers will tend toward thermodynamic equilibrium with the scatterers. Only if the configuration is limited in space or time, allowing accelerated particles to escape, can the production of a suprathermal tail occur. In this case the conditions favoring equilibrium are violated; the very few energized particles depart from the system carrying their excess energy with them. This turns out to be a very effective way for a plasma to get rid of excess free energy.

The application to thermal electrons is complicated by the fact that these particles typically have fast timescales and small gyroradii. The electromagnetic field in the interstellar and interplanetary medium generally varies over much larger time and spatial scales, and thermal electrons are generally thought to respond "only" adiabatically to the relatively small changes on the electron scales.

It is worth noting, however, that adiabatic changes can be sufficient for acceleration, *if* there is a way to "lock in" the gains made during the favorable phases of the adiabatic cycles. An example of this is acceleration in a converging magnetic trap[3]. This is essentially a first-order Fermi process in which an electron is trapped on a flux tube between two approaching magnetic mirrors. The electron gains energy on each encounter with one of the mirrors until it has gained enough parallel momentum to enter the loss cone of one of the mirrors. As it escapes through the loss cone the electron loses some energy, roughly the energy gained in its last approach to the mirror, but previous adiabatic gains are preserved. During each encounter with an approaching mirror, the electron is accelerated by the inductive electric field produced by the motion of the mirror.

THE SIMULATIONS

I choose to address this problem using the techniques of numerical simulation. A thermal population of particles is dropped into a distribution of randomly moving magnetic inhomogeneities, and its evolution is observed. Because the extraction process is posited to be inefficient, I have done *test-particle* simulations rather than fully self-consistent ones. There are two reasons for this choice, one physical, the other numerical. Physically, since the process is inefficient, the few particles that are extracted into the suprathermal tail do not greatly influence the evolution of the field inhomogeneities. Numerically, for every extracted particle one needs to compute thousands of particles that remain in the core of the distribution, and this makes self-consistent calculations expensive and impractical.

An example of this approach is given in Gisler and Lemons[3], where we presented simulations of electron acceleration in a closed flux tube with converging magnetic mirrors (*i.e.* first order Fermi acceleration), with application to bow shocks. In the present work I use a substantially upgraded version of that earlier code.

The magnetic scattering centers are represented as moving magnetic dipoles. The particles (which may be either electrons or ions) are followed in three dimensions along their orbits in the magnetic fields of these dipoles, using an adaptive Runge-Kutta scheme which amply resolves the particle gyrofrequency. The code is fully relativistic. There is no grid; the field sources all contribute to the field at any particle position. A particle-Hamiltonian approach is used for the equations of motion, so the acceleration of a particle is a result of the local inductive electric field at the particle position from the motion of nearby dipoles.

The code is implemented in THINK Pascal on the Apple Macintosh computer. Menus allow run-time adjustment of all important physics parameters in the code, and provide control of the real-time plotting and other outputs from the code.

Particles are chosen at random from a 3-dimensional thermal population, and dropped into a swarm of moving magnetic dipoles. In this paper, the dipole swarm also has a 3-dimensional Maxwellian velocity distribution, and flat distributions in position, dipole moment, and "hardness". Because of the finite size of the swarm and its thermal character, the configuration is always expanding, and particles almost always manage to escape. There are twenty dipoles in the swarm, and to avoid any statistical peculiarities caused by this small number, a new swarm from the same population is chosen after every hundred particles have been run. The dipoles are unaccelerated, that is, they are infinitely massive compared to the particles. Some individual swarms have been found to be much more effective than the average at producing suprathermal particles, and we have used these to try to understand how particles are selected for acceleration.

Since the full equation of motion is used, particles will scatter off the magnetic dipoles or not, according to the appropriate physics. Particle trapping can also occur, as follows. Two moving dipoles with opposing fields approach each other so that the field between them is temporarily reduced. A particle enters this low-field region on an open field line and becomes trapped in the field of one of the dipoles as the dipoles move apart. Subsequently the particle executes bounce orbits in the field of one of the dipoles, and is freed only by an encounter with another moving dipole. Three-body "recombination" events of this form are frequently observed in the simulations, but they result in no net energy gain for the particle. Somewhat less common are events in which a particle is trapped between two or more dipoles, but these events are exceedingly important for particle acceleration because when these traps are converging, very large energy gains can be realized.

We accumulate histograms of the initial and final particle energies, measure the amount of bulk heating, the index of the power-law tail (if any), the efficiency of boosting (defined as the number of particles with final energies greater than ten times the initial thermal energy divided by the total number of particles). We also examine initial and final energies of the boosted particles, to characterize the population from which energized particles are drawn. These statistics are summarized in Table I.

TABLE I. TABLE OF RUNS

Geometry	Swarm Speed	Number of Particles	Efficiency of Boosting	Index of power law tail	Bulk Heating
spherical	0.002	44729	0.29%	-1.76	1.02
spherical	0.006	62474	0.39%	-1.22	1.04
spherical	0.02	74503	0.44%	-1.51	1.14
spherical	0.04	45953	0.97%	-1.27	1.33
spherical	0.1	58744	1.18%	-1.15	1.80
spherical	0.3	45025	8.70%	-2.68	3.85
spherical	0.38	58235	13.32%	-2.27	4.66
disk, Bz	0.02	41230	0.19%	-1.27	1.07
disk, Bx	0.02	40950	0.21%	-1.56	1.06
disk	0.02	41079	0.23%	-1.09	1.09
optimal swarm*	0.02	10564	11.05%	-2.64	5.01
		523486			

* optimal swarm: one of the ~745 mirror swarms used in the spherical 0.02 case was found to be spectacularly successful at accelerating particles (13/100 compared to a Poisson distribution with the expected value of 0.44/100). This swarm was isolated and run, without reshuffling the swarm.

Results

When the swarm thermal speed is only a little less than the thermal speed of the particle population, as in Figure 1, the output spectrum looks very much like the initial one, just shifted to higher energy. The particle population has simply been heated by its contact with the dipole swarm.

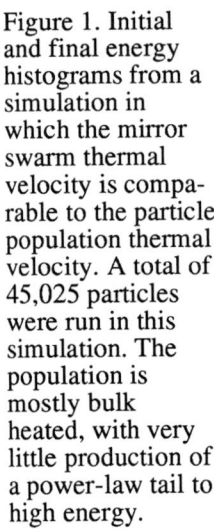 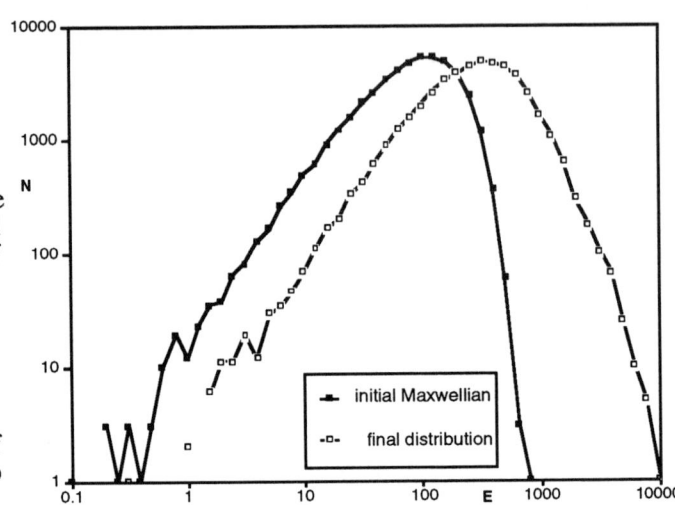

Figure 1. Initial and final energy histograms from a simulation in which the mirror swarm thermal velocity is comparable to the particle population thermal velocity. A total of 45,025 particles were run in this simulation. The population is mostly bulk heated, with very little production of a power-law tail to high energy.

On the other hand, when the dipole swarm thermal speed is small compared to the particle population thermal speed, as in Figures 2 and 3, there is no bulk heating, but instead the production of a modest suprathermal tail. About as many particles are dropped into a low-energy tail also. As expected, the acceleration process is relatively inefficient, and both the boosting efficiency and the slope of the power-law tail depend on the relative size of the swarm thermal speed.

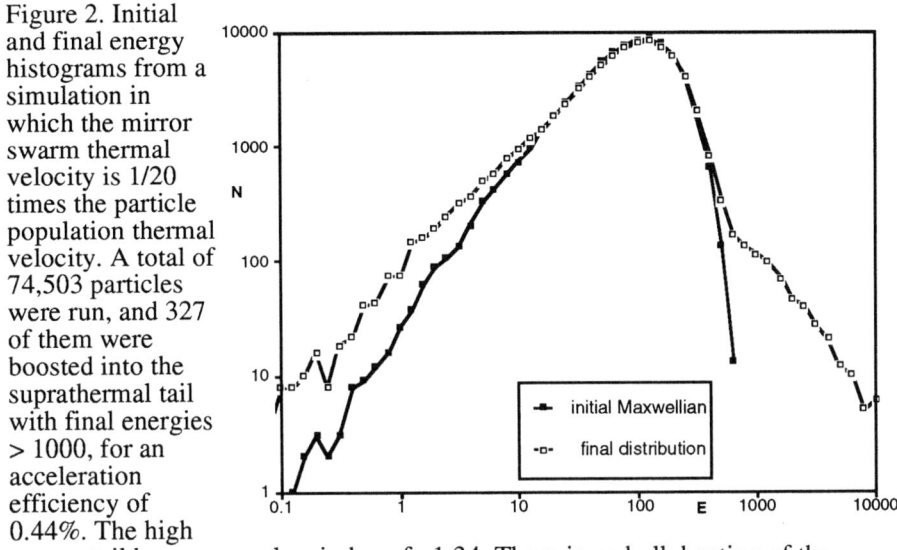

Figure 2. Initial and final energy histograms from a simulation in which the mirror swarm thermal velocity is 1/20 times the particle population thermal velocity. A total of 74,503 particles were run, and 327 of them were boosted into the suprathermal tail with final energies > 1000, for an acceleration efficiency of 0.44%. The high energy tail has a power-law index of −1.34. There is no bulk heating of the population.

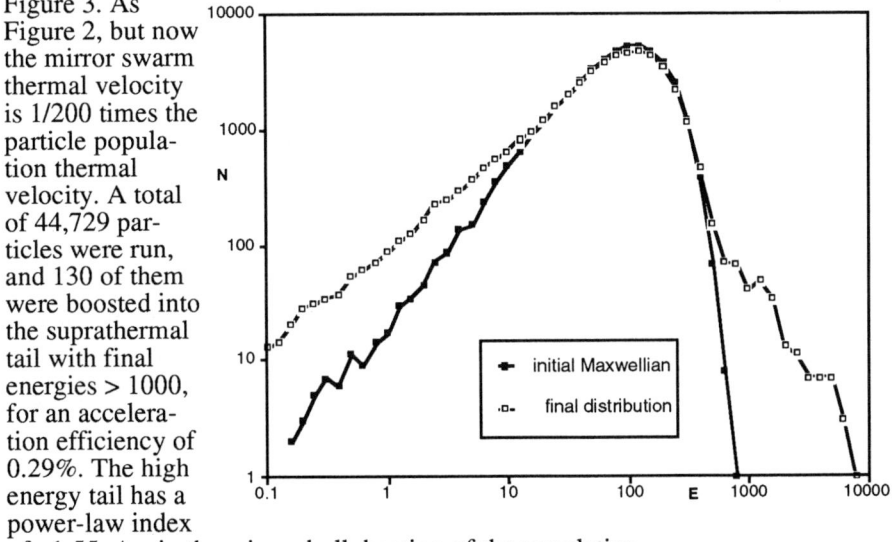

Figure 3. As Figure 2, but now the mirror swarm thermal velocity is 1/200 times the particle population thermal velocity. A total of 44,729 particles were run, and 130 of them were boosted into the suprathermal tail with final energies > 1000, for an acceleration efficiency of 0.29%. The high energy tail has a power-law index of −1.55. Again there is no bulk heating of the population.

An analysis of those particles in the suprathermal tail in all the runs of Table I yields the somewhat surprising result that most of them originate at energies well below the most probable energy in the original distribution (see Figure 4). In fact, because of the finite extent of the dipole swarm, particles that start out at higher energies tend to have fewer encounters with the mirrors before escaping, so have fewer opportunities to gain energy. The particles starting out at low energies suffer more encounters and hence are more likely to be involved in a trapping event. Single-dipole traps, as mentioned earlier, do not result in energy gains, but converging multi-dipole traps can produce enormous energy gains with the result that a particle is scattered far from its original energy.

The trapping of a particle in a converging multi-dipole trap is a relatively improbable event, so those particles that end up in the suprathermal tail are a very small fraction of the total number of particles studied.

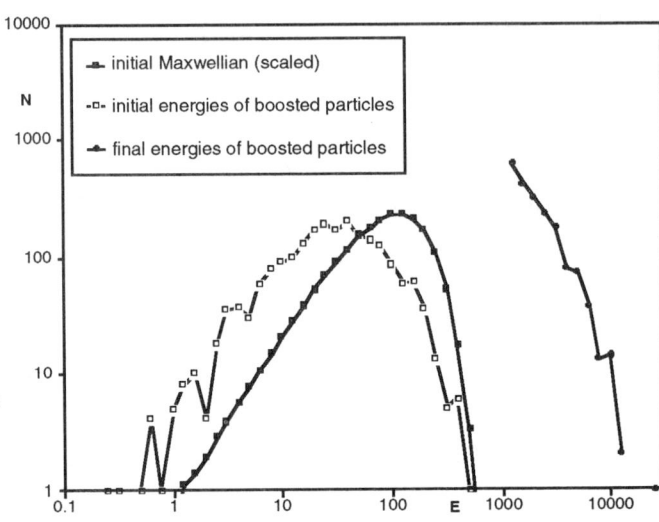

Figure 4. Comparison of initial and final energies for boosted particles from all runs. Boosted particles are defined as those particles with final energies greater than ten times the initial thermal energy. Also shown here is the shape of the initial Maxwellian, scaled to the total number of boosted particles. The boosted particles are found to originate in a distribution that is substantially cooler than the bulk of the population.

As mentioned before, the swarm of dipoles is undergoing, as a whole, a thermal expansion. One might expect that a typical particle in such a swarm would *lose* energy, since receding mirror velocities predominate. Indeed many particles lose a substantial amount of energy, as evidenced by the low-energy tails in Figures 3 and 4. But also occurring is the production of a power-law high-energy tail.

Thermal expansions are extremely common in astrophysics. Stellar atmospheres and winds, HII regions heated by hot young stars, supernova remnants, and galactic jets are but a few of the examples of such expansions. While the obvious result of a thermal expansion is an adiabatic cooling of the expanding gas, the acceleration of a few particles to high energies may also occur.

Acknowledgments

This work was done under the auspices of the U. S. Department of Energy's Office of Basic Energy Sciences.

References

1. M. A. Forman, R. Ramaty, and E. G. Zweibel, in *Physics of the Sun*, ed. P. A. Sturrock, T. E. Holzer, D. M. Mihalas, and R. K. Ulrich; D. Reidel, Volume 2, pp 249-289 (1986).
2. M. A. Miller, N. Guessoum, and R. Ramaty, *Astrophys. J.* **361**, 701 (1990).
3. G. Gisler and D. Lemons, *J. Geophys. Res.* **95**, 14925 (1990).

PROTON GYRORESONANCE WITH PARALLEL WAVES IN A LOW-BETA SOLAR FLARE PLASMA

Jürgen Steinacker[*,**] and James A. Miller[*]
Laboratory for High Energy Astrophysics, Code 665
NASA/Goddard Space Flight Center, Greenbelt, MD 20771

ABSTRACT

We consider the gyroresonant interaction of protons with parallel electromagnetic plasma waves. These waves have either right- or left-hand circular polarization, and include as a subset Alfvén and whistler waves. We identify three comoving gyroresonances, which can lead to divergences in the Fokker-Planck coefficients. Taking into account thermal damping, we calculate the Fokker-Planck coefficient D_{pp}, along with momentum diffusion coefficient $D(p)$ and the mean-free path $\lambda(p)$. Resulting acceleration time scales are compared with solar flare observations.

DIFFUSION COEFFICIENTS

We have previously derived[1] the Fokker-Planck equation describing the evolution of a particle distribution function due to gyroresonance with parallel transverse waves. We also specifically found that an electron can gyroresonate with a wave whose group velocity is equal to the parallel velocity of the particle along the magnetic field $\boldsymbol{B}_0$. We referred to this interaction as a comoving gyroresonance (since the particle is comoving with the energy flow), and showed that it leads to a divergence in the coefficients. The waves which can resonate with a particle of momentum p and pitch-angle cosine μ are shown in Figure 1 for the case of both electrons and protons, where we have taken $B_0 = 100\,\text{G}$ and a hydrogen density of $10^{10}\,\text{cm}^{-3}$. Here, "F" or "B" refer to wave propagation parallel (forward) or antiparallel (backward) to $\boldsymbol{B}_0$, respectively, while "R" and "L" indicates either right- or left-handed polarization; the preceding number is the number of resonant waves of this type. For example, if $p/m_p c = 0.1$ and $\mu = 0.4$, the proton can interact with one backward L wave and three forward R waves. The solid lines indicate values of p and μ where the particle resonates with a forward wave whose group speed is the parallel particle speed. (If $\mu \to -\mu$, "B" should be replaced by "F" and vice versa.)

We have numerically evaluated the Fokker-Planck coefficients for the power law spectral density given by equation (4.1) of reference 1. Taking the spectral index of the turbulence $y = 1$ and the normalization $A = 0.1$, we show the coefficient D_{pp} for protons in Figure 2. We have taken into account the thermal damping of the waves in a $10^7\,\text{K}$ plasma by introducing a maximum normalized wavenumber kc/Ω_p (Ω_p is the proton cyclotron frequency) in the spectral density equal to 4.6×10^3 for R waves and 95 for L waves. (Below these wavenumbers, the waves suffer negligible thermal damping on time scales of several seconds[2].)

---

* Universities Space Research Association
** Alexander von Humboldt-Stiftung

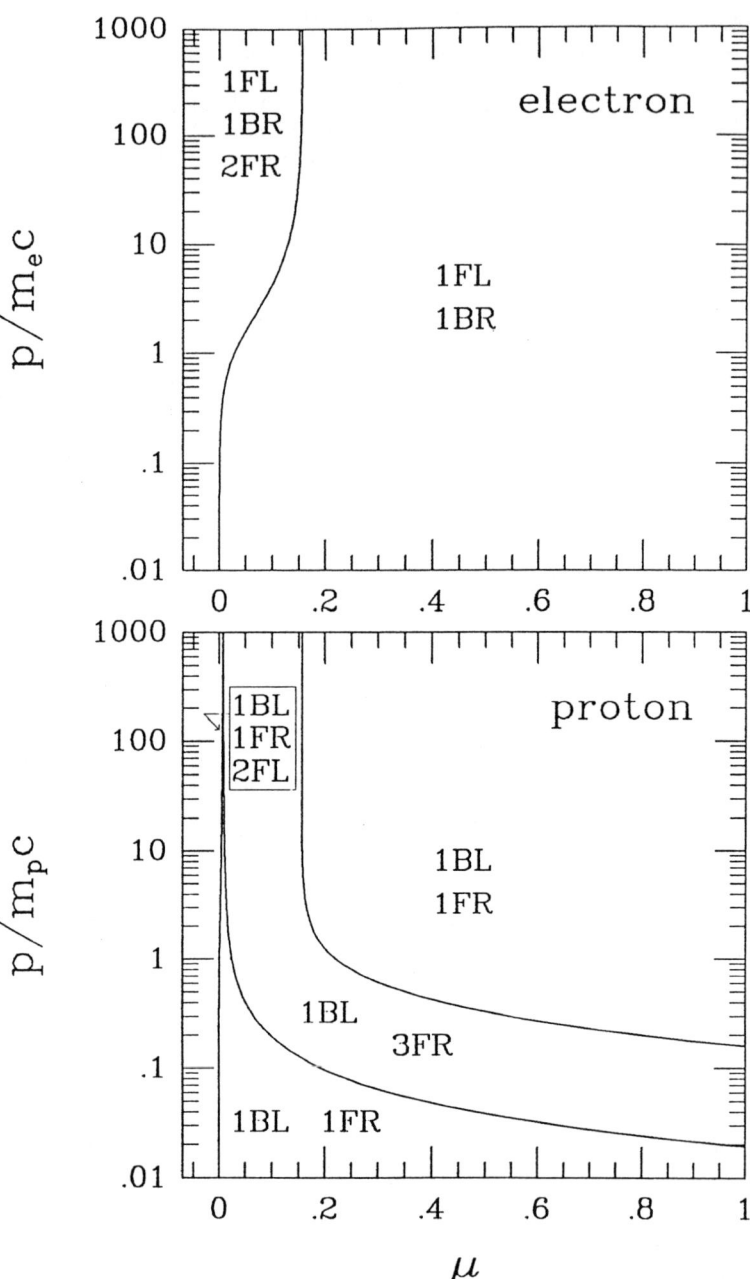

Fig. 1. Number, propagation direction, and polarization of waves in gyroresonance with electrons and protons as a function of μ and p.

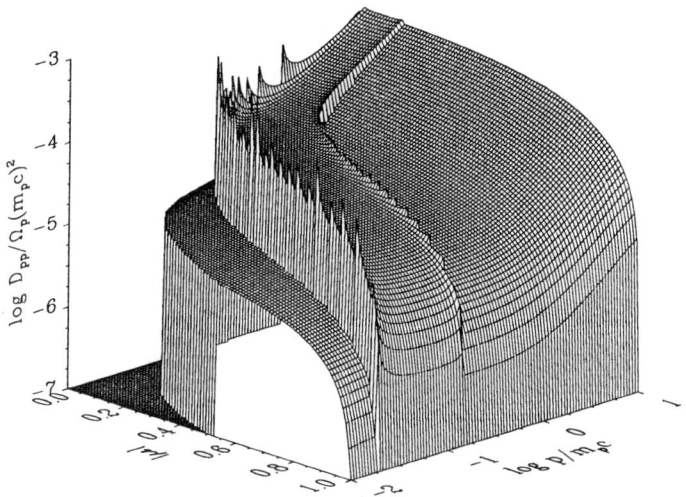

Fig. 2. Fokker-Planck coefficient D_{pp} as a function of p and μ for $B_0 = 100$ G, $n_H = 10^{10}$ cm^{-3}, and $T = 10^7$ K.

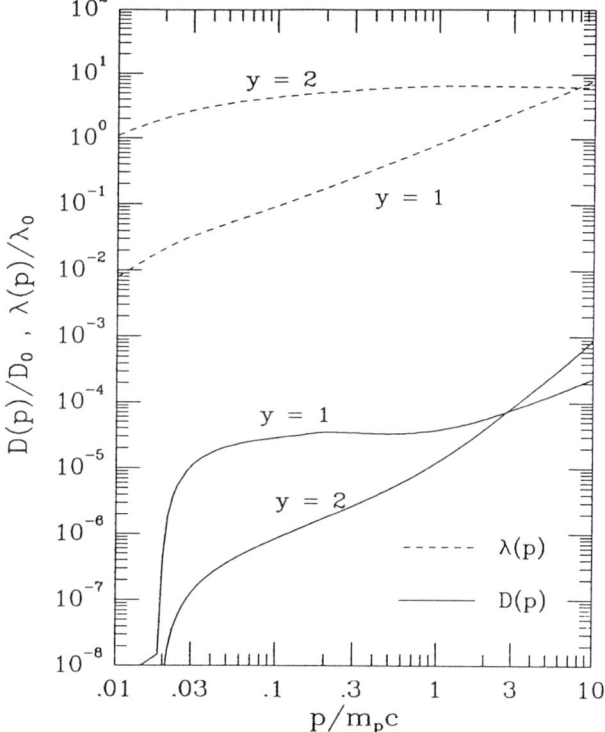

Fig. 3. Momentum diffusion coefficient $D(p)$ and mean-free path $\lambda(p)$ as a function of momentum for two values of y.

Thermal damping has the effect of introducing a threshold for resonance, which can be seen in this Figure as the momentum where D_{pp} becomes nonzero. This threshold is μ dependent and increases from $p/m_p c \lesssim 10^{-2}$ for $\mu = 1$ to about 0.3 for $\mu = 0$. The divergence resulting from the forward L wave comoving gyroresonance is not manifest, since the waves responsible are thermally damped and not present in the spectrum of turbulence. The forward low-frequency R wave comoving gyroresonance is important, however, and leads to a marked divergence. The last comoving gyroresonance (namely, that with a high-frequency R wave) does not yield a divergence, since these waves are also damped and not present. We point out that if waves of all frequencies are present, there are three lines of divergences instead of the one here.

The Fokker-Planck equation can be integrated over position along the magnetic field and averaged over μ to obtain a momentum diffusion equation[1,3]. This equation contains a momentum diffusion coefficient $D(p)$ and characteristic escape time $T(p) = 3L^2/(v\lambda)$, where $\lambda(p)$ is the mean-free path, v the particle speed, and L the scale length of the acceleration region. We show in Figure 3 $D(p)$ and $\lambda(p)$ for $A = 0.1$ and $y = 1$ or 2. The normalization factor $D_0 = \Omega_p(m_p c^2)(\Omega_e/\Omega_p)^{y-1}$ and $\lambda_0 = D_0/c$, where Ω_e is the electron cyclotron frequency. The effect of the comoving gyroresonances is to produce a significant flattening in $D(p)$ for $y = 1$, which was expected on the basis of Figure 2. This effect is not nearly as pronounced for $y = 2$.

The diffusion coefficient $D(p)$ can be used to derive an acceleration time scale τ_E. Following the procedure in reference 1, we find for $y = 2$ that

$$\tau_E \approx \begin{cases} 0.17 \frac{U_B}{W_t} \left(\frac{10^{-6} \text{ cm}^{-1}}{k_0} \right) \left(\frac{p}{m_p c} \right) \text{ s}, & \text{for } 0.04 \lesssim \frac{p}{m_p c} \lesssim 1 \\ 0.37 \frac{U_B}{W_t} \left(\frac{10^{-6} \text{ cm}^{-1}}{k_0} \right) \text{ s}, & \text{for } \frac{p}{m_p c} \gtrsim 1, \end{cases} \quad (1)$$

where W_t is the total magnetic energy density in the turbulence, $U_B = B_0^2/(8\pi)$, and k_0 is the minimum wavenumber in the spectral density. Acceleration to 30 MeV (1 GeV) thus occurs in less than about 0.4 s (4 s) for $W_t \approx 10 \text{ ergs cm}^{-3}$ and $k_0 \gtrsim 4 \times 10^{-6} \text{ cm}^{-1}$. These time scales are consistent with estimates from impulsive solar flares based upon the time profiles of nuclear deexcitation lines[4] and pion radiation[5]. The relatively short wavelength and high intensity of these waves rules out photospheric motion and cascading as their source; higher frequency motion or a microinstability (such as a proton beam[6]) is therefore necessary for their production.

REFERENCES

1. J. Steinacker, and J. A. Miller, ApJ, **393**, in press (1992).
2. J. A. Miller, and J. Steinacker, ApJ, submitted (1992).
3. R. Schlickeiser, ApJ, **336**, 243 (1989).
4. S. R. Kane et al., ApJ, **300**, L95 (1986).
5. J. A. Miller, R. Ramaty, and R. J. Murphy, Proc. 20th ICRC (Moscow), **3**, 33 (1987).
6. S. P. Gary et al., Phys. Fluids., **27**, 1852 (1984).

ACCELERATION OF PARTICLES BY AN ENSEMBLE OF SHOCK WAVES AND HIGH ENERGY EMISSION OF SOLAR FLARES.

L.G.Kocharov
State Technical University
St.Petersburg, 195251, RUSSIA

G.A.Kovaltsov
Ioffe Physico-Technical Institute
St.Petersburg, 194021, RUSSIA

ABSTRACT

An analysis of observational data on X- and gamma-emission and energetic particles from impulsive solar flares shows that particles are accelerated over a few seconds in a region where matter density $\sim 10^{11}$ cm^{-3} and magnetic field $\gtrsim 300$ G. We take the June 21 1980 flare as an example of an impulsive flare. For this flare one can obtain two different spectra of primary particles from the very same observations of high-energy neutral radiation. We consider the possibility of obtaining such a spectra of energetic ions as they are accelerated on an ensemble of shocks in the presence of short scale turbulence. This system of shocks may originate from a great number of points of explosive current loop coalescence.

INTRODUCTION

Observations carried out over the previous maximum of solar activity considerably improved our understanding of the processes of particle acceleration on the Sun. Analysis of the data on hard electromagnetic and corpuscular emissions of solar flares showed that there are two classes of events in solar energetic particles: events linked to the impulsive solar flares (with the duration of gamma burst $\leqslant 1$ min. and variabilities on the time scales of seconds) and those linked to gradual flares ($\gtrsim 5$ min)[1] (see also [2]). As was shown by the observations of flare gamma emission carried out aboard the SMM-satellite [3], in the impulsive solar flares acceleration of protons to energies several tens of MeV may occur almost simultaneously with acceleration of electrons up to ~ 1 MeV on a time scale of about 1s. The well-known event of 1980 June 07 is an example of such a flare. For such impulsive flares as was shown in paper [4] (see also [2,5]), it turns out possible to evaluate the matter density n_a and magnetic field strength B_a in the particle acceleration site basing on rather simple and general relations. As a result values $n_a=10^{11}-10^{12}$ cm^{-3}, $B_a=300-1000$ G were obtained. On the other hand data on >10 MeV gamma ray anisotropy over the previous 21st cycle of solar activity were an incentive for the development of models of nuclear emission generation in magnetic loops (see [6,7] and references therein). In its turn as a consequence there came the conclusion that not only in the impulsive but also in a number of gradual flares it is possible to interpret the gamma ray on the assumption that the whole lot of particles get accelerated only in the impulsive phase of a flare and their energy spectrum is a power law [7,8] (see also [9]). Acceleration of energetic particles in the impulsive phase of the flare will be discussed in Section 2 but earlier in Section 1 we consider an example of a flare with high energy emission to explicate the observational restrictions for theoretical models of acceleration.

1. NEUTRONS AND GAMMA-RAYS FROM 21.06.1980 FLARE

Interpreting the results of high-energy neutral radiation observations on 21.06.1980 Murphy et al. [10], Hua and Lingenfelter [11] assumed that neutrons and π^0-decay γ-rays were produced at the impulsive phase of the event (during 60 s) simultaneously with the 4-7 MeV nuclear γ-ray lines. The particle distribution in the emission region was supposed either isotropic [10] or parallel to the solar surface [11]. In this case the spectrum of accelerated ions could not be a simple power law and steepened abruptly with energy. However angular and temporal characteristics of the distribution of primary particles in the emission region strongly depend on the conditions of their propagation in the solar atmosphere, i.e. magnetic field structure, turbulence level etc. For instance if the generation of radiation occurs in the magnetic arched trap with particles scattering on MHD-turbulence, these observational data may be explained with a power law spectrum of primary particles [7,12]. The pitch-angle diffusion of particles leads to their precipitation from the magnetic trap down into dense regions of the solar atmosphere.

For this flare we consider generation of high-energy neutral radiation by particles confined after acceleration in a magnetic loop at different levels of MHD-turbulence. We assume that all particles were

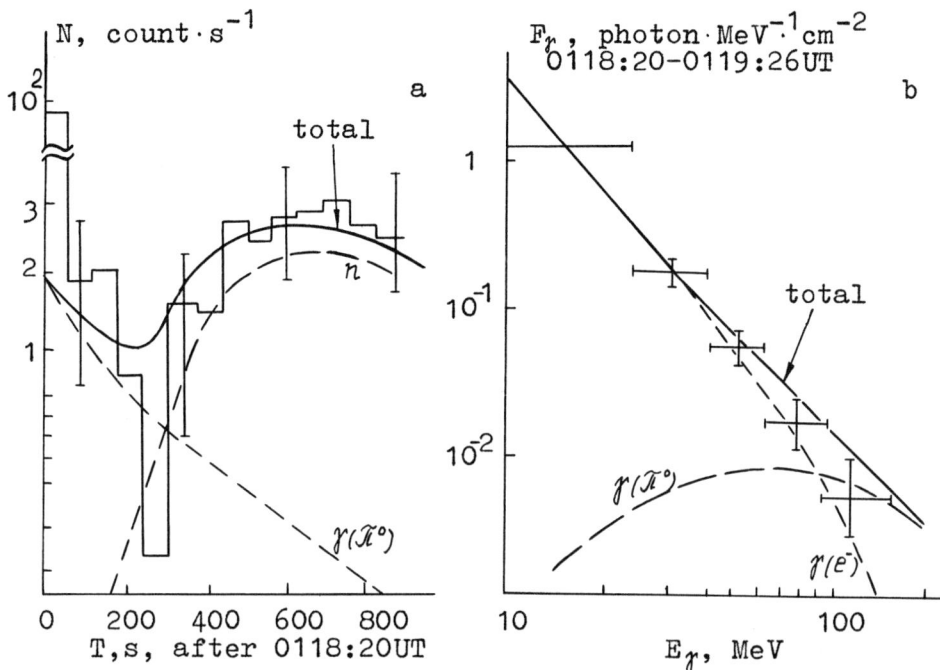

Figure 1. 1980 June 21 event: (a) SMM GRS count rate (>10 MeV) [13]; (b) Energy spectrum of γ-ray (>10 MeV) at the impulsive phase [14]. Curves are the results of our calculation. Model parameters: N_p(>30 MeV)= 6.6×10^{32}, $S_p=3.7$, N_e(>1 MeV)= 1.3×10^{34}, $S_e=3.6$, $R_A=10^9$ cm, $B_c=500$ G, $n_c=10^{11}$ cm^{-3}, (dB/dz)/B =1.25×10^{-8} cm^{-1}, $W/(B_c^2/8\pi) = 10^{-8}$.

accelerated during the impulsive phase of the flare and ion spectrum was a simple power law with a cutoff energy E_{max} (parameters of acceleration region were proposed different from the parameters of confinement region). For details of the applied model of arch structure see [6,7]. The main parameters of our model are: plasma densities n_c; coronal magnetic field B_c; arch radius R_A; magnetic field gradient in the chromosphere dB/dz; energy density of external turbulence W ; total number of accelerated particles $N_{p(e)}$; index of power law spectrum $S_{p(e)}$. Choosing model parameters we have to fit the following experimental data: 1. count rate of the high energy channel of GRS which detected neutrons (>50 MeV) and γ-ray (>10 MeV) [13]; 2. γ-ray (>10 MeV) spectrum at the impulsive phase [14]; 3. total flux of 4-7 MeV nuclear γ-ray line radiation - 98 photons cm^{-2} [15]. The power law spectrum of γ-ray (>10 MeV) at the impulsive phase places an upper limit to the generation rate of π^0-decay γ-ray emission over this phase.

In the case of weak pitch-angle diffusion these data can be fit by the results of calculation without a cutoff in the power law spectrum of accelerated particles. The region of possible parameter values turns out to be rather narrow: $W/(B_c^2/8\pi) = 10^{-8} - 10^{-7}$; $(dB/dz)/B > 10^{-8}$ cm^{-1} ; $N_p(>30$ MeV$) \approx 7 \times 10^{32}$; $S_p = 3.4-3.7$. The calculated emission fluxes for one of the possible sets of parameters are presented in Figure 1. At this turbulence level the calculated decay time τ of the generation rate of γ-ray line emission would be 50 s which is much higher than the observed value $\tau \approx 5$ s [16]. However accelerated protons propagating in flare loops can excite Alfven waves in the resonant region of these protons which leads to a more intense scattering of such protons and reduces the characteristic decay time of radiation [7,17]. It is required for an excitation of such Alfven waves that the cyclotron instability increment $\gamma = n_p v \omega_i /(n_c v_A \rho)$ be higher than collisional decrement of Alfven waves $\delta = 3.4 \times 10^{-2} n_c T^{-1.5}$ s^{-1} where n_p and v are density and velocity of accelerated protons, ρ is mirror ratio, v_A is Alfven velocity, T is plasma temperature in degrees, ω_i is the proton gyrofrequency. For $T=10^6$ K, arcade volume $V=10^{28}$ cm^3, $n_c = 10^{11}$ cm^{-3} and parameters of the arch and particles as above the condition $\gamma > \delta$ is valid for protons ~ 10 MeV responsible for the production of nuclear γ-ray lines, but it is not true for protons $\gtrsim 100$ MeV producing pions and high-energy neutrons. Therefore the characteristics of neutron and pion generation remain unchanged while the decay time of the nuclear γ-ray line generation rate decreases to $\tau_s \approx \pi R_A/(2v_A)$ due to the scattering by the self-generated turbulence. The value $\tau_s = 5$s is corresponded by the coronal magnetic field strength $B_c = 500$ G. In Figure 1b we present the calculated spectrum of relativistic electron bremsstrahlung in these conditions. We find that about 10^{34} electrons (>1 MeV) with $S_e = 3.6$ are required.

We come to another possible explanation of observational data at a high level of plasma turbulence: $W/(B_c^2/8\pi) > 10^{-4}$ when particles of all energies undergo intense scattering. In this case the time scale of both neutron and γ-ray production is several seconds. Then the number of accelerated protons should be $N_p(>30$ MeV$) \approx 10^{33}$ with $S_p = 2.8-3.0$ and cutoff energy $E_{max} = 300-500$ MeV. The spectrum of accelerated electrons should be the same as in the previous case. Note that in this case we obtain a spectrum of accelerated ions which is similar to that obtained in [10,11].

Thus there are two possibilities:
1) After acceleration particles are trapped in a loop with a very low level of external turbulence; the primary spectrum of particles is a power law in energy.
2) After acceleration particles propagate in the plasma with a higher level of turbulence; the energy spectrum of primary particles steepens abruptly with energy.

The principal reason of this uncertainty is that the differential energy spectrum of secondary neutrons is not measured. The observed time profile of GRS count rate may be fit either by softer spectra of neutrons emitted on a short time scale or harder spectra of neutrons emitted on a large time scale.

2. MODEL OF ACCELERATION.

Sharp fronts of γ-radiation in impulsive flares point out that the ion acceleration time <1s. If the acceleration of ions with such a characteristic time occurs on MHD-turbulence then the required density of turbulence energy must be close to the energy density of the regular magnetic field, that is the acceleration of ions should be considered under the condition of strong turbulence (system of shock waves). As for the source of such a turbulence it may be a "multi impulsive" energy release which occurs provided there are many reconnection regions of magnetic field. The number of such regions in a flare may be as high as 10^4 [18]. In this case there appear numerous shock waves and a short scale turbulence is generated. As was shown in [19] a three dimensional X-type current loop coalescence may go up like an explosion. Provided there are many coalescence points of the kind there may appear a rather complicated structure of hydrodynamical motion at different scales and amplitudes. Acceleration of particles in a similar situation for the galactic cosmic rays was investigated in [20]. Equation for the particle distribution function $f(p,x,t)$ in the acceleration region is the following:

$$\frac{\partial f}{\partial t} = \frac{1}{p^2}\frac{\partial}{\partial p}\left(p^2 D_p \frac{\partial f}{\partial p}\right) + \frac{\partial}{\partial x}\mathcal{x}\frac{\partial f}{\partial x} + \frac{div\,\vec{U_0}}{3}p\frac{\partial f}{\partial p}. \qquad (1)$$

Here $\vec{U_0}$ is the averaged hydrodynamical velocity (we propose that $U_0=0$, but $div\,\vec{U_0}$ may be not equal to zero). Spatial diffusion is proposed to be one dimensional due to the large scale regular magnetic field; x is the coordinate along this magnetic field. If the scattering path length on short scale turbulence $\Lambda << L_0 U/v$ (L_0 is the average distance between shocks, U^2 is the average squared hydrodynamical velocity on the scale L_0, v is the particle velocity), then the diffusion coefficient in the momentum space $D_p = \alpha p^2$ and coefficient of spatial diffusion $\mathcal{H} = $ const.

If $\Lambda \gg L_0 U/v$ then $D_p \approx (U^2/L_0)(p^2/v)$ and $\mathcal{H} \approx L_0 v$. Thus in this case for nonrelativistic ions D_p and $\mathcal{H}$ are proportional to p.

Here we propose that the time over which the acceleration exists is small: $t \ll L^2/(4\mathcal{H})$; L is the length of acceleration region. In this case particles do not leave the acceleration region before the acceleration mechanism stops operating. In this case the equation (1) has the form:

$$\frac{\partial f}{\partial t} = \frac{1}{p^2}\frac{\partial}{\partial p}p^2\left(b(t)p^n\frac{\partial f}{\partial p} + a(t)pf\right) - c(t)f, \qquad (2)$$

where n=2 for the strong scattering by small scale turbulence ($\Lambda \ll L_0 U/v$): n=1 for the case of weak scattering ($\Lambda \gg L_0 U/v$);

$$a = \frac{1}{3}\text{div}\,\vec{U_0}, \qquad b = D_p/p, \qquad c = \text{div}\,\vec{U_0}. \qquad (3)$$

Solutions of equation (2) may be found using a Green function the initial condition for which are:

$$G(p, p_0, t, t_0)\bigg|_{t=t_0} = (4\pi p_0^2)^{-1}\delta(p-p_0). \qquad (4)$$

For n<2 the Green function is

$$G(p, p_0, t, t_0) = \frac{(2-n)(pp_*)^{-(n+1)/2}}{4\pi \Delta_1(t)}I_R\left\{\frac{2(pp_*)^{(2-n)/2}}{\Delta_1(t)}\right\} \times$$

$$\times \exp\left\{-\frac{p^{2-n} + p_*^{2-n}}{\Delta_1(t)} - \int_{t_0}^{t}c(\tau)\,d\tau\right\}, \qquad (5)$$

where $I_R\{x\}$ is the modified Bessel function,

$$R = \frac{1+n}{2-n}, \qquad p_* = p_0 \exp\left\{-\int_{t_0}^{t}a(\tau)\,d\tau\right\},$$

$$\Delta_1(t) = \exp\left\{(n-2)\int_{t_0}^{t}a(\tau_1)\,d\tau_1\right\} \times$$

$$\times \int_{t_0}^{t}(2-n)^2 b(\tau_2)\exp\left\{(2-n)\int_{t_0}^{\tau_2}a(\tau_3)\,d\tau_3\right\}d\tau_2.$$

For n=2:

$$G(p, p_0, t, t_0) = \frac{\exp\left\{-\frac{9}{4}\Delta_2(t)\right\}}{(4\pi p p_*)^{3/2}\sqrt{\Delta_2(t)}} \times$$

$$\times \exp\left\{-\frac{(\ln p/p_*)^2}{4\Delta_2(t)} - \int_{t_0}^{t} c(\tau)\,d\tau\right\}, \quad (6)$$

where

$$\Delta_2(t) = \int_{t_0}^{t} b(\tau)\,d\tau.$$

In the latter case (for a=c=0 and constant b=α) proposing that particles with momentum p_0 are injected into the acceleration process during the whole acceleration time, one can obtain the distribution function at the end of acceleration [21,22]:

$$f = \frac{\pi^{-1}}{3\alpha t(4pp_0)^{3/2}} \left[\left(\frac{p_0}{p}\right)^{3/2} \text{erfc}\left\{\frac{\ln(p/p_0)}{2\sqrt{\alpha t}} - \frac{3}{2}\sqrt{\alpha t}\right\} - \left(\frac{p}{p_0}\right)^{3/2} \text{erfc}\left\{\frac{\ln(p/p_0)}{2\sqrt{\alpha t}} + \frac{3}{2}\sqrt{\alpha t}\right\}\right], \quad (7)$$

where

$$\text{erfc}\{x\} = \frac{2}{\sqrt{\pi}} \int_{x}^{\infty} \exp\{-y^2\}\,dy.$$

As is seen from calculations a close to a power law spectrum at E>10MeV can be obtained if αt=0.3 ; $p_0^2/(2m)$= 0.1 MeV. This case may be realized under the parameters: $L_0 \approx 3 \times 10^7$ cm, $L \approx 3 \times 10^8$ cm, $U \approx 10^8$ cm s^{-1}. In this case the acceleration time = $L_0/U \approx 0.3$s and spatial diffusion time $L^2/(4\varkappa) = 10$s. Then in order to create an accelerating system of shocks one needs $\approx (L/L_0)^3 \approx 10^3$ energy release points.

On the other hand for n=1 using (5) one can see that the accelerated particle spectrum steepens abruptly with energy.

CONCLUSIONS

It may be seen from above that the data at hand and considered estimates favour the hypothesis of proton acceleration by an ensemble of shocks which is formed as a result of "multi impulsive" energy release during a flare. This kind of flare model may be called a "statistical" one as was done by Vlahos [23].

Depending on the scattering path length on a short scale turbulence Λ one can obtain a nearly power law spectrum or a spectrum steepening with energy. Further investigations are needed to distinguish between those two possibilities.

REFERENCES

1. G.E.Kocharov, L.G.Kocharov, G.A.Kovaltsov, Proc.18th ICRC, Bangalore. 4, p.105 (1983).
2. L.G.Kocharov, G.A.Kovaltsov, Proc.Joint Varenna-Abastumani Int.School and Workshop on "Plasma Astrophys.", Sukhumi, ESA SP-251, p.101 (1986).
3. D.J.Forrest, E.L.Chupp et al.,Proc.17th ICRC, Paris, 10, p.5 (1981).
4. G.E.Kocharov et al., In "Energetic particles and photons from solar flares", Leningrad, p.46 (1984).
5. J.M.Loran, J.C.Brown, Astrophys.Space Sci., 117, 173 (1985).
6. L.G.Kocharov,G.A.Kovaltsov, Solar Phys.,125,67 (1990).
7. V.G.Gueglenko, G.E.Kocharov et al., Solar Phys., 125, 91 (1990).
8. V.G.Guglenko et al.,Astrophys.J.Suppl.,73,209 (1990).
9. J.M.Ryan, D.J.Forrest, W.T.Vestrand, Proc.21st ICRC, Adelaide, 5, 52 (1990).
10. R.J.Murphy et al., Astrophys.J.Suppl. 63, 721 (1987).
11. X.-M.Hua,R.E.Lingenfelter,Astrophys.J.,323,779 (1987).
12. V.G.Gueglenko et al., Izv.AN SSSR, ser.fiz., 55, 1908 (1991).
13. E.L.Chupp et al.,Proc.19th ICRC,La Jolla,4,126 (1985).
14. D.J.Forrest et al., Proc.19th ICRC, La Jolla, 4, 146 (1985).
15. E.W.Cliwer et al., Astrophys.J.343, 953 (1989).
16. D.J.Forrest, E.L.Chupp, Nature 305, 291 (1983).
17. V.V.Zaitsev, A.V.Stepanov, Solar Phys.99,313 (1985).
18. S.S.Bulanov et al., Proc XI Leningrad Seminar on Astrophysics, p.83 (1979).
19. J.I.Sakai, C.De Jager, Solar Phys. 123, 389 (1989).
20. A.M.Bykov, I.N.Toptyghin, Izv.AN SSSR, ser.fiz., 45, 474 (1981).
21. L.G.Kocharov, G.A.Kovaltsov, Proc.Flare 22 Workshop "Dynamics of solar flares",Observatoire de Paris,71 (1991).
22. G.A.Kovaltsov, L.G.Kocharov, Izv.AN SSSR, ser.fiz., 55, 1912 (1991).
23. L.Vlahos, Proc.Flare 22 Workshop "Dynamics of solar flares",Observatoire de Paris, 91 (1991).

THE RUNAWAY OF FAST ELECTRONS INTO TURBULENT
PLASMA OF SOLAR FLARES

Yu.E.Charikov and I.V.Kudrjavtsev

A.F.Ioffe Physical-technical Institute, St.Petersburg,Russia

Introduction

The consideration of the transport problem for a beam of fast non-relativistic particles in solar flare plasma is ordinarily carried out in the Coulomb collisions approximation(e.g.see[1,2]).However, it is not the only reason leading to a transformation of their distribution function. As is well known, beams of fast electrons are a source of plasma instabilities which in turn may scatter injected particles itselves Actually it is necessary to solve a self-consistent problem in order to obtain the distribution functions for the waves and particles. However, in the considered case of ion-sound turbulence generated by an return current [3] in plasma, the various groups of electrons generate the waves and are scattered on them.Actually it is possible to pose the problem as follows:a beam of fast particles fall into a layer of plasma with induced ion-sound waves and propagates inside the layer scattering by plasmons.We have obtained the solution for a turbulent plasma, and as an application considered two model cases: the non-thermal distribution of fast particles and the quasi thermal one which are discussing in interpretations of the emissions from solar flares [3]. Now we pass onto a more correct definition of the problem.

1.Statement of the Problem.

We assume that a beam of fast non-relativistic electrons with a given distribution in momentum are incident upon the boundary of a plane one dimensional layer z=0 of plasma with induced oscillations(energy density W of waves is given).We also assume that the ion-sound turbulence is isotropic, the particle beam is compensated in charge and current [4]. In this case the equation for the distribution function takes the form [5,6]:

$$\frac{\partial f}{\partial t} + \vec{v}\frac{\partial f}{\partial \vec{r}} = \frac{\partial f}{\partial p_i} D_{ij} \frac{\partial f}{\partial p_j} \qquad (1)$$

$t, \vec{r}$, and $\vec{p}$ are the ordinary variables in the coordinate and momentum spaces; $p_i = mv_i$ is the momentum of a non-relativistic particle. The process of scattering by plasmons is determined by diffusion tensor D_{ij}. In the case of isotropic turbulence it may be presented as a sum of longitudinal and transverse to the momentum components:

$$D_{ij} = D^t \frac{p_i p_j}{p^2} + D^l (\delta_{ij} - \frac{p_i p_j}{p^2}) \qquad (2)$$

taking into account that the velocity of fast electrons v is ordinarily much higher than the phase velocity of ion-sound in plasma

v_{is} ($v_{is}^2 = T_e/m_i$). in solar flares and that $D^t = D^2 v^2/v_{is}^2$ [5,6], we obtain

$$D^t \gg D^1 \qquad (3)$$

Therefore the angle diffusion is more important comparing with the diffusion on absolute value of momentum. As regards the diffusion coefficient D^t we take into account that it turns zero at the layer boundaries.

Applying (2) and (3) to (1) we obtain:

$$\sin\theta \frac{\partial f}{\partial x} = \frac{1}{\sin\theta} \frac{\partial}{\partial \theta} \left(\sin\theta \frac{\partial f}{\partial \theta} \right) \qquad (4)$$

where

$$x = \int D_3^t m/p^3 \, dz \qquad (4')$$

In (4) it is accepted that the stationary case is being considered, i.e. the propagation of the beam through the layer of plasma is settled and the transitional processes at the initial front of propagation are relaxed. It is known that $x \sim p^{-4} \varepsilon^{-2}$ where ε is the electron kinetic energy. The boundary values may be presented as $f(x, \theta=0, \pi)$ is finite; $f(x=0, \theta)$ for the case of $x \ll 1$ (very fast particles); for other x:

$$f(x, \theta=0, \pi) \text{ is finite; } f(x=0, \theta) = \begin{cases} z^1(\theta), & 0 \leq \theta \leq \pi/2, \\ z^2(\theta), & \pi/2 \leq \theta \leq \pi \end{cases} \qquad (5)$$

Functions z^1 and z^2 describe the particle distribution at the boundaries of the turbulent layer and are given. We consider certain peculiar cases of their definition below.

2. Solution of the Basic Equation

Taking into account the structure of (4), it is reasonable to present its solution as an expansion into Legandre polynomials $P_1(\cos\theta)$

$$f(x, p, \theta) = \Sigma \, F_1(x, p) P_1(\cos\theta) \qquad (6)$$

If we substitute now relation (6) in (4), multiply (4) by $P_1(\cos\theta)$ and integrate over θ making use of the recurrent relation and the normalization of Legandre function then we come to the set of differential equations regarding F_L functions, which impossible to solve in a closed finite form. However, one may limit the number N of equations and evaluate the accuracy of the solution. We limit N as 2k+1 where k is an integer value, obtaining a truncated system (7)

$$\begin{cases} \dfrac{\partial F_1(x)}{\partial x} = 0 \\ \dfrac{n+1}{2n+3} \dfrac{\partial F_{n+1}(x)}{\partial x} + \dfrac{n}{2n-1} \dfrac{\partial F_{n-1}(x)}{\partial x} = -n(n+1) F_n, \quad n=1,2..k \\ \dfrac{2k+1}{4k+1} \dfrac{\partial F_{2k}(x)}{\partial x} = -(2k+1)(2k+2) F_{2k+1}(x), \quad n=2k+1 \end{cases} \qquad (7)$$

and $F_n(x) = 0$ for any $n > 2k+1$. We seek a power series (over x) solution of the system (7):

$$F(x) = \Sigma \, c_m^n x^m, \qquad F_n(x=0) = c_0^m \qquad (8)$$

Now the problem of determining the functions $F_n(x)$ is reduced to the determination of the series coefficients C_n^m. A set of equations for this coefficients is obtained from (7) by means of a substitution of the expansion (8). The system of C_n^m coefficients can be solved for arbitrary k, while for k=1,2 series(8) can be easily summarized. We present the solution for k=1. Let $\mu = \cos\theta$

$$f(x,\mu) = (P_0(\mu)P_1(\mu)P_2(\mu)P_3(\mu))F(x) \qquad (9)$$

The matrix-column $F(x)$ is determination $\tilde{F}(x) = A\tilde{F}(x=0)$. Matrix $\tilde{A}$ is presented in [7]

3. Turbulent layer of a finite thickness

This case may be realized on the condition that particle path length with regard to the scattering by plasmons is less than the layer width. Condition $x \ll 1$ leads to the accounting for the particle escape backwards from the layer through the boundary $z=0$, i.e. applying of the boundary values (5). The solution of equation (4) for k=1 on the boundary of the layer z=L is the following:

$$F_0 = \frac{Z_0}{2} + Z_1 \frac{x}{x+2/3} \{ \frac{3ab}{5q}(1 - \frac{3x}{8(x+2/3)}) - 1\} + Z_2 \frac{ad}{8q} + Z_3 \frac{3ab}{5q} [4 - \frac{3x/2}{x+2/3}]; \quad F_1 = \frac{Z_1}{x+2/3}\{1 + \frac{x}{x+2/3}\frac{9ab}{40q}\} + Z_3 \frac{9ab}{10q(x+2/3)}; \qquad (10)$$

$$F_2 = -\frac{a}{q}\{Z_1 \frac{3x/2}{x+2/3} - Z_2 c + 6bZ_3\}; \quad F_3 = \frac{1}{q}\{Z_1 \frac{3x/2}{x+2/3} - Z_2 ad + 6Z_3\};$$

where

$$a = (\frac{1+ch(y)}{5} + \frac{\sqrt{70}sh(y)}{80})^{-1}; \quad b = \frac{2sh(y)}{\sqrt{70}} + \frac{ch(y)-1}{8}$$

$$d = \frac{75x+56}{80(x+2/3)}(ch(y)-1) + \frac{6}{\sqrt{70}}sh(y); \quad c = \frac{75x+56}{8\sqrt{70}(x+2/3)}sh(y) + 6(1+ch(y))/7;$$

$$q = \{12/35 - \frac{75x+56}{320(x+2/3)} + ch(y)(12/35 + \frac{75x+56}{320(x+2/3)}) + sh(y)(3/2\sqrt{70} + \frac{75x+56}{20\sqrt{70}(x+2/3)})\}a; \quad Z_n = \int Z_0^{(1)} P_n(\mu) d\mu, \quad Z_0^{(2)} = 0.$$

Let us make the boundary values (5) more concrete. We assume that $Z^1 = Z^0(\varepsilon)\mu^2$, $0 \leq \mu \leq 1$. Solution of (6–10) is presented in Fig.1. As a matter of fact, different particle momentum correspond the different x (x p^{-4}). Therefore x=0 is corresponded by p→∞, i.e. curve 1 describes the distribution of high energy particles which are only slightly scattered in the plasma layer, and the curve 1 coincides with the law of boundary distribution $\sim \mu^2$. However, the effect of scattering is significant for the particles with lower velocities. This is confirmed by distributions in cases 2,3 and a considerable deviation of the curves 2,3 from the law μ^2. So the turbulent layer is transparent for the high energy particles while for the low velocity ones it is "optically thick". This conclusion leads to an important result concerning the formation of beam-like particle distributions. Indeed the energy distribution of particles in a beam is usually $\varepsilon^{-\alpha}$ where $\alpha > 0$. Therefore it should be expected that a turbulent layer with the given parameters will "define" a certain region(boundary) of particle energy on the two sides of which there will be particles escaping the layer and confinement in it(see fig.2).Note that as a result of the truncated expansion of distribution function into Legandre polynomials solution (6),-(10) does not turn zero at $\theta \geqslant 90$, however,the accuracy is within 15%.

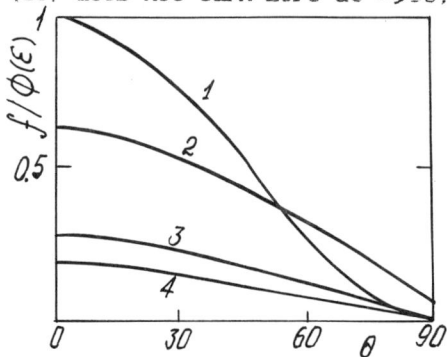

fig.1. Distribution function f of fast electrons via θ on the boundary of layer x=1 for energies X: 1.X=0; 2.X=0.1; 3.X=1.0; 4.x=2.0

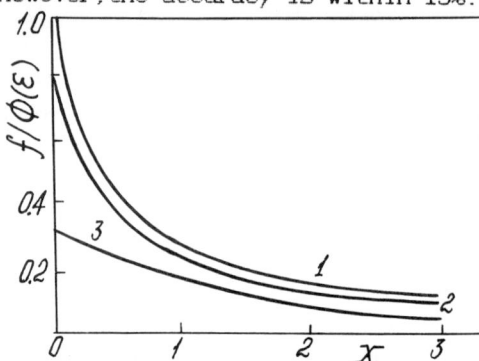

fig.2. Distribution function f of fast electrons via energies X on the boundary of layer x=1 for angles θ: 1.$\theta=0°$; 2.$\theta=\pi/6$; 3.$\theta=\pi/3$.

In order to carry out quantitative estimates of the geometrical length of the layer with ion-sound turbulence which seizes (lets through) fast electrons of given velocity v we pass from the dimensionless "thickness" x to z. We make use of (4') and the relation for diffusion coefficient for the ion sound [5].And, finally, it follows from (4'),[5,6]

$$\frac{x}{z} = \frac{\pi}{2\sqrt{2}} \omega_{pe} \frac{\sqrt{m_e}(kT_e)^{3/2}}{(2\varepsilon)^2} k_g d_e \frac{W^S}{nkT_e} \qquad (11)$$

Let us evaluate the parameter $\alpha = W^S/nkT_e$. The weak turbulence approximation implies $\alpha \ll 1$. One should keep in mind that turbulent energy must be such as to provide that path length with regard to the scattering on ion-sound plasmons be much less than the Coulomb length. i.e. $\lambda_k/\lambda_{is} \gg 1$. For λ_k from [8]

$$\lambda_k/\lambda_{is} = \frac{\sqrt{2\pi}}{32\sqrt{n} \, e^3 \Lambda} (kT_e)^{3/2} k_g d_e \frac{W^S}{nkT_e} \qquad (12)$$

where Λ is the Coulomb logarithm [8].Taking into account (12) and the condition that $\lambda_k/\lambda_{is} \gg 1$ we obtain:

$$\alpha = \frac{W^S}{nkT_e} \gg \frac{32}{\sqrt{2\pi}} \frac{\sqrt{n} \, e \, \Lambda^3}{(kT_e)^{3/2} k_g d_e} \qquad (13)$$

The right hand of (13) is a function of temperature, plasma number density and depends on the wavelength of generated ion-sound waves $2\pi k_g^{-1}$. For instance,the right hand of(24) for $T_e=10^6$K. $n=10^{10}$ cm^{-3}. $k_g d_e=0.1$ and $\Lambda=10$ equals 10^{-3}, i.e. $\alpha \gg 10^{-5}$. So the right hand of (13) cannot be significantly larger than the estimated value $\alpha \gg 10^{-5}$ and this wave energy density is quite real in the regions of solar flares. Analyses of hard X-ray emission of the solar flares gives us that the low energy band of energetic spectrum is approximately 16-20 keV. Therefore we estimate the length of the ion-sound fronts confining the electrons with this energy. For an efficient penetration the beam electrons with energies >20 keV it is necessary that $x(\varepsilon)$ is less than 1 which leads to the condition:

$$L < 1.3 \cdot 10^5 \alpha^{-1} cm. \qquad (14)$$

We obtain an estimate of the layer's thickness L starting from (14) and assuming $\alpha=10^{-3}$ which does not contradict (13): $L < 1.3 \cdot 10^8$ cm. It is known that density of thermal plasma energy in the region of solar flares $W \sim 1$ erg cm^{-3}, therefore at $\alpha=10^{-3}$, $W=10^{-3}$ erg cm^{-3}. Supposing that the turbulent plasma is confined in an arched structure of transverse size 10^8 cm, we obtain an estimate for the value of volume 310^{24} cm^3. The total energy of ion-sound waves in this case $W^s = 310^{24}$ erg which is much less than the flare energy. Taking into account this one should expect that the presence of turbulent layers of various configuration in the flare region will not lead to major changes in the scenario of secondary phenomena but will influence their details the identification of which will make it possible to judge about turbulence itself. According to the model concept, the ion-sound fronts divide the regions of hot ($T_e > 10^7$ K) and cold plasma in flare loops. In this case the thickness of the turbulent layer keeping thermal electrons with temperature $5 \cdot 10^2$ K is to be $L > 6 \cdot 10^8 \alpha$ cm. For the same value of $\alpha = 10^{-3}$ we obtain $L > 6 \cdot 10^6$ cm. Having carried out these estimates, we pass onto the consideration of the processes of hard X-ray generation by electrons which traversed a layer of turbulent plasma and reached denser regions.

4. Kinetics of Electrons in the Model of Hot Plasma.

An alternative model is that of hot ($T_e > 10^8$ K) flare plasma with ion-sound fronts [3]. The generation of secondary emissions takes place both in the region of hot plasma itself and beyond it due to "escaping" electrons. It is obvious that these electrons are to have high energy. Mechanism of "escaping" is determined by either thermal "escaping" [9] or scattering on ion-sound oscillations at the front of the hot region. For the description of electron flux leaving the hot region the model Manheimer function is ordinarily applied [10]

$$f_m = (\pi v_{Te}^2)^{-3/2} \exp\left(-\frac{m v^2}{2kT_e}\right)\{1 + g_T[(v_z/v_{Te})^3 - (v_z/v_{Te})3/2]\}, \qquad (15)$$

where $v_{Te} = \sqrt{2kT_e/m_e}$ is the thermal velocity of hot region electrons. T_e is the electron temperature of the hot region, $g_T = 8Q/(3mnv_e^3)$, Q is the thermal flux. However, in this model the change of the distribution function of fast electrons resulting from their interaction with ion-sound turbulence is not considered. In order to find such a function we use the solution (6)-(10). At the boundary z=0 we adopt:

$$f(z=0) = \begin{cases} f_n, & 0 < \theta < \pi/2 \\ f_u, & \pi/2 < \theta < \pi \end{cases} \qquad (16)$$

f_n is, as before, an unknown function describing the escape of fast particles backwards from the turbulent layer. At the boundary of the turbulent layer z=L we have the following distribution function:

$$f(z=L) = \Sigma \, F_L P_L (\cos\theta)$$

where coefficients F_L are determined from (10) and (15),(16). As was shown above, for the ion-sound turbulence $x \approx v^4$. Therefore one can write $x=A/(v/v_e)^4$. Then at $v=v_e$, $x=A$ and for an efficient trapping of electrons with velocities $v > v_e$ it is necessary that $A > 1$. In Fig.3 we present the distribution function of electrons of the beam which passed through the turbulent layer with the initial (at z=0) distribution (15) It follows from Fig.3 that the scattering on the front with ion-sound waves changes considerably the distribution function and not only quantitatively, but also qualitatively. Instead of a monotonous decrease with particle energy increase, in our case there appears a region of velocities for which the condition of an inverse distribution is true i.e. $\partial f/\partial v > 0$ appears. Such distributions are usually assumed in problems concerning the amplification of oscillations in astrophysical plasma. here we naturally obtained them.

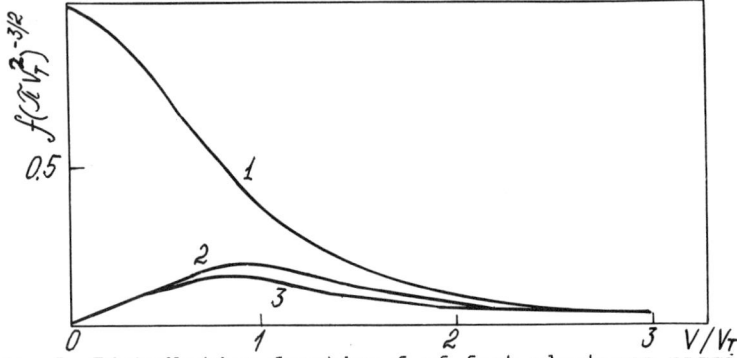

fig.3. Distribution function f of fast electrons passing through a layer of plasma with ion-sound turbulence in the thermal dissipative model for 1. z=0; 2. z=L and $\theta=0$; 3. z=L and $\theta=\pi/3$.

References

1. Leach J., Petrosian V., Astrophys. J., 1981, v.251, P.781.
2. Leach J., Petrosian V., Astrophys. J., 1981, v.269, P.715
3. Brown J.C., Melrose D.B. and Spicer D.S., Astrophys. J., 1979, v.228, P.592.
4. Vallis G., Zauer K., Zjunder E. et al., Uspechi phys.nauk, 1974, v.113, P.435.
5. Tsytovitch V.N., Nonlinear processes in plasma, M.:Atomizdat, 1971, 423P.
6. Tsytovitch V.N., The theory of turbulent plasma., M.:Nauka, 1967, 287P
7. Kudrjavtsev I.V., Charikov Yu.E., Astronom.J. (Sov.), 1991, v.68, P.628.
8. Spitzer L., The physics of fully ionized gas.M., 1957.
9. Nocera L., Skrynnikov Y.I., Somov B.V., Solar Physics, 1985, V.95, P.81.
10. Manheimer W.M., Phys. Fluids, 1977, V.20, P.265.

GAMMA-RAYS AND NEUTRONS AS A PROBE
OF THE PROTON SPECTRUM DURING
THE SOLAR FLARE OF 1988 DECEMBER 16*

P. P. Dunphy and E. L. Chupp
Physics Department and Space Science Center
Institute for the Study of Earth, Oceans and Space
University of New Hampshire
Durham, New Hampshire 03824, USA

ABSTRACT

We have previously reported on high-energy (> 10 MeV) γ-rays and neutrons from the flare of 1988 December 16 detected by the Gamma–Ray Spectrometer on the SMM satellite.[1] In this paper, we present results on γ-ray lines seen by the same detector during this flare. Together, these measurements constitute a powerful probe of the proton spectrum (> 10 MeV) that produces the flare neutrals. Analysis of the data suggests a Bessel-function proton spectrum with a shape parameter (αT) of 0.054 ± 0.004 and the number of protons above 30 MeV equal to $(9.0 \pm 0.9) \times 10^{32}$. The number of neutrons detected from this flare is much smaller than what is predicted from an isotropic distribution of the protons, indicating that the distribution may be non-isotropic.

OBSERVATIONS

We report on the observation of low-energy (0.3–10 MeV) γ-rays and high-energy (>10 MeV) γ-rays and neutrons from the flare of 1988 December 16. We apply the results from model calculations to the data to evaluate the proton spectrum that produced the γ-rays and neutrons.

The high-energy data are from the *SMM* GRS High Energy Matrix (HEM).[2] This mode of the GRS treats the seven 7.6 cm × 7.6 cm NaI(Tl) "main channel" scintillators as one layer of a two-layer detector. The second layer is a 7.5 cm thick × 25 cm diameter CsI(Na) "back shield." The HEM records energy-loss events in the range 10–100 MeV in broad energy-loss channels (~20 MeV wide).

Active region 5278 produced an X4.7/1B flare on December 16 at ~ 0830 UT. The flare was located at N27 E33 on the solar disk, corresponding to a heliocentric angle of 43°. Figure 1(a) shows the time history of the flare in the energy range 4.1–6.4 MeV, which is dominated by nuclear de-excitation line emission. Figure 1(b) shows the time history for "multiple" events in the HEM from the flare. The multiple events are "showering" events in both layers of the HEM and are produced primarily by γ-rays (> 25 MeV). The bulk of the high-energy γ-rays are produced in a peak that occurs well after (~ 7 minutes) the flare onset at lower energies.

Figure 1 in Dunphy et al.[1] showed a similar time history of the flare in the HEM for "singles" events (involving only one layer of the HEM). The emission that was seen to continue after the main peak is a signature of solar flare neutrons, delayed by their time of flight from the Sun.

---

* Work supported by NASA through grants NAG5-720 and NAGW-2755

254 The Solar Flare of 1988 December 16

DATA ANALYSIS

Murphy et al.[3] have shown that the spectrum of solar flare protons can be evaluated by using the ratios of various γ-ray emissions, namely: (1) the line at 0.51 MeV from positron annihilation, (2) the line at 2.22 MeV from deuterium formation, (3) the emission between 4 and 7 MeV from nuclear de-excitation, and (4) the broad peak centered at 68 MeV from neutral pion (π^0) decay. The GRS main channel spectra were used to determine the fluences of the 0.51 MeV, 2.22 MeV, and 4-7 MeV photons.

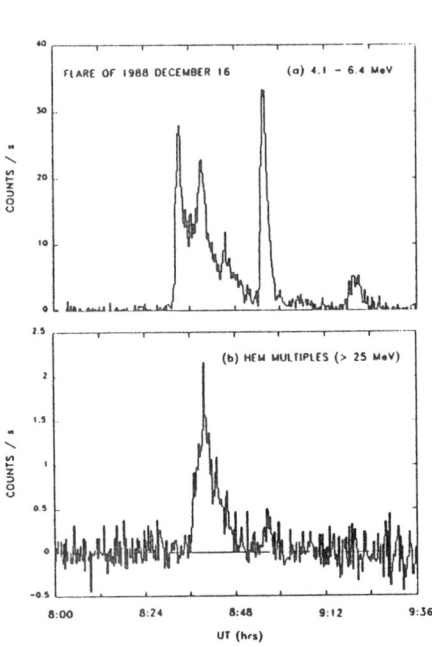

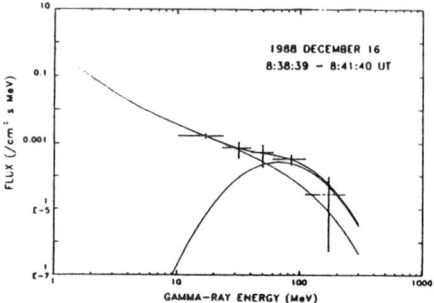

Figure 2. Gamma-ray spectrum in the GRS HEM during the peak of high-energy (> 10 MeV) γ-ray emission. A significant "bump" due to π^0-decay photons is present. The smooth curves are fits to the continuum (including the data < 10 MeV, which are not shown), to the π^0-decay peak, and to the total spectrum.

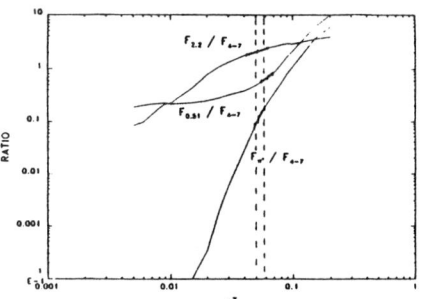

Figure 1. Time history of the 1988 December 16 flare in two GRS detector bands (background subtracted): (a) the 4.1-6.4 MeV band, sensitive to nuclear deexcitation γ-rays, and (b) the High Energy Matrix (HEM) multiple events, sensitive to γ-rays > 25 MeV.

Figure 3. The calculated dependence of γ-ray fluence ratios on the parameter αT for proton spectra with a Bessel-function shape.[4] The 1σ ranges in the ratios are indicated by the darkened sections of the curves. The corresponding range of αT (0.054±0.004) is also shown. The ratio $F_{2.2}/F_{4-7}$ depends on the heliocentric angle of the flare, which is 43°.

To determine the flux of the π^0-decay photons, we have used the GRS HEM data. In order to distinguish solar γ-rays from neutrons and to determine their energy spectra, we have developed an iterative fitting technique using the calculated response of the HEM. In principle, both the γ-ray and neutron fluxes can be determined independent of any model of the functional form for the differential energy spectra. In practice, the HEM response is not sensitive enough to neutron energy to generate a stable, well-defined neutron energy spectrum unless the spectral shape is restricted. Therefore, we constrained the spectrum by assuming that all of the neutrons were produced at the Sun in a single, short time interval (*i.e.*, a δ-function production model). Figure 2 shows a γ-ray spectrum in the HEM for one 180 s time interval with strong emission above 10 MeV. The spectrum shows a significant feature due to π^0 decay.

RESULTS

The relationship between γ-ray yields and solar proton spectral shapes has been calculated by Murphy and Ramaty [4] for the case of isotropic production in a thick-target model of the interaction environment on the Sun. We have used their results to calculate the expected fluence ratios, normalized to the the fluence in the 4-7 MeV band. Figure 3 shows the dependence of these fluence ratios on the parameter αT that describes proton spectra with a Bessel function shape. Also shown in Figure 3 are the ratios for the December 16 flare. These ratios were determined for the time interval 08:36-08:48 UT during which there was significant flux from π^0 decay.

The observed ratios are consistent with a Bessel-function proton spectrum with a shape parameter (αT) equal to 0.054 ± 0.004. The measured fluences can then be used to calculate the intensity of the proton spectrum. In this case, the number of energetic protons above 30 MeV ($N_P > 30$ MeV) is $(9.0 \pm 0.9) \times 10^{32}$.

Although the γ-ray data are consistent with a Bessel function proton spectrum, they are *not* consistent with a single, unmodified power-law spectrum. (The data might be consistent with a modified power law – e.g., a power law with a high-energy cutoff.)

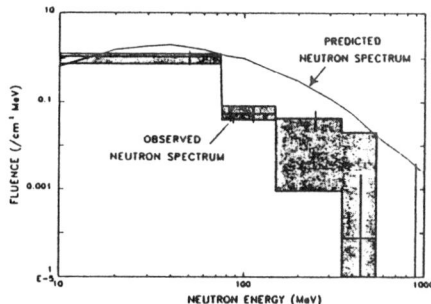

Figure 4. The time-integrated neutron spectrum observed by the GRS HEM, assuming δ-function neutron production at the time of peak γ-ray production > 10 MeV, is shown by the histogram. The shaded region encloses a range of spectra that depends on the range of assumed neutron production times (08:37-08:43 UT). The smooth curve is the neutron spectrum predicted from a Bessel-function proton spectrum interacting isotropically in a thick-target model.[5] The parameters of the proton spectrum, as determined from γ-ray line data, are $\alpha T = 0.05$ and $N_P (> 30$ MeV$) = 9 \times 10^{32}$.

The proton spectrum calculated from the γ-ray data can be used to predict the neutron spectrum for the same isotropic thick-target model. We have used the neutron spectrum predicted by Murphy et al. [3] for a Bessel-function proton spectrum with an αT of 0.05. This is shown in Figure 4 with the neutron spectrum observed by the GRS at 1 AU on December 16. In the range 10-75 MeV, the predicted fluence is reasonably close to the observed fluence, but above 75 MeV, the observed fluence falls below the predicted fluence by a factor of ~5. Since the predicted neutron spectrum does not account for γ-ray (and neutron) production outside the time interval 08:36-08:48 UT, the discrepancy is at least as large as is shown in Figure 4.

A similar analysis of the large 1989 March 6 flare [5] also showed an observed neutron spectrum that was below the predicted spectrum, in that case by a factor of ~2. It is important to determine whether these discrepancies are due to beaming effects, [6] since the present model assumes an isotropic distribution of energetic protons. Further observations of solar flare γ-rays and neutrons (for example, by the Compton Gamma Ray Observatory) would be valuable for such studies.

REFERENCES

1. Dunphy, P. P., Chupp, E. L., and Rieger, E., Proc. 21st Int. Cosmic Ray Conf. **5**, 75 (1990).
2. Forrest, D. J., et al., Sol. Phys. **65**, 15 (1980).
3. Murphy, R. J., Dermer, C. D., and Ramaty, R.,Ap. J. (Suppl.) **63**, 721, 1987.
4. Murphy, R. J., and Ramaty, R., Advances in Space Research (COSPAR) Vol. 4, No. 7, 127, 1985.
5. Dunphy, P. P., and Chupp, E. L., Proc. 22nd Int. Cosmic Ray Conf. **3**, 65 (1991).
6. Hua, X-M., and Lingenfelter, R. E., Ap. J. **323**, 779, 1987.

ARE CORONAL TYPE II SHOCKS PISTON DRIVEN?

N. Gopalswamy and M. R. Kundu
Astronomy Department, University of Maryland
College Park MD 20742

ABSTRACT

Flare blast waves and shocks piston driven by coronal mass ejections (CMEs) have been proposed to be responsible for generating type II radio bursts in the solar corona. The idea for piston-driven shocks came primarily from temporal association of shocks and CMEs. Our compilation of CME events with simultaneous radio observations with positional information supports idea of flare blast waves.

INTRODUCTION

Coronal mass ejections (CMEs) move through the solar corona with speeds ranging from a few kms^{-1} to more than 1000 kms^{-1}. If their speeds exceed local characteristic speeds, a bow shock is formed ahead of the CME. The flare explosions can send blast waves through the corona, moving at speeds exceeding local characteristic speeds[1]. Since solar radio bursts of type II are supposed to be produced by shock waves, both kinds of shocks have been proposed as likely candidates. There were three white light mass ejection events observed by the OSO-7 coronagraph that could be compared with type II radio bursts (with no positional information). Assuming a few time Newkirk density model using two events observed on Jan 11, 1973 Stewart et al[2,3] argued that the type II shock was ahead of or in the vicinity of the leading edge of the CME. For the Dec 14 1971 event, Kosugi[4] showed that the type II was too far ahead of the white light ejection and hence might be a blast wave, independent of the white light ejection. Thus, a controversy had started as to which excites a type II burst, the piston-driven shock or the blast wave ?

During the Skylab era, more coronal mass ejections were observed, but few simultaneous observations were made in radio with positional information. Using temporal association of CMEs and type II radio bursts, two statistical studies were undertaken: Gosling et al[5] found that 85% of fast (speed > 450 kms^{-1}) CMEs were associated with type II/Type IV bursts and that these bursts were associated only with fast (> 400 kms^{-1}) CMEs. Munro et al[6] found that all type II/type IV bursts occurring within 45° of the limb were associated with CMEs. These studies were taken as evidence for piston-driven shocks[7,8]. However, Gergely et al[9] were able to compare the type II burst position with that of the Oct 27, 1973 Skylab CME. From multi frequency observations, they could estimate the shock speed as 4900 kms^{-1} which is 6 – 7 times larger than the CME speed (670 kms^{-1}). Moreover, the shock trajectory made an angle of about 35° with the CME trajectory. Both the speed and the direction of propagation were inconsistent with the piston–driven shock model. In another CME event (April 27 1980) observed by the SMM – C/P instrument, Stewart et al[10] noted that the type II burst was generated within the CME loop. Robinson and Stewart[11] attempted a positional comparison between SOLWIND CMEs and Culgoora radio burst data. According to them, reasonable comparison could be made only in three cases. Among these, two events had type II bursts located well below the CME leading edge. In the third case, the lower frequency type II was clearly located below the CME leading edge, while the higher frequency one just coincided with the leading edge. Thus the picture started changing when more simultaneous observations became available: Wagner and MacQueen[12] proposed that the type II bursts are produced by shocks (blast waves) arising in flare explosion, thus supporting Uchida's[1] idea. Here we compile all the events with simultaneous positional information in optical and radio wavelengths to see which of the two shock mechanisms is supported by the data.

DATA AND RESULTS

In Table I, we list 12 events published with positional information for type II and CME, observed by SKYLAB, SOLWIND and SMM-C/P instruments in white light and by Culgoora and Clark Lake Radioheliographs. We have listed projected onset time of the CMEs (T_p), starting time of the associated flare (T_{Fl}), starting time of type II (T_{II}), type II and CME speeds (V_{shock} and V_{cme}) and type II and CME heights in units of solar radii (R_{shock} and R_{cme}). If the shocks are piston driven, we expect definite relations between the piston and the shock in terms of relative speed and location: for a piston driven shock, $V_{shock} \leq V_{cme}(1 + 1/\gamma)$ where γ is the ratio of specific heats. Similarly, The shock must be located at the stand-off distance determined by the radius of curvature of the piston and γ. Table I shows that out of the 12 events, 8 had type II locations well below the CME leading edge. In the remaining 4 cases, one can argue in favor of the blast wave model based on other characteristics of the association such as relative speed, trajectories and onset times.
1.**Oct 27 1973:** We have already discussed this event in the introduction. We note that the shock speed can not exceed 1072 kms^{-1} for $\gamma = 5/3$ if it is piston driven by the 670 kms^{-1} CME.
2. **April 12, 1980:** Although no type II positional information is available for this event, estimates of maximum and minimum height of the type II indicates that the type II onset coincides with the flare onset while the CME onset is 10 min earlier.
3. **June 27 1984:** The type II was ahead of the CME by about 0.5 $R_\odot$. It might seem to be a stand-off shock. However, the trajectory of the shock made an angle of about 45° with the CME trajectory. The CME was a single clump of material, 20 times smaller than the size of the type II burst.
4. **Feb 10 1986:** The type II started and ended even before the projected CME onset. If we extrapolate the shock height to the time when the CME was at a height of 3.0 $R_\odot$, it was 4.6 $R_\odot$, which is twice the stand-off distance. Thus the shock is too far ahead of the CME to be considered as piston driven.

Table I: Speed and Positional Comparison of CMEs and Type II Shocks

Date	T_p (UT)	T_{Fl} (UT)	T_{II} (UT)	V_{shock} (kms^{-1})	V_{cme} (kms^{-1})	R_{shock} (R_o)	R_{cme} (R_o)	Reference
10/27/73	1556	1554	1603	4900	670	2.05	1.40	9
5/4/79	0212	?	0221	?	800	2.25	2.50	11
9/2/79	0000	0033	0040	1100	750	1.67	2.59	11
4/9/80	2146	2229	2236	1300	300	1.83	2.30	13
4/12/80	2030	2040	2053	446	400	?	1.80	14
4/17/80	2240	?	2257	830	830	1.69	2.32	15
4/27/80	0226	0227	0239	580	750	1.60	1.90	10
6/29/80	0230	0233	0241	850	560	1.58	2.04	16
6/27/84	1752	1806	1822	1100	350	2.60	1.80	17
2/17/85	1855	2010	2025	1100	200	2.16	2.73	18
11/3/86	1917	1921	1946	600	750	2.30	2.80	19
2/10/86	2028	2019	2024	1900	1600	1.38	?	20

DISCUSSION AND CONCLUSIONS

We note that the available data is consistent with the blast wave model for coronal type II bursts. This is also supported by another statistical result that a class of type II bursts exists with no CME and conversely a class of fast CMEs exists with no associated metric type II[21,22]. The type IIs without CMEs have to be due to blast waves. According to these authors, the absence of type II bursts in some fast CME events meant either some fast CMEs did not reach the super-Alfvenic speeds in the lower corona or that some CMEs formed shocks that were unable to produce type II in the lower corona. A simpler explanation of course is that no blast wave was generated. Kahler et al[22] also found that the statistical properties derived from CME–type II association does not seem to conclusively support piston-driven or blast wave model. We feel that, the lack of positional information is a major source of uncertainty.

Cane[23] also considered Sheeley et al's[21] data and proposed a model in the framework of Wagner and MacQueen's[12] model where a shock is generated below the CME and produces a radio burst when it catches up with the CME. As she noted, this model cannot explain all the events because some type II bursts are located outside the CMEs also. If the shock wave is not related to the CME, then the type II burst can be located anywhere with respect to the CME and hence the question of relative location becomes irrelevant[15]. Thus we may conclude that the available data on simultaneous CME and radio burst data is consistent with a flare blast wave model for the generation of type II bursts. Since the dynamics of the blast wave is independent of the CME, type II position ahead of a CME is consistent with the blast wave model.

This research was supported by NASA grants Solar Maximum Mission Guest Investigator Program NAGW – 2760 and NAGW – 1541 and NSF grant ATM-9019893.

REFERENCES

1. Y. Uchida, Solar Phys. **28,** 495 (1973).
2. R. T. Stewart, R. A. Howard, F. Hansen, T. Gergely, and M. Kundu, Solar Phys. **36,** 219 (1974).
3. R. T. Stewart, M. K. McCabe, M. J. Koomen, R. T. Hansen, and G. A. Dulk, Solar Phys. **36,** 203 (1974).
4. T. Kosugi, Solar Phys. **48,** 339 (1976).
5. J. T. Gosling, E. Hildner, R. M. MacQueen, R. H. Munro, A. Poland, and C. L. Ross, Solar Phys. **48,** 389 (1976).
6. R. H. Munro, J. T. Gosling, E. Hildner, R. M. MacQueen, A. I. Poland, and C. L. Ross, Solar Phys. **61,** 201 (1979).
7. G. A. Dulk, in Radio Physics of the Sun (ed. M. R. Kundu and T. E. Gergely, D. Reidel, Hingham, 1980), p 419.
8. R. M. MacQueen, Phil. Trans. Roy. Soc. London, Ser. A, **297,** 605 (1980).
9. T. E. Gergely, M. R. Kundu and E. Hildner, Ap. J. **268,** 403 (1983).
10. R. T. Stewart, G. A. Dulk, K. V. Sheridan, L. L. House, W. J. Wagner, C. Sawyer, and R. Illing, Astron. Astrophys. **216,** 217 (1982).
11. R. D. Robinson and R. T. Stewart, Solar Phys. **97,** 145 (1985).
12. W. J. Wagner and R. M. MacQueen, Astron. Astrophys. **120,** 136 (1983).
13. T. E. Gergely, M. R. Kundu et al, Solar Phys. **90,** 161 (1984).
14. C. Sawyer, Adv. Space Res. **2,** 265 (1983).
15. W. J. Wagner, Adv. Space Res. **2,** 203 (1983).
16. D. E. Gary, G. A. Dulk et al, Astron. Astrophys. **134,** 222 (1984).
17. N. Gopalswamy and M. R. Kundu, Solar Phys. **111,** 347 (1987).
18. M. R. Kundu, N. Gopalswamy, S. M. White, P. Cargill, E. J. Schmahl, and E. Hildner, Ap. J. **347,** 505 (1989).
19. N. Gopalswamy and M. R. Kundu, Solar Phys. **122,** 91 (1989).
20. M. R. Kundu and N. Gopalswamy, in Eruptive Solar Flares (ed. Z. Svestka, B. V. Jackson and M. E. Machado, Springer Verlag, N. Y. , 1991), in press.
21. N. R. Sheeley, R. T. Stewart, R. D. Robinson, R. A. Howard, M. J. Koomen, and D. J. Michels, Ap. J. **279,** 839 (1984).
22. S. W. Kahler, E. W. Cliver, N. R. Sheeley, R. A. Howard, M. J. Koomen, and D. J. Michels, J. G. R. **90,** 177 (1985).
23. H. V. Cane, Astron. Astrophys. **140,** 205 (1984).

MHD Turbulence, Reconnection, and Test-Particle Acceleration

Perry C. Gray and William H. Matthaeus
Bartol Research Institute, University of Delaware

March 24, 1992

Abstract

We examine homogeneous MHD turbulence and turbulent magnetic reconnection as possible mechanisms for accelerating cosmic ray particles. Test particle calculations are performed using fields from MHD simulations, and initially Maxwellian particle distributions are shown to evolve into power-law distributions $N(E) \propto E^{-n}$ with $n \approx -2$ to -2.5. Simple estimates[1] for both the maximum energy attainable and the mean energies of the accelerated particles are fairly successful and are consistent with timescales for flares and cosmic rays.

1 Introduction

Magnetic reconnection has long been studied as a mechanism for converting magnetic energy into kinetic energy in solar flares and magnetospheric substorms. Most of the theoretical models for magnetic reconnection treat it as a steady-state process with magnetic fields that are essentially laminar. The reason for this is that MHD processes are commonly believed to be dominated by long-wavelength, low-frequency effects. From a theoretical standpoint, the major difficulty with these models is that the time spent by particles in the diffusion region, where the induced electric fields are greatest, is too short to account for significant energization of particles. The introduction of a turbulent component of the magnetic field in reconnection dramatically alters the physics of the above models. Recent test particle studies[1-3] have shown that the effect of MHD turbulence in reconnection alters the particle dynamics by trapping particles and retaining them in the diffusion region. Currently, we are studying this process in the context of homogeneous MHD turbulence, wherein reconnection plays an important role in the inverse cascade from small- to large-scale magnetic field structures. The preliminary results presented here show that this process can result in significant particle "heating," (i.e. an increase of the average test particle energy). Because the process produces power-law distributions of accelerated particles, it may be a viable mechanism for accelerating particles to cosmic ray energies.

2 Simulation of Turbulent Reconnection and Homogeneous Turbulence

The electric and magnetic fields used in the present studies are obtained from simulations of turbulent reconnection and homogeneous MHD turbulence. These simulations are two dimensional, such that $U_z = 0$ and $\mathbf{B} = (B_x(x,y), B_y(x,y), B_z)$, where B_z is a constant guide field in the ignorable direction. Furthermore, the flow is incompressible so that $\nabla \cdot \mathbf{U} = 0$ and $\rho = const$. The fields in this case are conveniently written in terms of the vorticity ω and the magnetic vector potential $\mathbf{a} = (0,0,a(x,y))$:

$$\frac{\partial \mathbf{a}}{\partial t} + \mathbf{U} \cdot \nabla \mathbf{a} = \mu \nabla^2 \mathbf{a},$$

$$\frac{\partial \omega}{\partial t} + \mathbf{U} \cdot \nabla \omega = \mathbf{B} \cdot \nabla \mathbf{j} + \nu \nabla^2 \omega$$

where $\mathbf{B} = \nabla \times \mathbf{a}$, $\omega = \nabla \times \mathbf{U}$, and μ and ν are the resistivity and kinematic viscosity. Because the vorticity, vector potential, and current density $\mathbf{j} = -\nabla^2 \mathbf{a}$ all have non-zero components in the z direction only and $\partial/\partial z = 0$, the equations effectively describe the evolution of those components as scalar fields in two dimensions. We also define the Reynolds number and magnetic Reynolds number as $R \equiv VL_0/\nu$ and $R_m \equiv 4\pi VL_0/\mu c^2$, where L_0 and V are the characteristic scale length and velocity of the system. Typically, V is taken to be the mean Alfvén velocity V_A. A truncated Fourier representation of the fields is advanced using an second-order accurate modified Euler method[3]. This method exactly conserves the three rugged invariants of two-dimensional MHD: the total energy density $E = E_u + E_b = (\langle U^2 \rangle + \langle B^2 \rangle)/2$, the cross helicity $H_c \equiv \langle \mathbf{U} \cdot \mathbf{B} \rangle/2$, and the mean square vector potential $A \equiv \langle a^2 \rangle/2$, where $\langle \ldots \rangle$ is a suitably defined averaging operation.

Previous studies of turbulent reconnection have used a sheet-pinch configuration[1-4]. As described in Matthaeus and Lamkin[3], the initial conditions for these simulations are described by a truncated Fourier representation of

$$j(x,y) \propto \delta(y - \frac{3\pi}{2}) - \delta(y - \frac{\pi}{2}),$$

which corresponds to two oppositely directed current sheets with magnetic field reversals at $y = \pi/2$ and $3\pi/2$. To initiate the reconnection process in this system, which slowly evolves by itself without spectral transfer, small-amplitude random-phase perturbations are added with a flat spectrum and finite bandwidth $1 \leq k^2 \leq 256$. The initial conditions are such that $E \approx A$ and $H_c \approx 0$.

For simulations of homogeneous MHD turbulence, the fields are initialized with $E_b \approx E_u$ and $H_c \approx 0$. The mode amplitudes are chosen to give an energy spectrum

$$E(k, t=0) \propto \frac{R_k}{2\pi k} \frac{1}{1 + (k/k_0)^{5/3}},$$

where R_k is a normally distributed random number of rms value 1. The parameter k_0 is the characteristic scale associated with the shift of the spectrum into the inertial range, which is consistent with a Kolmogorov omnidirectional spectrum at large k. For the results shown here, $k_0 = 6$, $R = R_m = 200$, and the calculations are carried out on a 64^2 grid ($k_x \times k_y$) for durations of several characteristic times. We are currently analyzing results for larger systems (256^2) and much longer times.

3   Test Particle Calculations

To examine how the turbulent magnetic and electric fields occurring in reconnection effect particle distributions, test particle calculations have been performed for two cases: the sheet pinch and homogeneous MHD turbulence. The results shown here for the sheet pinch are from previous simulations[1-3]. In those studies and the present one, the equations of motions for test particles are integrated in time as

$$\frac{d\mathbf{X}}{dt} = \mathbf{U},$$

$$\frac{d\mathbf{U}}{dt} = \alpha\big((\mathbf{U}\times\mathbf{B})+\mathbf{E}\big),$$

where $\alpha \equiv \Omega\tau$, where $\Omega = ZeB_0/mc$ and B_0 is the mean initial magnetic field strength. The electric and magnetic fields, $\mathbf{E}$ and $\mathbf{B}$, are calculated from the vector potential obtained from the simulations described in the previous section. The distances, velocities, and times are normalized in terms of characteristic length L_0, the initial mean Alfvén velocity V_A, and characteristic time $\tau \approx L_0/V_A$. The fields from the simulations are recorded every $\delta\tau = 1/512$ for the sheet-pinch calculation and $1/100$ for the homogeneous turbulence simulations. The magnetic field at the particle position is then approximated by linear interpolation, and the electric field is calculated as

$$\mathbf{E} = -\mathbf{U}\times\mathbf{B} + \mu\mathbf{j} = -\frac{\partial a_z}{\partial t}\hat{z}.$$

Hence, the electric field is held constant in the z direction between frames from the simulation. Because the system is two dimensional, the canonical z momentum $P_z = U_z + \alpha a_z(\mathbf{x},t)$ is a constant of motion. This fact was used by Ambrosiano et al.[1] to explicitly determine the z component of the particle velocity when integrating the equations of motion using a fourth-order Runge-Kutta scheme. For the present test-particle calculations in homogeneous turbulence, we have used the constancy of P_z as a diagnostic to determine the overall accuracy of our time integration, a leap-frog scheme with a conservative choice of time step. This choice of integration scheme was made in order to minimize the computational expense for the large number of particles (typically 10^5 per run) used in the MHD turbulence study.

4   Results and Discussion

Test particle calculations for the the sheet pinch[1] where carried out for $R = R_M = 400, 750,$ and 1000 using 2500 particles for each run. The particles were positioned initially in the middle of the current sheet with velocities of magnitude V_A and random direction. The left frame of Figure 1 shows the evolution of the current sheet and the position of sample particles at various times. The basic particle dynamics, as determined from earlier non-turbulent studies of reconnection[5], can be described as follows: The particles are convected into a reconnection region (an X line) and are accelerated in the z direction, after which they tend to escape along separatrices and end up in a magnetic island (an O line) where the reconnection electric field is weak. The major, distinguishing effect seen in turbulent reconnection is that the turbulent component of

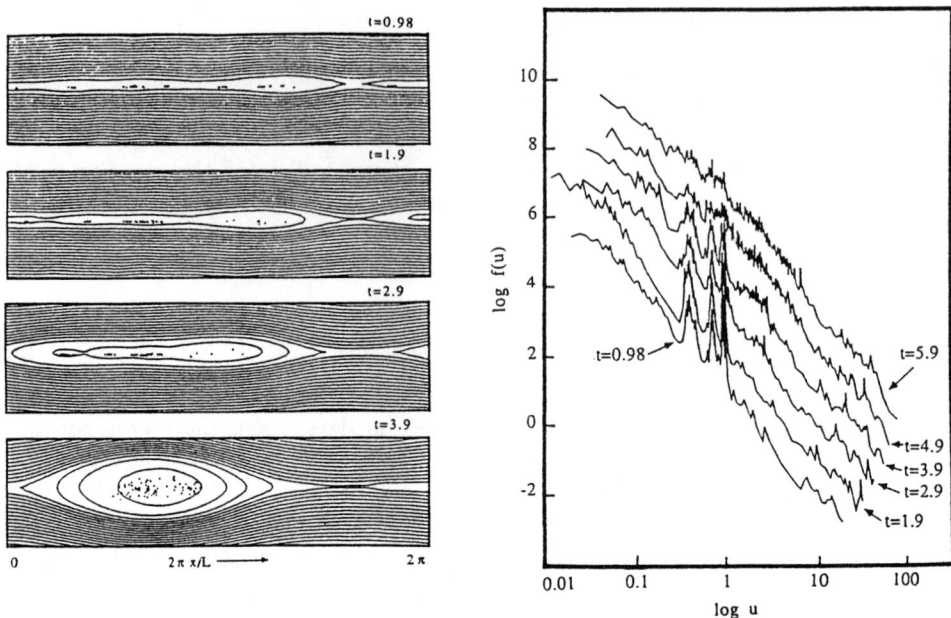

Figure 1: Magnetic field and particle positions (left) and distribution function (right) at various times for turbulent reconnection (adapted from Figs. 3 and 15 of Ambrosiano et al.[1]).

the fields cause the formation of "bubbles," which break up the reconnection region into multiple X points. Particles that are trapped in these bubbles are convected along the current sheet with them, thus spending more time in the reconnection region and gaining more energy than particles that escape along the separatrices. The maximum energy gain per particle can be estimated from the conservation of the z-momentum. This implies that $\Delta U_z = \alpha \Delta a_z$, where Δa_z is approximately the reconnection rate $\epsilon \approx 0.1$ multiplied by the time the particle spends in the reconnection region. The maximum energy attainable thus depends on the time required for a bubble to convect out of the reconnection region, which turns out to be about an Alfvén transit time τ. The result is $(\delta U_z)_{max} \approx \alpha\epsilon\tau$, which is in good agreement with model results. e.g., for $\alpha = 643$ the estimated value is $(U_z)_{max} = 64$ versus 60 from the simulation. The right frame of figure 1 shows the distribution of the test particle speeds at various times for the fields shown on the left. With the exception of the spike at $u = 1.0$ due to the initialization $u(t = 0) = V_A$, the distribution is well fitted by a power law with an exponent of approximately -2.5 .

Figure 2 illustrates the connection between reconnection and the processes involved in the formation of large-scale magnetic structures in MHD turbulence. Shown are the magnetic field lines and contours of current density in a two-dimensional simulation of homogeneous MHD turbulence (cf., Matthaeus and Lamkin[3] for details). At this late time in the simulation ($t = 9.76\tau$), only a few magnetic islands remain and these are merging together, as is apparent from the formation of current-sheet structures near the magnetic null points. Eventually, the system will relax to a state with two magnetic

 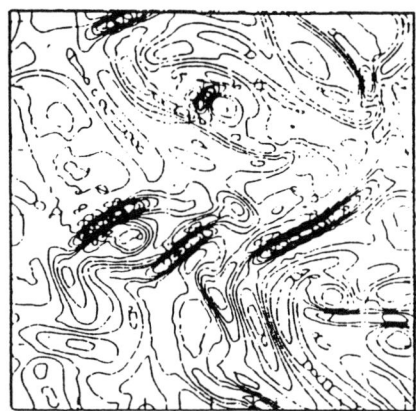

Figure 2: Magnetic field (left) and contours of current density (right) in for 2D simulation of MHD turbulence (adapted from Fig. 2c of Matthaeus and Lamkin[3]).

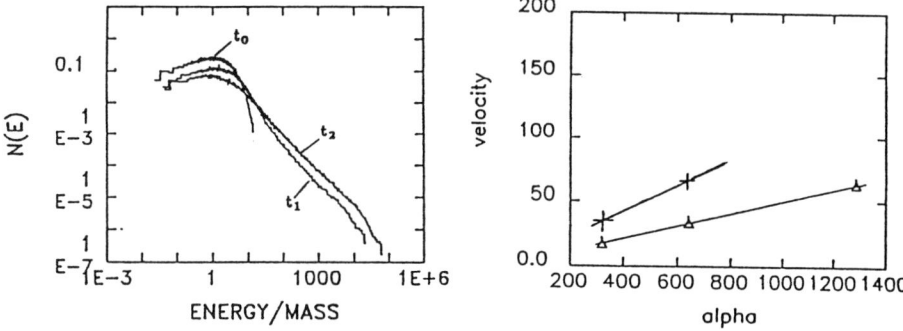

Figure 3: Distribution function (left) at $t_0 = 0$, $t_1 = 1$, and $t_2 = 4$, and (right) maximum velocity at the end of the run for test particles in turbulent reconnection (+), and mean velocity for test particles in MHD turbulence ($\triangle$).

islands corresponding to two oppositely directed current channels.

Figure 3 shows results from test particle calculations in turbulent fields generated in a periodic simulation with $R = R_m = 200$. Particles are initially distributed randomly in the 64^2 periodic simulation box with a Gaussian velocity distribution and mean thermal speed $v_t = V_A$. The left frame shows the distribution function calculated for $\alpha = 643$ at various times during the simulation. What is generally seen in these simulations is a evolution from the initial Gaussian to a power-law distribution that shifts to progressively higher energies as the simulation advances. At $t_2 \approx 4$, relatively few reconnection regions remain in the system. The right frame of Figure 3 shows the behavior of particle velocities, with various values of α, at the end of the test particle simulations. The values for the top curve are maximum velocities taken from sheet-pinch simulations[1,2], and the values for the lower curve are RMS particle velocities from simulations of MHD turbulence for the three cases $\alpha = 321, 643,$ and 1286. Evidently both the mean and maximum velocities scale according to the simple argument given previously[1,2] for the maximum obtainable velocity. Along with the power law distributions (cf. Fig. 3), this scaling suggests a statistical, and perhaps self-similar, underlying theory to account for energization of charged particles in MHD turbulence.

These preliminary results demonstrate that reconnection associated with the merging of magnetic islands in MHD turbulence is an effective mechanism for accelerating particles and is consistent with results from sheet-pinch calculations[1]. It remains to be determined whether the scaling demonstrated above extends to the extremely large values of α and the associated high particle energies expected in the cosmic ray parameter regime. We are currently extending these calculations to systems that more closely model the case for cosmic rays, an effort which will require larger two dimensional simulations, three-dimensional simulations, relativistic equations of motions and a supporting theoretical kinetic model.

Acknowledgments: We gratefully acknowledge previous collaboration and helpful discussions with D. Pontius, M. Goldstein, J. Ambrosiano, D. Plante, D. Montgomery, and L. Drury. Supported by NSF grant ATM-8913627 and the NASA Space Physics Theory Program. Computing was done at the San Diego Supercomputer Center.

References

[1] Ambrosiano, J. J., W. H. Matthaeus, M. L. Goldstein, and D. Plante, Test particle acceleration in turbulent reconnecting magnetic fields, *Jour. Geophy. Res.*, **93**, 14383, 1988.

[2] Matthaeus, W. H., J. J. Ambrosiano, and M. L. Goldstein, Particle acceleration by turbulent magnetohydrodynamic reconnection, *Phys. Rev. Lett.*, **53**, 1449, 1984.

[3] Matthaeus, W. H. and S. L. Lamkin, Turbulent magnetic reconnection. *Phys. Fluids*. **29**, 2513, 1986.

[4] Matthaeus, W. H. and D. C. Montgomery, Selective decay hypothesis at high mechanical and magnetic Reynolds numbers, *Annals New York Acad. Sci.*, **357**, 203, 1980.

[5] Wagner, J. S., P. C. Gray, J. R. Kan, T. Tajima, and S.-I. Akasofu, Particle dynamics in reconnection field configurations, *Planet. Space Sci.*, **29**, 391, 1981.

PARTICLE ACCELERATION AT COMETS

Tamas I. Gombosi
Space Physics Research Laboratory
Department of Atmospheric, Oceanic and Space Sciences
The University of Michigan, Ann Arbor, MI 48109

ABSTRACT

This paper compares calculated and measured energy spectra of implanted H^+ and O^+ ions on the assumption that the pick-up geometry is quasi-parallel and about 1% of the waves generated by the cometary pickup process propagates backward (towards the comet). The model provides a good description of the implanted O^+ and H^+ energy distribution near the pickup energies.

INTRODUCTION

Instruments at comets Giacobini-Zinner and Halley detected large fluxes of energetic particles[1-4]. A significant part of the observed energetic ion population was detected at energies considerably larger than the pickup energy indicating the presence of some kind of acceleration process acting on implanted ions. Velocity diffusion of lower energy implanted ions (near the pickup energy) has also been observed by the several instruments upstream of the comet Halley bow shock[5,6].

The acceleration of the implanted ions in the cometary upstream region has also generated considerable theoretical interest. This problem was first examined just before the Giacobini-Zinner encounter[7]. In a subsequent paper written shortly before the Halley encounters, Ip and Axford considered five potential mechanisms that can act to accelerate implanted ions[8]. They concluded that in cometary environments the second-order Fermi acceleration (slow velocity diffusion due to the interaction with propagating Alfven waves) was likely to play a dominant role in accelerating ions of cometary origin far upstream from the comets. Their conclusion was also endorsed by Gribov et al.[9]. Later Isenberg published an elegant analytic solution for a specific scenario, which took into account convection, adiabatic acceleration and velocity diffusion[10].

Shortly after Isenberg's analytic solution Gombosi developed a self-consistent, three-fluid model of plasma transport and implanted ion acceleration in the unshocked solar wind[11]. The model described convection, adiabatic and diffusive velocity change, as well as mass addition and charge exchange losses. Later, this model was extended to include the effects of first-order Fermi acceleration[12] and the predicted velocity distribution functions were compared with observations[13]. In this model a second-order Fermi mechanism accelerates ions to moderate energies in the cometary upstream region and then in the foreshock region (where the solar wind slows down from its ambient speed to about 0.8 times its upstream value) the superthermal implanted ions are further energized by a diffusive-compressive shock acceleration process (first-order Fermi acceleration)[12,13].

MODEL

A well developed cometary atmosphere extends to distances some six orders of magnitude larger than the size of the nucleus. The dominant neutral molecules in this extended exosphere are H_2O, CO_2, CO and their daughter products. Most of these neutral particles move with velocities of about 1 km/s with respect to the cometary nucleus and with a velocity of about $-\mathbf{u}$ ($\mathbf{u}$=solar wind velocity) with respect to the plasma flow. Pickup of cometary particles, ionized by photoionization, charge exchange or electron impact, is the main physical process whereby comets interact with the solar wind.

Freshly born ions are accelerated by the motional electric field of the high-speed solar wind flow. The ion trajectory is cycloidal, resulting from the superposition of gyration and $\mathbf{E} \times \mathbf{B}$ drift. The resulting velocity-space distribution is a ring-beam distribution, where the gyration speed of the ring is $v_\perp = u \sin\alpha$, (where u is the bulk plasma speed and α is the angle between the solar wind velocity and magnetic field vectors) and the beam velocity (along the magnetic field line) is $v_\parallel = u \cos\alpha$. The ring beam distribution has large velocity space gradients and it is unstable to the generation of low frequency transverse waves.

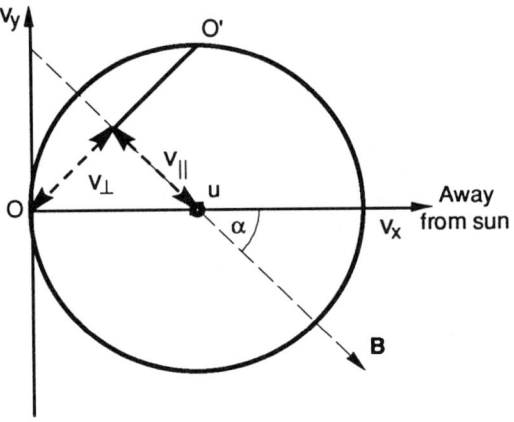

Fig. 1. Schematic representation of the implanted ion pickup geometry showing a velocity space diagram in the cometary frame of reference (the v_x axis points from the sun towards the comet). The bulk velocity of the ambient plasma distribution is marked by u. The initial velocity of a newly born cometary is marked by O. The thin dashed line indicates the direction of the magnetic field. The initial velocity-space distribution is a ring-beam distribution. The projection of the pickup ring beam to the (v_x, v_y) plane is denoted by the line OO'.

In a first approximation the newly ionized pickup particles interact with the low frequency waves (sometimes called Alfven waves) without significantly changing their energy in the average wave frame. As a result of this process the pitch angles of the pickup-ring particles are scattered on the spherical velocity space shell of radius u (see Fig. 1) around the local solar wind velocity. Observations indicate that this process does not lead to pitch-angle isotropy until very close to the cometary shock[14,15].

Our transport model is based on a series of recent papers discussing implanted ion acceleration processes in the upstream cometary region[11-13]. It takes into account advection, adiabatic acceleration, velocity diffusion (second order Fermi acceleration) and compressive-diffusive acceleration (first order Fermi acceleration) in a quasi-parallel geometry. Assuming that (i) the gyrophase distribution is totally random (this assumption, in effect means that diffusion across magnetic field lines is neglected), (ii) that the pitch-angle diffusion is a much faster process than energy diffusion, and (iii) the **u** and **B** vectors are parallel, one can derive the following Fokker-Planck equation for the pitch-angle averaged distribution function, $F(t,x,v)$ [13]:

$$\frac{\partial f}{\partial t} + u\frac{\partial f}{\partial z} - \frac{v}{3}\frac{du}{dz}\frac{\partial f}{\partial v} = \frac{1}{v^2}\frac{\partial}{\partial v}\left(v^2 D\frac{\partial f}{\partial v}\right) + \frac{\partial}{\partial z}\left(\kappa\frac{\partial f}{\partial z}\right) + \frac{Q_n \delta(u-v)}{(4\pi)^2 \lambda_n v^2 r^2}e^{-r/\lambda_n} \quad (1)$$

Here x is distance along the flow line (x increases towards the comet), r is cometocentric distance, v is the implanted ion velocity in the plasma frame, u is plasma bulk velocity, D_0 is the velocity diffusion coefficient, κ is the field aligned spatial diffusion coefficient, Q_n is the cometary gas production rate, and λ_n is the ionization scale-length of a cometary neutral species. $F(t,x,v)$ is the pitch angle averaged phase-space density of implanted ions (separate equations are used to describe protons and water group ions) having random velocity magnitude v at time t and position x.

The first two terms on the left hand side of Equation (1) are the convective time derivative of the distribution function, while the third term describes energization due to adiabatic compression (in the cometary environment du/dx<0, therefore this term results in energization of implanted particles). On the right hand side the first term describes energy diffusion in the plasma frame, the second term corresponds to spatial diffusion along magnetic field lines (in the present approximation this also corresponds to diffusion along flow lines), while the third term accounts for continuous production of cometary ions. The interplay between the adiabatic compression and spatial diffusion terms results in diffusive-compressive (first order Fermi) acceleration.

The spatial and velocity diffusion coefficients were obtained by using the quasilinear approximation. In this approximation the coefficients of spatial and velocity diffusions can be expressed in terms of the power spectrum of magnetic field fluctuations. However, one has to be careful, because different power spectra have to be used in the determination of these transport coefficients. When calculating κ, all magnetic field fluctuations have to be taken into account, regardless of their direction of propagation in the plasma frame of reference. When calculating the velocity diffusion coefficient, D_0, one has to recall that the wave field in the cometary environment is a superposition of fluctuations generated by the pickup process (these waves are predominantly low-frequency transverse waves propagating away from the comet along the magnetic field lines) and ambient waves in the solar wind (these waves move primarily away from the sun along the magnetic field lines). In the velocity diffusion coefficient an appropriately averaged magnetic field power spectrum has to be used to describe the effective power of randomly propagating (in both directions) magnetic field fluctuations. The details of this calculation can be found in our recent paper[13].

RESULTS AND DISCUSSION

Implanted ion distributions were calculated using the model described in the previous section. The model solved the transport equation along flow lines (which in this particular model are identical to magnetic field lines). For each flow line the calculation started at 2.5×10^6 km upstream from the point where the flow line intersected the Giotto trajectory, and it extended to a distance of 1.5×10^6 km downstream from the intersection point. The 4×10^6 km distance (along the curved flow line) was divided into 200 spatial grid points resulting in a 2×10^4 km spatial step size. The velocity interval extended from 0 to 3500 km/s (in the plasma frame) with a 25 km/s step size. There were free escape boundaries at both ends of the flow line, and there was no flux through $v=0$ and $v=3500$ km/s (this velocity was high enough so that only an insignificant number of particles was accelerated to this value).

The model predictions were compared to observations made by the IMS HERS and JPA IIS instruments onboard the Giotto spacecraft. Fig. 2 shows a detailed comparison between observed and calculated energy spectra at a cometocentric distance of $r=1.20 \times 10^6$ km along the inbound Giotto trajectory. The proton noise level is about $10^{-26.5}$ s^3/cm^6, therefore data points below this value should be ignored. The sharp rise of the proton phase space density function at low energies (below about 200 km/s) is due to the contribution of solar wind protons. The agreement between the calculated and observed spectra is very good. At the same time one should keep in mind that the plasma environment of comet Halley was a highly dynamic medium, with frequent changes of magnetic field configuration and flow patterns.

Finally, the sensitivity of our results to some of the simplifications has to be discussed. One of the fundamental simplification of the model is the assumption of a parallel geometry, i.e. that the flow velocity and magnetic field vectors are parallel. Recent numerical simulations showed that for quasi-parallel geometries (when the angle, α, between the flow velocity and the magnetic field vectors is less than about 60°) the backward propagating ratio of pickup generated waves is a couple of percent[17]. This means that our velocity diffusion coefficient is applicable for situations with $\alpha<60°$. In the case of quasiperpendicular geometry the effective wave power is much larger than in a quasi-parallel situation (because waves propagate in both directions with equal probability). The immediate consequence of the increased effective wave power is a large increase of the velocity diffusion coefficient. A temporary change from quasi-parallel to quasiperpendicular geometry might be the reason for the appearance of large energetic particle spikes observed in the cometary upstream region[3,4].

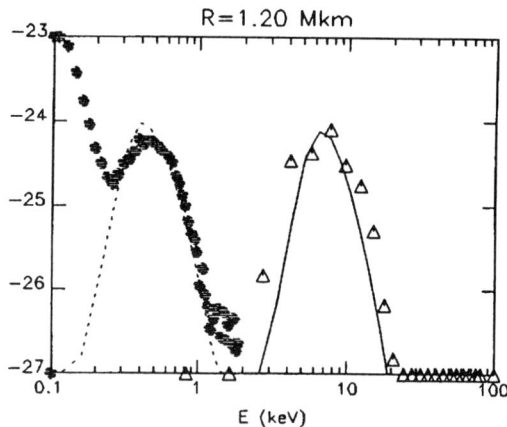

Fig. 2. Comparison of calculated and measured implanted ion phase space distribution functions at $r=1.20\times10^6$ km along the inbound part of the Giotto trajectory. The distribution functions are given in units of s^3/cm^6. The observed distribution functions are represented by filled circles (H^+) and triangles (O^+), while dotted and solid lines represent calculated distributions of implanted protons and oxygen ions, respectively. The sharp rise of the observed proton phase space density function at low energies (below about 200 km/s) is due to the contribution of solar wind protons.

SUMMARY

The results of model calculations describing transport and energization of cometary pickup ions were compared to observations made by the Giotto spacecraft. The model takes into account adiabatic compression, energy diffusion, field aligned spatial diffusion and mass addition along flow lines of decelerating plasma flow lines. The model used the quasilinear approximation of the velocity and spatial diffusion coefficients. The generalized transport equation for the velocity-space solid-angle-averaged distribution function was solved for individual flow lines. The implanted ion spectra at the intersection points of the flow line with Giotto's trajectory were compared with spectra observed at that particular spatial location. The theoretical model provides a good description of the implanted proton and water group ion energy distribution near the pickup energy.

ACKNOWLEDGEMENTS

This work was supported by NASA grants NAGW-1366 and NAGW-2162. Acknowledgement is also made to the National Center for Atmospheric Research sponsored by NSF, for the computing time used in this research.

REFERENCES

1. R.J. Hynds, et al., Science, 232, 361 (1986).
2. F.M. Ipavich et al., Science, 232, 366 (1986).
3. K. Kecskeméty et al., J. Geophys. Res., 94, 185 (1989).
4. S. McKenna-Lawlor et al., Ann. Geophys., 7, 121 (1989).
5. A.J. Coates et al., J. Geophys. Res., 94, 9983 (1989).
6. M. Neugebauer et al., J. Geophys. Res., 94, 5227 (1989).
7. E. Amata and V. Formisano, Planet. Space Sci., 33, 1243 (1985).
8. W.-H Ip and W.I. Axford, Planet. Space Sci., 34, 1061 (1986).
9. B.E. Gribov et al., Astron. Astrophys., 187, 293 (1987).
10. P.A. Isenberg, J. Geophys. Res., 92, 8795 (1987).
11. T.I. Gombosi, J. Geophys. Res., 93, 35 (1988).
12. T.I. Gombosi et al., J. Geophys. Res., 94, 15011 (1989).
13. T.I. Gombosi et al., J. Geophys. Res., 96, 9467 (1991).
14. A.J. Coates et al., J. Geophys. Res., 95, 4343 (1990).
15. M. Neugebauer et al., J.Geophys. Res., 94, 1261 (1989).
16. J. Gaffey et al., J. Geophys. Res., 93, 5470 (1988).
17. R.H. Miller et al., J. Geophys. Res., 96, 9479 (1991).

PICKUP IONS OBSERVED AT COMET HALLEY

M. Neugebauer
Jet Propulsion Laboratory, California Institute of Technology, Pasadena, CA 91109

ABSTRACT

Instruments on the Giotto spacecraft obtained data on the velocity distributions of pickup ions in the vicinity of comet Halley. Combination of data from two different instruments allows a comparative study of the pitch-angle scattering rates and velocity diffusion rates of protons and water-group ions. The key features of the data are reviewed.

INTRODUCTION

The vicinity of an active comet is an excellent laboratory for the study of wave-particle interactions associated with the nonequilibrium distributions of cometary material that becomes ionized in the solar wind. In its flyby of comet Halley in March, 1986, the Giotto spacecraft carried two plasma experiments -- the Johnstone Plasma Analyzer (JPA) and the Ion Mass Spectrometer (IMS) -- for measuring such ion distributions. This paper summarizes some of the findings of those two experiments that are relevant to the topic of ion pickup and acceleration.

In the outer region of the coma, principally upstream of the bow shock, protons were best measured by the high-energy-range spectrometer (HERS) of the IMS [1]. HERS was a magnetic mass spectrometer capable of mapping the velocity distribution as a function of ion mass/charge so the proton distributions could be studied without contamination or confusion by other species. HERS could not, however, detect ions with energy/charge greater than ~4 keV. In this region of the coma, the best water-group data were obtained by the JPA implanted ion sensor (IIS) [2,3] which is a time-of-flight mass spectrometer sensitive to ions in the energy/charge range 2.5-55 keV. An ion is considered to belong to the water group (WG) if it was observed in the IIS mass/charge bin that covered a nominal range of 6.5 to 20 amu/e. The relevant details concerning the data quality and methods of analysis have been summarized by *Neugebauer et al* [4,5] for IMS and by *Coates et al.* [6,7] for JPA.

A superthermal ion is created whenever an atom or molecule in the extensive cometary coma becomes charged, usually by photoionization or by charge exchange with solar wind ions. Immediately upon ionization, the pickup ions are located on a ring in velocity space. Their subsequent angular distribution depends on both the angle α between the solar-wind flow vector and the interplanetary magnetic field and on the degree to which the ions have been pitch-angle scattered since they were first picked up. Figure 1 shows pitch-angle (left) and velocity (right) distributions of protons upstream of the Halley bow shock for selected intervals when the direction of the interplanetary field, and hence α, were fairly steady. The velocity distributions $f(v)$ are calculated in the reference frame of the solar wind and the increasing value of f at low speeds is the high-velocity tail of the distribution of solar-wind protons. For each pair

© 1992 American Institute of Physics

274 Pickup Ions Observed at Comet Halley

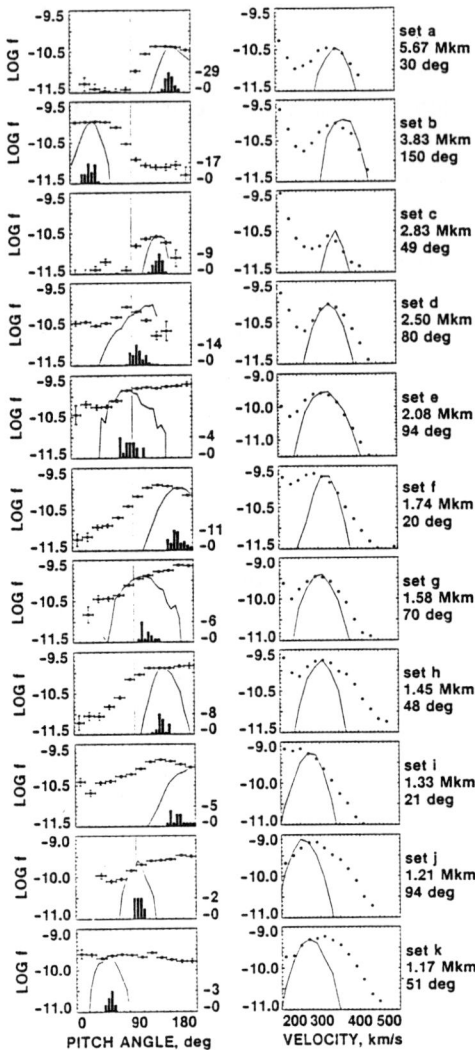

Figure 1. (Left) The circles are measured pitch angle distributions of picked-up protons; the histograms are the pitch angles expected for a perfect instrument due to variations in the direction of the interplanetary magnetic field; and the smooth curves are the calculated cold-ring distributions after the instrument response is convolved with the field variations. (Right) The circles are measured velocity distributions averaged over pitch angle and the smooth curves are the calculated cold-ring distributions after accounting for the instrumental resolution. The ordinates are $\log_{10} f$ in units of cm^{-3}km^{-3} s^3. From Neugebauer et al. [10].

of plots, the distance from the comet (in units of 10^6 km = Mkm) and the value of α are given in the upper right corner of the velocity plots. In these plots, the points are the measured values, the histograms indicate the range of variation of α during each interval, and the continuous curves indicate the calculated response of the HERS instrument to a cold ring of pickup protons, taking the distribution of α into account. Thus the amounts of pitch-angle scattering and velocity diffusion are indicated by the differences between the points and the curves.

PITCH-ANGLE SCATTERING

The spacecraft crossed the cometary bow shock immediately following interval k. Up until that interval, the pitch-angle distributions were broader than the initial ring distributions, but still very anisotropic. Far from the comet (intervals a, b, and c), α was either quite large or quite small (quasi-parallel pickup) and "half-shell" distributions were observed. At these times, the water-group ions also had half-shell distributions, as can be seen in Figure 9 of Coates et al.[8]. Yoon and Wu[9] have argued that a half-shell distribution is to be expected far upstream from the comet where the turbulence is weak and it is appropriate to use an angular diffusion coefficient that goes to zero at a pitch angle of 90°. They argue that it is only in regions of stronger fluctuations that resonance broadening allows particles to be pitch-angle scattered onto the other half shell. The observations are consistent with this point of view, with the half-shell effect less pronounced by quasi-parallel interval i and essentially gone by interval k.

Under quasi-perpendicular conditions, such as intervals d, e, g, and j in Figure 1, the pitch-angle scattering of both protons and WG ions was unexpectedly asymmetric. Neugebauer et al.[10] suggested that the asymmetric scattering may be a macroscopic or global rather than a local phenomenon; the picked-up ions might have been in a large-scale magnetic configuration that allowed escape in one direction but not the other. The global magnetic configuration was not uniform, with the field between the spacecraft and the comet being stronger and more turbulent than the field between the spacecraft and the Sun. Ions picked up with pitch angles near 90° might have been readily mirrored by either compressional waves or the increase in field strength as the solar wind was slowed down by mass loading near the comet. Intervals d, e, g, and j all had geometries consistent with such an explanation.

The comparison of the pitch-angle scattering rates of protons and WG ions can be put on a quantitative basis by considering the mean angular width

$$<\Delta p> = \int |p - <p>| f \, d^3v / \int f \, d^3v$$

where $f(p)$ is phase-space density averaged over the thickness of the shell of pickup ions and p is the pitch angle whose average is given by

$$<p> = \int p f \, d^3v / \int f \, d^3v.$$

For an isotropic distribution, $<p> = 90°$ and $<\Delta p> = 32.7°$. Figure 2 shows that there is a systematic dependence of $<\Delta p>$ on α and on the density of the picked-up ions. In this scatter plot of α vs ion density (n_{WG} or n_p), the data are coded according to whether the corrected mean width = $<\Delta p> - <\Delta\alpha>$ was high, moderate, or low, with the scales given in the lower left corners. The correction amounts to subtraction of the contribution of the fluctuations in the field direction from the observed $<\Delta p>$. The WG data are on the left and the proton data are on the right; note that the numbers are larger for the WG ions than for the protons, indicating greater pitch-angle scattering for the heavier ions.

The n_{WG} dependence shown in the left side of Figure 2 is certainly not surprising. At distances from the comet sufficiently great that the waves caused by the pickup ions are not saturated, one expects that more pickup ions will lead to more waves and more pitch-angle scattering. The dependence on α is more difficult to understand because both observations [11] and simulations [12] show greater wave amplitudes when α is small than when it is large. Thus the waves generated under quasi-perpendicular conditions (α near 90°) must be more effective scatterers than the greater amplitude waves generated under quasi-parallel conditions. Recent simulations by Gary et al. [13] support this point of view; they found that the O^+ pitch-angle scattering rate at perpendicular injections is approximately twice that in the quasi-parallel regime, while Yoon et al. [14] have shown theoretically, for a simplified wave field, that the α dependence of the initial diffusion rates for a ring-beam distribution are consistent with the observations.

The right-hand side of Figure 2 shows that the pitch-angle scattering rate of protons did not show a strong dependence on α.

VELOCITY DIFFUSION AND ION ACCELERATION

In Figure 3, the mean widths of the velocity peaks $<\Delta v>$ (i.e., the shell thicknesses) are plotted in a fashion similar to the $<\Delta p>$ data in Figure 2. The apparent shell thickness due to instrumental smear is ~24 km/s for both protons and WG ions. The proton shell thickness seems to correlate mainly with pickup proton density, while the WG shell thickness seems to correlate mainly with α.

The reasons for the differences in the pitch-angle scattering and shell thickening rates for protons and WG ions are not readily apparent. The answer must lie in differences in the wave spectra under quasi-parallel and quasi-perpendicular conditions. Tsurutani et al. [15] showed that during the flyby of comet Giacobini-Zinner, the ICE magnetometer observed qualitatively different waves for large and small α. Water-group ion cyclotron waves were evident for small α, while the fluctuations had higher frequencies (more likely to be resonant with protons) when α was large. Similarly, Glassmeier et al. [11] reported qualitatively different spectra for the two regimes, with spectral peaks near the WG cyclotron frequency being much more pronounced under quasi-parallel conditions than under quasi-perpendicular.

The plasma instruments on Giotto, whose measurements are the topic of this paper, had neither the energy ranges nor the geometric factors required to detect the particles in the high-energy tails of the ion distribution functions. Extrapolation of the JPA energy spectra to higher energies matched nicely with the higher energy data obtained by the Energetic Particle Analyzer on Giotto [16]. The bulk of the distribution of the picked up ions remained close to, and usually slightly below the local value of the solar wind speed, as can be seen on the right side of Figure 1. Upstream of the bow shock, the only place where the radius of the shell was greater than the local solar wind veloc-

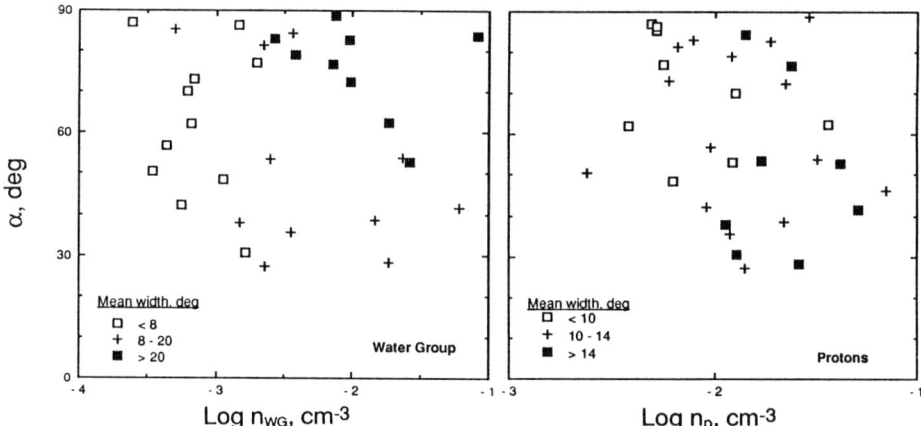

Figure 2. (Left) Scatter diagram of the angle α between the interplanetary field and the solar wind flow direction (converted to an equivalent angle <90°) versus the logarithm of the density of picked-up water group ions. Different symbols are used to denote small, intermediate, and large amounts of pitch angle scattering. (Right) A similar plot for picked-up protons. From Neugebauer et al.[16].

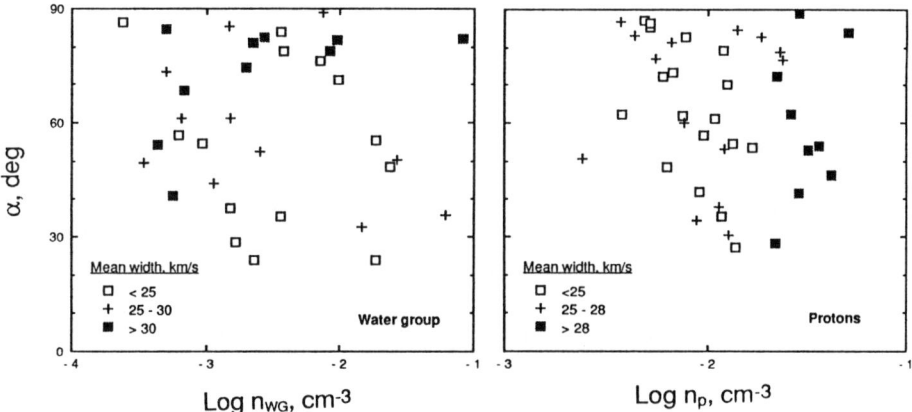

Figure 3. Same as Figure 2 except that the data are coded according to the level of energy diffusion (shell thickness) rather than pitch angle scattering. From Neugebauer et al.[16].

ity was very close to the shock (see intervals j and k in Figure 1). One might expect to see the radius of the shell increase in this region as a result of first-order Fermi acceleration and/or adiabatic deceleration associated with mass loading of the plasma. We cannot, however, demonstrate that either or both of these acceleration mechanisms played a role in the increased shell radius because just before the detection of the increased shell radius, the spacecraft crossed an interplanetary discontinuity that changed the solar wind speed and density; such discontinuities are quite commonly observed in the slow solar wind such as was present on the day of the Giotto encounter with comet Halley. With a single shock crossing, the ambiguities due to such effects cannot be removed.

Acknowledgements. This paper presents the results of one phase of research conducted at the Jet Propulsion Laboratory, California Institute of Technology, under contract with the National Aeronautics and Space Administration.

REFERENCES

1. H. Balsiger et al., J. Phys. E., 20, 759 (1987).
2. A.D. Johnstone et al., J. Phys. E, 20, 795 (1987).
3. B. Wilken et al., J. Phys. E., 20, 778 (1987).
4. M. Neugebauer et al., J. Geophys. Res., 94, 1261-1269 (1989).
5. M. Neugebauer, Geophys. Res. Lett., 16, 1261 (1989).
6. A.J. Coates et al., J. Geophys. Res., 94, (1989).
7. A.J. Coates et al., J. Geophys. Res., 95, 4343 (1990).
8. A.J. Coates, Cometary Plasma Processes (American Geophysical Union, Washington, D. C., 1991), p. 301.
9. P.H. Yoon and C.S. Wu, Cometary Plasma Processes (Amer. Geophys. Un., Washington, D. C., 1991), p. 241.
10. M. Neugebauer et al., J. Geophys. Res., 94, 5227-5239 (1989).
11. K.-H. Glassmeier et al., J. Geophys. Res., 94, 37-48 (1989).
12. S.P. Gary, K. Akimoto, and D. Winske, J. Geophys. Res., 94, 3513 (1989).
13. S.P. Gary, R.H. Miller, and D. Winske, Geophys. Res. Lett., 18, 1067 (1991).
14. P.H. Yoon and L. F. Ziebell, Phys. Fluids B 3, 2124 (1991)
15. B.T. Tsurutani et al., J. Geophys. Res., 94, 18 (1989).
16. A.J. Coates et al., J. Geophys. Res., 95, 20701 (1990).
17. M. Neugebauer, A.J. Coates, and F.M. Neubauer, J. Geophys. Res., 95, 18745 (1990).

PARTICLE ACCELERATION IN (BY) ACCRETION DISCS

J. I. Katz

Department of Physics and McDonnell Center for the Space Sciences
Washington University, St. Louis, Mo. 63130

ABSTRACT

I present a model for acceleration of protons by the second-order Fermi process acting on randomly scrambled magnetic flux arches above an accretion disc. The accelerated protons collide with thermal protons in the disc, producing degraded energetic protons, charged and neutral pions, and neutrons. The pions produce gamma-rays by spontaneous decay of π^0 and by bremsstrahlung and Compton processes following the decay of $\pi^\pm$ to $e^\pm$.

INTRODUCTION

The most remarkable property of AGNs is the appearance, in many cases, of much of their luminosity as the acceleration of nonthermal particles. Evidence for this consists of the polarized optical continuum with a power law spectrum found in some AGN, the great power required to supply radiating electrons to radio galaxies, and the gamma-ray luminosity of 10^{48} erg/sec of 3C279 recently discovered[1] by GRO.

This paper is a preliminary account of the calculation of a model[2] of particle acceleration in low density astronomical shear flows. Particles are accelerated by a second-order Fermi mechanism. They are assumed to be trapped in magnetic mirrors consisting of magnetic flux arches whose feet are pinned to the surface of a quasi-Keplerian accretion disc. The differential motion of points on the disc surface at differing radii accelerates the particles. The resulting forces on the disc (acting through the magnetic field) are described by a viscosity, and may be its chief dissipative process. This mechanism directly converts the gravitational power of black hole accretion to particle acceleration. Although this process is second order in the Keplerian velocity, in rapidly rotating inner discs it may be rapid.

CALCULATION

The greatest uncertainty of this model is the magnetic field configuration, particularly how flux tubes connect disc points at different radii. This uncertainly is not presently resolvable empirically or theoretically. In addition, it is not known how well the magnetic arches act as mirrors, nor is the density distribution of thermal gas within and above the disc known. The accelerated protons collide with the gas (the interaction of GeV protons with radiation is negligible, but Compton scattering prevents the acceleration of electrons). These uncertainties affect both the acceleration rate and the loss rate, and may be combined into a single parameter describing the ratio between these two rates.

The evolution of the isotropic volume-averaged proton distribution function in momentum space $n(p)$ is given by

$$\frac{\partial n(p)}{\partial t} = \vec{\nabla}_p \cdot [D(p)\vec{\nabla}_p n(p)] - \rho\sigma(p)v(p)n(p) + \rho\int v(p')\sigma(p',p)n(p')\,dp'; \quad (1)$$

on the right hand side the first term represents the momentum space diffusion, the second represents collisional losses in thermal matter of density ρ, and the third represents the contribution of collision products to the proton distribution. The velocity $v(p) = p/\sqrt{m_p^2 + p^2/c^2}$ and $\sigma(p)$ and $\sigma(p',p)$ are total and differential proton-proton cross-sections. If the momentum-space diffusion results from scattering by magnetic mirrors with uncorrelated speed u and has a mean scattering length ℓ, then

$$D(p) = \frac{4}{3}\frac{u^2}{\ell c^2}pE, \quad (2)$$

where the coefficient has an order-of-unity uncertainty resulting from the unspecified correlation between the directions of the incident and scattered particles. The parameter $\rho\ell$ describes the comparative importance of acceleration and collisional losses.

The scattering processes are

$$p + p \to p + p \quad (3a)$$
$$p + p \to p + p + \pi^0 \quad (3b)$$
$$p + p \to p + n + \pi^+ \quad (3c)$$
$$p + p \to \text{others}. \quad (3d)$$

A characteristic energy scale is set kinematically by $m_p c^2 = 938$ MeV, and also by the increase and saturation of the cross-sections for the inelastic processes (3b) and (3c) in the range 400–700 MeV. At laboratory energies > 1 GeV (3d) begins to replace (3b) and (3c), but was not included in these preliminary calculations (although the correct total inelastic cross-section was used).

The calculations reported here consisted of the evolution of equation (1), using (2), and the calculation of the spectrum of π^0 produced in (3b) and of the subsequent decay gamma-rays. The spectrum of π^+ was not calculated (although the cross-section of [3c] was included in equation [1]); this important process (its cross-section is about five times that of [3b]), as well as the subsequent decay e^+, and the effects of (3d) are presently being added to the code. The differential cross-sections used were scaled from the measurements of Bugg, et al.[3] at 970 MeV; this is one of the few experimental papers which give the laboratory distribution of the energies of the scattering products (most of the literature gives center-of-momentum energies, which are insufficient unless the scattering angle is also known).

RESULTS

The effect of (3a) is to multiply the number of energetic protons, while conserving their total kinetic energy. Processes (3b)–(3d) have a similar effect, although some kinetic energy is lost. The fate of the neutron produced in (3c) depends on the dimensions of the acceleration region; in regions larger than 10^{13} cm (appropriate to a massive black hole in an AGN) it decays and may be regarded as equivalent to a proton, while neutrons are lost from smaller regions (galactic X-ray sources).

Momentum space diffusion multiplies the energetic proton energy, while reducing their number (some diffuse to a Coulomb drag sink at zero momentum). The combination of diffusive acceleration and collisional loss will, for suitable values of $\rho\ell$, lead to an exponential (in time) runaway in proton number and energy density, with a stationary normalized spectrum. In reality such a runaway would saturate because the growing particle energy density would disrupt the confining magnetic field, or because the growing viscosity would deplete the accretion flow.

The following figures show the results of such a calculation. Particles up to 6 GeV were included, but cross-sections above about 1 GeV were inaccurate because of the exclusion of (3d). The calculated maximum in the gamma-ray spectrum at $m_{\pi^0}c^2/2$ was not observed in the data[1]. This may perhaps be explained by the contribution of gamma-rays from $e^\pm$ bremsstrahlung, as is the case for the Galactic gamma-ray spectrum. These processes will be included in the future.

REFERENCES

1. R. C. Hartman, et al., *Ap. J. (Lett.)* **385**, L1 (1992).
2. J. I. Katz, *Ap. J.* **367**, 407 (1991).
3. D. V. Bugg, et al., *Phys. Rev.* **133**, B1017 (1964).

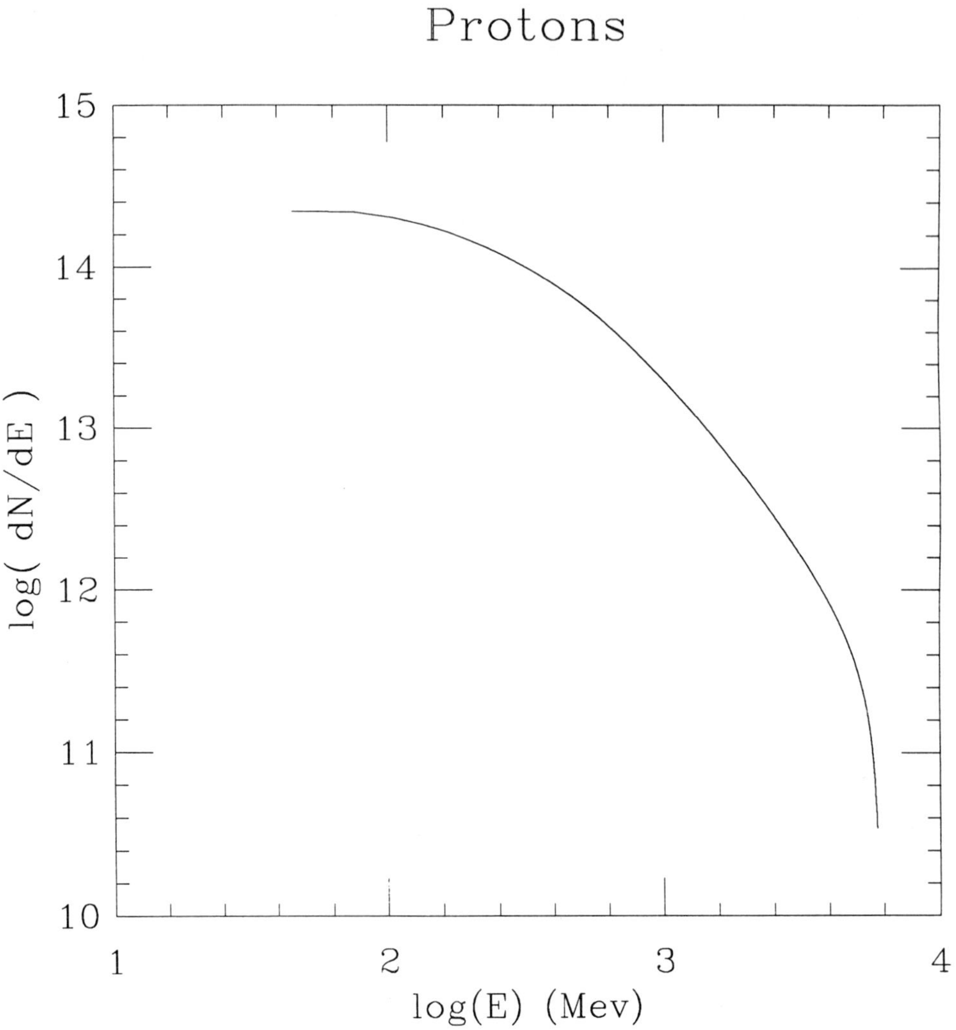

Figure 1: Proton spectrum.

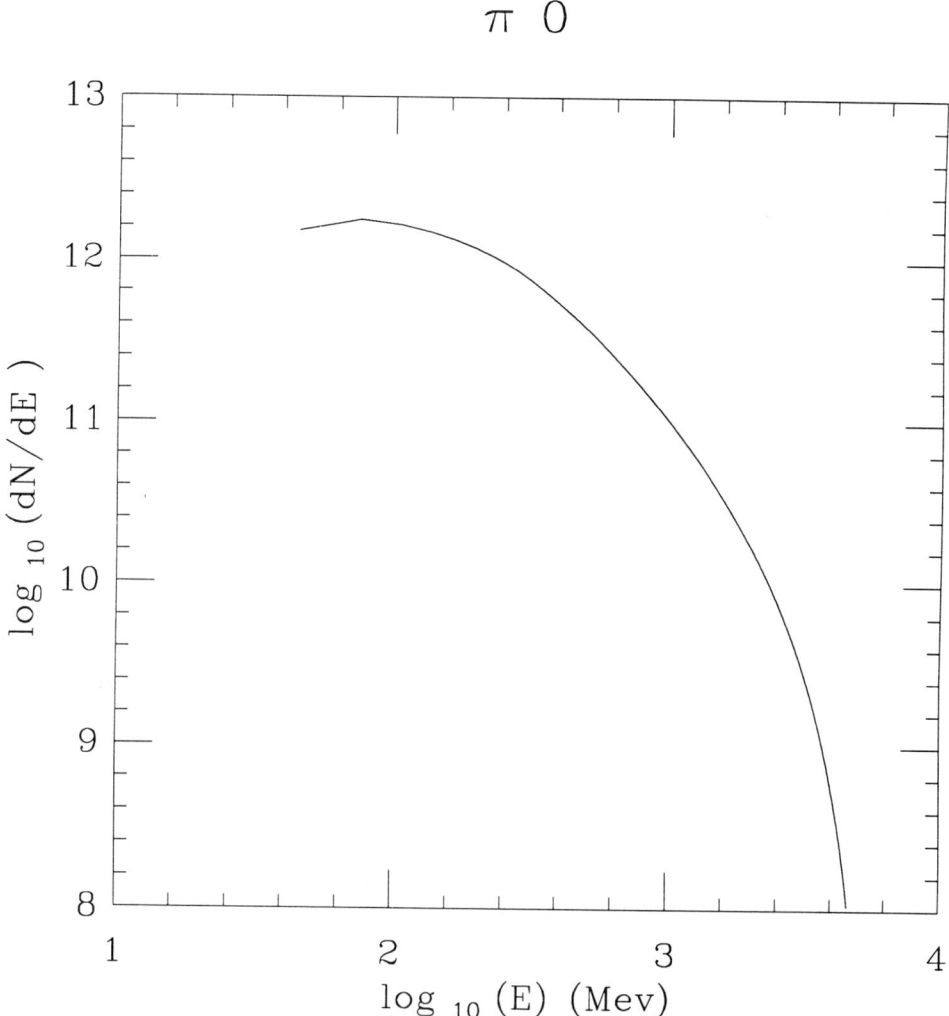

Figure 2: Neutral pion spectrum.

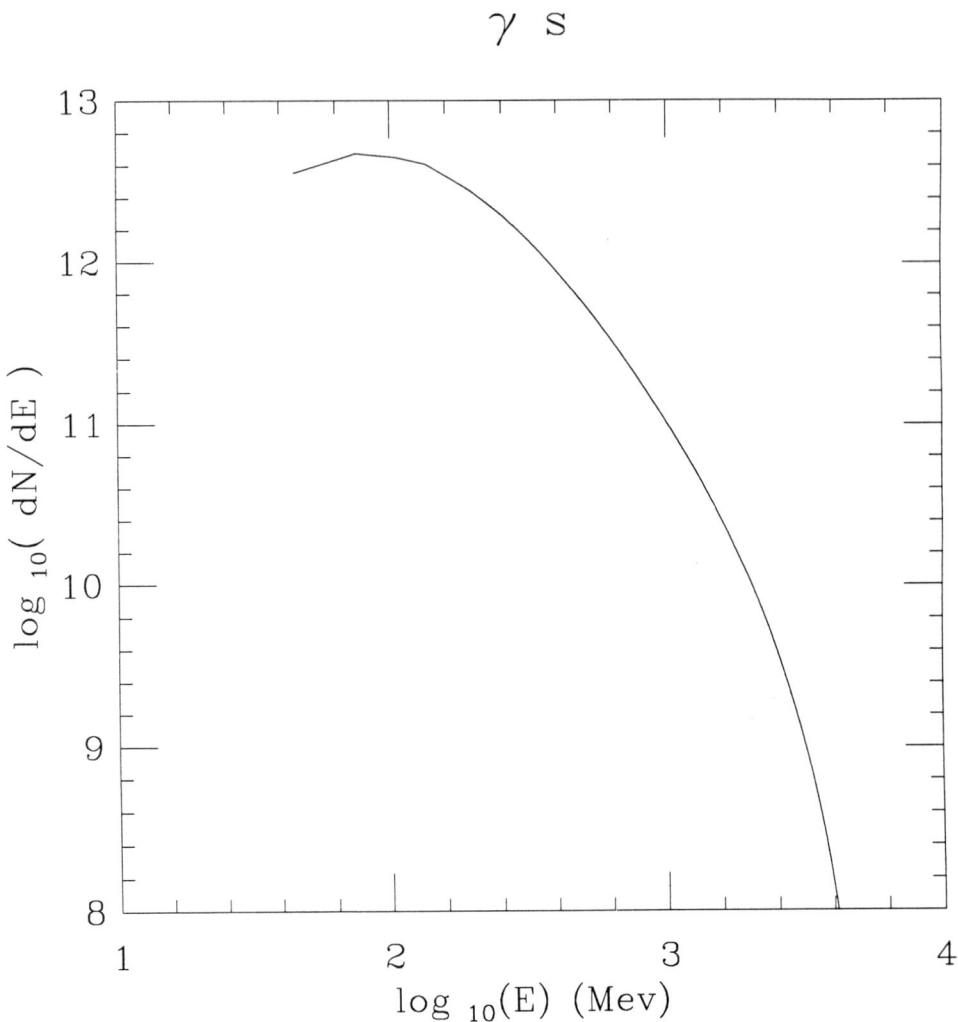

Figure 3: Gamma-ray spectrum from neutral pion decay.

Particle Acceleration in
Relativistic Flows and Shocks

ENERGETIC PARTICLE TRANSPORT IN RELATIVISTIC FLOWS

G.M. Webb

University of Arizona
Department of Planetary Sciences, Tucson Arizona 85721, U.S.A.

ABSTRACT

A discussion is given of pitch angle dependent and diffusive transport equations for cosmic rays applicable for both special relativistic and general relativistic flows, derived from the relativistic Boltzmann equation. As an example of particle transport in a curved spacetime we give a pitch angle dependent transport equation appropriate for radial accretion onto a Schwarzschild black hole. The roles of fluid shear, acceleration and compression on the energy changes of particles in the diffusive transport equations are emphasized. Also discussed are special flows (e.g. rigidly rotating flows) associated with a Killing vector for which a constant of the motion of the particles can be identified, and for which simplified transport equations can be constructed.

1. INTRODUCTION

In this paper we consider aspects of energetic charged particle transport in relativistic flows. The development of transport theory for both particles and photons in relativistic flows has a long history, with early seminal work by Thomas[14] on radiative transfer in special relativistic flows. The extension of the Boltzmann equation and associated transport theory to curved spacetimes expressed in geometric frame independent language was carried out by Lindquist[8] (see also Thorne[13]). Work on transport theory for cosmic rays in relativistic flows has been considered by Webb[15,16,17]; Kirk, Schlickeiser and Schneider[5] and Krülls[6].

In section 2, we give a brief introduction to the relativistic Boltzmann equation, including the effects of the Lorentz force, the radiation reaction four-force, and curved spacetime on the particle transport. We illustrate the effects of non-inertial forces by showing the role of fluid velocity acceleration, shear and expansion in determining the rate of change of the particle energy as measured in the background fluid frame. In section 3, we discuss pitch angle dependent transport equations for the energetic particles, for the special relativistic case with a flat background metric as well as the corresponding equation appropriate for radial accretion onto a Schwarzschild black hole. Section 4 considers diffusive transport equations and a discussion of special flows which admit a symmetry, and for which a constant of the motion for the particles can be identified, leading to simplified transport equations. Section 5 concludes with a discussion.

2. THE RELATIVISTIC BOLTZMANN EQUATION

In this section we briefly review aspects of relativistic transport theory associated with Liouville's equation that are useful in the description of cosmic ray transport in relativistic flows. We adopt Gaussian cgs units in the development. For the case of charged particle motion in electromagnetic fields, the particle motion is governed by the equations

$$m_0 \frac{dx^\alpha}{d\tau} = p^\alpha, \quad m_0 \frac{dp^\alpha}{d\tau} = -\Gamma^\alpha_{\beta\gamma} p^\beta p^\gamma + \frac{q}{c} F^\alpha{}_\beta p^\beta + m_0 G^\alpha, \quad (2.1)$$

where x^α and p^α denote the position and momentum four vectors of the particle, ($\alpha = 0, 1, 2, 3$); $\Gamma^\alpha_{\beta\gamma}$ are the affine connection coefficients; $F^\alpha{}_\beta$ is the Faraday tensor; q is the particle charge; m_0 is the particle rest mass; τ is the proper time (since $m_0 \neq 0$, and the particle speed $v \neq c$, we can use τ as the affine parameter), and G^α is the radiation reaction four-force. The form of the radiation reaction force G^α follows from a consideration of Larmor's formula for the power radiated by an accelerated charge in an electromagnetic field, and may be written as

$$G^\alpha = \frac{2q^2}{3m_0 c^2}\left(\delta^\alpha{}_\beta + \frac{p^\alpha p_\beta}{m_0^2 c^2}\right)\frac{p^\gamma}{m_0 c} F^\beta{}_{em;\gamma}, \qquad (2.2)$$

where

$$F^\beta_{em} = qF^\beta{}_\alpha p^\alpha/(m_0 c), \qquad (2.3)$$

is the Lorentz force on the particle. Equation (2.1) is known as the Abraham–Lorentz equation, or Lorentz–Dirac equation (e.g. Jackson[4]; Landau and Lifshitz[7]; Hakim[3]). The semi–colon in equation (2.2) denotes covariant differentiation, $\delta^\alpha{}_\beta$ denotes the Kronecker–delta symbol, and c is the velocity of light. The metric tensor $g_{\alpha\beta}$ is taken to have signature $-+++$.

In the consideration of energetic particle transport in a relativistic bulk flow with four velocity u^α, it is useful to introduce a local Lorentz frame Σ', moving with the fluid, with orthonormal basis vectors $\{e'_\alpha, \alpha = 0, 1, 2, 3\}$. Quantities measured relative to Σ' are labelled with a prime superscript. The metric in Σ' is the Minkowski metric $\eta_{\alpha\beta}$ [i.e., $\eta_{\alpha\beta} = diag(-1, 1, 1, 1)$].

From the equations of motion (2.1) and (2.2) for the individual particle, and applying Liouville's theorem we obtain the relativistic Boltzmann equation governing the particle phase space distribution function $F(\mathbf{x}, \mathbf{p})$ as

$$p^\alpha \frac{\partial F}{\partial x^\alpha} - \Gamma^\alpha_{\beta\gamma} p^\beta p^\gamma \frac{\partial F}{\partial p^\alpha} + \frac{q}{c}F^\alpha{}_\beta p^\beta \frac{\partial F}{\partial p^\alpha} = \frac{p'^0}{c}(C_{rad} + C_s). \qquad (2.4)$$

In equation (2.4),

$$C_{rad} = -\frac{\partial}{\partial p^\alpha}\left(\frac{d\tau}{dt}G^\alpha F\right), \qquad (2.5)$$

denotes the radiation energy losses of the particle (synchrotron losses), C_s represents the scattering of the energetic particles by the turbulent electromagnetic fields, and the Faraday tensor term on the left handside of equation (2.4) represents the Lorentz force on the particle owing to the mean background electromagnetic fields. The theoretical basis for incorporating synchrotron losses in equation (2.4) is discussed in detail by Hakim[3] (see also Kirk, Schlickeiser and Schneider[5]; Krülls[6]).

The frame Σ' is in general a non-inertial reference frame, and due account of non-inertial and gravitational forces on the particle motion are contained in the affine connection coefficients $\Gamma'^\alpha_{\beta\gamma}$. Some idea of the effects of non-inertial forces can be obtained by computing the rate of change of the comoving frame particle energy, $p'^0 = -p^\alpha u_\alpha$ along it's trajectory. From equations (2.1) we obtain

$$\frac{dp'^0}{d\tau} = -\frac{1}{m_0}\left(p^\alpha p^\beta u_{\alpha;\beta} + \frac{q}{c}u_\alpha F^\alpha{}_\beta p^\beta + m_0 u_\alpha G^\alpha\right), \qquad (2.6)$$

for the rate of change of p'^0 with proper time τ. Equation (2.6) can be reduced to a more suggestive form by using the Cauchy–Stokes formula (Mihalas and Mihalas[9]):

$$u_{\alpha;\beta} = -\dot{u}_\alpha u_\beta + \tfrac{1}{3}\Theta P_{\alpha\beta} + \tfrac{1}{2}\sigma_{\alpha\beta} + \omega_{\alpha\beta}, \qquad (2.7)$$

where $\dot{u}_\alpha$, Θ, $\sigma_{\alpha\beta}$, and $\omega_{\alpha\beta}$ denote respectively the acceleration vector, expansion, shear tensor and rotation tensor of the fluid and $P_{\alpha\beta}$ is the projection tensor perpendicular to u^α:

$$\dot{u}_\alpha = u^\beta \nabla_\beta u_\alpha, \quad \Theta = \nabla_\mu u^\mu, \quad P_{\alpha\beta} = g_{\alpha\beta} + u_\alpha u_\beta,$$
$$\sigma_{\alpha\beta} = \nabla_\alpha u_\beta + \nabla_\beta u_\alpha + \dot{u}_\alpha u_\beta + \dot{u}_\beta u_\alpha - \tfrac{2}{3}P_{\alpha\beta}\nabla_\mu u^\mu,$$
$$\omega_{\alpha\beta} = \tfrac{1}{2}(u_{\alpha;\beta} - u_{\beta;\alpha} + u_\beta \dot{u}_\alpha - u_\alpha \dot{u}_\beta). \qquad (2.8)$$

Decomposing the momentum four vector into components parallel and perpendicular to u^α:

$$p^\alpha = p'^0(u^\alpha + w^\alpha), \quad u_\alpha w^\alpha = 0, \tag{2.9}$$

(note that in Σ', $w^\alpha = (0, \mathbf{v}'/c)$ where $\mathbf{v}'$ is the particle velocity), and using the Cauchy Stokes formula (2.7) in equation (2.6) yields

$$\frac{dp'^0}{d\tau} = -\frac{(p'^0)^2}{m_0}[\dot{u}_\alpha w^\alpha + \tfrac{1}{3}(\nabla_\mu u^\mu)w_\alpha w^\alpha + \tfrac{1}{2}\sigma_{\alpha\beta}w^\alpha w^\beta] - \frac{p'^0}{m_0 c}q u_\alpha F^\alpha{}_\beta w^\beta - u_\alpha G^\alpha. \tag{2.10}$$

The first three terms in equation (2.10) in a flat spacetime, with no gravity represent non-inertial energy changes of the particle associated with the fluid acceleration $\dot{u}_\alpha$, divergence $\nabla_\mu u^\mu$, and shear $\sigma_{\alpha\beta}$, whereas the last two terms represent particle energy changes associated with the comoving frame electric field, and radiation losses respectively. A similar formula for photon transport has been derived by Novikov and Thorne[10]. For non-relativistic flows, equation (2.10) can be cast into the more familar form

$$\frac{dp'}{dt'} = -(\tfrac{1}{3}p'\nabla\cdot\mathbf{U} + p'^i A_i/v' + \tfrac{1}{2}p'\tilde{\sigma}_{ij}e'_{pi}e'_{pj}) + qE'_i e'_{pi} - (m_0 c/p')u_\alpha G^\alpha, \tag{2.11}$$

where p' is the magnitude of the particle 3— momentum; $\mathbf{U}$ is the 3-velocity of the fluid; $A_i = c^2 \dot{u}_i$ is the acceleration vector of the fluid; $\tilde{\sigma}_{ij} = c\sigma_{ij}$ is the usual fluid shear tensor; E'_i is the comoving frame electric field; $e'_{pi} = p'^i/p'$, and the indices $i, j = 1, 2, 3$. The results (2.10) and (2.11) indicate that the fluid acceleration, divergence and shear play a central role in transport equations using comoving frame momentum variables.

3. PITCH ANGLE DEPENDENT TRANSPORT EQUATIONS

If the gyro-radius of the particle is much less than the scale length of variation of the magnetic field B, and if the $q\mathbf{v} \times \mathbf{B}$ Lorentz force is dominant in determining the particle motion, then the distribution function is to lowest order independent of gyrophase. We consider both the special relativistic case and the case of radial accretion onto a Schwarzschild black hole.

(a) The Special Relativistic Case

For the case of a flat background Minkowski metric, the Boltzmann equation (2.4), averaged over gyrophase yields the pitch angle evolution equation for the particle transport (Skilling[12]; Webb[15]; Kirk Schlickeiser and Schneider[5]; Krülls[6]):

$$\gamma(1 + \mu' v' \mathbf{n} \cdot \mathbf{U}/c^2)\frac{\partial F}{\partial t} + \{\gamma \mathbf{U} + \mu'[v'\mathbf{n} + (\gamma - 1)v'\mathbf{n}\cdot\mathbf{UU}/U^2]\}\cdot \nabla F$$
$$+ p'\frac{\partial F}{\partial p'}\left[\tfrac{1}{2}(1 - 3\mu'^2)\tilde{\Theta}_\| - \tfrac{1}{2}(1 - \mu'^2)\tilde{\Theta} - \mu'\gamma^2 \mathbf{n}\cdot\mathbf{A}/v'\right]$$
$$+ \tfrac{1}{2}(1 - \mu'^2)\frac{\partial F}{\partial \mu'}\left[v'\nabla'\cdot\mathbf{n} + \mu'(\tilde{\Theta} - 3\tilde{\Theta}_\|) - 2\gamma^2 \mathbf{n}\cdot\mathbf{A}/v'\right] = C_s + C_{rad}, \tag{3.1}$$

where $\mathbf{n} = \mathbf{B}'/B'$ is a unit vector along the background magnetic field, and

$$\tilde{\Theta}_\| = (\gamma + 1)\mathbf{n}\cdot\frac{\partial}{\partial x'_3}\left(\frac{\gamma \mathbf{U}}{\gamma + 1}\right), \quad \tilde{\Theta} = \frac{\partial \gamma}{\partial t} + \nabla\cdot(\gamma\mathbf{U}),$$

$$\mathbf{A} = \frac{d\mathbf{U}}{dt} + \frac{\mathbf{U}}{\gamma(\gamma + 1)}\frac{d\gamma}{dt}, \quad \frac{d}{dt} = \frac{\partial}{\partial t} + \mathbf{U}\cdot\nabla,$$

$$\nabla'\cdot\mathbf{n} = \frac{\gamma\mathbf{U}}{c^2}\cdot\frac{\partial \mathbf{n}}{\partial t} + \nabla\cdot\mathbf{n} + (\gamma - 1)\mathbf{UU}\cdot\nabla\mathbf{n}/U^2,$$

$$\frac{\partial}{\partial x'_3} = \frac{\gamma \mathbf{n}\cdot\mathbf{U}}{c^2}\frac{\partial}{\partial t} + \mathbf{n}\cdot\nabla + (\gamma - 1)\mathbf{n}\cdot\frac{\mathbf{UU}\cdot\nabla}{U^2}, \tag{3.2}$$

In equations (3.1) and (3.2) **U** is the background fluid velocity, $\mathbf{n} = \mathbf{e}_3' = \mathbf{B}'/B'$ is a unit vector along the mean magnetic field, $\gamma = (1 - U^2/c^2)^{-1/2}$ is the Lorentz gamma of the background flow and $\mu' = \cos\theta' = \mathbf{p}'\cdot\mathbf{n}/p'$ is the cosine of the particle pitch angle. Equation (3.1) contains the effects of fluid compression in the parameters Θ and $\tilde{\Theta}_{\parallel}$, the effects of fluid acceleration **A**, and adiabatic focussing via $\nabla'\cdot\mathbf{n}$.

The term on the right handside of equation (3.1) contains the effects of particle scattering by waves travelling in the background flow (C_s) and synchrotron losses (C_{rad}). For the case of particle scattering by Alfvén waves propagating along the magnetic field (e.g., Schlickeiser[11]; Kirk Schlickeiser and Schneider[5]) we have

$$C_s = \frac{\partial}{\partial\mu'}\left(D_{\mu\mu}\frac{\partial F}{\partial\mu'}\right) + \frac{1}{p'^2}\frac{\partial}{\partial p'}\left(p'^2 D_{pp}\frac{\partial F}{\partial p'}\right) + \frac{\partial}{\partial\mu'}\left(D_{\mu p}\frac{\partial F}{\partial p'}\right) + \frac{1}{p'^2}\frac{\partial}{\partial p'}\left(p'^2 D_{p\mu}\frac{\partial F}{\partial\mu'}\right), \quad (3.3)$$

where the diffusion coefficients are

$$D_{\mu\mu} = \sum_{\epsilon=\pm 1}\mathcal{D}(\epsilon), \quad D_{\mu p} = D_{p\mu} = \sum_{\epsilon=\pm 1}\frac{\epsilon V_A p'}{(v' - \mu'\epsilon V_A)}\mathcal{D}(\epsilon),$$

$$D_{pp} = \sum_{\epsilon=\pm 1}\frac{V_A^2 p'^2}{(v' - \mu'\epsilon V_A)^2}\mathcal{D}(\epsilon), \quad (3.4)$$

and

$$\mathcal{D}(\epsilon) = \frac{\pi\Omega^2}{2}(1-\mu'^2)\frac{(1-\mu'\epsilon V_A/v')^2}{|v'\mu' - \epsilon V_A|}\frac{I^L(-k_{\parallel}) + I^R(k_{\parallel})}{B_0^2}. \quad (3.5)$$

Here $k_{\parallel} = \Omega/(v'\mu' - \epsilon V_A)$ is the resonant wave number, V_A is the Alfvén speed, Ω is the relativistic gyrofrequency, and I^L and I^R are the power spectra of left and right polarized waves respectively. The synchrotron loss term for a straight background magnetic field is

$$C_{rad} = \frac{1}{p'^2}\frac{\partial}{\partial p'}\left(\frac{(1-\mu'^2)p'\gamma_p'}{t_s}p'^2 F\right) - \frac{\partial}{\partial\mu'}\left(\frac{\mu'(1-\mu'^2)F}{\gamma_p' t_s}\right), \quad (3.6)$$

(Kirk, Schlickeiser and Schneider[5]), where $\gamma_p' = (1 - v'^2/c^2)^{-1/2}$ is the particle Lorentz gamma, and

$$t_s = \frac{4\pi m_0 c}{\sigma_T B_0^2}, \quad \sigma_T = \frac{8\pi}{3}\left(\frac{q^2}{m_0 c^2}\right)^2, \quad (3.7)$$

defines the synchrotron loss time scale t_s in terms of the Thomson cross section σ_T.

(b) Particle Transport in the Schwarzschild Geometry

For the case of radial accretion onto a Schwarzschild black hole with radial magnetic field geometry the generalization of the pitch angle evolution equation (3.1) is

$$\frac{\gamma}{A}(p'^0 + p'\mu'\beta)\frac{\partial F}{\partial x^0} + \gamma A(\beta p'^0 + p'\mu')\frac{\partial F}{\partial r} + G^p\frac{\partial F}{\partial p'} + G^\mu\frac{\partial F}{\partial\mu'} = \frac{p'^0}{c}(C_s + C_{rad}), \quad (3.8)$$

where

$$G^p = -p'^0 p'\left\{\frac{A\gamma\beta}{r}(1-\mu'^2) + \mu'^2\left[\frac{\partial}{\partial r}(\gamma\beta A) + \frac{\partial}{\partial x^0}\left(\frac{\gamma}{A}\right)\right]\right\}$$
$$- (p'^0)^2\mu'\left[\frac{\partial}{\partial x^0}\left(\frac{\gamma\beta}{A}\right) + \frac{\partial}{\partial r}(A\gamma)\right], \quad (3.9)$$

$$G^\mu = -\frac{(1-\mu'^2)}{p'}\left\{(p'^0)^2\left[\frac{\partial}{\partial x^0}\left(\frac{\gamma\beta}{A}\right) + \frac{\partial}{\partial r}(A\gamma)\right] - p'^2\frac{A\gamma}{r}\right.$$
$$\left. + p'p'^0\mu'\left[\frac{\partial}{\partial r}(\gamma\beta A) + \frac{\partial}{\partial x^0}\left(\frac{\gamma}{A}\right) - \frac{A\gamma\beta}{r}\right]\right\}. \quad (3.10)$$

In the above equations
$$A = (1 - r_s/r)^{1/2}, \qquad r_s = 2GM/c^2, \qquad (3.11)$$
where r_s is the Schwarzschild radius; M is the central gravitating mass; G is the universal gravitational constant; $x^0 = ct$; $\beta = U/c$ is the radial inflow speed and $\gamma = (1 - \beta^2)^{-1/2}$ is the Lorentz gamma of the flow. Further aspects of particle transport in the Schwarzschild geometry are discussed in Webb[19].

4. DIFFUSIVE PARTICLE TRANSPORT EQUATIONS

In this section we discuss a form of the diffusive particle transport equation for cosmic rays derived by Earl, Jokipii and Morfill[2] for non-relativistic flows, and by Webb[17] for relativistic flows that uses a BGK type collision term in the Boltzmann equation (2.4), i.e.,

$$C_s = -(F - F_0')/\tau_c, \qquad (4.1)$$

where τ_c is the collision time, and F_0' is the mean distribution function averaged over all momentum directions in Σ' (see also Berezhko and Krimsky[1]; Williams and Jokipii[20]). We neglect synchrotron losses. We also discuss particle transport in special flows that admit special symmetries.

From an analysis of the first three moments of the Boltzmann equation (2.4), and using the diffusion approximation one obtains the diffusive particle transport equation (e.g. Webb[17]):

$$\nabla_\alpha(cu^\alpha F_0' + q^\alpha) + \frac{1}{p'^2}\frac{\partial}{\partial p'}\left[-\tfrac{1}{3}p'^3 F_0'\nabla_\beta u^\beta - p'(p'^0)^2 \dot{u}_\alpha q^\alpha - \Gamma p'^4 \tau_c \frac{\partial F_0'}{\partial p'}\right] = 0, \qquad (4.2)$$

where

$$q^\alpha = -K^{\alpha\beta}\left[\nabla_\beta F_0' - \dot{u}_\beta \frac{(p'^0)^2}{p'}\frac{\partial F_0'}{\partial p'}\right], \qquad (4.3)$$

is the diffusive particle flux, including the relativistic heat inertia term ($\propto \dot{u}_\beta$). Here $K^{\alpha\beta}$ is the diffusion tensor including diffusion parallel and perpendicular to the mean magnetic field, as well as antisymmetric terms representing the Hall current or particle drift term. A more complete model should include the effects of second order Fermi acceleration and synchrotron losses in equation (4.2).

The first two terms in equation (4.2) represent particle transport via convection and diffusion, with the remaining momentum derivative terms representing particle energy changes. The third term proportional to the fluid velocity four divergence $\nabla_\beta u^\beta$ represents particle energy changes due to adiabatic compressions or expansions of the fluid. This term plays a major role in the acceleration of energetic particles at shocks in the first order Fermi mechanism. The fourth term in equation (4.2) dependent on the acceleration of the fluid represents non-inertial energy changes associated with the fact that Σ' is in general an accelerating frame. It may be thought as a gravitational redshift term and has the form $f_\alpha q^\alpha$ ($f_\alpha = -p'(p'^0)^2 \dot{u}_\alpha$) which is reminiscent of the form of the particle energization by a force field f_α. The remaining term in equation (4.2) represents particle energization by shear (i.e. cosmic ray viscosity: e.g. Berezhko and Krymsky[1]; Earl, Jokipii and Morfill[2]). The viscous energization coefficient Γ depends in general on both the shear tensor $\sigma_{\alpha\beta}$ of the fluid (equation (2.8)), the magnetic field geometry; and whether the scattering is strong ($\omega\tau \ll 1$) or weak ($\omega\tau \gg 1$), where $\omega = qB'c/p'^0$ is the particle gyrofrequency. For the case of strong scattering Γ depends only on the shear tensor of the fluid, and is given by $\Gamma = c^2 \sigma_{\alpha\beta}\sigma^{\alpha\beta}/30$. The detailed form of Γ for the general case is given by Webb[17].

(a) Special Flows

For special background fluid flows, both the Boltzmann equation moments, and the diffusive particle transport equation (4.2) can be reduced to purely spatial transport equations by the introduction of a new independent variable H of the form:

$$H = p'^0 R = -p^\alpha u_\alpha R, \tag{4.4}$$

to replace the particle momentum p'. In the terminology of Thorne[13], R is called the universal redshift function. We outline a method to determine these special flows below. Setting the radiation reaction force equal to zero in equation (2.1), and using equation (2.1) to compute $dH/d\tau$, we find

$$\frac{dH}{d\tau} = \frac{1}{m_0}\left[\frac{q}{c}\xi_\alpha F^\alpha{}_\beta p^\beta + p^\alpha p^\beta \xi_{(\alpha;\beta)}\right], \quad \xi_\alpha = -R u_\alpha, \tag{4.5}$$

for the rate of change of H along the particle's path. Hence H is a constant of the motion if the comoving frame electric field $E_\beta = u_\alpha F^\alpha{}_\beta \equiv 0$, and if $\xi_\alpha = -u_\alpha R$ is a Killing vector, i.e.,

$$\mathcal{L}_\xi(g)_{\alpha\beta} = \xi_{\alpha;\beta} + \xi_{\beta;\alpha} = 0, \tag{4.6}$$

where $\mathcal{L}_\xi(g)$ denotes the Lie derivative of the metric tensor g with respect the vector field ξ.

An example of such a flow is that of a rigidly rotating special relativistic flow, with flat Minkowski metric (Webb and Jokipii[18]). Using cylindrical coordinates $x^\alpha = (x^0, r, \theta, z)$ ($x^0 = ct$), and using holonomic basis vectors $e_\alpha = \partial x/\partial x^\alpha$, we have $u^\alpha e_\alpha = \gamma e_0 + (\gamma\Omega/c)e_\theta$, and $\dot{u}_\alpha \omega^\alpha = -(\gamma^2 \Omega^2 r/c^2)\omega^r$ for the fluid velocity four vector u^α and acceleration vector $\dot{u}_\alpha$, where Ω is the constant angular speed of the fluid about the z-axis; $\gamma = (1-\Omega^2 r^2/c^2)^{-1/2}$ is the Lorentz gamma of the flow, and $\{\omega^\alpha\}$ is the basis dual to $\{e_\alpha\}$. For the above flow, both $\sigma_{\alpha\beta} = 0$, and $\nabla_\alpha u^\alpha = 0$, so that the particle energy changes result solely from the acceleration vector term $\dot{u}_\alpha q^\alpha$ in equation (4.2). Taking R to be solely a function of r, we find from equation (4.5) that

$$\frac{dH}{d\tau} = \frac{p'^0 p'^r}{m_0}\left[\frac{dR}{dr} + \frac{\gamma^2 \Omega^2 r}{c^2} R\right]. \tag{4.7}$$

It follows from equation (4.7) that $dH/d\tau = 0$ is satisfied by

$$H_0 \equiv H = cp'^0/\gamma, \text{ and } R = c/\gamma. \tag{4.8}$$

Making a change of independent variables from (x^α, p') to (x^α, H_0) in equation (4.2), we obtain for the case of rigid rotation the purely spatial transport equation:

$$cu^\alpha \tilde{\nabla}_\alpha F_0' - \tilde{\nabla}_\alpha(K^{\alpha\beta}\tilde{\nabla}_\beta F_0') + [2+(p'^0/p')^2]\dot{u}_\alpha K^{\alpha\beta}\tilde{\nabla}_\beta F_0' = 0, \tag{4.9}$$

where $\tilde{\nabla}_\alpha$ denotes space-time covariant derivatives keeping H_0 constant. Note that there are no H_0 derivatives in equation (4.9), which is consistent with H_0 being a constant of the motion. The physical significance of H_0 is most clearly seen for the case of non-relativistic particles ($v'/c \ll 1$) in non-relativistic flows ($\Omega r/c \ll 1$), in which case H_0 approximates to

$$H_0 = \tfrac{1}{2}m_0 v'^2 - \tfrac{1}{2}m_0 \Omega^2 r^2 + m_0 c^2. \tag{4.10}$$

The result (4.10) is the Hamiltonian for a bead on a rigidly rotating wire and consists of the particle kinetic energy $\tfrac{1}{2}m_0 v'^2$ plus the centrifugal potential energy $-\tfrac{1}{2}m_0\Omega^2 r^2$ plus the rest mass energy $m_0 c^2$. Since the rest mass energy is constant it can be eliminated in a nonrelativistic treatment. Other examples of specialized flows for which purely spatial transport equations can be constructed by exploiting the constants of the motion are discussed by Thorne[13] (e.g., Friedmann cosmological models with expansion).

5. CONCLUDING REMARKS

In this paper we have discussed various aspects of cosmic ray transport and acceleration via diffusive and pitch angle dependent transport equations derived from the relativistic Boltzmann equation.

In the pitch angle transport equation description we indicated the effects of fluid compression, acceleration and adiabatic focussing on the particle transport (Webb[15]) as well as the effects of pitch angle and momentum dependent scattering of the particles by Alfvén waves propagating along the field, plus the effect of synchrotron losses (Kirk, Schlickeiser and Schneider[5]; Schlickeiser[11]; Krülls[6]). We also indicated the extension of these transport equations for the case of radial accretion onto a Schwarzschild black hole. This was followed by discussions of diffusive transport equations based on the BGK Boltzmann model (Berezhko and Krymsky[1]; Earl, Jokipii and Morfill[2]; Webb[17]; Williams and Jokipii[20]), and special flows for which a constant of the motion of the particles can be identified. Hydrodynamical versions of these equations may also be constructed (e.g. Webb[16]; Zank, Webb and McKenzie[21]).

ACKNOWLEDGEMENTS

This work was supported in part by NASA grant NSG 7101 and NSF grant ATM 8317701

REFERENCES

1. E.G. Berezhko, and G.F. Krymsky, *Soviet Astr. Letters*, **7**, (5), 352 (1981).
2. J.A. Earl, J.R. Jokipii, and G.E. Morfill, *Ap. J. Lett.*, **331**, L91 (1988).
3. R. Hakim, *J. Math. Phys.*, **8**, (7), 1379 (1967).
4. J.D. Jackson, *Classical Electrodynamics* (New York: Wiley 1975).
5. J.G. Kirk, R. Schlickeiser, and P. Schneider, *Ap. J.*, **328**, 269 (1988).
6. W.M. Krülls, Ph. D. Dissertation, Rheinischen Friedrich Wilhelms Universität, Bonn, F.R.G. (1990).
7. L.D. Landau and E.M. Lifshitz, *The Classical Theory of Fields*, Course of Theoretical Physics, Vol. 2, (Pergamon : Oxford 1971).
8. R.W. Lindquist, *Ann. Phys.*, **37**, 487 (1966).
9. D. Mihalas and B.W. Mihalas, *Foundations of Radiation Hydrodynamics* (Oxford: Oxford University Press 1984).
10. I.D. Novikov and K.S. Thorne, in *Black Holes*, Eds. Dewitt, C. and Dewitt, B.S., (Gordon and Breach, New York 1973) p. 343.
11. R. Schlickeiser, *Ap. J.*, **336**, 243 (1989).
12. J.A. Skilling, *M.N.R.A.S.*, **172**, 557 (1975).
13. K.S. Thorne, *M.N.R.A.S.*, **194**, 439 (1981).
14. L.H. Thomas, *Q. Jl. Math.*, **1**, 239 (1930).
15. G.M. Webb, *Ap. J.*, **296**, 319 (1985).
16. G.M. Webb, *Ap. J.*, **319**, 215 (1987).
17. G.M. Webb, *Ap. J.*, **340**, 1112 (1989).
18. G.M. Webb and J.R. Jokipii, *Proc. 21st Internat. Cosmic Ray Conf.*, Adelaide, Australia, **4**, 122 (1990).
19. G.M. Webb, *Proc. 22nd Internat. Cosmic Ray Conf.*, Dublin, Ireland, paper OG 9.1-22 (1991).
20. L.L. Williams and J.R. Jokipii, *Ap. J.*, **371**, 639 (1990).
21. G.P. Zank, G.M. Webb and J.F. McKenzie, *Astron. Astrophys.*, **189**, 338 (1988).

RELATIVISTIC, COSMIC-RAY MODIFIED SHOCKS

J. G. Kirk
Max-Planck-Institut für Kernphysik, W-6900 Heidelberg, Germany

ABSTRACT

The theory of first order Fermi acceleration at relativistic shock fronts was developed initially in the test particle picture. However, the efficiency of the mechanism depends on the backreaction of the particles on the fluid. The investigation of stationary solutions which include this effect can in principle provide a means of estimating the efficiency. In the nonrelativistic case, stationary solutions turn out not to be relevant indicators of efficiency, basically because realistic acceleration timescales are too slow. Arguments are presented which indicate that this problem does not necessarily apply to shocks which move relativistically.

PROPERTIES OF STATIONARY SHOCK FRONTS

As cosmic rays undergo first order Fermi acceleration at a shock front they automatically start to build up their pressure. In fact, the energetic particles derive their energy by forcing the incoming upstream plasma to do work in approaching the shock against this pressure gradient. Perhaps the best way of understanding this process is to consider the cosmic rays themselves as a fluid separate from the background plasma and with different properties [4]. It is a relatively simple matter to take the cosmic ray transport equation, multiply by particle energy and integrate over momentum space. The resulting equation contains quantities which are readily interpreted as cosmic ray pressure p_c and energy density e_c. There remain two formal difficulties: dealing with the diffusion term and establishing a functional relationship between e_c and p_c. Drury and Völk [4] circumvented the former by using an "effective diffusion coefficient" $\bar{\kappa}$. The latter difficulty is usually taken care of by introducing a new parameter, the adiabatic index γ_c, in analogy with the theory of ideal gases. Of course, as has been stressed by, amongst others, Jones and Ellison [9] there remains the physical (as opposed to formal) problem of identifying those particles which should belong to the gas of cosmic rays and those which should belong to the background fluid. This is really another way of formulating the "injection" problem. Although it is possible by varying the parameter γ_c to account for some of the effects of injection, there are certainly others which require a more detailed approach [17].

For a single fluid the conservation laws determine, via the jump conditions, the fluid flow through the shock once the upstream Mach number is given. In the case of two fluids, on the other hand, the conservation laws no longer suffice to determine the stationary state given only upstream parameters. A whole continuum of possibilities arises, and which of these is realised depends on the details of the interaction between the two fluids. Of course, the simplest approach is to require the interaction to be adiabatic in the sense that the background is not heated by the cosmic ray fluid (this would correspond to scattering off fluctuations which are fixed in the background fluid). But such a requirement does not suffice to lift the degeneracy: a continuum of final states is still permitted because an unspecified amount of energy can be transferred between the components as a result of accelerating or decelerating the background fluid. Under these circumstances, it is quite remarkable that a simple, physically plausible assumption, namely that $\bar{\kappa}$ is positive definite, is enough to restrict the possible flow patterns to a discrete number (up to 3) for any given set of upstream parameters. With this assumption, the energy transfer between the components in a stationary shock front must also take on one of a number of discrete values. Clearly, it is these values which might be used to estimate the efficiency of the acceleration process. A closer look at the assumption $\bar{\kappa} > 0$ confirms that only rather pathological microphysics could conspire to invalidate it [11].

In the case of relativistic shocks, the situation is similar. The conservation relations themselves do not provide enough information to determine the energy exchange between the cosmic ray fluid and the background in a stationary shock front; additional physical input is required. However, although the diffusion approximation – and therefore the parameter $\bar{\kappa}$ – cannot be used where the fluid moves relativistically, it is interesting to note that the requirement $\bar{\kappa} > 0$ need be applied only in those parts of the flow which have reached the subsonic downstream region [2]. Therefore, provided the diffusion approximation may be applied there, the degeneracy of the permitted solutions is once again lifted.

A detailed analysis of the stationary shock structures using the two fluid approach is, even in the nonrelativistic case, a complicated undertaking [4, 8, 1]. The relativistic case is not significantly simpler. In a recent investigation of stationary, relativistic solutions [2] it is found that, as in the nonrelativistic case, there exists more than one solution for a given set of upstream parameters. For high Mach numbers there exists a strong, cosmic ray modified shock in which the gas sub-shock disappears completely. Whether or not such a solution is realistic is unclear, since it requires continuous particle injection in the absence of the sub-shock. However, there exists an additional solution for the same upstream parameters in which the sub-shock is retained. An interesting aspect of the relativistic case is that the efficiency of this solution is quite large ($\sim 50\%$ when the upstream fluid has a Lorentz factor of 3), in stark contrast to the nonrelativistic case.

SELF-CONSISTENCY OF STATIONARY SHOCK FRONTS

The two fluid analysis does not require a knowledge of the momentum dependence of the spatial diffusion coefficient $\kappa(p)$. It also gives no information about the particle spectrum to be expected from a stationary shock front. For this, it is necessary to solve the cosmic ray transport equation. As a result, it is possible that solutions which appear perfectly reasonable in the fluid picture turn out to be possible only for certain restricted classes of $\kappa(p)$. If, for example, the time asymptotic solution of the cosmic ray transport equation is a power law at high

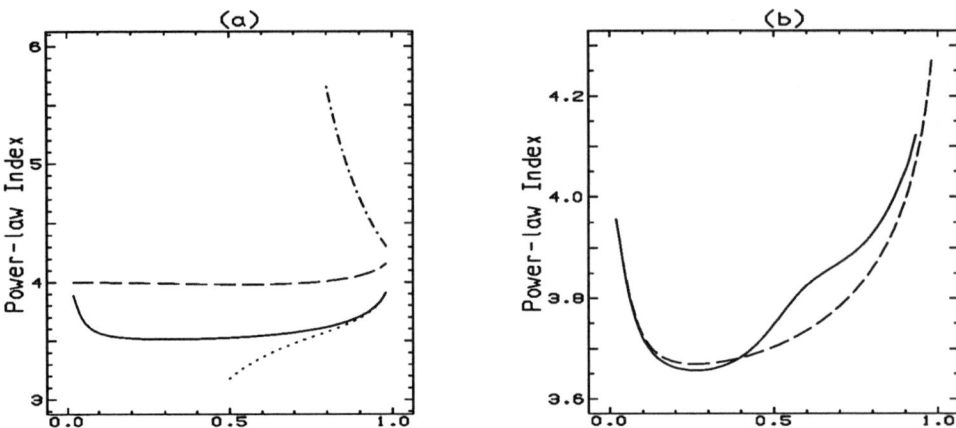

Figure 1: The power law index of a test particle distribution accelerated at a relativistic shock front of upstream speed u_- (in units of c). In (a), the background plasma is taken to be either a relativistic gas (dash-dotted line), or a gas which is cold upstream and has in the downstream part only ion pressure (dashed line), only electron pressure (solid line) or only pressure from electron/positron pairs (100 per proton) created at the shock front (dotted line). In (b) the gas is again cold upstream, and in full thermodynamic equilibrium downstream (solid line). The dashed line in this case shows the nonrelativistic formula $s = 3u_-/(u_- - u_+)$, where u_+ is the downstream speed.

momentum i.e., a distribution function $f(p) \propto p^{-s}$ and if, for a particular flow pattern and a particular $\kappa(p)$ it happens that $s < 4$, then the integrals determining both e_c and p_c diverge, although their ratio remains finite. It is then no longer possible to use the two-fluid equations, since solutions of these would not be self-consistent.

Such a situation is indeed to be expected for strong, nonrelativistic shocks, provided $\kappa(p)$ increases indefinitely as $p \to \infty$. This point has been emphasised by Ellison and Eichler [5, 6] and by Krymsky [12], who have advocated introducing a cut-off at high momentum to simulate the effect of the loss of particles of large mean free path through the system boundaries. In practise, however, first order Fermi acceleration at *nonrelativistic* shock fronts is such a slow process that the nature of the time asymptotic solutions is irrelevant (see, for example, Drury, Markiewicz and Völk [3]).

At relativistic shocks, on the other hand, the acceleration process is likely to be much more rapid [14, 7, 13], so that one may expect stationary solutions to provide a better guide to the overall efficiency of acceleration. However, stationary solutions are consistent only if the particle spectrum at large p is sufficiently steep ($s > 4$). Calculating the spectrum of a modified shock is rather involved in the relativistic case [15], but for $\kappa(p)$ an increasing function of p, the high p spectrum should return to that found in the test particle approximation, provided the overall compression ratio is adopted. Using the Synge equation of state [16] and a variety of compositions for the background fluid, the test particle results [10] are shown in Figure 1. It is interesting to note that highly relativistic shocks in general tend to satisfy the self-consistency test $s > 4$.

REFERENCES

1. Achterberg, A., Blandford, R.D., Periwal, V., Astron. Astrophys., *132*, 97, 1984.

2. Baring M.G. and J.G. Kirk, Astron. Astrophys., *241*, 329 1991.

3. Drury L.O'C., Markiewicz W.J. and Völk H.J., Astron. Astrophys., *225*, 179, 1989.

4. Drury, L.O'C., Völk H.J., Astrophys. J., *248*, 344, 1981

5. Ellison D.C. and Eichler D., Astrophys. J., *286*, 691, 1984.

6. Ellison D.C. and Eichler D., Phys. Rev. Letts. *55*, 2735, 1985.

7. Ellison D.C., Jones F.C. and Reynolds S.P., Astrophys. J., *360*, 702, 1990.

8. Heavens, A.F. Monthly Notices Roy. Astron. Soc., *204*, 699, 1983.

9. Jones F.C. and D.C. Ellison, Space Sc. Rev., *58*, 259, 1991.

10. Kirk J.G. Habilitationsschrift, 1988.

11. Kirk J.G. Astron. Astrophys., *239*, 404, 1990.

12. Krymsky G.F., Advances in Space Research *4*, 175, 1984.

13. Ostrowski M., in "Relativistic Hadrons in Cosmic Compact Objects", eds. A.A. Zdziarski, M. Sikora, page 121 (Springer-Verlag 1991).

14. Quenby J.J. and Lieu R., Nature, *342*, 654, 1989.

15. Schneider P. and Kirk J.G., Astron. Astrophys., *217*, 344, 1989.

16. Synge J.L. "The relativistic gas" (North Holland Publishing Company) 1957.

17. Zank G.P., these proceedings.

PARTICLE ACCELERATION IN DISORDERED MAGNETIC FIELDS

A.F. Heavens

University of Edinburgh, Royal Observatory, Edinburgh, United Kingdom

ABSTRACT

The acceleration of particles by the Fermi mechanism is widely invoked to explain the synchrotron emission from active galaxies and supernova remnants, and for the origin of cosmic rays. Previous studies have generally assumed that the scattering of the particles is effected by small irregularities in an otherwise uniform magnetic field. Motivated by polarisation measurements, we investigate particle acceleration in highly disordered magnetic fields, following particle trajectories numerically. We find that mildly relativistic shocks give rise to synchrotron emission with a spectral index ($S_\nu \propto \nu^{-\alpha}$) $\alpha \sim 0.5 - 0.7$, in excellent agreement with low-frequency optically thin synchrotron emission from active galaxies. For highly relativistic shock speeds ($u/c \gtrsim 0.9$) the spectrum steepens ($\alpha > 1$). This behaviour may be connected with the observed steepening of synchrotron spectra with radio power observed in powerful radio sources.

INTRODUCTION

The theory of particle acceleration in shock waves has undergone continual development since the first papers appeared in the late seventies[1-4]. The principle of the acceleration mechanism is that high-energy particles are assumed to pass unaffected through the shock, and scatter elastically off magnetic irregularities (usually Alfvén waves) in the fluid either side of the shock. Since, in the rest frame of a shock wave, the incoming (upstream) fluid speed exceeds the outgoing (downstream) fluid speed, the particles gain energy by the first-order Fermi process (see the reviews[5,6] for details). The acceleration process, coupled with the fact that particles in the downstream region have a chance of not returning to the shock, naturally gives rise to a power-law momentum distribution, with an expected synchrotron spectral index of $\alpha = 0.5$ (the differential electron energy distribution is $\propto E^{-(2\alpha+1)}$). This fits well with many shell-type supernova remnants[7] and some hotspots of radio galaxies, but there are extragalactic sources[8-10] with steeper spectra, in the range $0.5 \lesssim \alpha \lesssim 1.0$. These are quite difficult to explain by standard shock acceleration theory.

The generalisation to relativistic shocks[11,12] is not immediately helpful, as parallel shocks predict $0.35 \lesssim \alpha \lesssim 0.6$, and oblique shocks[13] have flatter spectra still $\alpha \to 0$. However, one possible effect which has not been extensively discussed is the disordering of the magnetic field. The effect of the magnetic field direction is much more important for relativistic shocks than in the non-relativistic case. The reason for this is that, if the angle between the magnetic field and the shock normal is too great, the point of intersection between the magnetic field and the shock normal moves at a speed greater than c, and the shock is said to be 'superluminal'. In this case, if the scattering process is unable to stop a particle being 'tied' to a particular field line, they are unable to recross the shock and gain more energy. This is especially severe for highly-relativistic shocks, where the magnetic field upstream must lie within an angle $1/\gamma$ to the shock normal, where γ is the Lorentz factor corresponding to the shock speed. Since this is rare unless the upstream field has a special configuration, the usual case will be a superluminal shock, where acceleration should take place[14], but a power-law does not naturally arise.

The evident importance of the magnetic field geometry in even mildly relativistic shocks provides the motivation for the work presented here, performed in collaboration with Keith Ballard. The purpose of this work is to investigate what happens if the upstream magnetic field is disordered, and has no preferred direction (an approximate study of a situation like this was done[15] for non-relativistic shocks). Our expectation is that this may be a more realistic description of the conditions in jets of active galaxies than the assumption of a large-scale uniform field, based on the fact that the polarisation levels of synchrotron emission rarely come close to the high values expected from uniform fields[8]. The compression of the magnetic field at the shock front does introduce a preferred axis, so high polarisations would be expected from hotspot emission in this model.

The spectral indices from shocks in disordered fields are encouragingly steep ($\alpha \sim$ $0.5-0.7$ for a wide range of shock speeds), and show a progressively steeper spectrum as the speed of light is approached. This qualitative behaviour is in accord (as is the range of spectral indices) with the observed correlation between radio power and spectral index[16]. However, the interpretation of these results is not entirely straightforward, so this statement is somewhat speculative.

MODEL ASSUMPTIONS

The flow is taken to be one-dimensional. We assume that, in the rest frame of the upstream fluid, the magnetic field is entirely disordered – there is no mean field, and the magnetic fluctuations are described in Fourier space by a specified power spectrum. The magnetic field in the downstream rest frame is obtained by compression of the tangential components of the upstream field.

The accelerated particles are test particles and the magnetic field is also assumed to be dynamically unimportant. The downstream flow speed and density are obtained using the hydrodynamic jump conditions for shocks in a proton-electron plasma, and thermal equilibrium between the protons and electrons is assumed for the downstream gas[12] The upstream gas is assumed to have zero pressure. Radiation losses are not taken into account, and the situation is assumed to be in a steady-state.

METHOD

To ensure that $\nabla.\mathbf{B} = 0$, the magnetic vector potential is generated on a cubic grid, and $\mathbf{B}$ is evaluated numerically from it. The field is specified by the Fourier power spectrum of the vector potential, which was taken to be a Kolmogorov spectrum between upper and lower cutoffs. The long-wavelength cutoff was set by the size of the cube (normally with 100 or 64 grid points per side). The short-wavelength cutoff was set at 4 grid points and the field interpolated on sub-grid scales[17]. The field is specified throughout space because of periodic boundary conditions.

Particles are injected some distance upstream, so that, in the upstream rest frame, they are monoenergetic, ultrarelativistic and isotropic. The magnetic field strength is chosen such that the gyroradius of the injected particles, in a field equal to the r.m.s. field in the box, is 0.5 grid points.

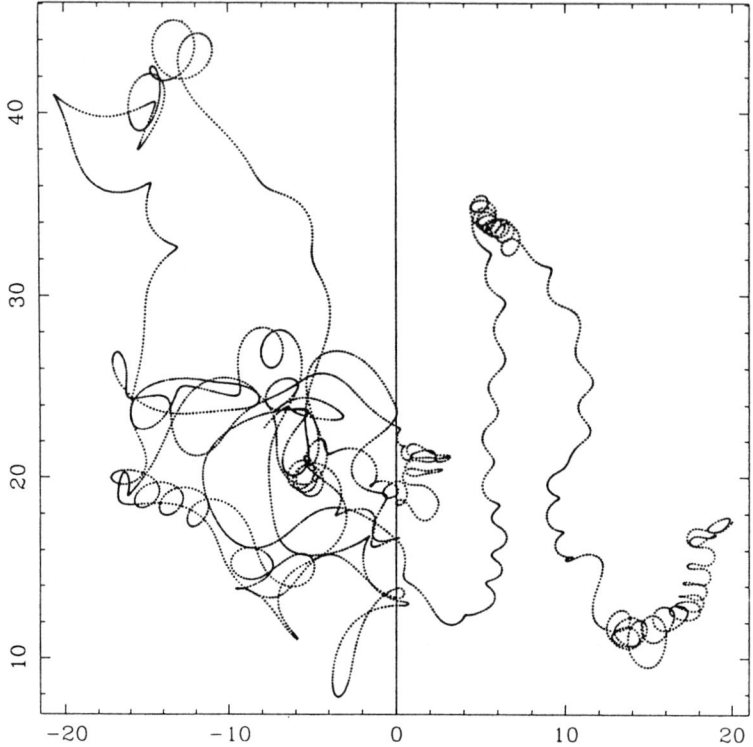

Figure 1. *A typical particle orbit in a disordered magnetic field.*

The transport of charged particles in disordered fields has been approximated in a number of ways, such as large-angle scattering[18,19] and Bohm diffusion. These approximations do not obviously apply, so we avoid difficulties by integrating particle orbits directly. Figure 1 shows a typical orbit; its complexity justifies our approach. Full details, including tests, may be found in a paper to be submitted to Mon. Not. R. astr. Soc. (Ballard & Heavens).

Two quantities are logged on shock crossing, for calculation of the particle momentum distribution and the anisotropy. These are the momentum in the rest frame in which the shock is at rest, and μ, the cosine of the angle between the momentum and the shock normal, also measured in the shock rest frame. Note that μ is not the pitch angle cosine. Many different realisations of the magnetic field are generated, and the distribution functions combined to give the final results.

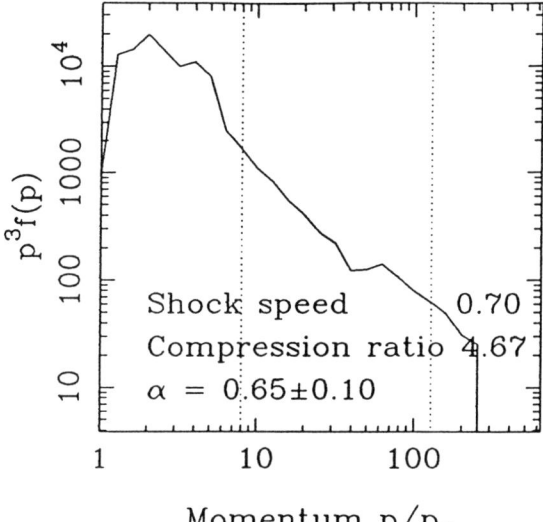

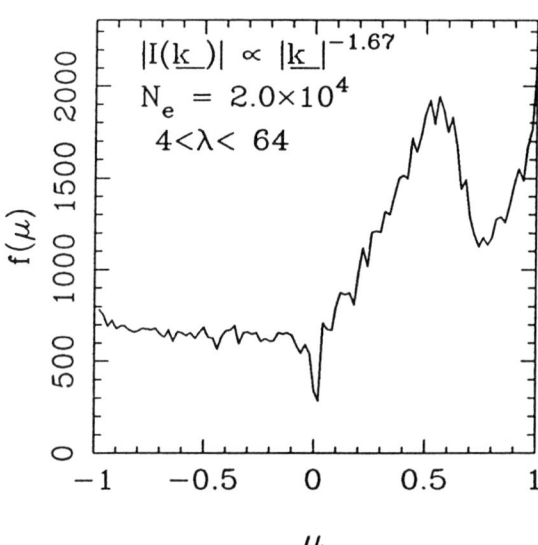

Figure 2. *The accelerated particle distribution function (multiplied by p^3) plotted against momentum p. Particles are injected at $1.2 p_0$ in the upstream rest frame. For dotted lines, see text. α is the best-fit spectral index for the synchrotron emission from electrons with momenta higher than the left dotted line. The lower panel shows the anisotropy of the particle distribution. Note that the peak at $\mu = 1$ arises from beamed injected particles (isotropic upstream) undergoing a single shock crossing.*

RESULTS

Figure 2 shows the distribution function and anisotropy of particles accelerated at a mildly relativistic shock of speed $u_1 = 0.7$ (units are chosen such that $c = 1$). The following features are apparent: there is a reasonable power-law established, up to a cutoff probably resulting from the large gyroradii of the high-energy particles. The dotted lines show the momenta of particles which have gyroradii equal to 4 grid points and the size of the box. If the fluctuations were small $(\delta B/B)^2 \ll 1$, these particles would be able to resonate with the upstream magnetic fluctuations present. Note that the peak in the angular distribution at $\mu = 1$ arises from the injected particles making their first shock crossing. These particles are more-or-less isotropic in the upstream frame when they cross first, and appear beamed in the shock rest frame.

The spectrum shows two interesting features. First, many of the injected particles undergo only a single shock crossing, and contribute most of the hump seen near $p/p_0 = 5$. The apparent acceleration arises from the change of frame, and is similar to the shock drift results[14]. Only a few of the particles are able to make a return crossing, but the ones that do are able to gain a lot of energy. These particles make up the power-law tail seen in the figure.

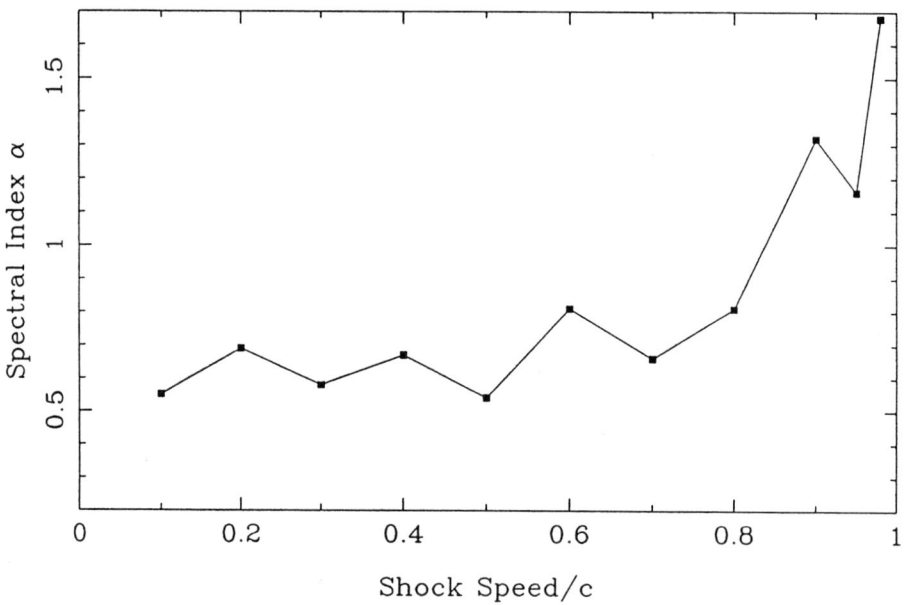

Figure 3. *The synchrotron spectral index from electrons accelerated in disordered magnetic fields against the shock speed.*

Figure 3 summarises the spectral indices for shock speeds between 0.2 and 0.98. The compression ratios are those appropriate for shocks in cold proton-electron plasma, where the downstream gas has equal proton and electron temperatures.

DISCUSSION

Shock acceleration in disordered magnetic fields gives a plausible explanation for the range of synchrotron spectral indices $0.5 \lesssim \alpha \lesssim 1$ which are often observed but which are difficult to account for in standard shock acceleration theory. We are able to reproduce the range of spectral indices seen in active galaxies and hotspot emission from radio galaxies with shocks travelling at mildly relativistic speeds. Shocks with speeds very close to c accelerate particles to a rather steep spectrum ($\alpha > 1$ for speed $> 0.9c$), and we therefore conclude that the shock speeds at hotspots are not highly relativistic.

An alternative explanation for spectral indices greater than 0.5 is that the shocks are weak[3]. The relatively small range of observed low-frequency spectral indices argues against this, as the predicted spectral index is a rather sensitive function of the Mach number of the shock. It is also difficult to reconcile this with the observed correlation between radio power and spectral index[16]. It seems unnatural for the higher power radio sources to arise from the weaker shocks. The explanation we have in this paper has a correlation between shock speed and spectral index which is in the same sense as the radio power–spectral index correlation observed. It is inappropriate to pursue this argument in more detail than this, as the single-frequency radio power (including no doubt a considerable proportion of radiatively-aged flux) may not be related in a straightforward way to the shock speed.

REFERENCES
1. G.F. Krymsky, Dok. Acad. Nauk. SSSR. 234, 1306 (1977).
2. W.I. Axford, E. Leer, & G. Skadron, Proc. 15th Int. Cosmic Ray Conf., 11, 132 (1977).
3. A.R. Bell, Mon. Not. R. astr. Soc., 182, 147 (1978).
4. R.D. Blandford & J.P. Ostriker, 1978. Astrophys. J., 221, L29.
5. L.O'C. Drury, Rep. Progr. Phys., 46, 973 (1983).
6. R.D. Blandford, & D. Eichler, Phys. Reports,154, 1 (1987).
7. D.A. Green, Astrophys. Spa. Sci., 148, 3 (1988).
8. K.I. Kellerman, & I.I.K. Pauliny-Toth, Ann. Rev. Astron. Astrophys., 19, 373 (1981).
9. R. Laing, In 'Hotspots in Extragalactic Radio Sources', Ed. K. Meisenheimer & H.-J. Röser, Springer-Verlag, Berlin (1989).
10. K. Meisenheimer, H.-J. Röser, P. Hiltner, M.G. Yates, M.S. Longair, R. Chini & R.A. Perley, Astr. Astrophys., 219, 63 (1989).
11. J.G. Kirk & P. Schneider, Astrophys. J., 315, 425 (1987).
12. A.F. Heavens & L.O'C. Drury, Mon. Not. R. astr. Soc., 235, 997 (1988).
13. J.G. Kirk & A.F. Heavens, Mon. Not. R. astr. Soc., 239, 995 (1989).
14. M.C. Begelman & J.G. Kirk, Astrophys. J., 353, 66 (1990).
15. A. Achterberg Mon. Not. R. astr. Soc., 232,323 (1988).
16. J.A. Peacock & S.F. Gull, Mon. Not. R. astr. Soc., 196, 611 (1981).
17. W.H. Press, B.P. Flannery, S.A. Teukolsky & W.T. Vetterling, Numerical Recipes, Cambridge University Press (1989).
18. J.G. Kirk & P. Schneider, Astrophys. J., 322, 256 (1987).
19. D.C. Ellison, F.C. Jones & S.P. Reynolds, Astrophys. J., 360, 702 (1990).

SHOCK ACCELERATION IN A RADIATION FIELD

A.P. Szabo and R.J. Protheroe
Department of Physics and Mathematical Physics
University of Adelaide, Adelaide, South Australia 5000, Australia.

ABSTRACT

We consider the case of shock acceleration of protons taking place in the radiation field of an active galactic nucleus (AGN). We use an heuristic approach to devise a simple model which reproduces the overall features of shock acceleration and can be used to simplify numerical calculations. Two possible forms of the radiation field based on observations of the infrared to hard X–ray spectrum are considered. In the first of these, the bulk of the infrared emission is assumed to come from reprocessing of UV photons by dust far outside the assumed acceleration region, while in the second the observed infrared emission is assumed to be mainly of synchrotron origin. In both cases, we find that a significant fraction of the total energy used to accelerate a proton goes into particle production in interactions with photons occurring during the acceleration process. We obtain the energy spectra of neutrons, neutrinos and electromagnetic radiation produced both during acceleration, and subsequently as a result of interactions of the accelerated protons after leaving the accelerator. We compare the spectra of the particles produced for the two possible forms of the radiation field and give the expected neutrino flux from 3C273 and 3C279.

INTRODUCTION

Shock acceleration may be responsible for the energetic particles and radiation in active galaxies, possibly at a shock in a spherical accretion flow onto a super-massive black hole[1,2]. High energy particles are produced when accelerated protons interact via pair production and pion photoproduction with photons. Neutrinos are produced from the decay of the charged pions and may result in a substantial neutrino flux from AGN's[1,3-6]. Energetic radiation produced in the central region of an active galactic nucleus can not easily escape because of photon–photon pair production and other absorption processes. However, neutron production in the central regions of AGN gives a way of transporting energetic particles out to regions from which high energy radiation can escape[7,8].

A simple idealised picture of shock acceleration in the absence of energy losses is as follows. A constant fractional energy gain occurs for every shock crossing. Particles crossing the shock from downstream to upstream will eventually be convected back to the shock, whereas particles crossing from upstream to downstream have some probability of escaping downstream, never to return to the shock (see ref. 9 for a very readable account of shock acceleration). Thus, we may visualise the shock accelerator as a leaky box. A consideration of the energy gain per shock encounter using the test particle appoximation and diffusive (assuming $D \propto E_p$) and convective transport towards and away from an infinite plane shock shows that the particles are accelerated at a constant rate, i.e. $dE_p/dt = a$. This results from the timescale for one cycle of the acceleration process (i.e. upstream$\rightarrow$ downstream$\rightarrow$upstream) being proportional to the energy of the particle (e.g., ref. 10). The probability of a particle injected with energy E_o reaching energy E_p without having escaped is then $P_{surv}(E_p) = E_o/E_p$. For a particle injected at E_o one finds that the distribution of energies on escaping from the box is $n(E_p) = E_o E_p^{-2}$ for energies above E_o.

Where the acceleration takes place in the presence of the radiation field of an AGN, the maximum proton energy achievable is governed by particle production (e.g., ref 11). Hence the maximum proton energy E_{max} may be adopted as a parameter, and the acceleration rate set to equal the loss rate for a given energy density. A simple order of magnitude estimate of the energy per injected particle going into particle production during acceleration is $E_{max} P_{surv}(E_{max}) = E_o$, and into particles escaping is $\int n(E_p) E_p dE_p = E_o \log(E_{max}/E_o)$. The fraction going into particle production during acceleration is therefore not insignificant.

In this paper we will present the spectra of high energy particles produced during and after shock acceleration in two possible radiation fields appropriate to AGN's. As well as the spectrum of neutrinos produced, we give the spectrum of the electromagnetic component produced in the central region, and the spectrum of neutrons escaping from the central region. This is of particular imposrtance as escaping neutrons may be responsible for producing an observable flux of very high energy γ-rays from AGN's. We use our results together with observations of the infrared to hard X–ray continuum to calculate the expected neutrino flux from 3C273 and the optically violent variable 3C279.

MODELLING THE OBSERVED A.G.N. CONTINUUM

Observations of the continuum emission from AGN's show that in many objects approximately equal amounts of energy are emitted per decade from the infrared to X–rays, with the exception of the UV excess (or 'bump'). This excess relative to the underlying power law spectrum was first observed to be a black body component by Malkan and Sargent[12], and is thought to come from the inner regions of an accretion disk. Edelson and Malkan[13] fitted blackbody spectra to the UV bump and found a mean temperature $\overline{T} = 26,000 \pm 4000$ K. Saunders et al.[14] found UV bump temperatures in the range 10^4–10^5 K for galaxies in the Palomar–Green (PG) survey.

Much debate has centred on the nature of the various components which go into making up the infrared continuum, in particular on whether the infrared emission is dominated by thermal or non–thermal processes. Carleton et al.[15] identified the infrared spectrum of 'bare' (dust–free) AGN as being non–thermal in nature, and fitted a power law spectrum $\varepsilon dn/d\varepsilon \propto \varepsilon^\alpha$ with index $\overline{\alpha}_{IR} = -1.0 \pm 0.1$. In these objects the infrared emission is probably produced by synchrotron radiation[13] which will occur at relatively small radii. However, Saunders et al.[14] found that for radio–quiet and steep–spectrum radio–loud galaxies in the PG survey the infrared seems to be thermal in nature. In this latter case, the emission appears to result from the absorption of UV photons by dust at large radii and subsequent re–emission in the infrared[16].

There are strong arguments suggesting that the X–ray emission of an AGN is produced in the central region[17]. Wilkes and Elvis[18] found power–law X–ray spectra with mean spectral indices $\overline{\alpha}_X = -1.0$ and $\overline{\alpha}_X = -0.5$ for radio–quiet and radio–loud quasars respectively. They did, however, find a wide range of spectral indices. For hard X–ray selected objects, Turner and Pounds[19] found that in the 2–10 keV energy range the spectra are well represented by a power–law with index $\overline{\alpha}_X = -0.7 \pm 0.17$.

Clearly to adopt any generic AGN spectrum is an over-simpification. However, many of the obseved spectra display characteristics which may be roughly split into two categories, a spectrum with negligible energy density in the infrared component from the central region of the AGN (appropriate, for example to radio-quiet and steep-spectrum radio–loud galaxies in the PG survey), and a flat quasar–like spectrum which has roughly equal energy per decade from the infrared to hard X-rays, except in the UV region. Hence in the present work

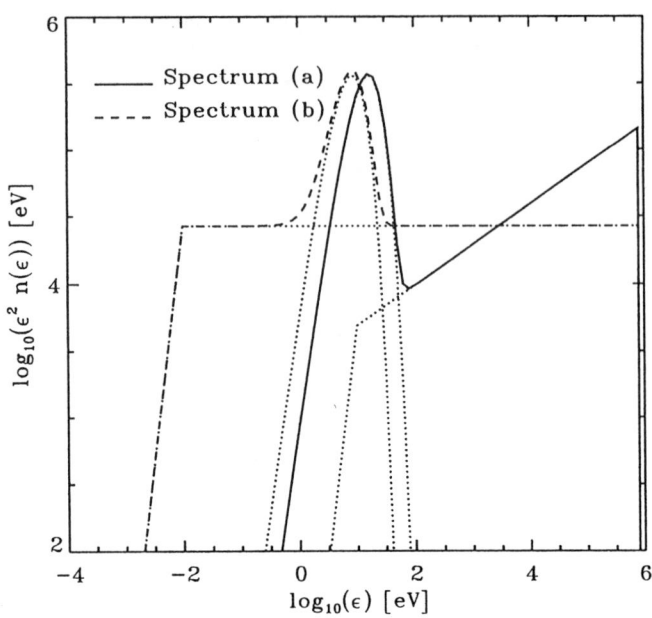

Figure 1: Possible AGN continuum spectra.

we consider two possible spectra: (a) a power–law of the form $dn/d\varepsilon \propto \varepsilon^{-1.7}$ extending from 10 eV to 1 MeV, plus a dilute black body spectrum of temperature $T = 50,000$ K; and (b) an ε^{-2} power–law from a low energy turn-over at 10^{-2} eV (below which the spectrum is assumed typical of synchrotron self–absorption, i.e. $\varepsilon^{1.5}$) to a high energy cut-off at 1 MeV, plus a dilute black body spectrum of temperature $T = 26,000$ K. In both cases equal energy density is assumed to be contained in the power–law and black body components. Both possible radiation fields are shown in Figure 1.

METHOD OF CALCULATION

The Monte Carlo method is used to follow a proton as it accelerates and to model its interactions. We have calculated the energy loss rate for protons,

$$\frac{dE_p}{dt} = -\frac{cE_p\bar{\kappa}(E_p)}{\bar{\lambda}(E_p)}$$

where $\bar{\kappa}(E_p)$ is the mean inelasticity for the process (obtained using a Monte Carlo simulation), and $\bar{\lambda}(E_p)$ is the mean free path length for the interaction on the photon field. The energy loss rates of protons, divided by the proton energy and total energy density in the radiation field, are shown in Figures 2(a) and 2(b) for spectrum (a) and spectrum (b) respectively. Contributions of pair production and pion production are shown separately for interactions with the power–law component (dashed lines) and black body component (dotted lines) of the field; curves for the lower energy threshold are for pair production. For the case of the flat quasar–type spectrum (spectrum b), the energy loss rates for pair production and pion photoproduction are nearly equal. However, pair production occurs much more frequently as the energy lost per interaction is a factor of ~ 100 less than in photoproduction. Hence, excessive computing time would be required to calculate the spectra of particles produced via the straightforward approach of following each proton until it is lost catastrophically. So, except when calculating the spectra of electrons from pair production, we treat pair production as a continuous energy loss process. Notice that for spectrum (a), pion photoproduction on the power–law component dominates, in agreement with the work of Begelman et al.[20]

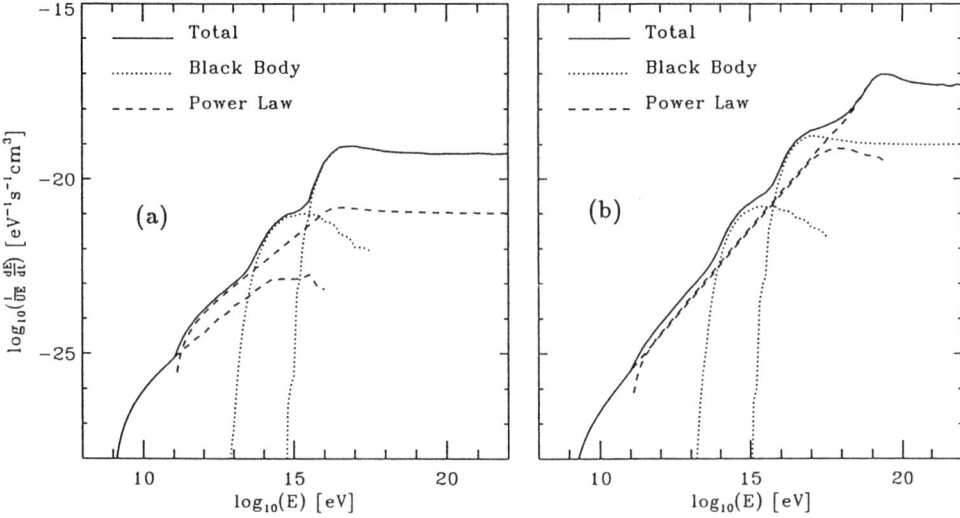

Figure 2: Energy loss rates of protons.

For detailed modelling of pair production interactions we use the differential cross sections given by Motz et al.[21] for the case of negligible nuclear recoil. An energy and direction for the positron is sampled in the proton rest frame using the rejection technique, and then Lorentz transformed to the LAB frame. The energy of the electron and the final energy of the proton are obtained by energy and momentum conservation. For photoproduction interactions, we use exclusive data of Genzel et al.[22] near threshold, and inclusive data of Moffiet et al.[23] at higher energies. Exclusive interactions are modelled exactly in the centre of momentum frame using the rejection method. For inclusive interactions we assume Feynman scaling to be approximately valid.

We inject protons with energy $E_o = 1$ GeV into the accelerator. Because protons are assumed to be constantly accelerating, and because of the energy dependence of the cross sections, we need to solve a differential equation to find the distance travelled before interaction. Thus, we sample a dimensionless interaction length, τ, from the exponential distribution,

$$p(\tau) = \exp(-\tau)$$

and solve

$$\frac{d\tau}{dx} = \overline{\lambda}[E_p(x)]^{-1}$$

to obtain the point of interaction x (see ref. 24). We do this numerically for photoproduction and pair production and decide which interaction takes place on the basis of whichever occurs first. The same approach is adopted when we treat pair production as a continuous energy loss process, although in this case we only sample an interaction length for photoproduction and the 'effective' acceleration rate will be lower and energy dependent.

Since the probability of a proton of energy E_p not having escaped from the acceleration region is E_o/E_p, we simply assign each proton a weight W equal to this value. Thus, the weight of a proton with energy E_p just before its first interaction is E_o/E_p. If n interactions have already taken place before a proton reaches energy E_p, the weight is

$$W = \frac{E_o}{E_1'} \frac{E_1}{E_2'} \frac{E_2}{E_3'} \cdots \frac{E_n}{E_p}$$

where E_n' and E_n are the energies of the proton immediately before and after the nth interaction. We model the interaction fully in three dimensions using the Monte Carlo method (see above), and the particles produced are binned in energy with weight W. If the proton survives the interaction, then we sample new interaction lengths, and the above procedure is repeated. If, however, there is no final–state proton (an interaction such as $p + \gamma \rightarrow n + \pi^+$ has taken place), then the simulation for this initial proton is complete and we inject another proton with energy E_o. For a full description of the calculation see Szabo and Protheroe[6,25].

RESULTS AND DISCUSSION

We present results for $E_{max} = 10^{17}$eV (i.e. we set $-\frac{dE}{dt}|_{loss} = \frac{dE}{dt}|_{accn}$ at 10^{17} eV; this does not preclude acceleration to energies slightly above E_{max}). Results for other E_{max} values will be given later[25]. The spectrum of protons escaping from the accelerator and the spectra of n, $\pi^{\pm}$ and π^o produced during the acceleration (dashed lines) are shown in Figures 3(a) and 3(b) for spectrum (a) and (b) respectively. We also show the spectra of particles produced after the acceleration (dotted lines) as a result of the protons interacting in the radiation field. The total spectrum produced in each case is shown by a solid line. In all cases, the spectra shown in Figure 3 result from the injection of one proton with an initial energy of 1 GeV. Notice that for the maximum energy chosen here the resulting spectra are very similar for the two radiation fields. This is not surprising if one compares the energy loss rates in Figures 2(a) and 2(b) for energies below 10^{17} eV. The major difference, due to the extension of the power–law spectrum down to infrared energies in spectrum (b), occurs mainly for proton energies above 10^{17} eV.

The spectra of particles produced by the escaping protons reflect the changing dominance of each loss process with energy. Examination of Figures 2(a) and (b) shows that for both radiation fields photoproduction is the dominant energy loss process for proton energies above 10^{11} eV except for the region $3 \times 10^{13} - 3 \times 10^{15}$ eV where pair production is dominant. This region where pair production dominates results in a reduction in photoproduction interactions relative to higher and lower energies and gives rise to the broad minima in the spectra of pions and neutrons shown in Figures 3(a) and 3(b).

We model the decay of the pions and muons, and the resulting neutrino, γ–ray, and electron spectra are shown in Figures 4(a) and 4(b) for spectrum (a) and (b) respectively.

Note that the electron spectrum contains contributions from both pair production and $\pi - \mu - e$ decay. We should point out that processes which depend on the details of the AGN model, such as proton–proton collisions (the matter density is model dependent) and convection onto the black hole at the heart of the AGN, may modify the spectra at low energies[20].

For our results to be useful, we need to be able to scale them to AGN flux observations. We assume the infrared to X-ray continuum is due to cascading in matter and radiation fields of the $e^{\pm}$ and γ-rays produced during and after acceleration. The total energy going into this component per injected proton is $W_{e\gamma} \simeq 8$ GeV for both spectrum (a) and spectrum (b) for $E_{max} = 10^{17}$ eV. Dividing the infrared to hard X-ray luminosity by this gives the rate at which protons are injected,

$$\dot{N}_p = L_{IR-X}/W_{e\gamma}.$$

The expected neutrino flux is then obtained from the neutrino spectrum n_ν in Figure 4 using

$$E^2 F_\nu(E) = \dot{N}_p E_1^2 n_\nu(E_1)/4\pi d_L^2$$

where $E_1 = (1+z)E$ is the neutrino energy on production, z is the redshift, and d_L is the luminosity distance. Hence, we obtain

$$E^2 F_\nu(E) = (F_{IR-X}/W_{e\gamma})E_1^2 n_\nu(E_1)$$

where F_{IR-X} is the obgerved infrared to hard X-ray energy flux.

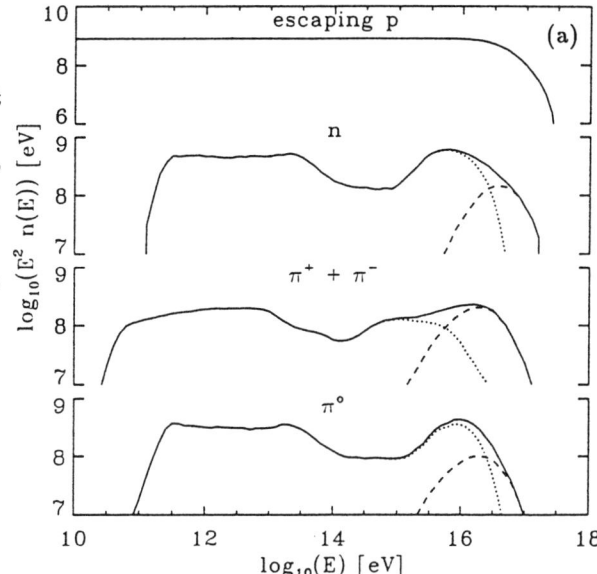
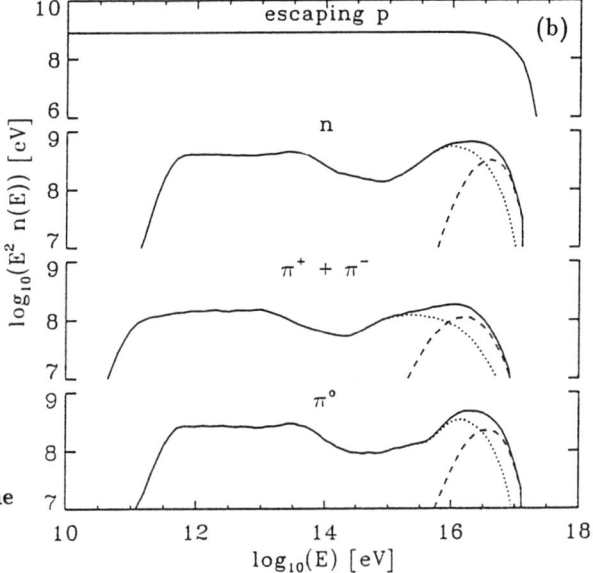

Figure 3: Spectra of hadrons per injected proton.

We do not include interactions of neutrons in calculating the total energy going into the cascade as neutrons will probably escape to the outer regions of the AGN where they may produce an observable flux of VHE γ-rays[8]. The neutrino flux resulting from interactions of the neutrons and protons from neutron decay will probably be the same order of magnitude as that from interactions of the protons escaping from the accelerator. Here we take this into account in an approximate way simply by multiplying by two the neutrino flux calculated as described above. Our final results[25] will include a full calculation.

100 MeV γ-rays have been observed from two AGN's, the quasar 3C273 (Bignami et al.[26]) which has a redshift of $z = 0.158$, and the more distant optically violent variable 3C279 (Hartman et al.[27]) with a redshift of 0.538. Both these objects are therefore prime candidates for neutrino astronomy. 3C273 has a continuum which may be approximated by spectrum (b) and a total luminosity of about 10^{47} erg s^{-1}. For this object the total infrared to X-ray energy flux is about 2000 eV cm^{-2} s^{-1}; if the 100 MeV γ-ray flux were included this would increase to about 5000 eV cm^{-2} s^{-1}. These values of F_{IR-X} are based on the summary by Bezler et al.[28]

Unfortunately 3C279 has a continuum which has features of both spectrum (a) and spectrum (b) and is not exactly described by either (see ref. 29 for a representation of the continuum).

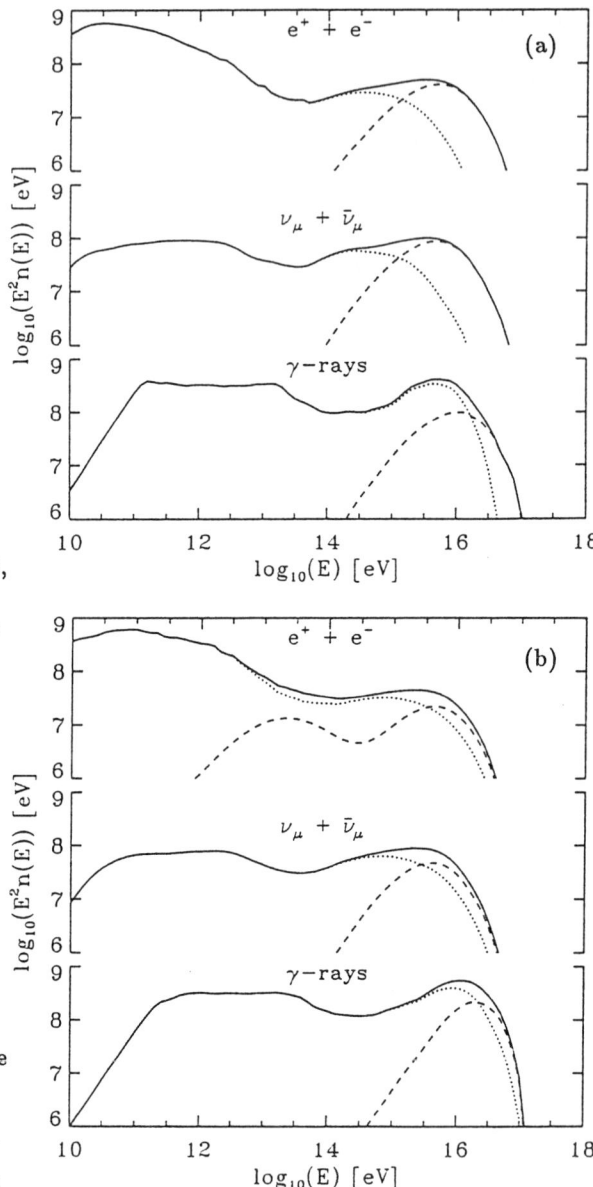

Figure 4: Spectra of leptons and photons per injected proton.

However, as noted above, for the maximum proton energy we adopt, the neutrino spectrum is approximately the same for both AGN continuum spectra assumed. The total luminosity of 3C279 during its quiescent state is $L_{tot} = 3 \times 10^{46}$ erg s^{-1} (Impey and Neugebauer[30]) and it has an X-ray luminosty (2-20 keV) of $L_X^{2-20} = 6 \times 10^{45}$ erg s^{-1}.

During outburst, the X-ray luminosty has been observed to increase to $L_X^{2-20} = 3 \times 10^{46}$ erg s^{-1} with 20% fluctuations occuring in $\sim$ 45 minutes. The γ-ray luminosity (100 MeV–10 GeV) was found to be $\sim 10^{48}$erg s^{-1} (assuming isotropic emission) during recent observations with the EGRET instrument on GRO[27]. The total infrared to X-ray energy flux from 3C279 varies from about 90 eV cm^{-2} s^{-1} in its low state to about 600 eV cm^{-2} s^{-1} in its active state. If the recently observed 100 MeV flux is also due to cascading of the high energy electromagnetic component in the source then the energy flux used should be increased to about 3000 eV cm^{-2} s^{-1}.

Our predicted neutrino flux from 3C279, scaled as described above, is shown in Figure 5(a). The two solid curves show the flux for the quiescent and active states based on only the observed infrared to X-ray flux. The dashed line shows the result of scaling to the energy flux assuming the γ-ray flux observed with GRO was also the result of cascading of the $e^{\pm}$ and γ-rays produced during and after acceleration. The flux of neutrinos from 3C273 is shown in Figure 5(b). Again, predictions based solely on the observed infrared to X-ray flux are shown by the solid line, and if γ-rays are included by the dashed line.

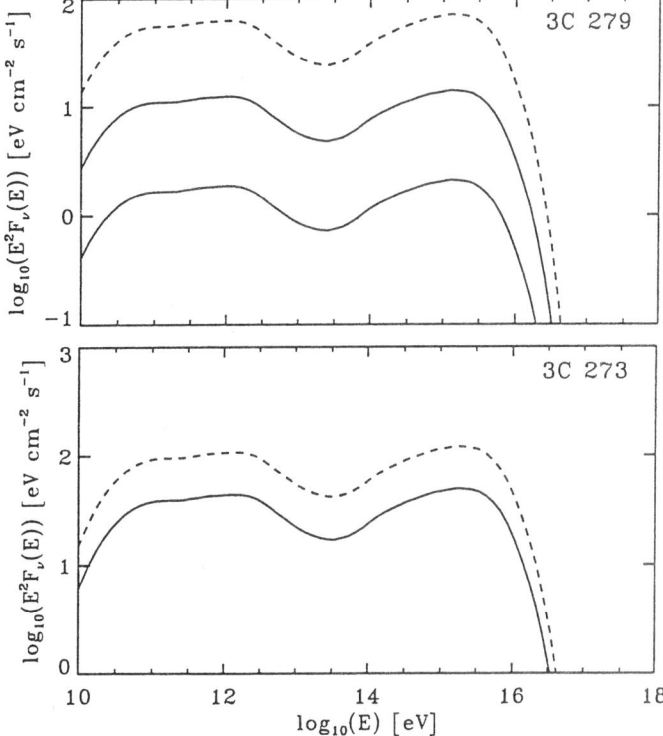

Figure 5: Predicted Neutrino flux from 3C279 and 3C273.

CONCLUSION

We have examined some of the consequences of shock acceleration of protons in two radiation fields appropriate for AGN. We find that a significant fraction of the total energy in high energy particles may be produced during the acceleration process. We have calculated the spectrum of neutrons escaping from the central regions of AGN's, as well as the spectrum of neutrinos, and the energy deposited in an electromagnetic cascade per low energy proton injected into the accelerator. We have described how to use the observed infrared to hard X-ray continuum to normalize the predicted neutrino spectrum in order to give the expected neutrino flux. We give, as examples, predictions for two objects, 3C273 and 3C279, from which 100 MeV γ-rays have been observed. While the predicted flux of neutrinos from either either of these objects is not observable with current neutrino telescopes 3C273 might just be observable with DUMAND II, as might 3C279 during outburst phase. Calculations of the expected signals are in progress and will be reported later[25].

REFERENCES

1. Protheroe, R.J., and Kazanas, D., Ap.J., **265**, 620 (1983).
2. Kazanas, D., and Ellison, D.C., Ap.J., **304**, 178 (1986).
3. Eichler, D., Ap.J., **232**, 106 (1979).
4. Berezinsky, V.S., and Ginzburg, V.L., M.N.R.A.S., **194**, 3 (1981).
5. Stecker, F.W., et al., Phys.Rev.Lett., **66**, 2697 (1991).
6. Szabo, A.P., and Protheroe, R.J., 22nd Int. Cosmic Ray Conf., Dublin, paper OG 9.1–20, in press (1991).
7. Kirk, J., and Mastichiadis, A., Astron.Ap., **211**, 75 (1989).
8. Mastichiadis, A., and Protheroe, R.J., M.N.R.A.S., **246**, 279 (1990).
9. Gaisser, T.K., 'Cosmic Rays and Particle Physics' (Camb. Univ. Press) (1990).
10. Drury, L.O'C., Rep.Prog.Phys., **46**, 973 (1983).
11. Sikora, M., et al., Ap.J., **320**, L81 (1987).
12. Malkan, M.A., and Sargent, W.L.W., Ap.J., **254**, 22 (1982).
13. Edelson, R.A., and Malkan, M.A., Ap.J., **308**, 59 (1986).
14. Saunders, D.B., et al., Ap.J., **347**, 29 (1989).
15. Carleton, N.P., et al., Ap.J., **318**, 595 (1987).
16. Chini, R., Kreysa, E., and Biermann, P.L., Ap.J., **219**, 87 (1989).
17. Rees, M.J., Begelman, M.C., and Blandford, R.D., *Tenth Texas Symp. on Relativistic Astrophysics*, ed. R. Ramaty and F.C. Jones (N.Y.Acad.Sci.), 254 (1981).
18. Wilkes, B.J., and Elvis, M., Ap.J., **323**, 243 (1987).
19. Turner, T.J., and Pounds, K.A., M.N.R.A.S., **240**, 833 (1989).
20. Begelman, M.C., Rudak, B., and Sikora, M., Ap.J., **362**, 38 (1990).
21. Motz, J.W., Olsen, H.A., and Koch, H.W., Rev.Mod.Phys., **41**, 581 (1969).
22. Genzel, H., Joos, P., and Pfeil, W., Landolt–Bornstein, **8**, ed. H. Schopper, (Berlin--Heidelberg–New York: Springer Verlag) (1973).
23. Moffiet, K.C., et al., Phys.Rev.D, **5**, 1603 (1972).
24. Protheroe, R.J., M.N.R.A.S., **246**, 628 (1990).
25. Szabo, A.P., and Protheroe, R.J., in preparation (1992).
26. Bignami, G.F., et al., Astron. Astrophys., **93**, 71 (1981).
27. Hartman, R.C., et al., Ap.J.Lett. in press (1991).
28. Bezler, M., et al., Astron. Astrophys., **136**, 351 (1984).
29. Makino, F., et al., Ap.J., **347**, L9 (1989).
30. Impey, C.D., and Neugebauer, G., Astron.J., **95**, 307 (1988).

RELATIVISTIC SHOCK WAVES AND THE EXCITATION OF PLERIONS

Jonathan Arons
Departments of Astronomy and of Physics,
University of California at Berkeley
and
Institute of Geophysics and Planetary Physics,
Lawrence Livermore National Laboratory

ABSTRACT

The shock termination of a relativistic magnetohydrodynamic wind from a pulsar is the most interesting and viable model for the excitation of the synchrotron sources observed in pulsar driven ("plerionic") supernova remnants. I describe results on the structure of relativistic magnetosonic shock waves in plasmas composed of electrons and positrons plus heavy ions as a minority constituent by number. Relativistic shocks in symmetric pair plasmas create fully thermalized distributions of particles and fields downstream. Therefore, such shocks are not good candidates for the mechanism which converts rotational energy lost from a pulsar into the nonthermal synchrotron emission observed in plerions. However, when the upstream wind contains heavy ions which are a minority constituent by number density, but carry the bulk of the energy density, much of the energy of the shock goes into a downstream, nonthermal power law distribution of pairs with energy distribution $N(E)dE \propto E^{-s}$. In a specific model presented in some detail, $s = 1.7$, with approximately 20% efficiency in converting upstream flow energy into downstream energy in nonthermally accelerated particles. These characteristics are close to those assumed for the pairs in macroscopic MHD wind models of plerion excitation. The acceleration mechanism is collective synchrotron emission of elliptically polarized extraordinary modes by the ions in the shock front at high harmonics of the ion cyclotron frequency, with the downstream pairs absorbing almost all of this radiation, mostly at their fundamental (relativistic) cyclotron frequencies. The spatial structure of the shock includes compressional overshoots of the magnetic field strength above what would be expected from the MHD jump conditions, at places where the ions are reflected. I outline a new model of the inner regions of the Crab Nebula, in which I propose that the shock structure is angularly resolved and the optical "wisps" are surface brightness enhancements created in the compressional overshoots of the magnetic field within and just down stream from the shock. I also briefly outline the application of these results to the relativistic shocks believed to terminate the winds from pulsars in compact binaries.

INTRODUCTION

The excitation of diffuse, nonthermal astrophysical synchrotron sources by energy lost from central compact objects has long been a puzzle in high energy astrophysics. Pulsars and their surrounding plerionic nebulae form the nearest at hand and best studied examples of this problem. Other examples include the excitation of extragalactic radio sources by jets, and possibly the emission from some active galactic nuclei themselves. The Crab Nebula is the best studied of the plerions, and I will take the physical problems it poses to be characteristic of the class. It is typical of what I shall call "diffuse" plerions, in which the synchrotron radiation and adiabatic expansion dominate the energy losses from relativistic electrons and positrons. The termination of the outflow from a pulsar in a compact binary, as the relativistic wind encounters the mass lost from the companion star, may form an example of a "compact" plerion, the difference from the diffuse case being in the competition between synchrotron losses and the acceleration process itself.

The X-ray and gamma-ray emission from the Crab requires continuous energy input, which is aptly explained by the energy lost from a central, rotating magnetized neutron star[1]. Indeed, the identification of this idea with the subsequently discovered pulsars[2] is one of the pillars of the reasoning by which we identify the rotating neutron star model with the observed pulsar phenomenon. However, the physics of the energetic link has remained an unsolved problem. The Crab has features similar to a wide variety of astrophysical synchrotron sources (power from a central compact object, a large scale magnetic field, diffuse, optically thin nonthermal synchrotron emission, dense filaments of thermal emission line plasma embedded in or adjacent to the diffuse nonthermally emitting gas). Therefore, an understanding of the physical mechanisms through which energy is carried, without emitting many photons, from the central compact object to the surrounding environment, then degraded into the observed nonthermal emission, has wide astrophysical significance. In addition, the nature of this conversion mechanism may provide constraints on models of the pulsar magnetosphere.

Relativistic, magnetohydrodynamic (MHD) models of the coupling between pulsar and nebula are the most successful at the present time. Building on earlier ideas[3-5], Kennel and Coroniti[6,7] showed that such a wind from the pulsar, terminated by a standing, transverse magnetosonic shock wave, could give a good account of the nebular dynamics and of the nebular spectrum of high energy photons (near infrared, optical, X-rays and gamma rays), which are an instantaneous probe of the pulsar's outflow. The prominent results of their model are several. 1) The post shock flow fits into the slowly expanding nebula only if the upstream flow is almost entirely dominated by kinetic energy, with

$$\sigma \equiv \frac{B_1^2}{4\pi m N c^2 \gamma_1} \simeq 0.003. \tag{1}$$

2) If the upstream flow is entirely composed of pairs (an assumption), and if the post-shock energy distributions are power laws with $N(E) \propto E^{-s}$ (another assumption), then the synchrotron emission from the downstream flow can be fit to the observed X- spectra of the Nebula, taken as an unresolved object, if $s \approx 2.3$. Once this fit is made, the optical and near infrared spectra of the nebula (again taken as an unresolved object) are *predicted* (more or less) correctly by the model.

This theory is sufficiently successful as a macroscopic explanation of the phenomenology to provide a setting for investigation of the microscopic physics. Several questions of basic physics are implied by the macroscopic flow model. Does a relativistic, magnetosonic shock in a pair plasma actually create the assumed power law distributions downstream? If so, what is the mechanism of acceleration in such shocks? Are there other observable consequences of this mechanism, other than the desired particle acceleration, which allow one to test the shock hypothesis directly? I emphasize that one cannot appeal to the ever popular ideas of diffuse shock acceleration proposed for particle acceleration by interplanetary and nonrelativistic interstellar shocks[8,9] even as extended to relativistic shocks with magnetic field parallel to the flow[10], since the termination shocks of pulsar winds must be fairly close to having the magnetic field perpendicular to the flow upstream when the plerion surrounds the pulsar — indeed, for the Crab, a laminar Archimedean spiral in the outflow from the pulsar would have Θ_{Bn}, the angle between the flow direction and the magnetic field in the shock frame where we observe the system, differing from 90° only by one part in 10^9!

In relativistic shocks, the mechanisms of high energy particle acceleration and of basic thermalization can be one and the same — accelerating relativistic particles can be part of the problem of shock structure itself. In fact, the highest energy electrons and positrons in the shock theory I outline below are those which have Larmor radii comparable to the gyro-radii of the heavy ions which control the shock thickness, and are precisely the $e^{\pm}$ which give rise to the highest energy synchrotron photons ($\varepsilon \sim 100$ MeV) observed in the *nebular* spectrum. Thus, the problems of shock structure and of nonthermal particle acceleration *are* one and the same in this case.

SHOCK STRUCTURE AND PARTICLE ACCELERATION

Pulsar outflows must have extremely high Mach number. The Alfven Mach number in the flow is $M_A = \beta_1 \gamma_1 / \sqrt{\sigma} \cong 1.5 \times 10^8$! Under these circumstances, magnetic reflection of particles from the shock front plays an essential role in the shock dynamics. Relativistic, magnetosonic solitary waves with unidirectional flow exist only for[11-13] $M_A < 1 + 1/\gamma$. At higher Mach number, incoming particles reflect from the enhanced magnetic field in the wave front and are set into Larmor gyration, leading to a new kind of magnetosonic soliton with reflected particles self-consistently incorporated[12]. This reflection process requires all the momentum of the incoming plasma to be temporarily stored in a magnetic overshoot, with $B_{peak}/B_1 = (1 + \sigma^{-1})^{1/2}$ in the wave frame of a soliton in a pair plasma.

While it is possible to construct solitary wave models with magnetically reflected particles in pair plasmas[12], unraveling the instabilities and thermalization mechanisms in this highly inhomogeneous environment is a formidable task for analytical theory. In a collaboration with Y. Gallant, M. Hoshino, A.B. Langdon and C.E. Max, I have made use of fully non-linear, fully self-consistent particle-in-cell numerical simulations in order to "experimentally" uncover the relevant physics. The basic method is to integrate the equations of motion of many charged particles in their *self-consistent* fields. Suppose at some time t_n, one knows the electromagnetic field on a spatial grid, and one knows the positions and momenta of all the particles $\{x_j(t_n), p_j(t_n)\}$ (they are not on the grid). One then finds the Lorentz force at each particle's position and uses Newton's law (in relativistic form) to advance the particles to new phase space positions $\{x_j(t_n + dt), p_j(t_n + dt)\}$. One then calculates new fields at time $t_n + (dt/2)$ using the charge densities and currents at time t_n, with the particles distributed on the grid with a suitable weight function. These new fields are used to advance the positions of the particles once again, with the time step set by various stability considerations. These techniques are quite well understood and have been in use for 20 years[14, 15].

We modeled shock waves by setting up a one-dimensional spatial grid in the computer, with length anywhere from 10 to 40 Larmor radii based on the magnetic field and particle energy in the upstream flow. The initial magnetic field points exactly transverse to the flow velocity. At the initial time, the cold plasma fills the computational box, with flow Lorentz factor $\gamma_1 \gg 1$ and a magnetic field polarized across the flow carried with the plasma at the fluid speed. The ratio of magnetic energy density to flow energy density upstream σ is the only parameter of significance. As expected from theoretical considerations, simulations with the same value of σ but differing γ_1 yield identical results, when $\gamma_1 \gg 1$. We have surveyed parameter space with these spatially 1D models from $\sigma = 13$ down to $\sigma = 0.001$.

At the injection point (the "left wall"), the particles are injected all with the same upstream velocity, and the magnetic field is carried with the fluid, by requiring the boundary electric field to be that of a perfectly conducting fluid. The opposite end of the grid is represented as a conducting rubber wall — particles bounce off the wall, which acts as a perfect conductor as far as the fields are concerned. It would be better to have outflow boundary conditions on the plasma and radiation, but so far a useful algorithm to implement these in the dense downstream plasma has not been constructed. Because the magnetic field points across the flow, the wall has no influence on the shock structure, once the point of particle reflection is more than a few Larmor radii away from the wall. In the low σ shocks of interest here, the wave energy is never more than a few percent of the thermal and kinetic energy of the plasma, so the conducting wall, which forces the downstream waves to be standing modes, has little influence on the shock once it its well upstream of the wall.

The details of the pair simulations are reported elsewhere[16, 17]. The essential results are that one finds an intense, quasi-coherent electromagnetic precursor travelling in front of the shock, while downstream, a few per cent of the total energy is turned into a linearly polarized Rayleigh-Jeans spectrum of electromagnetic waves with temperature close to what one expects from the jump conditions. The incoming plasma stream is refelcted from the shock front, followed by rapid thermalization of the Larmor gyration. The downstream spectra of electrons and positrons (not shown) are almost perfect relativistic Maxwellians. Thus, transverse shocks in a pure pair plasma create a fully thermalized downstream medium, contrary to a basic assumption of the MHD model of plerionic supernova remnants. We have done a few preliminary calculations with the magnetic field oblique to the flow, using what is otherwise the same 1D model. So long as $\Theta'_{Bn} > 50°$, the results are semiquantitatively the same as for the purely transverse case, as one might suspect since the structure is dominated by the reflected particles and the rapid dissipation. Smaller obliquities may be different, but for technical reasons this conclusion is not yet warranted.

The nature of the thermalization mechanism gives a clue to a shock model which does have nonthermal particle acceleration. As seen in the shock frame, the reflected particles form a gyrating ring with small momentum dispersion in the leading edge of the shock. Such rings are unstable — they form synchrotron masers[18–21]. Calculations of the growth rate for a homogeneous pair plasma, along with estimates of the maximum harmonic number that can be generated, suggest that the dissipation observed in the pure pair plasmas is due to the formation of synchrotron masers, with consequent plasma heating as the extraordinary modes radiated are absorbed.

Now imagine what happens if the upstream flow contains heavy ions, as well as pairs. Since all the species flow into the shock with the same speed (the upstream Larmor radii are small compared to the overall flow scale in which the shock is embedded), $ZN_{1i} + N_{1+} = N_{1-}$. The ions have a rest mass large compared to that of the $e^{\pm}$. Therefore, they will contribute the dominant kinetic energy in the flow, if $N_{1i} > (m_{\pm}/m_i)(N_{1+} + N_{1-})$. As the plasma encounters the shock, the lighter pairs, which have Larmor radii a factor $m_{\pm}/m_i$ smaller than the ions, form a leading leptonic shock almost identical to the shock found in a symmetric plasma, if $N_{1i} \ll (N_{1+} + N_{1-})$. The more massive ions have no chance to respond, since the pair shock develops on the pair cyclotron time. They plow on through until they too are magnetically reflected in the enhanced magnetic field of the shock front. An electrostatic field in the direction of flow will be associated with this spatial separation of the ions from the leptons, with electrostatic potential having magnitude $Ze\Phi \sim m_i\gamma_1 c^2$. Once the ions are set into gyration by the reflection from the shock front, they too are synchrotron maser unstable, as they gyrate in the much denser background of shock heated pairs.

Simulations of shock flows with ions[22] support these conjectures. In Figure 1, I show the electromagnetic structure from a simulation done in the same manner as the simulations of the purely leptonic shocks, but now containing ions with a charge to mass ratio appropriate to protons, at $12\omega_{ci}^{-1}$ after the plasma bounced off the simulation wall. In order to keep the computer time finite, the mass ratio is small, $m_i/m_\pm = 20$, and the density ratio is $N_{1i}/(N_{1+} + N_{1-}) = 0.2$, chosen as a compromise to have $N_{1i} \ll (N_{1+} + N_{1-})$ but still have most of the kinetic energy density in the upstream flow. In this case, 80% of the plasma flow energy is in the ions. The upstream flow is almost entirely dominated by the plasma kinetic energy flux — $\sigma = 0.005$. While the choice of γ_1 matters only to the relative scale of the flow, in the interests of making the problem be of direct relevance to plerions and pulsars, we chose $\gamma_1 = 10^6$, corresponding to protons being accelerated through a few percent of the polar cap potential of the Crab pulsar. Note the large overshoot in the magnetic field and the long wavelength oscillations downstream. The first overshoot occurs because as the ions reflect from the shock front, their flow momentum is stored in an increase in the magnetic pressure and in the thermal pressure of the pairs. Behind the overshoot, the ions continue gyrating quasi-coherently, creating compressional enhancements at each turning point, until their gyration finally thermalizes. These oscillations are spaced roughly one ion Larmor radius $r_{L2} = m_i c^2 \gamma_1 / e B_2$ apart.

In Figure 2 I show zoomed views of the $p_x - x$ projections of ion phase space for a region of the simulation near the shock front itself, at 12 and 13 ion cyclotron times after the simulation began. One sees the quasi-coherent reflection of the ions, followed by their rapid thermalization. The figure also shows the unsteady character of the shock structure. On this scale, the structure of the pair shock, with its initial thermalization to Maxwellian distributions, is infinitesimal, achieved as the ions have barely begun to gyrate. Figure 3 shows the spectra of the downstream electrons and positrons. The positrons are no longer everywhere Maxwellian in energy space, but now have a power law distribution with $N(E) \propto E^{-1.7}$ at energy/particle exceeding $\gamma_1 m_\pm c^2$, and have energy density comparable to the ions' kinetic energy density in the upstream flow. In this simulation, the electrons also show some sign of power law behavior at energies exceeding $\gamma_1 m_\pm c^2$. The positron spectrum is Maxwellian at energies below $\gamma_1 m_\pm c^2$, and cuts off at positron energy equal to that of the upstream energy/particle of the protons, $\gamma_1 m_i c^2$. Thus, a relativistic, low σ magnetosonic shock, composed of ions, electrons and positrons, with $N_{1i} \ll N_{1+} + N_{1-}$ but with the upstream flow energy dominated by the ions, can accelerate the downstream positrons into a power law distribution with high efficiency, with the slope of the power law being very close to what is desired for the macroscopic model. This particular simulation is highly resolved, with 64 particles per grid cell per species in the initial state. In other calculations of the shock structure[22] with various values of σ and of γ_1, and in simulations of the synchrotron maser process in isolation in a uniform medium[20], we have found the slope to be be of order 2. Figure 4 shows the spectral slopes found as a function of σ. For $\sigma = 0.0034$ and $\sigma = 0.005$, simulations were done for a variety of numerical parameters (number of particles/cell, number of grid cells per ion Larmor radius, ...). The dispersion in the values of s

obtained defines the "error" in these numerical experiments in deterimining the form high energy tail of the distribution function. The relatively small mass ratios and the upper limits to the number of particles per cell form the main sources of the variability in s found in the various simulations with the same physical parameters. Figure 4 also shows the fraction of the downstream energy density contained in the nonthermal part of the particle spectra. For all the simulations, this efficiency is between 10% and 20%, with the latter being the value appropriate to the physical conditions inferred in the Crab Nebula.

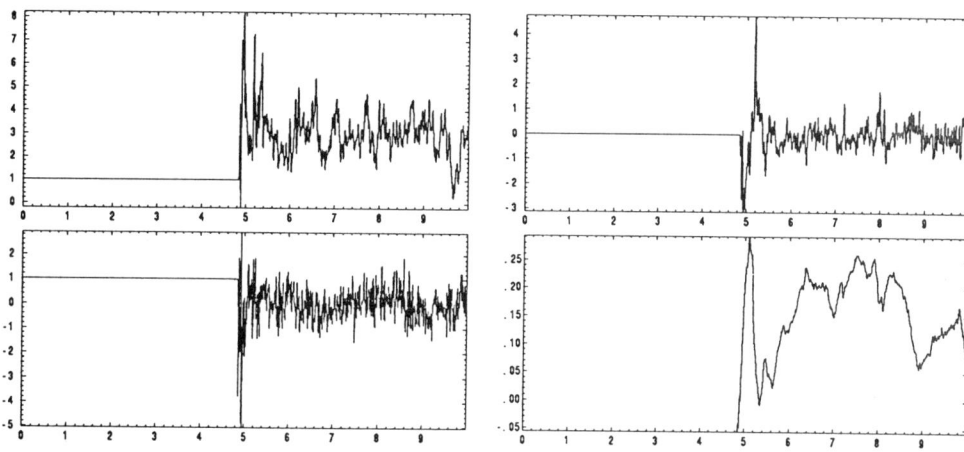

Figure 1: Electromagnetic structure of a relativistic magnetosonic shock in an electron–positron–proton plasma, at a time when the shock has crossed 50% of the simulation grid. Upper left panel: Magnetic field B_z, with upstream field amplitude equal to unity. Lower left panel: Transverse electric field E_y, with $E_y = (v_1/c)B_z$ at the injection wall. Upper right panel: Electrostatic field E_x. The downstream waves have electrostatic fields with amplitudes roughly one-half the magnitude of the electromagnetic field components, indicating strong elliptical polarization of the post-shock extraordinary modes. Lower right panel: Electrostatic potential in units of $\gamma_1 m_i c^2/e$. Since the potential never exceeds 0.3, the electrostatic field in the shock front acts to decelerate the ion stream, but the reflection is magnetic, as in the shocks in pure pair plasmas. The unit of length is the ion Larmor radius based on upstream prameters.

The physics of this acceleration is simple. In this non-symmetric plasma, the extraordinary modes are no longer linearly polarized electromagnetic waves but have a noticeable electrostatic component as they propagate across the magnetic field. The ion extraordinary modes behave like magnetosonic waves in the pair plasma, and have *elliptical* polarization with the electric vector rotating in the left handed sense with respect to the magnetic field. This is the same sense as the Larmor gyration of the positrons and of the ions. As a result, a positron can linearly absorb the magnetosonic modes at their cyclotron resonance (as well as at higher harmonics) and can gain energy. Because the polarization is elliptical, there is some power in right circularly polarized fields, so the electrons also can gain energy nonthermally from the ions. Because of the small mass ratio used in the simulations, the power in left handed circularly polarized fields emitted by the ions is much larger than in right handed circular polarization, so the positron acceleration is much stronger than the acceleration of electrons. Under realistic conditions, we believe the acceleration of electrons and positrons will be much more comparable[22].

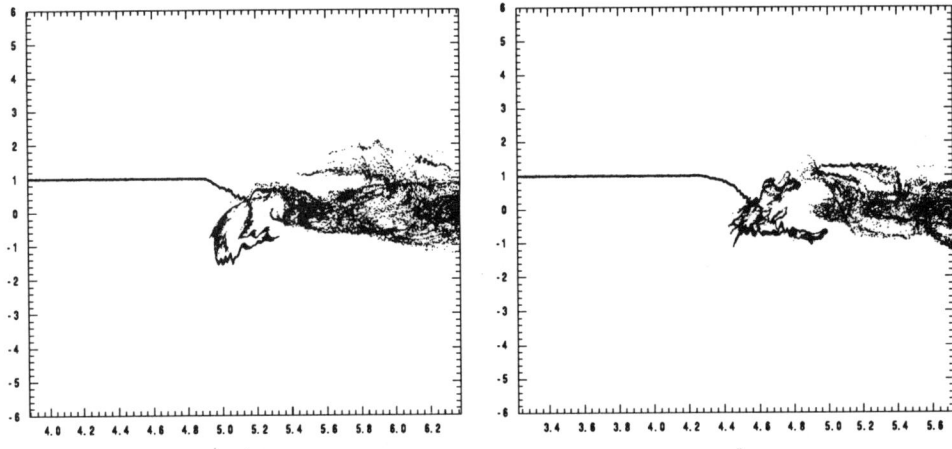

Figure 2: Ion phase space near the shock front. Left panel: Flow momentum in units of $10^6 m_i c$ versus x for ions near the shock front at $t = 12/\omega_{ci1}$. The unit of length is the ion Larmor radius based on upstream parameters. Note the initial deceleration of the ion stream in the electrostatic field, followed by the rapidly dissolving loop in the magnetic field. The main body of an ion Maxwellian distribution is produced within 2–3 Larmor times after the ions begin decelerating in the shock front. Right panel: ion flow momentum versus x at $t = 13/\omega_{ci1}$ near the shock front. These snapshots illustrate the unsteady character of the shock.

That this heating process can and should result in power law positron spectra can be shown by elementary "quasi-linear" arguments. If one computes the response of a positron or an electron to a given ensemble of small amplitude magnetosonic modes in a relativistically hot electron-positron plasma, one finds a quasi-linear rate of gain of a particle's energy with the Fermi-like form $\dot{\gamma} \propto \gamma(8\pi k U_k / B_0^2)$ when the magnetosonic wave spectrum has spectral energy density $k U_k \approx k |E_k^2|/4\pi \propto k^{-1}$ (Arons, in preparation), as is found in our simulations. When the relativistic ion cyclotron frequency is the fundamental frequency in the magnetosonic wave spectrum, as is the case in our simulations, the proportionality constant is approximately equal to the ion cyclotron frequency itself. Numerical investigation of a particle's response to finite amplitude spectra suggests the Fermi-like acceleration saturates, with $\dot{\gamma}/\gamma \approx \Omega_{ci}(\gamma_{ion})$, when $8\pi k U_k / B_0^2 > 1$. This fast acceleration is observed in the particle-in-cell simulations, where the energy density in field fluctuations does exceed that in the background magnetic field when $\sigma \ll 1$.

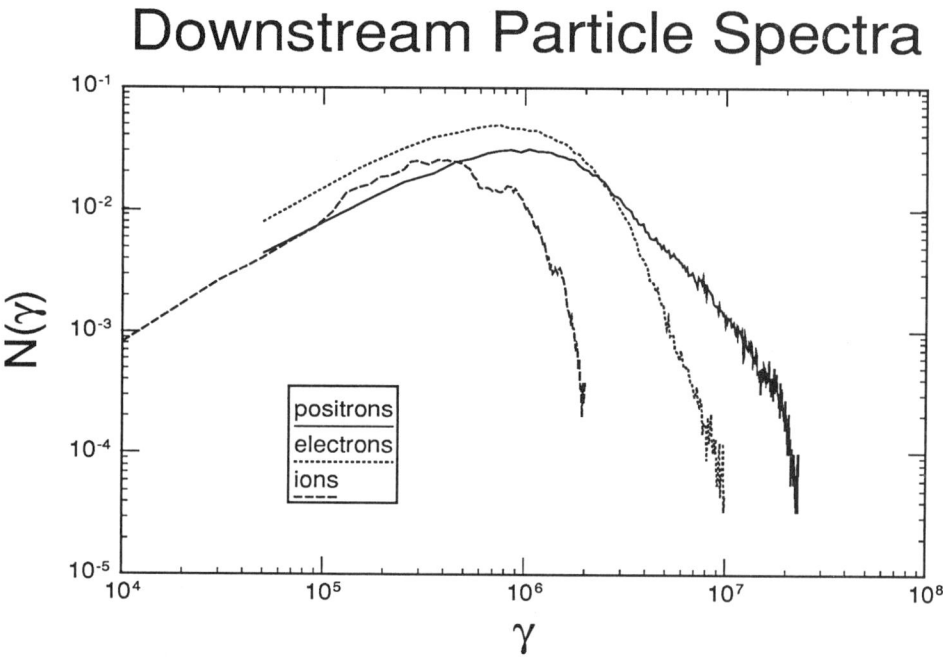

Figure 3: Post-shock distributions. The positrons have a high energy power law spectrum with slope s = 1.7, while the electrons show evidence for a power law supra-therma distribution with slope s = 3.7. The upper cutoff occurs where the energy/particle of the pairs equals the upstream energy/particle of the ions, as is expected from cyclotron resonant absorption of extraordinary modes emitted at and above the upstream ion cyclotron frequency.

APPLICATION TO THE CRAB NEBULA AND PULSARS IN COMPACT BINARIES

Models of astrophysical systems using these results are in the midst of construction. One can get the flavor of these from the following simple considerations. Suppose the magnetosphere of the Crab pulsar emits a dense flow of pairs, with $\dot{N}_\pm \sim 10^{38} \dot{N}_{\pm,38}$ coming out in the broad sector around the rotational equator where the pulsar must emit the wind required to feed the torus of X- and gamma-ray emission in the Nebula[23, 24]. A number of magnetospheric models suggest such a flow is possible[25, 26]. In addition, suppose the magnetospheric electrodynamics causes an

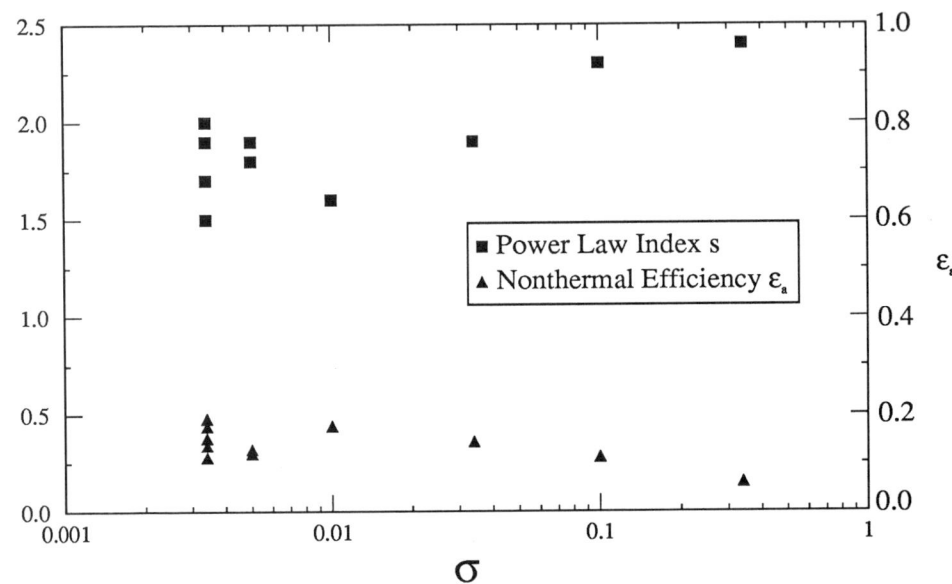

Figure 4: Squares show the the index s (left ordinate axis) of the power law part of the downstream positron spectra, as a function of the total σ of the upstream electron-positron-proton flow. The triangles show the nonthermal particle acceleration efficiency of these shocks (right ordinate axis), defined as the ratio of the downstream energy in the nonthermal part of the particle spectrum to the total downstream energy density.

outflow of heavy ions to be pulled up from the stellar surface and accelerated to high energy in the outflow around the magnetic equator[25]. The particle flux is constrained by the electrodynamics to be close to the Goldreich-Julian flux, $\dot{N}_{GJ} 2\Omega^2 \mu / Zec$ for this pulsar, where Z is the charge on the ions. The resulting ratio of the mass densities in the wind is

$$\kappa_i \equiv \frac{m_i n_i}{m_\pm(n_+ + n_-)} = 1836 \frac{A}{Z} \frac{\Omega^2 \mu}{ec\dot{N}_\pm}, \qquad (2)$$

where A is the atomic number of the ions, μ is the pulsar's magnetic moment and Ω is its angular velocity. Solving for Z yields $Z \approx 3(A/56)/\dot{N}_{\pm,38} \kappa_i$ in the case of the Crab; the model works for partially ionized iron, the primary constituent of the crust, while if the heavy ions are protons or α particles, an ion flux greatly in excess of the Goldreich-Julian rate would be required.

These ions might be accelerated through a fair fraction of the total polar cap potential $\Phi_{cap} \sim 10^{17}$ Volts, corresponding to $\gamma_1 \sim 5 \times 10^7 (Z/A)(\Phi/\Phi_{cap})$. The equatorial wind with these accelerated ions carries a fraction Φ/Φ_{cap} of the spindown luminosity of the pulsar. If the shock really puts essentially all the energy into the accelerated pairs, as our simulations suggest, and if these pairs radiate essentially all of this in X-rays and gamma-rays, as is indicated by the macroscopic flow models, then one would conclude, from the hard photon luminosity of the nebula, that the ions experience an acceleration potential $\Phi/\Phi_{cap} \sim 0.3/I_{45}$, where I_{45} is the moment of inertia of the neutron star in units of 10^{45} cgs.

We can go much further. From momentum conservation, one expects the shock to occur roughly $10^{17.5}$ cm from the pulsar[4]. Optical observations reveal the presence of the curious "wisps" in the surface brightness in this region[27], a series of enhancements in the surface brightness elongated in the direction perpendicular to the flow lying to the Northwest of the pulsar; there is also a fainter wisp to the Southeast[28]. The broad band synchrotron emissivity in the magnetic overshoot and downstream magnetic compressions shown in Figure 3 is readily found to be proportional to the $B^{11/3}$, assuming the downstream pitch angle scattering has largely isotropized the pairs. The essential new astrophysical idea is to identify the enhanced surface brightness expected in these compressions with the observed wisps.

From this identification, one can extract a host of conclusions. The details of these will be presented elsewhere. However, several qualitative statements can be made immediately. 1) A reasonable model, with $\sim 10^{38}$ pairs/s flowing out in the wind and 10^{33} iron ions/s with charge $Z = 3$ or 4 yields both a roughly correct optical surface brightness for the thin wisp and wisp 1, the observed spacing between the thin wisp and wisp 1, and the thickness of wisp 1 itself, if $\gamma_1 \sim 10^7$. This is a particle outflow rate and flow energy/particle *roughly* consistent with the values inferred above for the gross energetics of the nebular emission. 2) The series of downstream, large amplitude compressions of the plasma are spaced roughly one ion Larmor radius apart. This spacing is similar to that seen between the first few wisps in van den Bergh and Pritchett's image if $\gamma_1 \sim 10^7$. 3) The spectrum of wisp 1 should be close to that of Maxwellian synchrotron emission. The spectrum for a uniform medium is well known[29]. With $B \sim 10^{-4}$ Gauss and pair temperature $\sim 10^6 m_{e\pm} c^2$, the characteristic exponential decline of the spectrum from a relativistic Maxwellian should appear around ~ 1 eV. 4) Successive wisps further from the ion overshoot should show a progressively more nonthermal spectrum, as the ions transfer their energy to the positrons. 5) Because the pair Larmor radius is very small (~ 10 AU for $B \sim 10^{-4}$ Gauss and $\gamma_1 \sim 10^7$), we identify the observed thickness of the thin wisp [$\sim 10^{16.5}$ cm in the image obtained by van den Bergh and Pritchett[28] as being due to the projection of an infinitesimally thin sheet projected on the plane of the sky, as one expects from the flow geometry implied by the Aschenbach and Brinkmann[23] model of the X-ray emission in the nebula. 6) The flow velocity of the pairs throughout this region is between 0.3c and 0.5c, which provides a Doppler asymmetry to the luminosity just about right to explain the brightness ratio seen between wisp 1 and the one faint wisp observable to the pulsar's southeast. Possible partial anisotropy of the pairs with respect to the local magnetic field enhances this conclusion. 7) The shocks observed in our simulations are unsteady, as is shown in Figure 2. Translated into the shock frame, which is where we observe the Crab Nebula, the unsteady shock velocity becomes oscillation of the position of the various ion overshoots on a time scale of several months, a phenomenon consistent with time variability of the wisps reported long ago by Scargle[27]. Such variability should be uncorrelated with pulsar glitches and other rotational anomalies. In addition, such unsteady behavior may imply unsteady nebular emission at energies exceeding 1 MeV, as has been occasionally reported. 8) Finally, the parameters derived from the wisp model are appropriate to those expected in the diffuse X-ray source formed further downstream. The detailed modeling in progress is designed to test whether these rough numerical conclusions stand up to careful analysis of the flow and predicted surface brightness distribution.

Therefore, I suggest that the structure of a relativistic shock in a relativistic wind composed of pairs plus heavy ions has its spatial structure revealed to us in the detailed morphology of the wisp region in the Crab Nebula. That this may be so is a consequence of the very high rigidity of the outflows expected from pulsars, an expectation based on the enormous voltages developed in the magnetospheres of these objects combined with the negligible radiation drag on the flow near these photon poor compact objects.

The same ideas can be applied to the millisecond radio pulsars recently discovered in compact binaries[30, 31], whose relativistic winds interact with the mass lost from the "normal" stellar companion[32-34]. If one assumes the same fraction of the potential drop on open field lines goes into the energy/particle of these winds as appears in the wind from the Crab pulsar, one expects $\gamma_1 \sim 5 \times 10^5$, while if σ is as small as it appears to be in the Crab pulsar's wind, the magnetic field behind the termination shock of the relativistic wind is on the order of 10 Gauss, since the shock occurs less than one solar radius from these pulsars, at least along the line of centers between the pulsar and the star. In this strong field, synchrotron losses limit the energies of the accelerated pairs to values much less than $\gamma_1 m_1 c^2$, and convert almost all of the energy of the part of the pulsar's wind shocked within the binary into high energy radiation. The pairs accelerate to energies on the order of $10^7 m_{\pm} c^2$ and much of the wind energy is given up as 1–10 MeV synchrotron photons, whose flux at earth might be as high as 10^{-4} photons/cm^2–s. Most of the pulsar's wind goes into blowing the nebular bubble observed in Hα emission around the binary[35], but because of the much stronger magnetic field within the close confines of the binary, most of the high energy emission comes from within the binary rather than from the nebular bubble[36].

ACKNOWLEDGMENTS

The research described here on relativistic shock waves was supported in part by NSF grant AST-9115093 and by IGPP-LLNL grant 92-23, both to the University of California at Berkeley. Part of the work was performed under the auspices of the U.S. Department of Energy at the Lawrence Livermore National Laboratory under contract W-7405–Eng-48.

REFERENCES

1. F. Pacini. Energy emission from a neutron star. *Nature*, **216**:567, 1967.
2. T. Gold. Rotating neutron stars and the nature of pulsars. *Nature*, **221**:25, 1968.
3. J.H. Piddington. The crab nebula and the origin of interstellar magnetic fields. *Australian. J. Phys.*, **10**:530, 1957.
4. M.J. Rees and J.E. Gunn. The origin of the magnetic field and relativistic particles in the crab nebula. *Mon. Not. Roy. Astron. Soc.*, **167**:1, 1974.
5. W. Kundt and E. Krotscheck. The crab nebula — a model. *Astron. and Ap.*, **83**:1, 1980.
6. C.F. Kennel and F.V. Coroniti. Confinement of the crab pulsar's wind by its supernova remnant. *Astrophys. J.*, **283**:694, 1984.
7. C.F. Kennel and F.V. Coroniti. Magnetohydrodynamic model of crab nebula radiation. *Astrophys. J.*, **283**:710, 1984.
8. A.R. Bell. The acceleration of cosmic rays in shock fronts. *Mon. Not. Roy. Astron. Soc.*, **182**:147, 1978.
9. J. Quenby and R. Lieu. Enhanced shock acceleration in relativistic jets and cosmic ray origin in active galactic nuclei. *Nature*, **342**:654, 1989.
10. D.C. Ellison, F.C. Jones, and S.P. Reynolds. First-order fermi particle acceleration by relativistic shocks. *Astrophys. J.*, **360**:702, 1990.
11. C.F. Kennel and R. Pellat. Relativistic nonlinear waves in a magnetic field. *J. Plasma Phys.*, **15**:335, 1976.
12. D. Alsop and J. Arons. Relativistic magnetosonic solitons with reflected particles in electron-positron plasmas. *Phys. Fluids*, **31**:839, 1988.
13. T. Chiueh. Relativistic solitons and shocks in magnetized $e^- - e^+ - p^+$ fluids. *Phys. Rev. Lett.*, **63**:113, 1989.
14. A.B. Langdon and B.F. Lasinski. Electromagnetic and relativistic plasma simulation models. In B. Alder, S. Fernbach, and M. Rotenberg, editors, *Methods in Computational Physics*, volume **16**, page 327, New York, 1976. Academic Press.
15. C.K. Birdsall and A.B. Langdon. *Plasma Physics via Computer Simulation*. McGraw-Hill, New York, 1985.

16. A.B. Langdon, J. Arons, and C.E. Max. Structure of relativistic magnetosonic shocks in electron-positron plasmas. *Phys. Rev. Lett.*, **61**:779, 1988.
17. Y.A. Gallant, M. Hoshino, A.B. Langdon, J. Arons, and C.E. Max. Structure of relativistic magnetosonic shock waves in electron-positron plasmas. *Astrophys. J. (in press)*, 1992.
18. V.V. Zheleznyakov and E.V. Suvorov. Results and problems in the investigation of the synchrotron instability. *Ap. and Space Sci.*, **15**:24, 1972.
19. P. Yoon. Amplification of a high-frequency electromagnetic wave by a relativistic plasma. *Phys. Fluids B*, **2**:867, 1990.
20. M. Hoshino and J. Arons. Differential heating and acceleration of positrons by synchrotron maser instabilities. *Phys. Fluids B*, **3**:818, 1991.
21. J. Arons, M. Hoshino, and Y.A. Gallant. Synchrotron instability, absorption and suprathermal particle acceleration in relativistic, magnetosonic shock waves. *Astrophys. J.*, to be submitted, 1992.
22. M. Hoshino, J. Arons, Y.A. Gallant, and A.B. Langdon. Relativistic, magnetized electron-positron-proton shock waves in synchrotron sources: Shock structure and non-thermal acceleration of positrons. *Astrophys. J. (in press)*, 1992.
23. B. Aschenbach and W. Brinkmann. A model of the x-ray structure of the crab nebula. *Astron. and Ap.*, **41**:147, 1975.
24. R.M. Pelling, W.S. Paciesas, L.E. Peterson, K. Makashima, M. Oda, Y. Ogawara, and S. Miyamoto. A scanning modulation collimator observation of the high energy x-ray source in the crab nebula. *Astrophys. J.*, **319**:416, 1987.
25. J. Arons. Electron-positron pairs in radio pulsars. In M.L. Burns, A.K. Harding, and R. Ramaty, editors, *Proc. Workshop on Electron-Positron Pairs in Astrophysics*, page 163, New York, 1983. American Institute of Physics.
26. V.S. Beskin, A.V. Gurevich, and Ya.N. Istomin. Electrodynamics of pulsar magnetospheres. *Zh. Eksp. Teor. Fiz.*, **85**:401 (*Soviet Physics—JETP*, **58**:235), 1983.
27. J.D. Scargle. Activity in the crab nebula. *Astrophys. J.*, **156**:401, 1969.
28. S. van den Bergh and C.J. Pritchett. The crab synchrotron nebula at $0.5''$ resolution. *Astrophys. J.*, **343**:L69, 1989.
29. T.W. Jones and P.E. Hardee. Maxwellian synchrotron radiation. *Astrophys. J.*, **228**:268, 1979.
30. A.S. Fruchter, D.R. Stinebring, and J.H. Taylor. A millisecond pulsar in an eclipsing binary. *Nature*, **333**:237, 1988.
31. A.G. Lyne, R.N. Manchester, N. D'Amico, L. Stabeley-Smith, S. Johnston, J. Lim, A.S. Fruchter, W.M. Goss, and D. Frail. An eclipsing millisecond pulsar in the globular cluster terzan 5. *Nature*, **347**:650, 1990.
32. W. Kluzniak, M.A. Ruderman, J. Shaham, and M. Tavani. Nature and evolution of the eclipsing millisecond binary pulsar psr1957+20. *Nature*, **334**:225, 1988.

33. E.S. Phinney, C.R. Evans, R.D. Blandford, and S.R. Kulkarni. Ablating dwarf model for eclipsing millisecond pulsar 1957+20. *Nature*, **333**:832, 1988.
34. M.A. Ruderman, J. Shaham, and M. Tavani. Accretion turnoff and rapid evaporation of very light secondaries in low mass x-ray binaries. *Astrophys. J.*, **336**:507, 1989.
35. S.R. Kulkarni and J.J. Hester. Discovery of a nebula around psr 1957+20. *Nature*, **335**:801, 1988.
36. J. Arons and M. Tavani. High energy emission from the relativistic wind of psr 1957+20. *Astrophys. J.*, submitted, 1992.

Simulations

MONTE CARLO SIMULATION OF ELECTRON ACCELERATION IN MODIFIED RELATIVISTIC SHOCKS

Donald C. Ellison
Department of Physics
North Carolina State University
Raleigh, NC 27695-8202

Abstract We give a brief review of Monte Carlo simulations of nonlinear Fermi shock acceleration and then give new results on electron acceleration in SNRs and in relativistic parallel shocks. The acceleration of low energy electrons in shocks is poorly understood, but even when energetic electrons are considered, where electron and proton scattering should be qualitatively similar, dramatic differences result between electron and proton acceleration in relativistic shocks. If the shocked plasma is a mixture of electrons and protons, the electrons are accelerated much less efficiently than protons. We predict that only $e^- - e^+$ pair dominated plasmas can produce significant radio emission in relativistic flows if the standard Fermi mechanism operates in parallel shocks.

1. Introduction

Analytic techniques have been quite effective in studying first-order Fermi particle acceleration at shocks and, in particular, have shown that Fermi acceleration is expected to be extremely efficient in high Mach number shocks (see Drury 1983; Blandford and Eichler 1987; Jones and Ellison 1991 for reviews). However, most analyses based on the diffusion-convection equation assume that particle distribution functions are isotropic to first order, and since this condition is never satisfied for thermal particles at shocks, the viscous shock structure itself, together with thermal particle injection into the acceleration process, cannot be studied if this approximation is used. If higher order terms are retained in formulating the diffusion-convection equation, analytic difficulties become severe and solutions, when they are obtainable, are cumbersome and limited to small areas of phase space.

Large-scale plasma computer simulations of collisionless shocks (normally hybrid, i.e., particle protons and fluid electrons), on the other hand, give detailed information on the microphysics of shock dissipation and structure, but have not yet been able to simulate particle acceleration to relativistic energies (see Goodrich 1985 for a review of hybrid simulations of quasi-perpendicular shocks and Leroy 1984, Burgess 1987, and Quest 1988 for reviews on quasi-parallel plasma simulations). Hybrid simulations do show, however, that an important relationship exists between shock structure, dissipation, and particle acceleration, and particle acceleration may, in fact, be essential for dissipation in high Mach number shocks, particularly quasi-parallel ones (e.g., Jones and Ellison 1991). By parallel we mean that the magnetic field is parallel to the shock normal. Of primary importance for particle acceleration is the observation from quasi-parallel hybrid simulations (first reported by Quest 1988) that some ions are back-scattered from downstream to upstream with a resulting energization. This acceleration process can continue with successive reflections between the converging upstream and downstream waves, as has been confirmed in recent work by Scholer et al. (1992) (for alpha particles as well as protons) and Giacalone et al. (1992), where some thermal particles were accelerated by more than a factor of 100 in energy.

Thus hybrid simulations clearly show that the first-order Fermi process applies to thermal particles as well as superthermal ones and confirms the suggestion by Eichler (1979) and Ellison, Jones, and Eichler (1981) that first-order Fermi acceleration and parallel shock dissipation are two limits of the same process as long as the scattering is controlled by nearly stochastic, large amplitude turbulence.

The fact that Fermi acceleration is an intrinsic part of parallel shock formation means that the injection of thermal particles into the acceleration mechanism must be treated self-consistently, a requirement severely limiting analytic treatments. In addition, both theoretical and observational evidence suggests that the acceleration can be quite efficient, implying that test-particle approximations are poor and the influence of the highest energy particles on the shock structure must also be included self-consistently. To achieve a realistic description of collisionless shock structure coupled with particle acceleration to cosmic ray energies, some compromise must be reached between the detailed modeling of the shock microphysics and the overall effects of particle acceleration on shock structure. In the past decade we have developed a Monte Carlo description of parallel shocks which, we believe, keeps enough of the shock plasma physics to be realistic, while at the same time is computationally efficient enough to model the acceleration of particles to extremely high energies, including the nonlinear modifications to the large-scale shock structure that these accelerated particles produce. In this paper, I will briefly review the Monte Carlo techniques and present some new results on electron acceleration in shocks with flow speeds at an arbitrary fraction of the speed of light.

2. Monte Carlo Nonlinear Shock Simulations

We have developed Monte Carlo simulation techniques to model parallel shock structure coupled with Fermi acceleration. Our steady-state simulation describes the thermal gas shock in a simplified but self-consistent way, while at the same time, by not treating the scattering in detail, we are able to follow the evolution of individual ions long enough to model acceleration to high energies. Our basic assumption is that the same scattering processes responsible for producing energetic particles can also be applied to thermal particles producing the gas subshock. We assume that all particles scatter elastically and isotropically off some background magnetic turbulence with a mean free path, λ, that is some function of momentum. These assumptions have received direct support from spacecraft observations of diffuse ions at the quasi-parallel Earth bow shock (e.g., Ellison and Möbius 1987; Ellison, Möbius, and Paschmann 1990) and from large-scale plasma simulations of quasi-parallel shocks (e.g., Quest 1988; Scholer *et al.*, 1992; Giacalone *et al.*, 1992), and there seems to be no compelling physical reason to treat thermal and superthermal populations differently at least as far as acceleration is concerned.

The scattering mean free path for all energy particles, λ, is taken to be

$$\lambda = \lambda_o R^\beta / \rho, \qquad (1)$$

where $R = pc/(Ze)$ is the particle rigidity, ρ is the plasma density, e is the electron charge, Z is the charge state number, and λ_o is the mean free path of a thermal proton. As just mentioned, the scattering is not determined self-consistently from the particle distribution function and the background magnetic field, but the rigidity dependence of the mean free path can be varied to model pitch-angle scattering off hydromagnetic turbulence with various power spectra. Particles scatter in three dimensions but we have thus far treated only plane, parallel shocks, so fluid quantities vary only in x (see Baring and Ellison, this volume, for a discussion of Monte Carlo simulations of oblique shocks). The background magnetic field, which is assumed to produce the particle scattering via unspecified wave-particle interactions, lies along x and does not contribute to the jump conditions and we assume the scattering centers are at rest in the local fluid frame.

If particles convect and diffuse according to (1), some shock heated downstream ions scatter back across the subshock into the upstream region and become Fermi accelerated. The slowing and heating of the upstream flow, mandated by the presence of accelerated ions ahead of the shock, produces a smooth shock and strongly influences thermal ion injection: the more the unshocked gas is slowed, the smoother the shock and the fewer injected particles. In the simulation, the smoothing of the shock structure, or flow velocity profile, is found by iteration until the mass, momentum, and energy fluxes are conserved across the shock. We are able to make quantitative predictions for the absolute injection and acceleration efficiencies and calculate the complete particle distribution functions over all energies.

The model includes the escape of energetic particles at either an upstream free escape boundary (FEB) or a maximum energy cutoff, $E_{\max}$. The FEB phenomenologically models a finite shock size and/or the lack of sufficient scattering far upstream to turn particles around; ions which cross the FEB decouple from the shock system. The loss of energy flux at the FEB or $E_{\max}$ is strongly nonlinear, and, in steady-state shocks at least, will result in an increase in the overall compression ratio. Our simulation determines the shock compression ratio self-consistently including escape (see Ellison and Jones 1991).

The Monte Carlo simulation can model large-angle scattering or pitch-angle diffusion, but in the work considered here, only large angle scattering is used, and our results for relativistic flows would change if we had used pitch-angle diffusion instead.

The code has been generalized to include relativistic particle energies and relativistic flow velocities (Ellison, Jones, and Reynolds 1990; Ellison 1991a,b). Since the diffusion approximation never applies in a system with relativistic flow speeds, analytic treatments of relativistic shocks are extremely difficult and the approximate solutions which result are unwieldy (e.g., Webb 1985; Kirk and Schneider 1987a; see Kirk 1988 for a review of relativistic shock acceleration work). In contrast, relativistic generalizations are quite straightforward in the Monte Carlo simulation and once they are included, the simulation treats relativistic and nonrelativistic shocks identically except that now the acceleration depends on the details of the scattering. Test particle results for relativistic shocks (e.g., Kirk and Schneider 1987b; Ellison, Jones, and Reynolds 1990), suggest that these shocks tend to produce flatter spectra than nonrelativistic ones and this implies that nonlinear effects will be even more important than in nonrelativistic shocks. In addition, as we show below, the relativistic kinematics produce dramatic differences in the way shocks treat electrons and protons. In this paper, we will concentrate on the differences between electron and proton acceleration in shocks modified by accelerated protons.

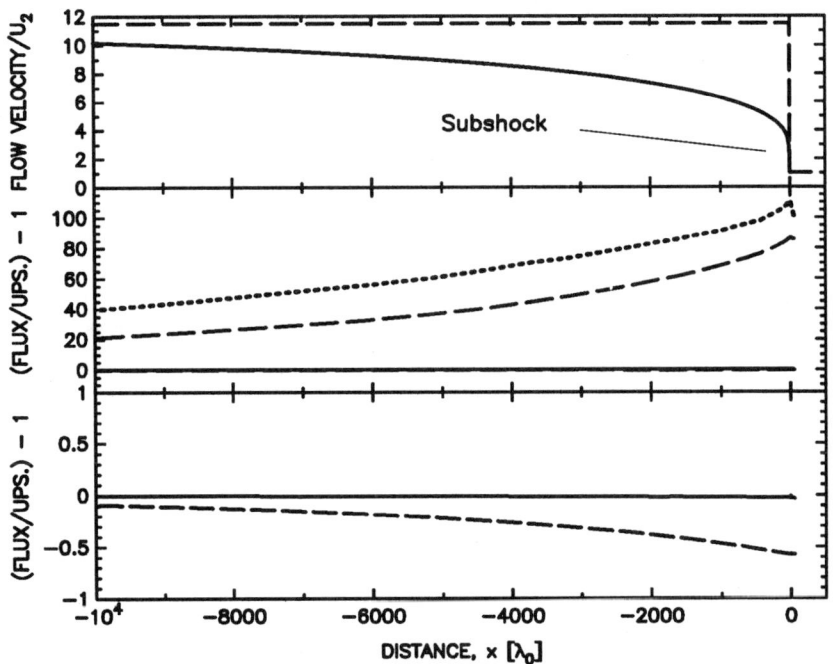

Fig. 1. Shock profiles (upper panel) and momentum and energy fluxes (center and lower panel). Note that the bottom solid line in the center panel contains two nearly flat curves overlaid. For this illustration, the sharp profile (dashed curve, upper panel) has the correct final $r = 11.5$. In an actual calculation, the correct r would not be known for this initial sharp shock run. The bottom panel shows the momentum flux (solid line) and energy flux (dashed line) on an expanded scale (zero corresponds to the far upstream value).

3. The Acceleration of Protons in Modified Shocks

We first consider nonrelativistic flows. Figure 1 shows an example of our results (taken from Ellison and Reynolds 1991) for the case where the shock velocity is $u_1 = 2 \times 10^4$ km s^{-1}, the upstream temperature is $T_1 = 1 \times 10^6$ K, and the Mach number is $M_1 = 170$. Only protons are considered in this example. We have used a maximum energy cutoff of $E_{\max} = 1 \times 10^6$ keV. The smoothed shock profile (solid line in upper panel) was determined in an iterative fashion with both the shape of the profile and the overall compression ratio being iterated simultaneously (see Ellison and Reynolds for a detailed discussion). We find that a unique r and shock profile exist which allow simultaneous conservation of momentum and energy fluxes from far upstream through the shock to the downstream region.

The momentum and energy fluxes are shown in the center and lower panels of Figure 1. The upper dashed and dotted lines in the center panel are the momentum and energy fluxes, respectively, calculated with the discontinuous shock profile of the upper panel. The lower solid curves in the center panel (the lower line contains two curves) and the curves in the bottom panel of Figure 1 show the final momentum and energy fluxes produced by the smooth profile with $r = 11.5 \pm 2$. (The "uncertainty" quoted here and elsewhere is the range within which convergence is obtained to $\pm 10\%$.) The bottom panel shows the final fluxes on an expanded scale and no deviations from the far upstream value greater than $\pm 10\%$ are seen for the momentum flux (solid line), while the energy flux (dashed line) falls below the far upstream value due to particle escape. In this example, $\sim 55\%$ of the energy flux is lost through $E_{\max}$. When this energy flux is included, the total energy flux is conserved across the shock. The escaping particles also carry away number and momentum; however, these are generally much smaller, as fractions of the upstream values, than the escaping energy flux.

The shock structure we determine, plotted as the flow velocity versus position, describes all of the plasma with no distinction between the thermal gas and the accelerated component. One important aspect of the smooth profile is the presence of a 'subshock' with a compression ratio of about three and a length-scale on the order of an upstream thermal particle convection length. This subshock heats the incoming cold gas and produces a hot downstream proton population some fraction of which diffuses back across the shock and becomes Fermi accelerated; the accelerated population emerges smoothly from the thermal gas with no separate seed population. We find, in contrast to analytic two-fluid results (see Jones and Ellison 1991 for a review), that a subshock exists regardless of the acceleration efficiency as long as only thermal ions are injected and accelerated by the shock.

Once the shock profile is determined for protons, electrons or heavier ions can be treated as test particles and scattered off the shock profile determined by protons.

4. Acceleration of Electron in Modified Shocks

Understanding the acceleration of electrons is of crucial importance for high energy astrophysics. Nonthermal processes are most often inferred from radio synchrotron emission from relativistic electrons and, with the exception of γ-ray sources, superthermal ions are not associated with particular objects. Therefore, an essential step in bringing shock acceleration theory closer to observations is developing the ability to calculate the spectra of accelerated electrons in self-consistent shock profiles. It turns out, however, that the shock acceleration of electrons is poorly understood compared to the acceleration of ions.

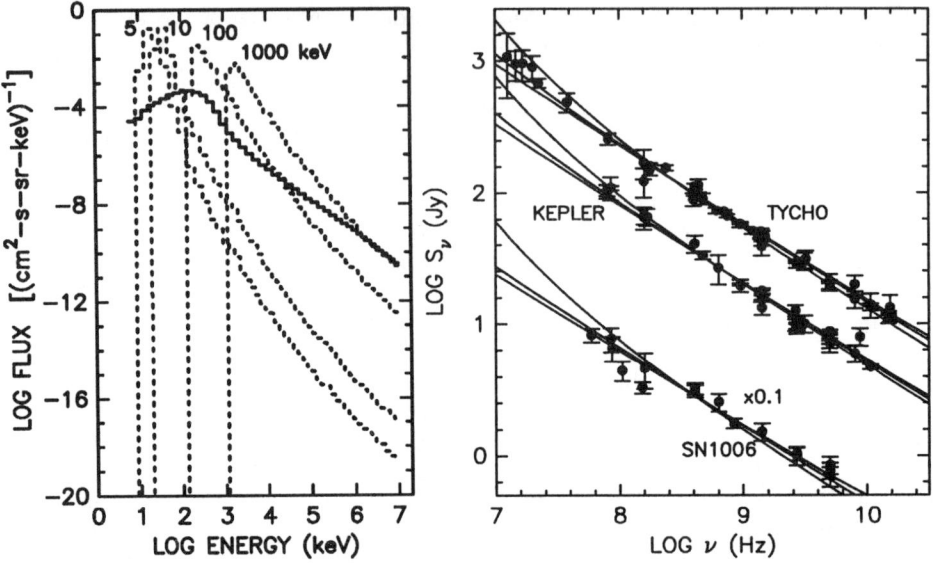

Fig. 2. Electron spectra (dotted lines) for various injection energies in keV. The test-particle electron spectra are calculated with the same shock structure determined self-consistently for protons and which yields the proton spectrum shown with the solid line. All spectra are calculated in the shock frame, at a downstream location, and are normalized to one entering particle per cm^2 per second.

Fig. 3. The data is a compilation from many sources (see Reynolds and Ford 1992) and gives the most complete and consistent representation of the fluxes from these young SNRs. There is a slight indication that the observed fluxes have an upward curvature in accord with the predictions of nonlinear Fermi acceleration. The solid lines are model fits to the data for various magnetic field values. For Tycho and Kepler, the fields go (from top to bottom) from 10^{-2} to 10^{-3} to 10^{-4} G. For SN1006, the fields are 10^{-2} to 10^{-4} to 10^{-6} G. We note that if the theoretical *proton* spectral shape had been used to model the data as is normally assumed, the fits would have been in strong disagreement with the observations.

The fundamental reason for this is the difficulty in formulating the electron diffusion coefficient at low energies. The apparent lack of magnetic fluctuations of sufficiently high frequency and of low phase velocity makes it difficult to estimate the acceleration of low energy electrons. On the other hand, at energies above where electron gyroradii are comparable to those of the thermal protons in the shock, the electrons should scatter off of the same magnetic turbulence as the protons and no fundamental scattering problem remains. However, even if Alfvén-like scattering centers are available for thermal electrons, another difficulty intervenes. As discussed above, efficient ion acceleration will smooth the shock on the proton diffusion-length scale. Low energy electrons, if they have short mean free paths, could not scatter far enough to sample substantial velocity differences, and would not be accelerated by the converging flows in the shock.

Since, in any case, the electron energy density is negligible compared to that of the ions, we have restricted ourselves to treating superthermal test-particle electrons by inserting them in the shock profile determined self-consistently for protons. Several important results have emerged from this work and are described in Ellison and Reynolds (1991). In particular we showed that the normalization of electrons, compared to protons, depends strongly on the injection energy and mean free path assumed for the electrons. If electrons are injected at energies much below 1 MeV, they are strongly depressed relative to protons (see Figure 2). We also showed that electron spectra generally share the same concave-upward curvature typical of protons accelerated by nonlinear shocks but have quite different spectral shapes in the energy range of interest for synchrotron radio emission (typically 3 – 30 GeV). The predicted radio synchrotron spectra from young supernova remnants (SNRs) do, however, match observations extremely well as shown in Figure 3.

The models shown in Figure 3 are tentative, however, because they are for steady-state shocks. In an evolving SNR shock, the maximum energy is set by the finite shock age rather than a FEB or $E_{\max}$. In any event, we expect the distinctive concave shape, which depends only on the fact that the scattering mean free path increases with energy, to occur in time-dependent shocks as well.

We now investigate, still in the confines of the parallel geometry, steady-state approximations, the differences between electron and proton acceleration as the shock velocity is increased and becomes relativistic. In Figure 4 we show proton spectra, all calculated in the shock rest frame, having the same upstream temperature, $T_1 = 1 \times 10^6$ K and $E_{\max} = 10^{11}$ keV but with shock speeds ranging from $u_1 = 2,500$ km s^{-1} to $u_1 = 0.9c$. The downstream thermal peak, consisting mainly of protons that have crossed the shock once, shifts to higher energy as u_1 increases from purely kinematic reasons, i.e., the Lorentz transformation from the u_1 frame to the shock rest frame increases with increasing u_1. The spectral shapes above the thermal peak remain quite similar even though several nonlinear effects are occurring. First of all, all spectra shown in Figure 4 were calculated in shocks where the shock profile and the compression ratio were determined self-consistently as described above. The compression varies because the flux of energy out $E_{\max}$ varies and because the effective ratio of specific heats, γ_{eff}, varies as the fraction of total pressure contributed by relativistic particles increases with increasing u_1. Our simulation includes a self-consistent determination of γ_{eff} (see Jones and Ellison 1991).

We now use the same shock profiles determined in the five cases shown in Figure 4 to accelerate test-particle electrons. We inject these electrons at 1 MeV, an energy where their gyroradii are well above those of thermal protons, and assume that the electrons and protons both scatter with mean free paths depending only on the particle gyroradius (with the same proportionality constant). The electron spectra are shown in Figure 5. Note that the vertical scales of Figures 4 and 5 are quite different and that the electron spectra in the relativistic shocks are suppressed by many orders of magnitude relative to the protons. Two effects cause the large difference in normalization: the first comes from the fact that electrons of a given energy (as long as it is below the proton

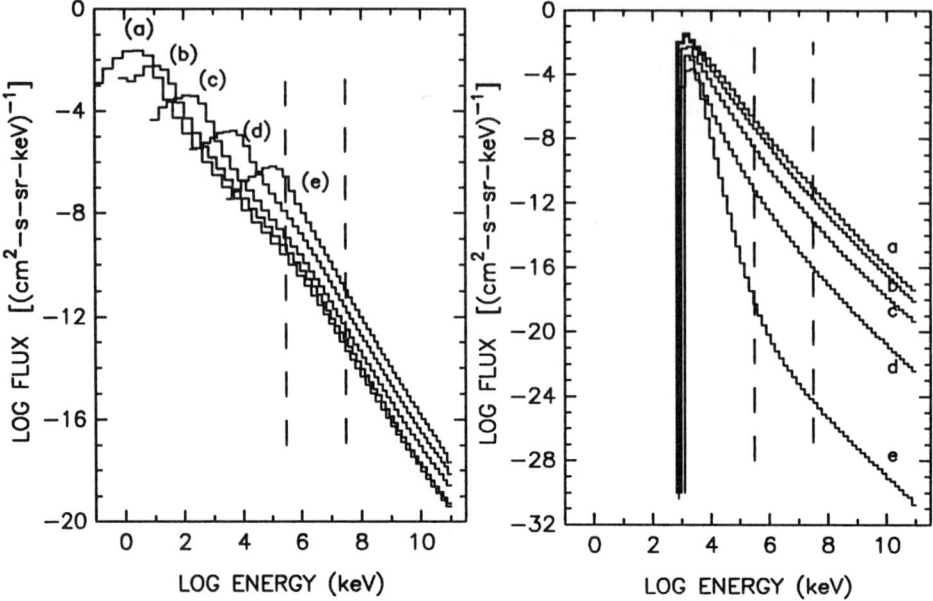

Fig. 4. Proton spectra for varying shock velocities: (a) 2,500 km s^{-1}, (b) 5,000 km s^{-1}, (c) 20,000 km s^{-1}, (d) 0.3c, and (e) 0.9c. The self-consistent shock structure, r, and $\gamma_{\rm eff}$ have been determined in each case. The range in r is from $r \simeq 7$ for $u_1 = 2,500$ km s^{-1} to $r \simeq 5$ for $u_1 = 0.9c$, and $\gamma_{\rm eff}$ ranges from ~ 1.45 for $u_1 = 2,500$ km s^{-1} to ~ 1.40 for 0.9c. The vertical dashed lines show roughly where the radio emitting energies (for electrons) lie. The energy flux escaping at $E_{\rm max}$ ranges from approximately 10-15% for $u_1 = 2,500$ km s^{-1} to less than 5% for $u_1 = 0.9c$.

Fig. 5. Electron spectra for the same shocks shown in Fig 4. The electrons are injected as test-particles at 1 MeV. Note the large difference in vertical scales between Figs 4 and 5.

rest mass energy) have a smaller diffusion length than protons of the same energy and will, therefore, not scatter as far upstream. These electrons will feel a smaller compression ratio and will be accelerated less efficiently then like-energy protons. This occurs regardless of the relativistic nature of the flow velocity and is just the effect that produced the results shown in Figure 2. The second effect depends purely on kinematics and becomes extremely important when the flow speed approaches c. To see this, consider the Lorentz transformation from one frame, O, (the upstream plasma frame) to another, $O_{\rm sk}$ (the shock frame), moving at speed u_1 relative to O. If a particle has a kinetic energy, E, in frame O, it will have a maximum kinetic energy $E_{\rm sk} = \gamma \left(E + m_o c^2 + p u_1 \right) - m_o c^2$ in frame $O_{\rm sk}$, where $\gamma = \sqrt{[1 - (v^2/c^2)]}$ and p and v are the particle momentum and velocity in O. This is a maximum because we have assumed that the particle crosses the shock along the normal. If heavy and light particles undergo this transformation, the heavy particles will gain more energy unless all particles are fully relativistic. In relativistic flows, the difference in energy gain can become quite large. For electrons and protons in a shock where $u_1 = 0.9c$, if $E_e = 1$ MeV and $E_p = 1$ keV, typical of the results shown in Figures 4 and 5, we find $E_{\rm sk,p}/E_{\rm sk,e} \simeq 200$ after a single shock crossing. Until the protons have become fully relativistic, they gain more energy in each shock crossing then do the electrons. This is reflected in the electron and proton spectra in Figures 4 and 5.

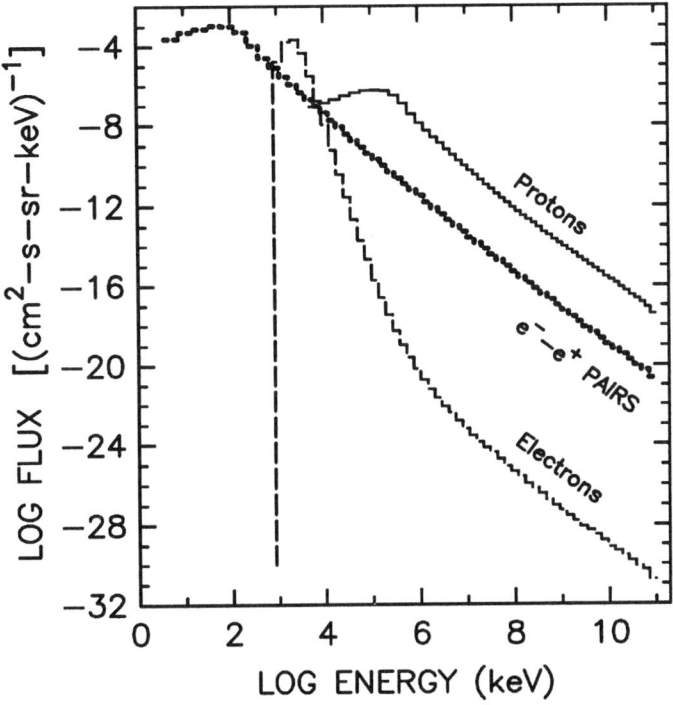

Fig. 6. The spectra labeled 'Protons' and 'Electrons' are the $u_1 = 0.9c$ results from Figures 4 and 5. The spectrum labeled 'Pairs' is a self-consistently determined result where only electrons are present. In the pair plasma case, the shock structure is determined by the light component injected at $T_1 = 10^6$ and we find that $r \simeq 4.5$ and $\gamma_{\rm eff} \simeq 1.41$.

In Figure 6 we compare directly the results for $u_1 = 0.9c$. The electrons are suppressed by almost 13 orders of magnitude compared to protons above $\sim 10^8$ keV. The entire difference in normalization occurs at energies below $\sim m_p c^2$, above this energy, the shock treats electrons and protons, both of which are now fully relativistic, identically. Even though our model makes a number of approximations and is restricted to plane-parallel, steady-state shocks, the normalization difference comes from two fundamental effects; shock smoothing and the kinematics of the transition between upstream and downstream reference frames. We expect this prediction will survive more detailed models.

Our results clearly suggest that relativistic, parallel shocks cannot produce significant radio fluxes. Since radio fluxes are seen from objects like extragalactic radio jets, where the flow velocities are not well known, we predict that either the plasma flows are nonrelativistic or that the heavy proton component is dynamically unimportant compared to electron-positron pairs. If no heavy component is present, the shock structure will be determined by the e^--e^+ pairs and the electron (or positron) spectrum will be essentially the same as the proton spectrum. In Figure 6 we have included the spectrum produced by a shock where only pairs are present.

5. Conclusions

The important role played by particle reflection and acceleration in the dissipation processes of collisionless shocks, combined with the expectation that shocks are extremely efficient particle accelerators, implies that realistic solutions of Fermi particle acceleration must cover the entire energy range from thermal to relativistic energies self-consistently. We have developed Monte Carlo techniques that keep, we believe, enough of the essential plasma physics to give a sensible description of the viscous subshock, and are still computationally efficient enough to model acceleration to extremely high energies.

Our simulation also allows us to model test-particle electrons and treat relativistic flow speeds. We have found that electrons, even if injected at high enough energies where their scattering properties are expected to be essentially the same as protons, are dramatically less efficiently accelerated in relativistic shocks compared to protons. This effect is quite fundamental, coming from (a) the difference in kinematics between relativistic electrons and nonrelativistic protons, and (b) the fact that the heavy protons determine the shock structure, which is then unfavorable for electron acceleration. We expect this result will survive more detailed models. Our results imply that shock acceleration in objects, such as extra-galactic radio jets, which produce radio synchrotron emission (almost certainly coming from accelerated electrons), would be the result of nonrelativistic flows if protons are a dynamically significant component of the plasma. If the flows are relativistic, shock acceleration would only produce significant radio fluxes if the plasma was pair dominated.

We caution that our model does contain a number of simplifying assumptions, most importantly those of a parallel shock with a simple scattering law (equation (1)) for all particles. Diametrically opposite predictions for *perpendicular* shocks have been made by Hoshino et al. (1992) using plasma simulations. These authors find that electrons will only be significantly accelerated in perpendicular relativistic shocks if protons are present, a result which depends on the detailed wave-particle interactions determined with the plasma simulation. We do not yet understand if the differences between our results and those of Hoshino et al. depend on the differences in shock geometry or on the differences in scattering properties. In any event, phenomena such as radio jets, pulsar winds, and electron acceleration in SNRs, makes understanding electron acceleration in shocks an extremely important problem in astrophysics.

Acknowledgements The author wishes to thank S. Reynolds for helpful comments and A. Ford for his diligence in gathering the SNR data. Partial support for this work came from NASA grants NAG 5-1042, NAG 5-1131, and NAGW 2001 and NSF grant AST-88-17567. Much of the computing was performed on the North Carolina Supercomputing Center's Cray Y-MP.

REFERENCES

Blandford, R. D. and Eichler, D.: 1987, *Physics Reports* **154**, 1.

Burgess, D.: 1987a, in Proc. of the Int. Conf. on Collisionless Shocks, (Balatonfüred, Hungary), Omikk-Techoinform, p. 89.

Drury, L. O'C.: 1983, *Rep. Prog. Phys.* **46**, 973

Eichler, D. 1979, *Ap. J.*, **229**, 419.

Ellison, D.C., 1991a, in *Relativistic Hadrons in Cosmic Compact Objects*, eds. A.A. Zdziarski and M. Sikora, Springer-Verlag, p. 101.

Ellison, D.C., 1991b, in *Astrophysical Aspects of the Most Energetic Cosmic Rays*, eds. M. Nagano and F. Takahara, World Scientific, p. 281.

Ellison, D.C., and Jones, F.C., 1991, *Proc. 22nd Int. Cosmic Ray Conf. (Dublin)*, **2**, 320.

Ellison, D. C., Jones, F. C., and Eichler, D.: 1981, *Journal of Geophysics*, **50**, 110.

Ellison, D. C., Jones, F. C., and Reynolds, S. P. 1990, *Ap. J.*, **360**, 702.

Ellison, D. C., and Möbius, E. 1987, *Ap. J.*, **318**, 474.

Ellison, D. C., Möbius, E., and Paschmann, G. 1990, *Ap. J.*, **352**, 376.

Ellison, D.C., and Reynolds, S.P., 1991, *Ap. J.*, **382**, 242.

Giacalone, J., Burgess, D., Schwartz, S.J., and Ellison, D.C., 1992, *Geophys. Res. Lett.*, **19**, 433.

Goodrich, C. C. 1985, in *Collisionless Shocks in the Heliosphere: Reviews of Current Research*, AGU Monograph Vol. 35, p. 153, ed. by B.T. Tsurutani and R.G. Stone, AGU, Washington, D.C.

Hoshino, M., Arons, J., Galland, Y. A., and Langdon, A. B., 1992, *Ap. J.*, in press.

Jones, F. C., and Ellison, D. C., 1991, *Space Sci. Rev.*, **58**, 259.

Kirk, J.G.: 1988, Thesis for Dr. rer. nat. habil., Ludwig-Maximillians-Universität, München.

Kirk, J. G., and Schneider, P. 1987a, *Ap. J.*, **315**, 425.

Kirk, J. G., and Schneider, P. 1987b, *Ap. J.*, **322**, 256.

Leroy, M. M.: 1984, *Adv. Space Res.*, **4**, 231.

Quest, K. B. 1988, *J. Geophys. Res.*, **93**, 9649.

Reynolds, S.P., and Ford, A., 1992, in preparation.

Scholer, M., Trattner, K. J., and Kucharek, H., 1992, *Ap. J.*, in press.

Webb, G. M., 1985, *Ap. J.*, **296**, 319.

ION ACCELERATION AT COLLISIONLESS SHOCK INTERACTIONS

Peter J. Cargill
Department of Astronomy, University of Maryland, College Park, Maryland 20742

ABSTRACT

When two collisionless shocks collide, a population of high energy ions is produced. Using hybrid numerical simulations, it is shown that when the two shocks are quasi-perpendicular, ions with energies up to 15 E_0 (where E_0 is the initial kinetic energy of upstream ions in the shock frame) are produced by direct electric field acceleration. In the quasi-parallel regime, energies as high as 80 E_0 can be obtained. These ions are accelerated by both scattering off the approaching shocks and subsequently in the intense turbulence left behind by the shock collision.

INTRODUCTION

Theory and experimental results have shown that collisionless shocks are a very effective means of producing energetic particles. Although the details of the acceleration process differ between the quasi-parallel and quasi-perpendicular regimes, (where quasi-parallel [quasi-perpendicular] shocks propagate approximately parallel [perpendicular] to the ambient magnetic field), in each case a distinct population of high energy electrons and ions results. In the quasi-parallel case, diffusive (1st order Fermi) acceleration is widely believed to be the most effective process[1]. Very high energy ions (many MeV) can be readily produced by the systematic scattering of these ions off hydromagnetic turbulence on either side of the shock. The turbulence is in turn produced by low frequency plasma instabilities involving the accelerated ions[2]. At quasi-perpendicular shocks, a different process is believed to be operative. In this case, ions interacting with the shock drift along its surface, picking up energy from the shock electric field. In the absence of turbulence, this drift acceleration only gives significant energies when the shock is very close to perpendicular[3].

Numerical simulations can shed only a limited amount of light on shock acceleration, as can be clearly seen from a glance at the parallel shock. Since the energetic ions are responsible for the hydromagnetic turbulence, they must be treated consistently (as opposed to test particles). However, higher energy particles travel ever further away from the shock before being scattered, so that a very large simulation box is needed. In addition, the timestep must be decreased to accurately follow the orbit of these particles on the grid. It is clear that numerical simulations of particle acceleration to the energies seen in astrophysical plasmas are many years (or even decades) in the future. However, numerical simulations are most useful in the regime where analytic theories (such as diffusive acceleration) have the most problem, namely in understanding how particles are initially extracted from the thermal population. Leroy et al.[4], and Quest[2] showed how such a process operated at perpendicular and parallel shocks respectively and subsequent simulation work has made progress in the understanding of quasi-parallel shocks[5].

This paper examines another form of acceleration of ions from the seed population. However, instead of concentrating on acceleration at a single shock as other workers have done, we will examine how ion acceleration is modified by the presence of multiple shocks. It is clear that the presence of a second shock can modify ion acceleration significantly. When ions escape from one shock after being accelerated there, they can then interact with the other shock and undergo further acceleration. In other words, the introduction of another scattering center can make the acceleration process more effective. Such a process can be important in a number of astrophysical situations. The best known example is in the solar wind, where forward and reverse shocks generated by stream-stream interactions can collide beyond 1 A. U..[6] Other locations of shock interactions can be in astrophysical plasmas with multiple energy release sites. A recent model for solar flares invokes such a process[7]. In addition, such interactions can be important when solar wind shocks meet the bow shock of the earth and other planets.

RESULTS

Particle acceleration at the interaction of collisionless shocks has been investigated extensively in recent years[8-12] using hybrid numerical simulations (massless electrons, kinetic ions[13]). Rather than use most of this section to discuss these results in depth, we will simply provide a brief summary here, with the rest of the paper presenting some new results concerning the details of the acceleration process.

Acceleration at perpendicular shocks was investigated by Cargill and Goodrich[9]. They showed that when two supercritical high Mach number ($M_A = 8$: where M_A is the Alfven Mach number defined as V_{shock}/V_A) shocks collided, some of the thermal ions were accelerated to energies > 10 E_0, where E_0 is the kinetic energy of the unshocked plasma in the shock frame of reference. It was subsequently shown that this process was optimized when the two shocks has roughly equal strengths[12]. In the quasi-parallel regime, the acceleration process produced higher energy ions (up to 35 E_0) and was much more efficient[12], although the efficiency fell off as the shocks tended to exactly parallel. This can be attributed to the fact that hot ions can easily leak upstream from the quasi-parallel shock and interact over an extended period (many Ω_i^{-1}, where Ω_i is the ion cyclotron frequency) with the other shock before the shock fronts themselves collided. For perpendicular shocks, ions are strongly tied to the shock front, so that acceleration could only occur within a small fraction of Ω_i^{-1} when the shock fronts themselves collided. This work emphasized detailed energy spectra of the accelerated ions at the expense of examining how exactly the shock electric fields increased the ion energies. We now present some new diagnostics that make the acceleration process much more transparent.

The first hybrid simulation is of the collision between two $M_A = 8$ shocks. In the unshocked plasma, $\beta_e = \beta_i = 0.5$ and $\theta_{BN} = 75°$ (Here θ_{BN} is the angle between the shock normal and the upstream magnetic field). In order to have a reasonable initial state, the two shocks were run for several Ω_i^{-1} before being allowed to collide. The shocks were then put into a numerical box with length 80 c/ω_i (ω_i is the ion plasma frequency) and 400 cells. A timestep of 30 ω_i^{-1} was chosen. Initially there were 25,000 ions in the simulation, but this increases in time as more ions are injected through the boundaries. By the end of the simulation there are roughly 70,000 ions present.

Figure 1 shows (v_x, x) phase space before (1a) and after (1b) the shocks collide. In Fig 1a, the familiar features of quasi-perpendicular shocks are present, including ion reflection[4]. Fig. 1b shows that the shock collision has led to the production of a small population of energetic ions, which can be seen around x = 40 c/ω_i. The two shocks have been transmitted through each other and are at x = 32 and 48 c/ω_i. Figure 1c shows the energy spectrum at this time. There is an extended tail with energies up to 16 or 17 E_0, corresponding to the accelerated ions.

Figures 1d,e shows a revised version of some diagnostics presented in Cargill and Goodrich[9] which shed light on the details of the acceleration process. The ions

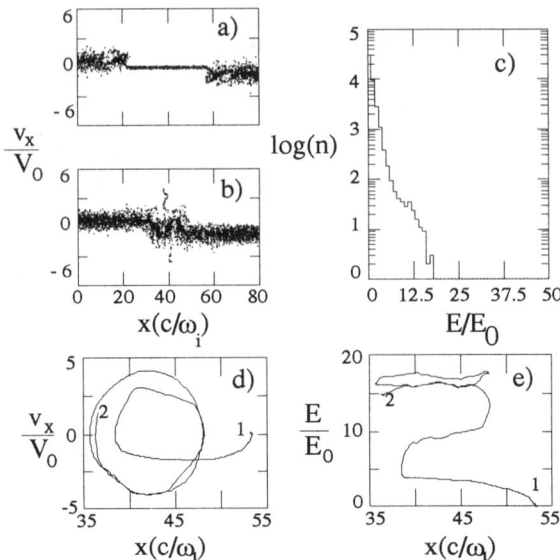

Figure 1 1a and 1b show (v_x, x) phase space at $\Omega_i t = 0$ (1a) and 3(1b). 1c shows the energy spectrum at $\Omega_i t = 3$. 1d and 1e show velocity (1d) and kinetic energy (1e) of a tracer particle as a function of x. Distance is normalized to the unshocked ion inertial length and energy to the kinetic energy of an unshocked ion in the shock frame.

that are most readily accelerated lie on the very tail of the distribution function of the unshocked plasma in Figure 1a. A population of the ions were labelled as tracers so that we could follow their evolution throughout the simulation. Figs 1d and 1e show the history of v_x and the kinetic energy of one of these ions throughout the simulation. In these plots, the point labelled "1 (2)" marks the position of the ion at the start (end) of the simulation. At point 1, this ion has just been reflected by the incoming (right) shock and so gyrates in front of that shock. Before the magnetic field can turn the ion around and send it back downstream of the shock, it encounters the other (left) shock, at around x = 39 c/ω_i. Its energy at that time is roughly $4E_0$. In the left shock, this ion is directly accelerated by a D.C. electric field, increasing its energy to 10 E_0, and is sent back toward the right hand shock. The ion encounters the right hand shock at x =45 and is subject to a further D.C. acceleration, taking the energy up to $16E_0$. Subsequent to this, the two shocks themselves collide and the energetic ion gains no further energy. It continues to orbit in the twice-shocked plasma left behind by the two shocks and, in the present case appears to have lost a small fraction of the energy.

This acceleration mechanism is reminiscent of both Fermi and drift processes. The energy gain itself is due to the D.C. field in each shock, similar to the case of drift acceleration, but the presence of the two shocks as scattering centers, which confine the ion to the acceleration region, is more reminiscent of Fermi acceleration. It turns out[9] that the only ions to be accelerated to high energies are those reflected by the incoming shocks just before they collide. The interaction time of ions from one shock with the other is thus very short (or order Ω_i^{-1}), so that the production of a large population of energetic ions by this process appears to be unlikely. One must enter the quasi-parallel regime to see really significant energies being produced.

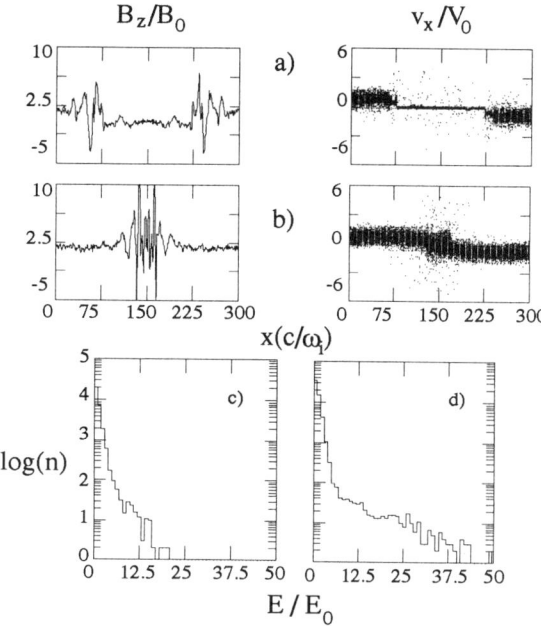

Figure 2 2a,b show B_z (left column) and (v_x, x) phase space (right column) at $\Omega_i t = 0$ (2a) and 24 (2b). The magnetic field is normalized with respect to the total field unshocked field strength. 1c,d show the energy spectra at $\Omega_i t = 0$ and 24 respectively.

Figures 2 and 3 show the results of the simulation of the collision between two $M_A = 8$ shocks with $\theta_{BN} = 30°$ and $\beta_{i0} = \beta_{e0} = 0.5$. The simulation box is now 300 c/ω_i long with 600 cells, the timestep is 70 ω_i^{-1} and there are 25,000 ions initially. As in the quasi-perpendicular case, the individual shocks are simulated for some time before they are allowed to collide: for this case they are run for 30 Ω_i^{-1} which allows a significant population of leaked upstream ions to develop.

Figures 2a and 2b show B_z (left column) and (v_x, x) phase space at $\Omega_i t = 0$ (2a) and 24 (2b). Figs. 2c and 2d show the energy spectra at $\Omega_i t = 0$ and 24 respectively. The initial conditions clearly show that a population of energetic ions exist upstream of each shock before they collide. The maximum energy of these ions is roughly 15 E_0 (Fig 2c). Unlike the quasi-perpendicular case, these ions are present over the entire 150 c/ω_i that separates the two shocks. In addition the shock magnetic fields show the strong wavelike structure expected in quasi-parallel shocks[14]. After the shocks collide, there is a population of very energetic ions located around the site of the shock collision, with energies as high as 70 E_0. (Fig 2d only shows energies up to 50 E_0, since these higher energy ions are very few in number and cannot be viewed as statistically meaningful.) The two shocks are transmitted through each other and are at x = 130 and 170 c/ω_i. The main difference with the quasi-perpendicular case is the increase of both the number and energy of the accelerated ions. This is attributable to two things. The leakage and/or scattering of hot ions from the pre-collision shocks leads to much higher energy ions interacting with the other shock. This makes any D.C. acceleration much more effective due to the stronger motional electric field that the ion will experience. Also, the presence of

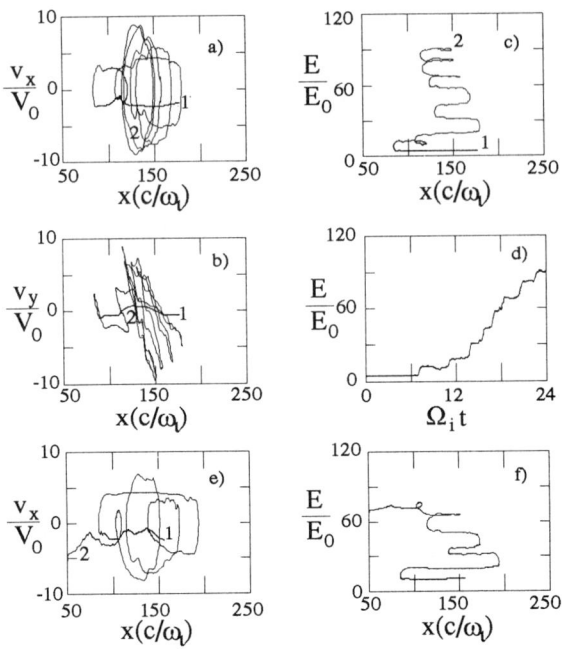

Figure 3 3a-d show the velocity as a function of x (3a,b) and the kinetic energy as a function of x (3c) and time (3d) for a chosen tracer particle. 3e,f show the x velocity and kinetic energy as a function of x for another tracer ion, chosen because it exits the simulation box before the simulation ends.

upstream ions means that the interaction between the two shocks now extends over many Ω_i^{-1} so that many more ions can be accelerated.

Figure 3 presents two sample ion orbits. Again tracer ions were selected with the knowledge that ions upstream of the right hand shock moving with a negative v_x were most likely to gain large energies. Figs. 3a-d show one ion which starts upstream of the right hand shock and eventually attains an energy of 90 E_0 (by far the most energetic ion in this simulation). Figs. 3e,f show another ion that gets an energy of 70 E_0 before escaping from the simulation box. Returning to Figs. 3a-d, the ion initially starts of with a small energy ($< 5E_0$) well upstream of the right hand shock. It moves to the left, encounters the left hand shock, and is accelerated to 6 E_0 by a process similar to that operating in the quasi-perpendicular case. The two shocks themselves collide at around $\Omega_i t = 10$, so that this ion does not per se interact with the shocks again. However, it is after the shocks collide that it gains most of its energy. This can be attributed to the presence of very large amplitude turbulence in the twice shocked plasma. Figure 3b shows that $\delta B/B \gg 1$ and Fig. 4c suggests that the ion is systematically scattered of these fluctuations.

The ion shown in Figures 3e,f is accelerated by a similar process to that discussed above. However, it has a different pitch angle from the other ion and its larger v_x sends it off toward $x = 0$. This ion is thus lost to the acceleration process with a smaller energy gain. Loss of ions is, in fact, not a particularly important feature for quasi-parallel shocks and only becomes really significant for θ_{BN} close to $0°$. This can be understood by noting that the local value of θ_{BN} in the twice shocked plasma is 68° here due to the compression of B_z in the collision[12]. The energetic ions are, in general, trapped on field lines lying in the twice shocked plasma.

DISCUSSION

The production of energetic ions at two colliding shocks has been examined. Energies up to 15 E_0 (> 50 E_0) arise for M_A = 8 quasi-perpendicular (parallel) shocks. The acceleration is dominated by D.C. electric fields for the perpendicular case and a combination of D.C. electric fields and strong MHD turbulence for the quasi-parallel case. A limitation of this process is that when two shocks are very far apart, the energetic ions streaming away from one shock will generate low frequency waves and be scattered back toward that shock[2,5,14] so that if the shock separation exceeds a certain distance, the two shocks will propagate as independent entities. If the growth time of the low frequency waves[15] is of order 50 Ω_i^{-1}, and assuming that the upstream ions move away from the shock at the shock speed, the shocks do not contaminate each other if they are separated by more than $50 M_A$ c/ω_i.

It is instructive to calculate some of the energies that can be attained in some real life scenarios. Only quasi-parallel shocks will be considered. In the solar wind, two shocks travelling at 800 km/s could give ions up to 200 KeV. In the solar corona, shocks travelling at a few times the Alfven speed could give energies up to 1 MeV and even stronger shocks found in astrophysical applications could give even higher energies. However, such energies are well short of the highest typically found in each of these examples, so no claim can be made that this mechanism can produce particles at such energies. This process can produce energetic ions that can serve as a seed population for further acceleration and should be viewed in that light.

ACKNOWLEDGMENTS

I am grateful to C. Goodrich and K. Papadopoulos for their input to the early stages of this work which was supported by NASA grant NAG5-1101. Computer support was provided by San Diego Supercomputer Center.

REFERENCES

1. M. A. Lee, J. Geophys. Res., **87**, 5063, (1982).
2. K. B. Quest, J. Geophys. Res., **93**, 9649, (1988).
3. R. B. Decker and L. Vlahos, Astrophys. J., **306**, 710, (1986).
4. M. M. Leroy, D. Winske, C. C. Goodrich, C. S. Wu and K. Papadopoulos, J. Geophys. Res., **87**, 5081, (1982).
5. M. Scholer and T. Terasawa, Geophys. Res. Lett., **17**, 119, (1990).
6. Y. C. Whang and L. Burlaga, J. Geophys. Res., **90**, 221, (1985).
7. A. Anastasiadis and L. Vlahos, Astron. Astrophys., **245**, 271, (1991).
8. P. J. Cargill, C. C. Goodrich and K. Papadopoulos, Phys. Rev. Lett., **56**, 1988, (1986).
9. P. J. Cargill and C. C. Goodrich, Phys. Fluids, **30**, 2504, (1987).
10. P. J. Cargill, Phys. Fluids, **B2**, 2294, (1990a).
11. P. J. Cargill, J. Geophys. Res., **95**, 20731, (1990b).
12. P. J. Cargill, Astrophys.J., **376**, 771, (1991).
13. D. Winske, Space Sci. Revs., **42**, 53, (1985).
14. D. Burgess, Geophys. Res. Lett., **16**, 345, (1989).
15. D. Winske and M. M. Leroy, J. Geophys. Res., **89**, 2673, (1984).

ACCELERATION OF DIFFUSE IONS UPSTREAM FROM PARALLEL SHOCKS

Kevin B. Quest
University of California, San Diego, Ca. 92093

ABSTRACT

We review the preliminary results of diffuse ion generation at parallel shocks using a combination of hybrid simulation and particle splitting techniques. Our basic motivation is to obtain improved statistics relative to earlier studies, and to examine in detail the spatial and velocity space distribution of the diffuse ion population. We find that for a parallel shock with Alfvén Mach number of 5 and plasma β of 2, approximately 2% of the incoming ion flux is reflected and scattered back upstream. This diffuse population has a shallow density maximum at a velocity corresponding to specular reflection and has a distribution that is approximately exponential in energy. Further, the velocity space distribution of these ions is limited to shock normal velocities less than the speed of the plasma core. We interpret these results as being due to the finite length of the simulation system, and the low level of upstream waves excited by the backstreaming ions. Implications for particle acceleration and cosmic ray generation are discussed.

INTRODUCTION

It is now generally accepted that a plausible model for the generation of cosmic rays is first order Fermi acceleration at collisionless shocks. Given an initial seed population of moderately energetic ions, a convecting, low frequency electromagnetic wave spectrum is generated that pitch angle scatters the ions in the wave frame. In the shock frame, the particles experience a net gain in energy as they reflect back and forth between the converging scattering centers. The resulting energy spectrum is relatively independent of the details of the scattering process, and if no particles are lost upstream from the shock, takes the form of a power law[1,2,3].

A central question concerning shock acceleration has been the nature and origin of the seed population. It was argued by Eichler[4] that the Fermi process could be used to generate the seed population from the thermal population as well as accelerate energetic ions. The spectrum of waves should accelerate some ions out of the thermal background, eliminating the need for a separate mechanism of creation. This concept was tested via Monte-Carlo simulation[5], where a simple energy dependent scattering law obtained energetic distributions and whose spectral forms agreed with available particle data[6]. Quest[7,8] performed a series of particle-in-cell simulations in which the electrons were taken to be a charge-neutralizing fluid, and concluded that collisionless parallel shocks require the creation of a steady stream of reflected/scattered ions to generate waves that in turn are essential for the process of shock dissipation. He also showed via particle orbit tracing that a few of these ions were accelerated to high energies via a Fermi-like acceleration process confined to within 20 to $30 c/\omega_{pi}$ of the shock. This analysis was extended to quasi-parallel shocks[9,10] where it was found that the majority of the energetic ions seen in the simulation are accelerated via a similar Fermi-like mechanism. This point is contested by Lyu and Kan[11], who concluded in a separate study that the diffuse ions are provided by diffusion from the region downstream from the quasi-parallel shock.

A limitation of previous simulation work has been the particle statistics for the scattered-reflected component. This arises because the number density of this population is so small, on the order of a few percent of the background ions. In order for the energetic population to be adequately resolved, several thousand macro-particles per spatial cell are required. One way to increase the statistics is to split the macro-particles, as is in the Monte-Carlo simulations. When a macro-particle reaches a certain energy, or is otherwise determined to be part of the diffuse population, the ion is split into several new particles whose sum equals the original particle weight. These split particles are then given a slightly different velocity or position relative to each other. This latter step is necessary, as otherwise all the particles would follow identical orbits. Thus, energy and/or momentum cannot be exactly conserved, although in practice the deviation is small. In the simulation results that follow, particles identified as diffuse ions are split in two, and then a random speed of 0.2 times the upstream thermal speed is added/subtracted. The particles are split again at later times until they reach 1/64 of their original weight. Thus, on average the particles will be artificially scattered by about a thermal speed, which is small compared to the ram velocity. More details on this method will be provided in a forthcoming publication.

The following figures are from a one-dimensional hybrid simulation of a parallel shock. The ions are injected at the left-hand boundary, and are specularly reflected at the right hand wall. More details on the methods of quasi-parallel shock simulation can be found in a number of review articles[12,13]. All the figures are taken at late time ($\Omega_{ci} t = 185$, where Ω_{ci} is the far upstream ion cyclotron frequency, and t is the time). In Figure 1 the number density normalized to its far upstream value, and the magnetic field B_y (solid line) and B_z (dotted line) normalized to B_x are displayed as a function of position (C/ω_{pi} is far upstream ion inertial length). Only a portion of the simulation box is shown for the magnetic field in order to enhance the scale of the shock associated waves. Lowpass filtering has been used on the density and field profiles to eliminate short wavelength oscillations. We define the diffuse population as consisting of ions separated by more than one shock speed relative to the incoming ion core. A small number (relative to the core) are just upstream from the shock (dashed line), decreasing slowly in number in the negative x direction. A more dramatic drop is seen as we cross from just upstream to just downstream from the shock. The magnetic field profile shows the associated waves. Note that only one or two large amplitudes oscillations are seen (as opposed to the trailing wavetrain observed in earlier work). This is likely a consequence of setting the resistive diffusion length in the electron momentum equation to a non-zero value ($0.05c/\omega_{pi}$ in this simulation).

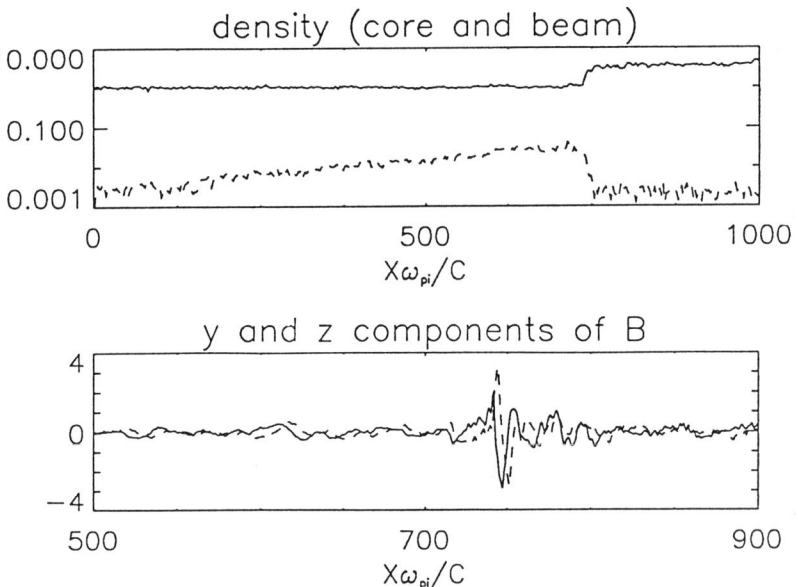

Fig 1. Total ion density (solid line) and diffuse ion density (dashed line) is shown in top panel. Transverse magnetic field components B_y (solid line) and B_z (dashed line) is shown in bottom panel. Both panels are plotted as a function of position $X\omega_{pi}/C$.

In Figure 2 we show a 3-dimensional wire diagram of the number density (on a decimal logarithmic scale, arbitrary units) as a function of position and of the x component of velocity. The velocity is normalized to the shock speed V_s, and $V_x = 0$ corresponds to the shock stationary frame. The incoming ion core is the narrow high wall that shifts and broadens in velocity at the position of the shock. The diffuse ion population is also clearly visible as a low density component with a broad spectrum in velocity. A striking feature of this diagram is the relative insensitivity to the spatial slope for a given velocity (where it finally does change rapidly owes more to the lack of statistics at those energies as opposed to a physical cause). If the upstream population reaches steady state, one expects an increasing decay length with increasing energy owing to the gyro-resonance condition. Note also that there is a significant number of ions escaping out the left hand end of the box, an indication that the scattering mean free paths and the simulation system size are comparable in size.

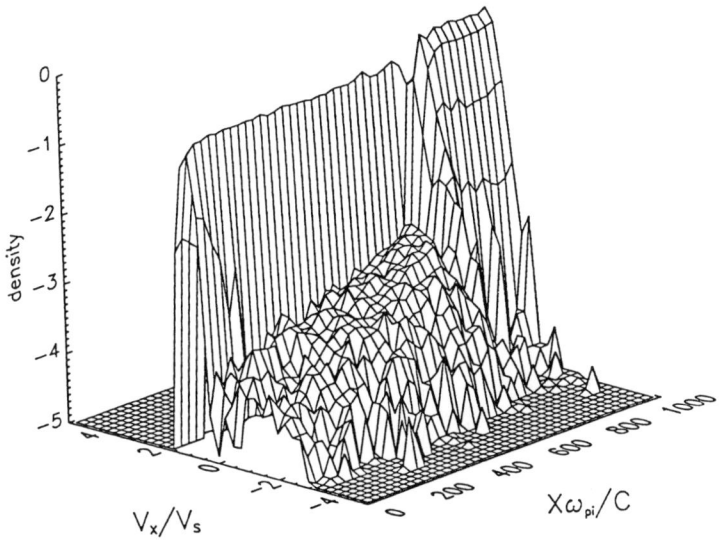

Fig 2. Perspective plot of the logarithm of the number density as a function of V_x/V_s and $X\omega_{pi}/C$. Numerals on vertical axis refer to decades.

In Figure 3 we display the number of diffuse ions found in a given energy band in a shock stationary reference frame (arbitrary units). The energy is normalized to the shock ram energy $E_R \equiv M_i V_s^2/2$. The resulting spectrum is seen to approximate an exponential at sufficiently high

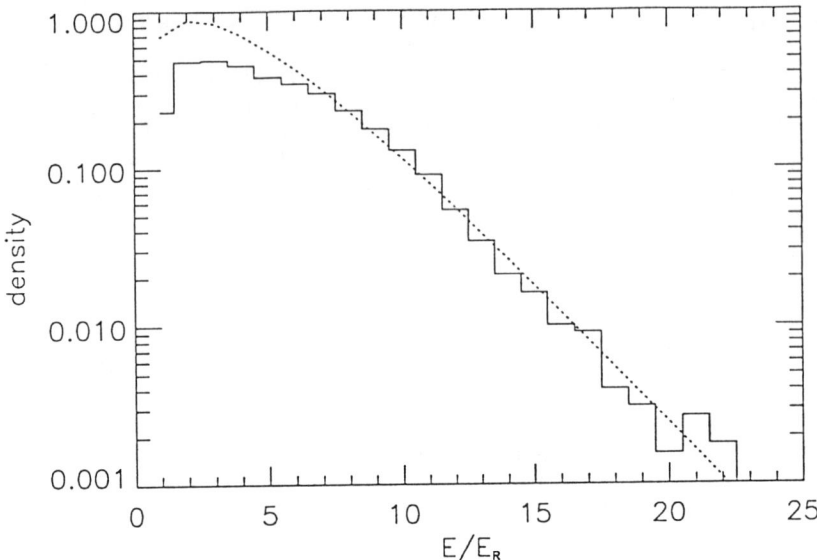

Fig 3. Density as a function of energy for the diffuse ions upstream from the shock.

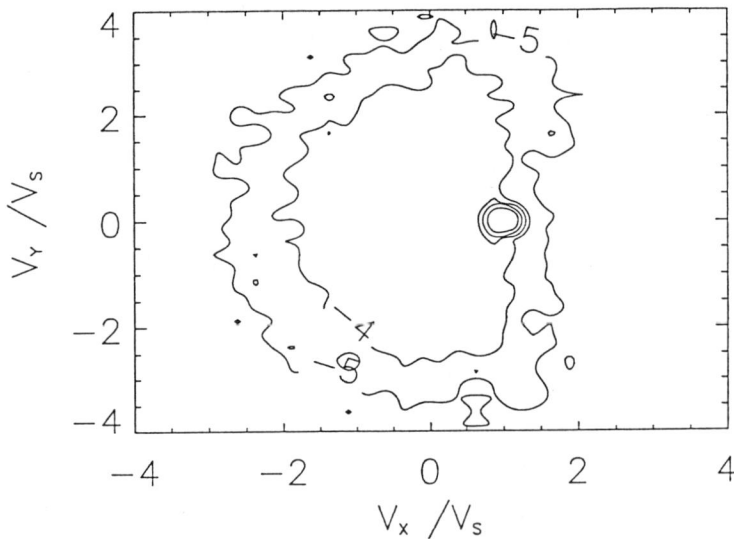

Fig 4. Iso-density contours of upstream diffuse ions

energies (dashed line) of the form $E\exp(-E/E_o)$, where E_o is $2.2E_R$ for the present shock.

Finally, in Figure 4 we display the iso-density contours of the V_x, V_y phase space for all ions upstream from the shock. The contours are logarithmic, and represent a one decade change from one contour to the next. The majority of the ions are focused near the ram speed. The diffuse ion population forms a nearly symmetric distribution for V_x velocities roughly less than the core, but is essentially zero for larger V_x values.

Conclusions

First, ions are being accelerated by the upstream waves as it is difficult to explain the effective temperature of the diffuse ions (over twice that of the downstream ions) in terms of simple leakage. What is unclear is whether the dominant method of energization is from small angle pitch angle scattering with respect to the upstream waves, or by the Fermi-like mechanism identified in earlier simulation studies. The exponential density profile as a function of energy is consistent with particle energization models assuming free escape, but may also be influenced by the form of the injected flux near the shock. Finally, the near isotropy of the particle distribution function for $V_x < V_s$ is strongly suggestive that there is limited pitch angle scattering past $90°$, consistent with the predictions of standard quasi-linear theory[14]. The overall results yielded by particle splitting are encouraging and the method will doubtless be used extensively in future studies of particle acceleration.

References

1. W. I. Axford, E. Lear, and G. Skadron, Proc. 15th Int. Cosmic Ray Conf. (Plovdiv 1977), 11, 132.
2. A. R. Bell, M.N.R.A.S. 182 (1978) 147.
3. R. D. Blandford and J. P. Ostriker, Astrophys. J. (Letters) 221 (1978) L29.
4. D. Eichler, Astrophys. J. 229 (1979) 419.
5. D. C. Ellison, F. C. Jones, and D. Eichler, Journal of Geophysics 50 (1981) 110.
6. D. C. Ellison and D. Eichler, Astrophys. J. 286 (1984) 691.
7. K. B. Quest, Proc. Sixth Int. Solar Wind Conf., II (NCAR, Boulder, 1987), p. 503.
8. K. B. Quest, J. Geophys. Res. 93 (1988) 9649.
9. M. Scholer, Geophys. Res. Lett. 17 (1990) 1821.
10. H. Kucharek and M. Scholer, J. Geophys. Res 96 (1991) 21195.
11. L. H. Lyu and J. R. Kan, Geophys. Res. Lett. 17 (1990) 1041.
12. D. Burgess, Proc. of the Int. Conf. on Collisionless Shocks (ed. K. Szegö, Central Institute ofr Physics of Hungarian Academy of Sciences, Budapest, 1987), p. 89.
13. K. B. Quest, Collisionless Shocks in the Heliosphere: Reviews of Current Research (AGU, Washington DC, 1985), p. 185.
14. F. Skillings, Mon. Not. R. Astr. Soc. 172 (1975) 557.

SIMULATIONS OF COLLISIONLESS SHOCKS: SOME IMPLICATIONS FOR PARTICLE ACCELERATION

D. Burgess

Astronomy Unit, Queen Mary and Westfield College, London E1 4NS, UK

ABSTRACT

The role of self-consistent plasma simulations is discussed with reference to collisionless shock structure and the extraction of thermal particles to supra-thermal energies. Examples are given from quasi-perpendicular and parallel shock geometries. The cyclic reformation behaviour of the quasi-parallel shock, as revealed by simulations, is detailed, and some implications given. Finally, some recent advances are described in the techniques of simulation of strong particle acceleration.

SIMULATIONS AND COLLISIONLESS SHOCKS

Understanding the operation of collisionless shocks is one of the most interesting and challenging problems of astrophysical plasma physics. The definition of the problem, namely the fact of maintaining a shock without collisions, entails a complex web of questions and implications. Of course the fact that shocks can be sites of particle acceleration is one consequence, but the central problem is always determining how there can be dissipation (i.e, thermalization) without collisions. In trying to solve this riddle the structure of the shock must be invoked so that anomalous processes may operate, whether they be instabilities or dynamical effects. The situation is complicated by the influence of a large number of parameters, such as the plasma beta, Mach number, and the angle between the upstream magnetic field and the shock normal. This angle, θ_{Bn}, is fundamental in controlling the shock structure, to such an extent that the major division of collisionless shocks is between quasi-parallel and quasi-perpendicular (with the dividing line usually chosen at $\theta_{Bn} = 45°$).

Plasma simulations have proved an invaluable tool for helping to understand collisionless shocks. The best example of their utility has been in the study of the high Mach number (i.e, super-critical) quasi-perpendicular shock. Satellite observations had revealed the importance of ion reflection at these shocks, but simulations were able to explain the connection between ion reflection and the structure of the shock itself (this whole topic is covered by several reviews[1,2]). Plasma simulations are of several different types, but the simulations used in the study of the quasi-perpendicular shocks were of the so-called "hybrid" method. Several excellent papers exist describing this model[3], but, briefly, the hybrid model treats ions kinetically using simulation macro-particles, and the electron response is based only on a fluid description. The advantage of this method is that long time and large length scales can be explored, but of course at the expense of an accurate electron response. The hybrid model has had its greatest successes with high Mach number phenomena, which are prevalent in astrophysical plasmas, and for which the hybrid method is well suited because the ions dominate the energy and momentum balance. One obvious problem with applying the hybrid simulation method to shock acceleration is that it might be useful to determine shock structure and acceleration of ions, but it fails to tell us anything about electron acceleration.

In this paper I am going to give an account of the use of hybrid simulations in modelling the acceleration of thermal ions, and I will finish by highlighting the problems of simulating particle acceleration to high energies, and the steps recently made towards that goal.

SIMULATING INJECTION

As I have pointed out above, the main use of simulations is to study thermalization, which means describing the behaviour of thermal particles. An understandable consequence is that the main application of self-consistent simulations to particle acceleration has been to model the extraction of thermal particles into what might be called supra-thermal energies. This is the so-called "injection problem," or problem of the formation of a "seed" population of energetic particles. It is worth pointing out that the terms "seed" and "problem" are the result of theories of particle acceleration which hit the headlines first. Theories are developed to quantify mechanisms, but a plasma's behaviour only depends on mechanisms, and a numerical simulation is similar, albeit within its mundane restrictions. Eventually one might expect a simulation of particle acceleration which replicates the observations. The job of the simulationist is then to describe what is happening within the simulation, which might, or might not, be in terms of prior theories.

Shock Drift Acceleration at Low Energies

The best observations of suprathermal ions are from the Earth's foreshock. *In situ* observations have enabled a detailed taxonomy of the various suprathermal populations, and their associated waves[4]. In particular, the field aligned beams (FABs) have energies between two and fives times the incident plasma energy. They are collimated along the magnetic field flowing away from the shock, with fractional densities of about 1%. They are predominantly observed coming from the bow shock where $40° < \theta_{Bn} < 75°$.

Such properties would tend to indicate that a suitable mechanism would be shock drift acceleration (SDA). But usually SDA is studied in the limit of initial energies high enough to treat the shock as a discontinuity[5]. Such an assumption can not be justified *a priori* when the initial energy of the particles is expected to be roughly thermal. This problem was approached using the hybrid simulation code[6,7]. Initial simulations[8] indicated that FAB ions could be produced at shocks with $\theta_{Bn} \sim 45°$, but not above, contrary to the observations. However, this result was limited by the statistics of the simulation, which only revealed the densest beams. By first examining the usual SDA situation[6] (i.e., arbitrary initial distribution) the efficacy of SDA at low initial energies was demonstrated. It then had to be shown that beam densities similar to those observed could be reproduced when the initial distribution was a Maxwellian. For this purpose it was necessary to improve the particle statistics in the simulation. Usually about 30-50 simulation particles are used in each grid cell, and this makes it difficult to detect with certainty subpopulations with densities at and below the 1% level. One way to improve the statistics is to crudely increase the total number of particles. However in the case of FABs this would be extremely wasteful, since it was shown that the ions which ended up in the FABs started in the wings of the initial distribution. So, in order to study the production of FABs extra simulation particles were used *only* in the wings of the initial distribution. It was found[7] that the FAB density decreased with increasing θ_{Bn}, and that at the same time the ions which were turned back upstream came increasingly far from the centre of the thermal distribution. The orbits of the FAB ions at the shock revealed that it was necessary to properly account for the structure of the shock, because all the ions which became backstreaming were, on first encounter with the shock, specularly reflected, i.e, reversed their normal component of velocity within the shock layer. At the quasi-perpendicular shock it is the specularly reflected ions which end up downstream which are responsible for most of the shock thermalization, and, indeed, the structure of the shock. The close relationship between FAB production and shock thermalization is emphasized by the results of a simulation study which showed that, for all quasi-perpendicular shocks, the gyrating reflected ions did not come from the core of the thermal distribution[9].

Most of the properties of FABs can be reproduced with the hybrid simulation. Further support for the operation of SDA from thermal energies has been given by simulations which included an alpha particle thermal distribution[10]. It was found that alpha particles could be found in FABs, but their fractional density was much lower than for the protons, i.e, the beams were deficient in alpha particles when compared with the solar wind. This result, important because FABs are often invoked as a seed population for other acceleration mechanisms, is in substantial agreement with observations[11].

One implication of the study of FAB production is that, because energetic particles are only a minor contribution to the number density, it is generally not possible to simply take a standard plasma simulation and expect to observe energetic particles, unless special steps are taken to improve the inherently bad particle statistics (given present computational constraints).

Quasi-Parallel Shocks and Cyclic Reformation

To the wider community of those interested in particle acceleration it has been (historically, if not logically) the parallel and quasi-parallel shock which has been the focus of attention, simply because a first-order Fermi acceleration mechanism is more feasible, and such a mechanism results in a power law spectrum in energy for the accelerated particles. The simulations of Quest[12] were of major importance for the understanding of the *exactly* parallel shock. Quest demonstrated that it was possible to extract some particles from the thermal population, and that their trajectories consisted of multiple traversals of the shock surface, as in the usual first-order Fermi model. However, there remained the suspicion that the exactly parallel shock appeared as some singular case, mainly since the simulation results didn't look like the observations of the Earth's bow shock.

Some progress had been made towards understanding the quasi-parallel shock when observations of upstream gyrating ions[13] appeared consistent with partial specular reflection of the incident solar wind. Such a picture was reinforced by simulations[8], motivated by the observations, which showed reflected-gyrating ions in front of a quasi-parallel shock. These simulations showed that the reflected-gyrating ions had an appreciable fractional density, several per cent of the incident density. An opportunity was missed, in retrospect, because it now seems obvious that such an appreciable beam should have some effect on the region immediately upstream of the shock. One factor may have been that the simulations were only carried out longer enough to see the reflection of the ions, but not their coupling to the upstream flow. It was only after the simulations by Quest of the exactly parallel shock, that simulations, comparable in scale, were attempted for the quasi-parallel shock. These simulations[14], confirmed by later work[15,16], revealed a new type of thermalization mechanism: cyclic reformation.

The simulations revealed a shock structure which alternates between a sharp transition in the field (and correspondingly the thermalization) and a longer, more gradual transition. This alternation repeats itself, and is linked to the convection of upstream waves (generated by foreshock ions) into the shock layer. Furthermore, periodic bursts of reflected ions (with densities of up to 20% the incident density) are observed within the shock structure. These reflected ions are specularly reflected from the incident distribution, and correspond to the upstream reflected-gyrating ions of earlier simulations. But in the current simulations, the spatial domain is large enough and the simulation followed long enough for a "realistic" foreshock to develop, and it is the interaction of the foreshock waves, as they are swept into the shock, with the reflected ions, which forms one part of the the reformation cycle mechanism.

One way to look at the cyclic reformation mechanism is that the presence of reflected-gyrating ions immediately upstream of an abrupt field transition (i.e., the nominal shock position) effectively means a large kinetic temperature "upstream." But the whole point of a shock is that the thermalization is to achieve a *downstream* state. And one way to move into the downstream region all the effective thermalization due the reflected ions is to simply move the shock. So. in effect, the upstream reflected ions, because they are at such large densities, constitute a pressure pulse immediately upstream of the nominal shock, which is strong enough to relaunch the shock from the new position.

The quasi-parallel reformation cycle is illustrated in Figure 1. Here I plot the total magnetic field in the form of a grey scale plot of B at successive times in the simulation against position in the (one dimensional) simulational domain. This representation shows the time evolution of the simulation across the whole simulation domain. the shock has $\theta_{Bn} = 30°$, and $M_A = 6.5$. The shock is launched of the rigid right hand boundary of the simulation box, and so the shock propagates leftwards, but the plot has been sheared by the appropriate amount so the average shock position is fixed. This representation graphically illustrates the behaviour of the shock. one can identify a major reformation cycle between $T = 42\Omega_{ci}^{-1}$ and $T = 54\Omega_{ci}^{-1}$, where the initial "shock" (i.e., abrupt transition) moves backwards, and is replaced by another "shock" front ahead of the nominal shock position. This representation also reveals that there are "mini" reformation cycles (e.g., at $T = 62$), which previous "stack plot" representations fail to show. An important conclusion from this figure is that there is no one unique shock surface. This means that although the shock front sometimes has a fairly short scale length, in fact the scale of the shock transition is really that associated with the reformation mechanism.

The implications of the cyclic reformation mechanism for particle acceleration are, firstly, that, because the downstream state is a mixture of plasma with different thermalization histories, it is not possible to model the downstream state by a simple Maxwellian for the purpose of determining the density of ions leaking into the upstream region, as in the model of Edmiston et al.[17] Secondly, it has been possible with the hybrid simulations to study the extraction of ions from the thermal population in detail and to show that their distribution upstream is diffuse-like[18,22]. The question of alpha particle injection has also been addressed[18]

Naturally enough, such a model would not be believed without the support of observations. Observations at the earth's bow shock are extremely detailed, but pose great problems of interpretation because they are only point measurements, so it is difficult to unambiguously identify the overall behaviour of the shock, especially in the case of the quasi-parallel shock. Nevertheless, much work has been done,[19–21] and the observations lend support for the model of a time-varying shock structure. I should note that I have been talking here exclusively about the high Mach number quasi-parallel shock. Low Mach number shock simulations show a different behaviour.

SIMULATING SHOCK ACCELERATION

As described above, self-consistent plasma simulations have been, until recently, only used to study the first step of acceleration out of the thermal distribution. But with the increasing power of computers, and the interesting physics to investigate, it is now possible to begin simulations of particle acceleration, starting with thermal energies and following the entire process of particle acceleration. There are some basic problems with trying to simulate particle acceleration. Firstly, as we have mentioned already, the particle spectrum falls away with energy, so that in a standard simulation there would be hardly any particles with very high energy. Recently, work has been presented[22,23] which uses the method of simulation particle "splitting." In this method when a particle crosses a given energy threshold (of which there are several at increasing energies) it is split into two simulation particles, each with half the simulation weight (to obey conservation laws), but separated one from another by a small amount in velocity space. This has the effect, by introducing new simulation particles, of increasing particle statistics at higher energies. The algorithm is efficient because extra particles are introduced only when acceleration occurs; there is no need to guess in advance which parts of the thermal distribution will eventually contribute to the accelerated particle distribution.

The work of Giacalone et al.[22] uses an additional technique to allow a full simulation of particle acceleration. Usually in plasma shock simulations the upstream boundary is "quiet." However in the case of parallel and quasi-parallel shocks we know that an extensive foreshock eventually develops. But, in a self-consistent simulation the foreshock may take a considerable amount of time to develop, and until the foreshock is mature, the particle acceleration cannot be considered properly developed. So, in order to hasten the formation of the foreshock, Giacalone et al. use an upstream source of seed turbulence, on which the self-consistent turbulence, driven by the energetic upstream particles, can grow.

Using these techniques to specially adapt the one dimensional hybrid code to study particle acceleration, we have examined ion acceleration at the parallel shock. We find[22,23] an enhanced high energy tail in the upstream particle distribution extending to over one hundred times the plasma flow energy, and also a prominent shoulder in the downstream distribution function, which has a slope similar to that predicted by the standard theory of first-order Fermi acceleration. It should be pointed out that the details of the acceleration are not just as would be expected from the usual theories of first-order Fermi acceleration. Because of the reformation structure of the shock there are times when θ_{Bn} is locally far from its nominal value of zero, and this leads to drift acceleration at the shock front.

Much work remains to be done: the limitations of a one dimensional simulation have to be explored, as well as the details of individual trajectory types. In terms of other related problems, the question of acceleration of electrons to high energies is of great importance, and it will only be a question of time before simulation methods are used, and hopefully found successful.

Acknowledgements. The author would like to thank J. Giacalone, S. J. Schwartz, D. C. Ellison and M. Scholer for useful discussions. This work was supported by SERC grant GR/H09454 and an SERC Advanced Fellowship. Computations were carried out at the Atlas Centre (RAL, UK) and the North Carolina Supercomputing Center.

REFERENCES

1. Goodrich, C. C., Collisionless Shocks in the Heliosphere: Reviews of Current Research, *Geophys. Monogr. Ser.*, vol. 35, (AGU, Washington, D. C., 1985) p. 153.
2. Burgess, D., Proceedings of the International Conference on Collisionless Shocks, (Central Research Institute for Physics of Hungarian Academy of Sciences, Budapest, 1987) p. 89.
3. Winske, D., *Space Sci. Rev.*, *42*, 53, 1985.
4. Thomsen, M. F., Collisionless Shocks in the Heliosphere: Reviews of Current Research, *Geophys. Monogr. Ser.*, vol. 35, (AGU, Washington, D. C., 1985) p. 253.
5. Decker, R. B., *J. Geophys. Res.*, *88*, 9959–9973, 1983.
6. Burgess, D., *J. Geophys. Res.*, *92*, 1119–1130, 1987.
7. Burgess, D., *Ann. Geophys.*, *5*, 133–145, 1987.
8. Leroy, M. M., and D. Winske, *Ann. Geophys.*, *1*, 527–536, 1983.
9. D. Burgess, W. Wilkinson, and S. J. Schwartz, *J. Geophys. Res.*, *94*, 8783–8792, 1989.
10. D. Burgess, *Geophys. Res. Lett.*, *16*, 163–166, 1989.
11. Ipavich et al., *Geophys. Res. Lett.*, *15*, 1153, 1988.
12. Quest, K. B., *J. Geophys. Res.*, *93*, 9649, 1988.
13. Gosling, J. T., et al., *Geophys. Res. Lett.*, *9*, 1333–1336, 1982.
14. D. Burgess, *Geophys. Res. Lett.*, *16*, 345–348, 1989.
15. Scholer, M., and T. Terasawa, *Geophys. Res. Lett.*, *17*, 119, 1990.
16. Thomas, V. A., D. Winske, and N. Omidi, *J. Geophys. Res.*, *95*, 18809, 1990.
17. Edmiston, J. P., C. F. Kennel, and D. Eichler, *Geophys. Res. Lett.*, *9*, 531–534, 1982.
18. Kucharek, H., and M. Scholer, *J. Geophys. Res.*, *96*, 21,195, 1991.
19. Thomsen, M. F., et al., *J. Geophys. Res.*, *95*, 957, 1990.
20. Onsager, T. G., et al., *J. Geophys. Res.*, *95*, 1261, 1990.

21. Schwartz, S. J., D. Burgess, W. P. Wilkinson, R. L. Kessel, M. Dunlop, and H. Lühr, *J. Geophys. Res.*, in press, 1992.
22. Giacalone, J., D. Burgess, S. J. Schwartz and D. C. Ellison, *Geophys. Res. Lett.*, in press, 1992.
23. Giacalone, J., et al., this volume.

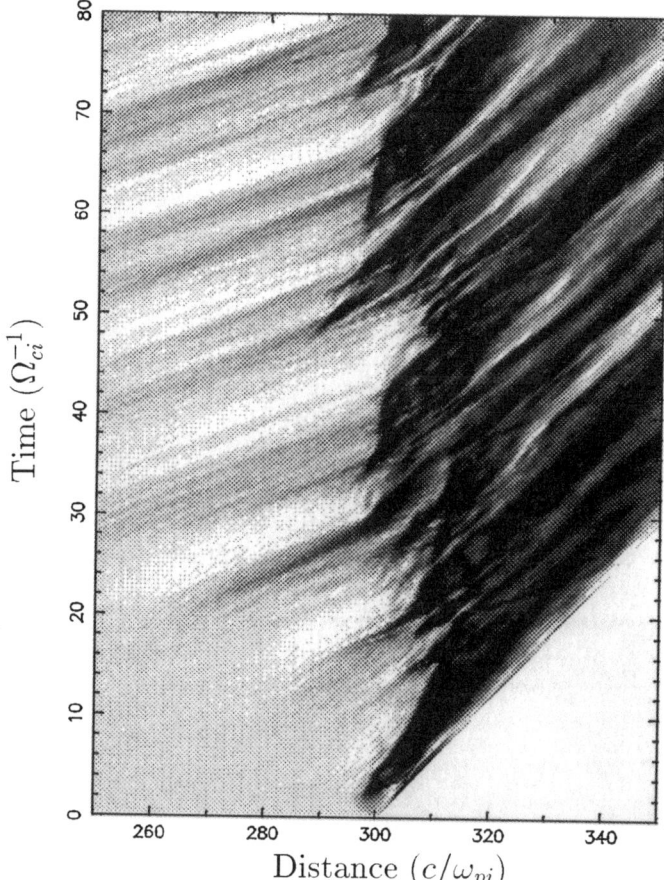

Fig. 1. A representation of the time evolution of a high Mach number ($M_A = 6.5$) quasi-parallel ($\theta_{Bn} = 30°$) shock. Plotted is a grey scale of total magnetic field, with white at $0.9B_1$ and black at $3.5B_1$, where B_1 is the nominal upstream value. The plot is in a frame where the shock is, on average, stationary at $x = 300$. Distance is in units of c/ω_{pi}, i.e., the nominal upstream ion inertial length, and only a portion of the simulation domain is shown. Time is plotted in units of Ω_{ci}^{-1}, where Ω_{ci} is the nominal upstream ion cyclotron frequency. The upstream is to the left, and the diagonal striations represent upstream waves convected into the shock by the plasma flow. Each reformation cycle consists of a period of sharp transition in B (e.g., $T \sim 36 - 42$), which then moves downstream as a region of low upstream B (i.e., local θ_{Bn} less than nominal) convects in ($T \sim 42 - 48$). Then a new sharp rise in B forms at, or upstream of the nominal shock position, replacing the previous shock front ($T \sim 48 - 50$). There is an initial "start up" period until about $T = 25$, associated with the formation of the shock, before the upstream waves are well developed.

SIMULATIONS OF ION ACCELERATION AT PARALLEL SHOCKS

J. Giacalone, D. Burgess, S. J. Schwartz
Astronomy Unit, Queen Mary and Westfield College, London E1 4NS, UK
D. C. Ellison
Physics Department, North Carolina State University, Raleigh, NC 27695, USA

ABSTRACT

Using very large scale hybrid plasma simulations of parallel shocks, we have studied the acceleration of ions to high energies. The energetic particle component of the (self–consistent) distribution function is effectively modelled by utilizing the technique of particle splitting, which has been used successfully in Monte Carlo simulations, and also by introducing a source of upstream turbulence which acts to efficiently isotropize the ions backstreaming from the shock.

INTRODUCTION

Until recently, self–consistent plasma simulations have received little attention in regards to particle acceleration at shocks. The reason for this is that simulations of this type have been used almost exclusively for studying the detailed structure of the collisionless shock wave which has a scale length that is usually neglected in models of ion acceleration. In the past few years, several authors have reported results from hybrid plasma simulations in which a small fraction of the incident plasma is energized to moderate energies (typically up to 50 keV)[1-4]. In the present study, we continue along these same lines, however, we introduce two techniques to aid in the development of the energetic particle component of the distribution function. These methods allow us to obtain a stastically valid energy spectra covering a wide range in energy and reduces the limitations imposed by the time–consuming, self-consistent treatment of the plasma. Further discussion on the application of the hybrid simulation model to particle acceleration studies can be found in Giacalone, *et al.*[5]

SIMULATION MODEL

We use a 1–D hybrid simulation with massless fluid electrons and kinetic ions. This approach is discussed in greater detail elsewhere in this volume [6-10]. In our simulations, a parallel shock is launched by a rigid piston and propagates from right to left with an average shock Mach number 6.4. Other simulation parameters are: ion and electron beta of 0.5, time step $0.02\ \Omega_i^{-1}$, grid cell size $0.5\ c/\omega_i$ and $\omega_i/\Omega_i = 5000$. Since we wish to investigate ion acceleration to high energies we use a large simulation box, $2000\ c/\omega_i$, which is approximately the gyroradius of a 50 MeV proton.

We superimpose a spectrum of Alfvén waves on the ambient, upstream magnetic field which lies along the x direction. The waves are an equal mixture of forward and backward propagating (in the plasma frame), left and right hand circularly polarized, coherent Alfvén waves. The waves have a power law spectrum with spectral index 1.5, wavelength range from 5 to 200 c/ω_i, and a total integrated energy density of $0.75 B_1^2/8\pi$. This gives a set of fluctuations with a maximum $\Delta B/B \sim 2$ initially, which suffers ion damping and reaches $\Delta B/B \sim 1$, shortly after the start of the simulation. The spectrum persists to provide the (time-dependent) upstream boundary condition but is not maintained artificially in the simulation domain. The actual wave field in the simulation is not merely the injected wave spectrum, but a complex combination of the shock generated turbulence with the injected spectrum. An additional, computationally advantageous, effect of the seed turbulence is that the shock forms much faster.

Since only a small fraction of the incident population reaches the energies of interest here, we improve the statistics by "splitting" particles, a technique used successfully in Monte Carlo simulations[11]. The details of this method and how it is employed in our simulation model is discussed in Giacalone, et al.[5]

RESULTS AND DISCUSSION

In Figure 1 we display the upstream and downstream energy spectra taken at the end of the simulations. Two distinct populations are clearly seen: the thermal core and a high energy diffuse (see Figure 2) component which extends to more than 200 times the plasma ramming energy, E_p. The increased variability in the upstream spectra in the region connecting these two parts of the distribution function suggests that there are few particles there. This indicates that there is a coherent mechanism for the injection of particles upstream with an energy of about 5 E_p. In the case with input seed waves, the downstream distribution is noticeably flatter above the shocked solar wind point. This is due to particles crossing the shock from the upstream region after being scattered by the upstream waves. The diffusive theory predicts that in steady state, such a process would yield a power law energy spectrum with an index that depends on the properties of the shock [12]. We have indicated the predicted power law slope by a dashed line in the right panel of Figure 1, which reveals that the self-consistent energy spectrum is in agreement with the diffusive theory. The distribution falls off severely for $E > 80 E_p$ (approximately exponentialy) which is due to a significant flux of ions with these energies escaping the left edge of the simulation box. There is no such feature in the no waves case because there is not enough power in the upstream resonant waves (generated by the particles themselves) to efficiently scatter them back across the shock. We expect that if the system were large enough and the simulations were run for much longer, the spectrum would begin to look like the one for the with waves case.

In Figure 2 we plot the shock frame $v_\perp$-$v_\parallel$ phase space diagrams of ions with $E > 2E_p$ at the end of the simulation, averaged over regions 100 c/ω_i far upstream (x_s-500 c/ω_i), near upstream (x_s-100 c/ω_i) and downstream of the shock (located at x_s). The left panels are for the with waves case, while the right panels are for the case without waves. This figure illustrates the effect of the waves on the isotropy of the energetic particles. Clearly, when waves are continually injected from the left boundary, the energetic ions are efficienty isotropized and a large number of them are scattered back to the shock front. In contrast, in the (initially) no waves case, the flux of the energetic ions is directed away from the shock.

In Figure 3 we plot the partial densities of the energetic ions, normalized to the total upstream plasma density. The energy corresponding to each profile is shown in the upper left. At low energies ($2 < E/E_p < 5$), there are no particles ahead of the shock and a noticeable spike at the shock front, while at larger energies, the density falls off exponentially upstream. The latter is a predicted feature in the steady state diffusive theory where the e–folding distance is a function of the diffusion coefficient. The former implies that particles are being trapped at the shock front and are being accelerated locally. This suggests that there is a coherent injection mechanism which we have found to be shock drift acceleration[13]. This occurs when the local θ_{Bn} at the shock front is different from zero[3].

In Figure 4 we display the e–folding distance, λ_e, of exponentially fit density profiles upstream of the shock as a function of energy. This plot shows a dependence upon energy which is approximately power law until $E \simeq 80$ E_p where the distribution function begins to turn over. At this point, λ_e is approximately equal to the distance from the shock to the left end of the simulation box and the fits become very dubious. From this diagram, the e–folding distance, for $10 < E/E_p < 80$, is approximately proportional to E.

CONCLUSIONS

We have investigated the acceleration of ions to large energies by a numerically simulated parallel collisonless shock wave. We have used the hybrid simulation model with the electrons treated as a fluid and the ions treated kinetically. Although such an approach is computationally time consuming, it is esential for studying ion acceleration since the energetic particles, and their effect on the shock, are treated self–consistently. We have introduced the technique of particle splitting to obtain better statistics at high energies, significantly resolving the distribution function to energies more than a few hundred times the plasma ramming energy. We have also introduced a source of seed turbulence in the upstream region at the start of the simulation and maintained their injection at the left boundary. This approach is designed to efficiently scatter the ions backstreaming from the shock so that multiple shock interactions are possible, as in the classic picture of first order Fermi acceleration. We expect that if simulations without such waves are run for very long times and using very large domains, such a spectrum would evenutally be the result.

We find that including the waves:

1. makes the shock form faster;
2. produces a power law in the downstream ion distribution function between $5 < E/E_p < 80$ with a slope that is in agreement with the steady state diffusive theory;
3. effectively isotropizes the backstreaming upstream energetic ions.

The energetic particles have been the primary focus of attention in this study. However, the question as to what effect the waves have on the detailed structure of the shock needs to be addressed and is the subject of a future investigation.

Acknowledgments We gratefully acknowledge helpful discussions with the participants of this workshop and in particular, R. Decker, M. Scholer, K. Quest, and H. Kucharek. This work was supported by SERC grant GR/H09454 and an SERC Advanced Fellowship (DB). Computations were carried out at the Atlas Center (RAL, UK) and the North Carolina Supercomputing Center.

REFERENCES

1. Quest, K. B., *J. Geophys. Res.*, *93*, 9649, 1988.
2. Scholer, M. *Geophys. Res. Lett.*, *17*, 1821, 1990.
3. Kucharek, H., and M. Scholer, *J. Geophys. Res.*, *96*, 21195, 1991.
4. Trattner, K. J., and M. Scholer, *Geophys. Res. Lett.*, *18*, 1817, 1991.
5. Giacalone, J. , D. Burgess, S. J. Schwartz, and D. C. Ellison, *Geophys. Res. Lett.*, in press.
6. Burgess, D., this volume.
7. Cargill, P. J., this volume.
8. Kucharek, H., this volume.
9. Quest, K. B. this volume.
10. Scholer, M., this volume.
11. Ellison, D. C., this volume.
12. Drury, L. O'C., *Rep. Progr. Phys.*, *46*, 973, 1983.
13. Decker, R. B., this volume.

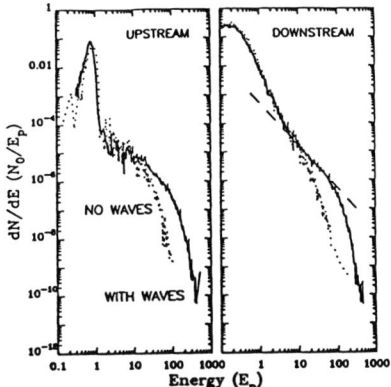

Fig. 1. Energy spectra taken at $t = 600\ \Omega_i^{-1}$. The distributions are averaged over 60 c/ω_i upstream (left) and downstream (right) of the shock. The ordinate axis is the number of particles per unit energy, normalized to the total number of upstream solar wind ions, N_0, contained within a region having the same size. The solid (dotted) lines are for a simulation with (without) injected seed turbulence.

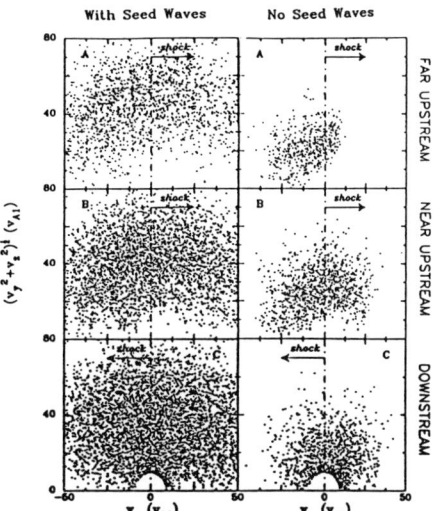

Fig. 2. Velocity space configuration of ions with $E > 2\ E_p$ at different locations with respect to the shock. The direction to the shock is indicated in each panel.

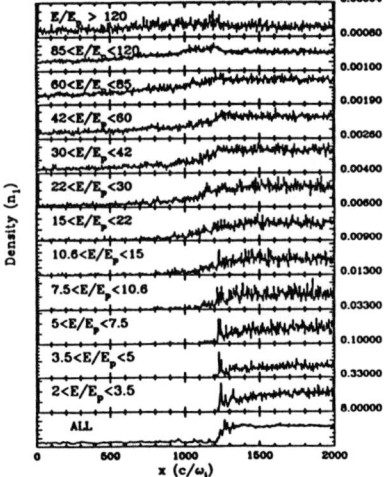

Fig. 3. Density of ions with energies indicated in the upper left of each panel. The number in the right margin, at the top of each panel, represents the maximum value of the plot and is different for each panel.

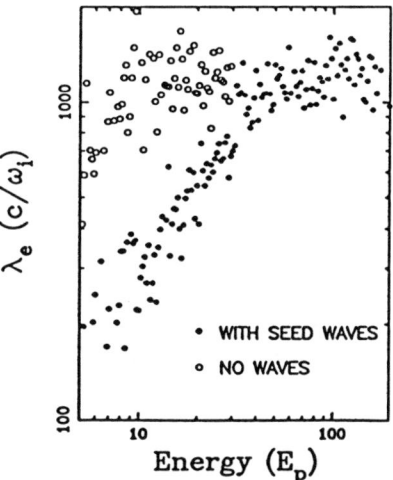

Fig. 4. The e–folding distance, λ_e, of the density profiles, illustrated in Figure 3, as a function of energy for simulations with input waves (solid circles), and without (open circles).

ACCELERATION OF DIFFUSE IONS AT QUASI-PARALLEL SHOCKS: SIMULATIONS

H. Kucharek and M. Scholer
Max-Planck-Institut für extraterrestrische Physik
8046 Garching, Germany

ABSTRACT

Large-scale hybrid simulations have been performed in order to study the acceleration of thermal particles to high energies at quasi-parallel shocks. The results show that most of the backstreaming ions are not due to leakage, but are directly accelerated at the shock out of the thermal population. These superthermal ions constitute a seed particle population for further acceleration by a first order Fermi process. The simulations show that ions after the initial acceleration from thermal energies reach high energies by multiple traversals of the shock. These ions are scattered by upstream waves of their own making. The relative averaged amplitude of the upstream magnetic field $\langle \delta B \rangle / B$ is of the order of 0.6.

INTRODUCTION

Hybrid simulations of quasi-parallel shocks have shown that thermal particles of the incoming solar wind are self-consistently accelerated to superthermal energies by a more or less coherent process and constitute a population of backstreaming ions[1,2]. In an initial process the acceleration to a possible seed particle population for further acceleration occurs when these particles stay close to the shock for a relatively long period of time. Due to a $\nabla \vec{B} \times \vec{B}$-drift these ions gain energy by the tangential electric field $(\vec{v} \times \vec{B})$ until they leave the shock into the upstream as well as in the downstream direction. A large fraction of the backstreaming particles leaves the shock during a reformation cycle. At this time large temporal changes of the magnetic field apparently contribute to energy increase of the particles. During this process the ions reach velocities up to twelve times of the shock ram velocity. These particles can be scattered back toward the shock by ultra low frequency upstream waves of their own making and become accelerated by multiple traversals of shock.

In order to study this subsequent acceleration process a hybrid simulation in a large simulation domain is performed. The time development of the system is followed up to about fifty ion gyro-periods. The upstream wave field and the spectra in the upstream as well as in the downstream region are analyzed. Since downstream heated ions may leak upstream and can thus contribute to backstreaming ions, we will furthermore investigate more closely what part of diffuse ions is due to leakage.

SIMULATION METHOD

A one-dimensional hybrid code is used which treats the electrons as a massless fluid and the ions as macro-particles [3]. All fields and variables depend only on one direction x, which is the shock normal direction. The shock is initiated by a continuous plasma flow from the left hand side (at $x = 0$) which is reflected at the end of the simulation domain (at $x = x_{max}$) by a rigid wall. A ion/ion beam instability between the incident and the reflected plasma causes the build-up of a shock which propagates to the left ($x = 0$) of the simulation frame.

The following dimensionless units are used throughout in the paper. The time is expressed in units of the inverse of the ion gyrofrequency ($\Omega_{ci} = eB_o/m_i c$, where c is the speed of light, e the electron charge, B_o the upstream magnetic field, m_i the proton mass). The distances are in units of the proton inertial length ($\lambda_p = c/\omega_{pi}$, where ω_{pi} is the proton plasma frequency). A grid size of $\Delta x = 0.5 c/\omega_{pi}$ and a time step of $\Omega_{ci}\Delta t = 0.025$ is used. The density and the magnetic field are normalized to their upstream values. The shock is characterized by the Alfvén Mach number M_{Ao} and the angle Θ_{Bn} between the upstream magnetic field and the shock normal. Energies are expressed in units of Alfvén energy $E_A = mv_A^2/2$, where v_A is the Alfvén speed.

For the results presented below, two simulation runs with the same upstream conditions were carried out. In the first simulation (run A) a large domain of $1600 c/\omega_{pi}$ was used and the shock was followed until $t = 375\Omega_{ci}^{-1}$. For the second simulation (run B) a smaller system was chosen: the size of the simulation region was $500 c/\omega_{pi}$ and the system was followed up to $t = 125\Omega_{ci}^{-1}$. In both cases the position of the shock x_s was determined as the position of the largest density gradient following the point where the density exceeds a critical value (three times the upstream density). In order to improve statistics of the diffuse ions in run A the method of particle splitting was used. Each incident solar wind ion carried a identification flag. This flag was changed when the particle encountered the shock. In case such a particle was found in the upstream region beyond the boundary $x_B = x_s - 15$ it was identified as a diffuse ion. Such a diffuse ion was then substituted by 40 new ions with the same three velocity components and was randomly distributed around the position of the original particle. When calculating the moments their contribution was weighted properly. In order to investigate the origin of the diffuse ions run B was made in two steps: at $t_1 = 75 c/\Omega_{ci}^{-1}$ the shock was well matured and particles which have been upstream as well as downstream were identified and given different flags. Then the run was continued until $t_2 = 105 c/\Omega_{ci}^{-1}$. At this time t_2 the position of particles which were upstream as well as downstream, respectively, at time t_1 is checked.

SIMULATION RESULTS

Under typical solar wind conditions at 1 AU the simulation domain in run A corresponds to $\sim 25 R_E$ and the time development of the system was followed up to a time of $t_s = 375\ \Omega_{ci}^{-1}$, which is about 8 minutes in real time. At this time there exists still a region of $x_s = 1000 c/\omega_{pi}$ upstream of the shock, which corresponds to about $\sim 16 R_E$.

In the simulation an Alfvén Mach number of $M_{Ao} = 6.6$ and a global $\Theta_{Bn} = 5°$ is used. At each time step we have calculated the average relative amplitude $\langle \delta B \rangle / B$ within $X = 200 c/\omega_{pi}$ immediately upstream of the shock. The result is displayed in Figure 1. $\langle \delta B \rangle / B$ is a stongly fluctuating, but continuously increasing function which reaches values of about 0.6 at $t = 300\Omega_{ci}^{-1}$. This result clearly demonstrates that quasi-parallel shocks are able to create self-consistently highly nonlinear upstream waves via an ion/ion beam instability between the incident solar wind and the diffuse ions.

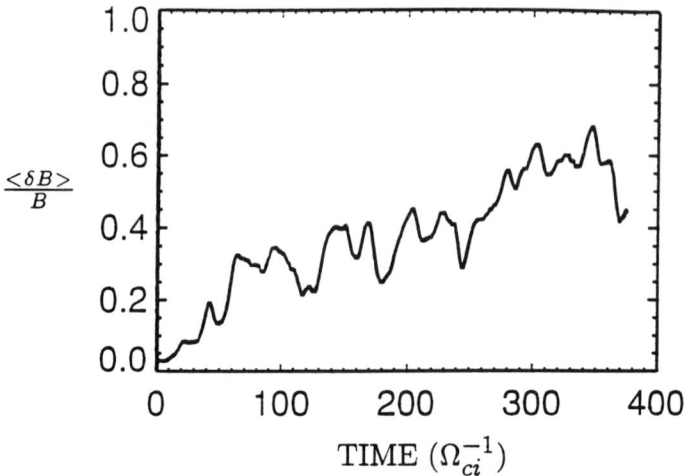

Fig. 1. The average relative amplitude of the magnetic field $\langle \delta B \rangle / B$ within $X = 200 c/\omega_{pi}$ immediately upstream versus time.

Further analysis of the upstream wave field shows that there is a dominant wave with a wave lenght of $\sim 50 c/\omega_{pi}$.

The power in this mode increases as a function of time and reaches a saturation level at a time of about $t = 300\Omega_{ci}^{-1}$. The superthermal seed particles are scattered by these waves in the upstream region and may eventually interact again with the shock.

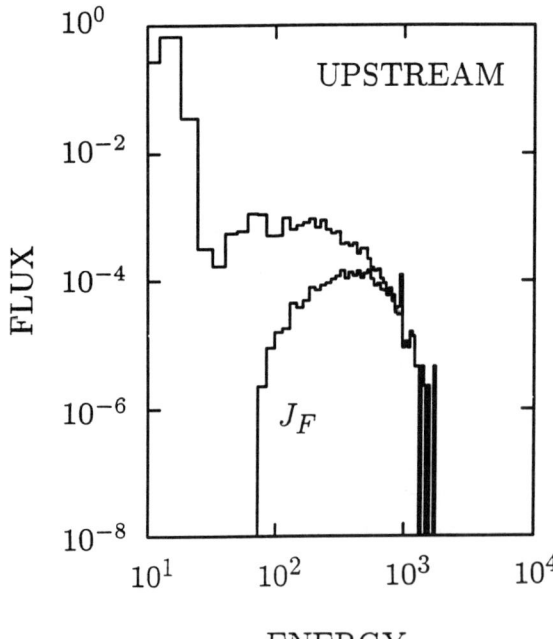

Fig. 2. The differential flux as a function of energy. J_F indicates the flux of those ions which have encountered the shock more than once.

Figure 2 shows the upstream flux (number of particles within a velocity bin) as a function of energy in a double logarithmic representation in the reference frame of the shock. The spectrum exhibits a peak at low energies corresponding to the thermal solar wind distribution and a broad superthermal tail extending up to $\sim 1\times 10^3 E_A$. Indicated by J_F is the differtential flux of those upstream ions which have been scattered back by upstream ULF waves and have encountered the shock at least a second time. A comparison of the spectrum labeled by J_F with the total upstream spectrum indicates that below $\sim 500\ E_A$ the ions are accelerated in a one step process at the shock wheras above this energy multiple interaction with the shock leads to further acceleration. The energy of $500\ E_A$ corresponds in our case to $12\ E_R$, where $E_R = mV_R^2/2$ and V_R is the shock ram velocity. Regarding the spectrum in the downstream regime (see Figure 3) one can distinguish between three parts: the flux of the thermal downstream population (up to $10\ E_A$) and at higher energies ($10\ E_A - 100\ E_A$) the contribution of the superthermal particles.

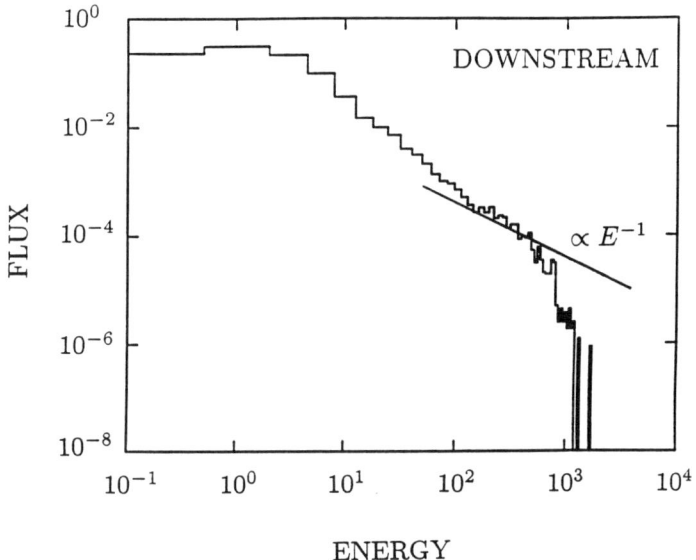

Fig. 3. The differential flux in the downstream region as a function of the energy. The straight line indicates the prediction of the first order Fermi theory.

These are solar wind particles which have been accelerated by staying close to the shock for an extended period of time and eventually go downstream. At the high energy end of the spectrum there are the Fermi accelerated ions. The straight line indicates the decrease of the spectrum predicted by the first order Fermi theory ($f \propto E^{-1}$). According this theory the energetic particle distribution function at the shock and in the downstream region is in the steady state a power law in momentum $f(p) \sim p^\alpha$, where $\alpha = 3\Delta N/(1-\Delta N)$ and ΔN is the density jump at the shock ram. This results for a density jump of 4 in $j = (dN/dv) = (dN/dE)(dE/dv) \propto E^{-1}$. As one can see the slop of the differential flux from $\sim 100\ E_A$ to $\sim 800\ E_A$ agrees with the diffusive acceleration

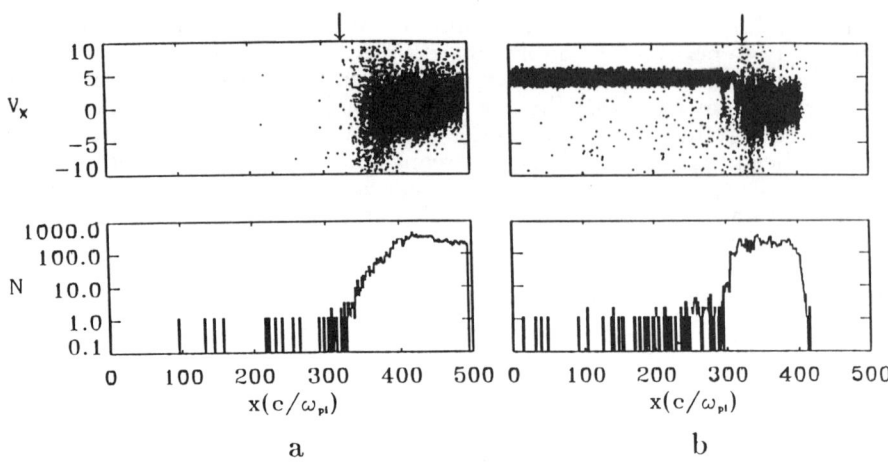

Fig. 4. v_x versus x and the number of particles as a function of x at t_2. Figure 4a shows those ions which have been downstream at t_1. The lower panel displays the corresponding number of particles. Figure 4b shows those ions which have been upstream at t_1. The particle number of the diffuse ions at $t = t_2$ is shown in the lower panel.

theory. The steep fall-off above $\sim 800\ E_A$ is due to the finite size of the upstream region and due to the fact that at high energies a steady state is not yet reached.

One possibility for backstreaming particles is leakage of heated downstream ions across the shock into the upstream region. Kucharek and Scholer[1] have argued that diffuse particles originate by an acceleration process at the shock, wheras Lyu and Kan[4] presented evidence for such a leakage process. In order to investigate this question in more detail we have performed the following simulation (run B). We used a simulation region of $x_{max} = 500 c/\omega_{pi}$ and the shock was followed up to $t_1 = 75\Omega_{ci}^{-1}$. At this time the shock was well developed and reached a position $x_s = 376 c/\omega_{pi}$. All ions were labeled by different identification flags corresponding to their position upstream or downstream, respectively, of the shock. The simulation was then continued until $t_2 = 105\Omega_{ci}^{-1}$.

Figure 4 shows v_x versus x as well as the number of particles as a function of x at the final state of the simulation (t_2). All ions which have been downstream at t_1 are displayed in Figure 4a. The lower panel of Figure 4a shows the corresponding number of particles as a function of x. In Figure 4b all ions which have been upstream at t_1 are shown. The lower panel of Figure 4b shows the number of particles of the backstreaming ions as well as of the heated downstream ions. In both figures the arrow indicates the shock position at t_2. As one can see from these figures, only very few of the ions downstream at $t = t_1$ are upstream at $t = t_2$ and contribute to the backstreaming population. More precisely, at $t = t_2$ only 2% of the upstream diffuse ions have been downstream at $t = t_1$ and have leaked upstream.

SUMMARY

The acceleration of ions at quasi-parallel collisionless shocks has been studied by hybrid simulations. In addition the upstream waves have been investigated. The results can be summarized as follows:

The diffuse ions are accelerated out of the thermal population by a "one step" process up to ~ 12 times the shock ram energy. Above this energy the ions interact more than once with the shock, i. e., a first order Fermi process is necessary for further acceleration. The shape of the downstream spectrum (omnidirectional differential flux) near the high energy end is close to E^{-1} as predicted by the Fermi theory. We have also shown that for quasi-parallel shocks the average relative amplitude, $\langle \delta B \rangle / B$, of the upstream waves reaches values of 0.6. The result of the simulations also show that the upstream diffuse ions are not leakage ions, but predominantly originate from direct acceleration of the particles out of the incident solar wind distribution at the shock.

REFERENCES

1. H. Kucharek and M. Scholer, J. Geophys. Res. 96, 21195 (1991).
2. M. Scholer, Geophys. Res. Lett. 17, 1821 (1990).
3. D. Winske and M. M. Leroy, in Computer Simulation of Space Plasma (ed. H. Matsumoto an T. Sato), p. 255, (1984).
4. L. H. Lyu, J. R. Kan, Geophys. Res. Lett. 17, 1041 (1990).

Composition and Sources of
High Energy Cosmic Rays

The UHE Cosmic Ray Spectrum

Pierre Sokolsky
Physics Department
University of Utah
Salt Lake City, Utah 84112

February 12, 1992

1 Introduction

This paper reviews the current experimental understanding of the UHE cosmic ray spectrum. The gross features of the spectrum are: the spectrum obeys a power law with index -2.6 from GeV to PeV(10^{15}eV) energies and then steepens to an index of between -3.0 and -3.1. This feature is known as the knee. The new power law continues up to an energy of 10 EeV(10^{19}eV). Beyond that the data suggest another flattening (known as the ankle). The spectrum must eventually terminate, either by the exaustion of the acceleration mechanism or due to propagation effects. K.Greisen and Zatsepin and Kuzmin[1] pointed out in 1966 that if the spectrum in the region of the ankle is composed of extragalactic protons and the sources of these protons are greater than 100 Mpc, then we should observe a spectral cut-off between 50 and 100 EeV.

The physical process underlying this GZK cutoff is the onset of inelastic photoproduction. Protons with energies of 100 EeV or greater will interact inelastically with the 2.7 deg black body photons and produce secondary pions, protons and neutrons. The secondary protons will have lower energies and generate first a pile up and then a steepening or cut-off in the observed protonic spectrum. There will also be associated gamma ray and neutrino fluxes from the subsequent decays of pions and muons. Since the effective interaction length for protons of this energy is about 6 Mpc, relatively close extragalactic sources will not exhibit such a cutoff.

There have been a number[2] of recent calculations of the GZK effect incorporating such refinements as cosmological evolution of sources and the black body radiation, following the reinteraction of secondary particles in a full transport calculation, etc. These calculations are in qualitative agreement. They predict a well-defined GZK cut-off below 100 EeV if cosmic ray sources extend beyond 100 Mpc (as in the universal cosmic ray origin hypothesis) and a less well-defined steepening above 100 EeV if sources are quasi-local (such as sources in the Virgo supercluster). Observation of a GZK cutoff below 100 EeV would be strong evidence for the universal origin theory of ultra high energy cosmic rays as well as direct confirmation of the universality of the black body radiation. For this reason the search for the GZK cutoff is the centerpiece of UHE cosmic ray studies.

Of course, if galactic sources can produce cosmic rays above 10 EeV, then the spectrum from these sources will not show a GZK cut off. However, if such sources are distributed in the galactic plane, the resultant protonic component of the spectrum will show strong anisotropy at these energies. Models, such as galactic wind acceleration models would have smaller anisotropies but would generate a predominantly heavy (Fe rich) spectrum[3]. Cutoffs from such local sources would be due to termination of the injection spectrum and not propagation effects.

It is clear that a definitive search for UHE sources requires measuring the spectrum, composition and anisotropy of the UHE cosmic ray flux.

2 Experimental Techniques

There are two basic techniques for studying the ultra-high-energy cosmic ray spectrum: ground arrays and the Fly's Eye detector.

2.1 Ground Arrays

At PeV and EeV energies, primary cosmic rays entering the atmosphere produce extensive air showers (EAS) of secondary particles many of which reach the ground. Ground arrays use large scintillation counters spaced by hundreds of meters to kilometers over areas of tens of square kilometers to sample these surviving secondaries. The core of the EAS is found by a fit to the lateral particle density and the zenith angle of the incident primary is determined by measuring the relative delays between adjacent scintillation counters. Determining the energy of the primary particle is done using the parameter $\rho(600)$. It has been found thru Monte Carlo studies, that the best measure of the primary energy is the charged particle density at about 600 meters from the shower core. This parameter, $\rho(600)$, is approximately linearly related to the incident energy and is not strongly dependent on the composition of the primary particle or the interaction model used[4].

Recent data on the UHE cosmic ray flux comes from three ground arrays:Haverah Park in the U.K. (which was turned off in 1988)[4], Yakutsk in the Soviet Union[5], and Akeno in Japan[6]. All these arrays have effective collecting apertures of 10 to 20 $km^2 ster$ and an approximately 100 percent duty cycle. They have been accumulating data for between six and 25 years.

2.2 The Fly's Eye

The Fly's Eye[7] is a unique detector that utilizes the scintillation properties of atmospheric N_2 molecules to detect the EAS produced by cosmic ray primaries. As the EAS develops in the atmosphere, the secondary charged particles excite N_2 molecules which subsequently emit scintillation light in the near UV. This light is collected by the Fly's Eye mirrors (67 1.6 meter diameter mirrors at one site and 36 at another site 3.5 km distant) and focused on PM tubes which give pulse height and relative timing information. The EAS can be thought of as a point source of light moving with the speed of light thru the atmosphere and growing first brighter and then dimmer as the shower comes to its maximum and then dissipates. Relative tube timing and stereo information allow the impact parameter and zenith and azimuth angles of the the track of the shower to be reconstructed. Once the geometry is known, pulse height information can be used to reconstruct the longitudinal shower development curve as a function of depth in the atmosphere. The primary energy is the shower profile integral times the mean charged particle energy loss in the atmosphere. The position of shower maximum X_{max} is also an indicator of the cosmic ray composition. The energy determination is much more direct than in the case of ground arrays, the energy scale being set by the N_2 scintillation efficiency (a measured number) and the detector optical efficiency. The geometrical reconstruction however is more complex and the detector aperture is an increasing function of the particle energy, going from tens of $km^2 ster$ at .1 EeV to 1000 $km^2 ster$ at 50 to 100 EeV. Since this is an optical technique, the duty factor is 10 percent leading to an effective aperture at 100 EeV of a few hundred $km^2 str$.

3 Experimental Results

Fig 1 shows the results on the UHE cosmic ray spectrum for the ground arrays and the Fly's Eye. Note that the spectrum is multiplied by E^3 to more easily see structure in a steep power law. The spectra from different experiments are qualitatively similar showing a slope of approximately -3.1 to -3.2 up to 10 EeV and some evidence of flattening to a slope of -2.5 above this energy. In no case is this evidence for flattening stronger than a 3 sigma effect, however. There is no evidence for a GZK cutoff below 50 EeV. With the exception of Haverah Park which reports four candidates, no reliable events have been reported above 100 EeV.

The energy resolution of ground arrays and the F.E. is on the order of 25 percent. The energy scale for ground arrays depends on the hadronic model and the composition assumed in the monte carlo relating $\rho(600)$ to primary energy and could shift, in the H.P. case, by 40 percent[4]. For the Fly's Eye, the energy scale depends fundamentally on the N2 scintillation efficiency and could shift by 20 percent. Much smaller shifts than these will bring the spectral normalizations of all experiments into substantive agreement. The presence of non-gaussian tails in the energy resolution function at the 10 to 20 percent level can wash out a cut-off at 60 EeV. Such tails can not presently be ruled out by any experiment. We can thus only say the the combined data are not inconsistent with a GZK cutoff between 50 and 100 EeV but do not prove that such a cut-off exists.

In the case of the Fly's Eye, a significant fraction of events seen by one eye is also seen by the second one. These stereo events have more precise reconstruction and the energy resolution can be checked since two independent measurements are performed. The resultant spectrum, although poorer in statisitics, has well controlled errors. It is in good agreement with the monocular Fly's Eye data but has insuficient statistics to explore the GZK cutoff region.

A better understanding of the presence or absence of the GZK cutoff will have to await the new generation of UHE cosmic ray detectors. For now, we can say that there is accumulating evidence for a flattening of the cosmic ray spectrum above 10 EeV, which may be a precursor to the long awaited cut-off.

The Fly's Eye group[8] has recently published results on the cosmic ray composition in the region .3 to 10 EeV based on high quality stereo data. The technique for establishing composition is to measure the distribution of the EAS shower maximum as a function of depth in the atmosphere. This so called Xmax distribution should reflect the cosmic ray composition because heavy nuclei such as Fe will interact earlier and

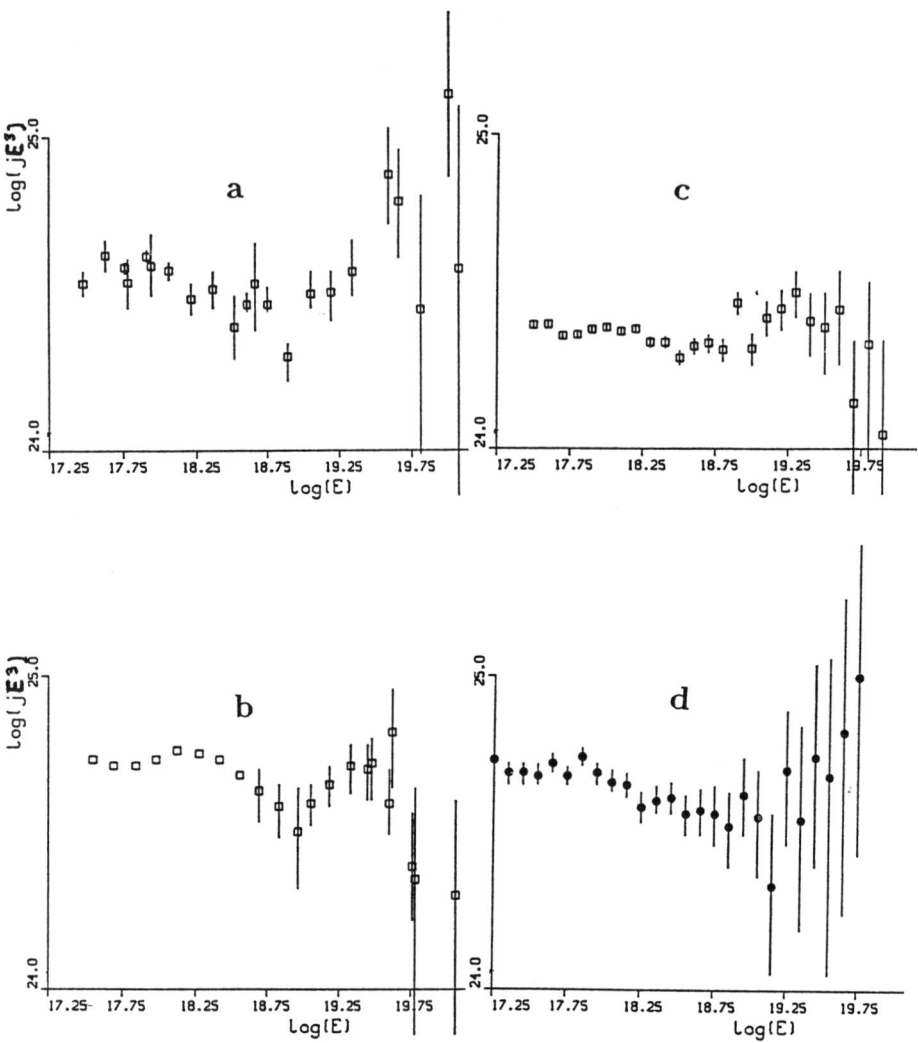

Figure 1: Spectra measured by ground arrays and Fly's Eye. a. Haverah Park final spectrum; b. Yakutsk spectrum; c. Fly's Eye spectrum; d. Akeno spectrum.

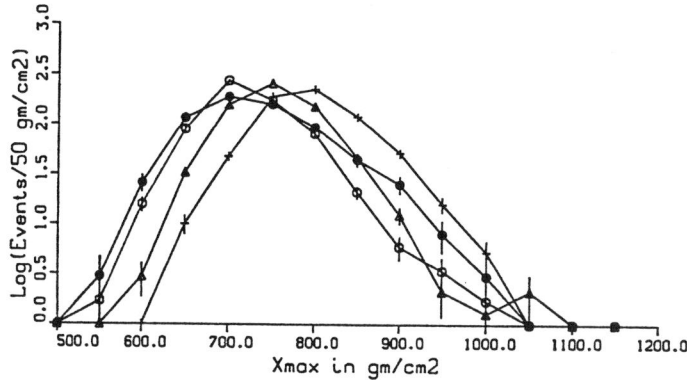

Figure 2: Xmax distribution from 1.0 to 50 EeV:•-data,o-Fe, ▷-C,†-p.

thus have Xmax's higher in the atmosphere than protons.

The overall Xmax distribution integrated from .5 to 10 EeV is shown in Fig 2. together with the expectations for pure Fe, pure C and pure p. The hadronic model used in predicting the Fe, C and p Xmax distributions is discussed in ref[9], but details of the model do not affect the general conclusion: a pure flux of any element cannot account for the shape of the observed distribution. The best fit requires roughly equal contributions of heavy and light nuclei. There is no significant data yet on this question above 10 EeV, where the composition is expected to change to predominantly protons if the flux becomes extragalactic.

4 New Detectors

A number of much larger detectors are now either coming into operation or are in the prototype and planning stages. We describe them briefly below.

4.1 The AGASA array

The Akeno array is currently being expanded again and will be known as AGASA (Akeno Giant Air Shower Array). The design calls for an expansion following the pattern of the 20 km^2 array i.e. a 1 km grid of 2.2 m^{-2} scintillators. One hundred and eight such detectors will bring the total area to 100 km^2. The stations are interconnected with a fiber optic network.

4.2 EAS-1000

A new large array is being planned for a site in the Kazakh Republic of the USSR, at an elevation of 500 m a.s.l.. An area of 900 km^2 is to be instrumented by 1 m^2 scintillator detectors on a 500m grid. In the center of this area another array of similar detectors will be built, covering an area of 25 km² with an detector spacing of 70m. To complete the electron detection section of the experiment, a compact array of 900 1 m² scintillators will be spaced on a 20m grid over an area of 0.36 km². The aim is to provide sensitivity and a large data rate over the energy range from 0.3 PeV to 1000 EeV.

4.3 High Resolution Fly's Eye

The Fly's Eye group has proposed the construction of a second-generation Fly's Eye detector, to be known as the High Resolution (HiRes) Eye. There are two principal goals: Increasing by an order of magnitude the aperture for well-reconstructed events above 10 EeV and increasing the resolution of the shower profile measurement so that $\sigma(X_{max}) \simeq 20$ g cm^{-2}. The present stereo Fly's Eye has an X_{max} resolution of 60 g cm^{-2}. The detector will accumulate 200 events per year above 10 EeV.

These goals are to be accomplished by constructing three stations, situated on the vertices of an equilateral triangle with 15 km sides. Each station will have 56 mirrors (2 meters in diameter), each with a cluster of 256 phototubes having one degree apertures. The increase in detector acceptance is achieved primarily as a result of reducing the solid angle of each tube by 1/25 relative the the Fly's Eye phototube apertures. That reduces the sky noise in each phototube and allows weaker light signals to be detected. Improvement in X_{max} resolution is due to increased shower profile sampling and the improved geometric reconstruction obtained from stereoscopic measurement of all showers.

References

[1] K.Greisen,P.R.L.**16**,748,1966.;G.T.Zatsepin,V.A.Kuzmin,JETPh Letters,**4**,78,1966.

[2] G.T. Hill and D.N. Schramm,Phys. Rev. D,**31**,564,1985;V.S. Berezinsky and S.I. Grigor'eva, Astron. Astrophys, **199**,1,1988;F.A. Aharonian, B.L. Kanevsky and V.V. Vardanian, Astrophys. Space Sci.,**167**,93,1990;S. Yoshida and M. Teshima, ICCR-Report-221-90-12,1990.

[3] J.R.Jokipii and G. Morfill,Astrophys. J.**312**,170,1987.

[4] N.N. Efimov,in *Proc. of ICRR International Symposium on Astrophysical Aspects of the Most Energetic Cosmic Rays*,Kofu,Japan,1990, to be published.

[5] M.A. Lawrence et al., J.Phys G.:Nucl. Part. Phys.,**16**,1991.

[6] M. Teshima,in *Proc. of ICRR International Symposium on Astrophysical Aspects of the Most Energetic Cosmic Rays*,Kofu,Japan,1990, to be published.

[7] R.M. Baltrusaitis et al.,Nucl. Inst. Meth. **A240**,410,1985.

[8] G.L. Cassiday et al.,Astrophys.J.**356**,669,1990. A240,410,1985.

[9] T.K. Gaisser et al., in *Proc. 22nd ICRC*, Dublin,1991,Vol.4,p 413.

COSMIC RAY COMPOSITION AT THE KNEE

Todor Stanev, Bartol Research Institute
University of Delaware, Newark, DE 19716

1. Introduction

Most of the participants of this meeting are familiar with the data on the chemical and isotopic composition of the cosmic rays in the GeV energy range which has been used to derive many basic properties of the acceleration and propagation of cosmic rays. This talk deals with the composition of cosmic rays at much higher energies where such information is restricted by the decreasing statistics and the ambiguity in the interpretation of detected events. We want to show, however, that the investigation of the composition in a wide energy range should reveal important clues on the acceleration mechanisms in general, and even a limited data set could meaningfully restrict the wide variety of acceleration models that are considered today.

Fig. 1 shows data on the cosmic ray spectrum above 10^3 GeV/nucleus with the flux multiplied by $E^{2.75}$. The solid points represent direct measurements, while the open ones refer to measurements derived from air shower data. Because the air shower measurements estimate the total amount of energy carried by a cosmic ray nucleus we have chosen to discuss the composition in the appropriate terms of energy/nucleus. The data set includes two regions of special interest in the spectrum: the *knee* region, where the slope of the spectrum increases by $\Delta_\gamma =$ 0.3-0.4, and the *ankle* where the spectrum might be flattening again. Although the representative error bars seem to indicate a consensus on the magnitude and shape of the spectrum, a warning is due. These data points are extracted from the measurements of the air shower size spectra using procedures that assume a cosmic ray composition and a good knowledge of high energy inelastic nuclear interactions in the atmosphere. Different input on the composition or on the interaction properties might change the magnitude of a point by a large factor.

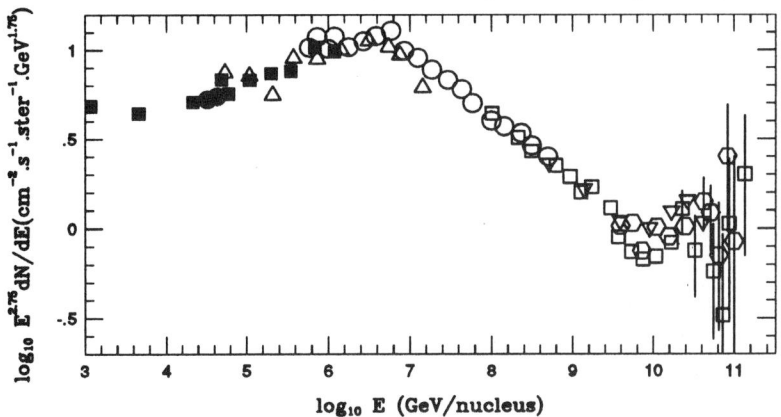

Fig. 1. The cosmic ray spectrum.

© 1992 American Institute of Physics

Specific questions that might have a serious impact on the validity of different acceleration models, such as *(i) Is the bump preceding the steepening of the spectrum real?* and *(ii) Is the break as sharp as some data sets indicate?* are mostly related to the spectral features in the *knee* region.

In this talk we briefly review the recent results of the direct measurements, that are now partially overlapping with shower data points. Apart from the statistical uncertainty all available direct data sets are in good agreement and show a strong trend of increasing fraction of heavy nuclei in the cosmic ray flux. Qualitatively this trend is in line with the expectations from shock front acceleration models and rigidity dependent escape from the Galaxy. The limited statistics at the approach to the *knee* does not allow us to draw major conclusions from the available data and the air shower data becomes very important. We describe how the composition can be extracted from different types of air shower data and discuss the difficulties associated with such procedures.

2. Results from the direct measurements.

There are three data sets that provide the link between the GeV region and the air shower meaurements. In order of increasing energy these are known as the HEAO[1], CRN[2] (Cosmic Ray Nuclei Space Shuttle experiment) and JACEE[3] (Japanese American Collaboration on Emulsion Experiments). Figure 2 compares the most recent results published by CNR and JACEE for three groups of primary nuclei. The CRN results[4] are in a very good agreement with HEAO data. It is not straightforward to compare exactly CRN and JACEE data although it is easy to see the general agreement between them. The Fe group data points show a smooth transition with a possible flattening of the spectrum above 10^{12} eV/nucleus.

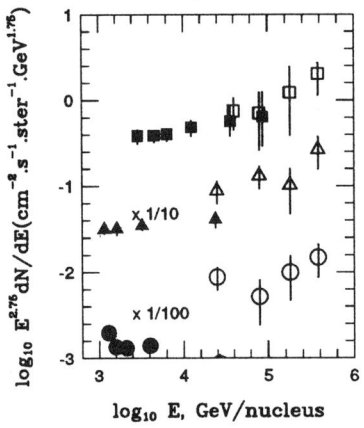
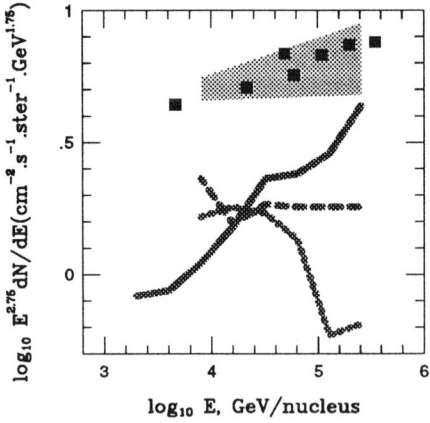

Fig. 2. Comparison of CNR (full symbols) with JACEE (open symbols) data for Fe - squares, O (C-O) - triangles and Mg+Si (Ne-S) - circles.

Fig. 3. H (dots), He (dashes) and heavy (Z>6, solid) components from direct measurements. The sum yields the shaded area. Data points are from Ref. 6.

JACEE does not have good enough statistics to give separately the carbon and oxygen spectra and CRN is not fully sensitive to carbon. It looks, though, that within the expected factor of $\sim$2 the experiments agree again and show the same trend. The comparison for the heavy group is the most difficult, because the data refers to a vastly different selection of nuclei – Si+Mg for CRN and everything between and including Ne and S for JACEE, and exhibits a large gap in energy. The composition of the JACEE set, however, suggests that a factor of two or three in magnitude is justified by the selection criteria, and leads us to conclude that both data sets agree over the whole mass range and that all cosmic ray nuclei show the trend of a flattening spectrum above 10^{12} eV/nucleus.

Having reached these conclusions we combine the data sets and fit the measurements with the solid line shown on Fig. 3. Unfortunately we can not make the same comparison for the important components of hydrogen and helium and have to rely exclusively on JACEE[5], which has good statistics for both components. These are shown respectively with a dotted and a dashed line on Fig. 3. Pretending that the three components thus derived represent the total cosmic ray flux, we have proceeded to add them together and obtain the shaded area on top of the graph that includes a rough estimate of all error bars. The data points in the same region show the results of the old soviet experiment Proton-4[6]. To our satisfaction we can conclude that a straightforward (and maybe unwarranted) treatment of the worldwide data set exposes a more than qualitative agreement. On Fig. 3 we can see how the fraction of protons in the cosmic ray flux declines, while being more than compensated by the increasing fraction of heavy nuclei. The helium component seems not to change much in the energy range from 10^{12} to 10^{14} eV.

Is this a surprising fact? Not being an expert on either cosmic ray acceleration nor on cosmic ray propagation, I am not impressed by it. It has been shown that the *standard* cosmic ray accelerators should be exhausted in the 10^{14} eV $\times$ Z region, and correspondingly higher Z nuclei should start dominating at about that energy. A simple rigidity dependent escape from the Galaxy might strengthen this effect. We could even use the current data set to set limits on some basic and general models. It has been shown by Tom Gaisser[7], that the rigidity dependent escape model of Peters[8] can fit the cosmic ray spectrum *below* and *above* the *knee* (but not at the knee itself) with an overal rigidity cutoff R_0 of $\sim 3 \times 10^{15}$ V applied to the form and composition of the GeV cosmic ray spectrum. The newest JACEE proton data[5] contradicts such a model. Fig. 4 shows the JACEE proton spectrum compared to rigidity dependent escape models with R_0 from 10^{13} to 3×10^{15} V. The two highest energy data points (that contain altogether 7 events) obviously favour very low R_0. The statistics is low indeed but the disagreement with the $R_0 = 3 \times 10^{15}$ V model is very significant.

We cannot leave unnoticed still another fact that does not follow from the simple arguments of the previous paragraph. These would suggest that the H, He, etc. components of the cosmic ray flux would fade, while the higher Z components will stay *constant* in the energy range in consideration. This is not what data shows - the heavy nuclei component of the cosmic ray flux seems actually to increase in terms of a flatter spectral index. It almost looks as if an acceleration mechanism that is more effective for heavy nuclei is taking over the conventional acceleration on the supernova blast wave in the interstellar medium. This feature in the composition is independently supported by the bump in the all particle spectrum as measured by air showers.

We can also ask ourselves the question what do direct observations tell us about the particle composition at the *knee* and generally at higher energy. To obtain a rough answer of this question I have constructed a primitive toy model of the cosmic ray acceleration and propagation, which nevertheless incorporates many features of our current understanding of these processes. Cosmic rays are accelerated in two processes which contribute to the GeV region with a ratio of 10:1. Process 1 has a differential spectral index $\gamma_1 = 2.3$ and an exponential cutoff above $E_{max} = 10^5 \times Z$ GeV while process 2 fails above $E_{max} = 10^8 \times Z$ GeV and $\gamma_2 = 2.2$. In addition the total flux of accelerated cosmic rays is modified at propagation as $(1. + \frac{E}{(10^5 \times Z)GeV})^{-0.4}$. At GeV energies the cosmic ray composition is as measured, i.e. the relative fraction of nuclei with Z=1, 2, 6-8, 9-16 and >17 is respectively 0.55, 0.21, 0.12, 0.10 and 0.02. Figure 5 shows the fractions of protons and heavy nuclei (Z>10) for energies above one TeV.

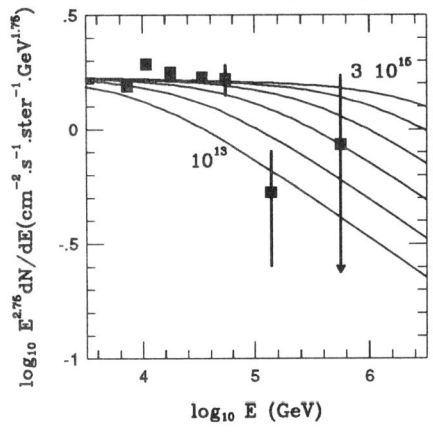 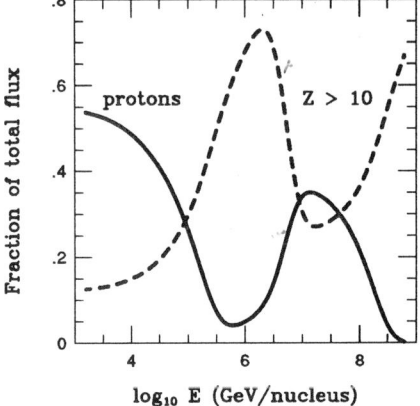

Fig. 4. Rigidity dependent cut-off with $R_0 = 10, 10^{1.5}...10^{3.5}$ TV compared to the JACEE proton spectrum.

Fig. 5. Energy dependence of the fraction of H and heavy (Z>10) nuclei from the toy model described in the text.

Being at least a five parameter model, this toy construction could have even given a decent fit of the measured cosmic ray spectrum. This was not my intention and *as is* the toy model produces a *knee* too early, at energies below 10^6 GeV. The resulting composition, however, exposes the main consequence of modeling the acceleration of cosmic rays with two shock acceleration models that intrinsically have Z dependent cutoffs. Protons dominate at low energy until the acceleration fails at 10^5 GeV. At higher energy, when the second class of accelerators prevails, the fraction of protons increases before it is cut-off again by the failure of process 2. The second maximum never reaches the magnitude of the first one because of the energy dependent leakage out of the Galaxy. The composition is constantly changing.

With Fig. 5 I want to emphasize the importance of air shower measurements of the composition. It has been shown in various proposals that a direct measurement of the cosmic ray composition at the *knee* would require several years exposure of a heavy calorimeter type instrument, presumably at the Space Station. While that wonderful scenario would be ideal, it seems more realistic to approach systematically the numerous bits and pieces of air shower data, that are recently becoming more and more reliable.

3. Composition Measurements with Air Showers.

Air showers are nuclear and electromagnetic cascades initiated in interactions of high energy cosmic ray nuclei high in the atmosphere. Since the prevailing majority of the secondary particles are highly relativistic, the shower geometry is that of a relatively flat pancake that propagates through the atmosphere with the speed of light. Apart from techniques that measure the light generated by charged particles in the atmosphere, air showers are usually detected on the ground, being identified by the coincidences between many individual counters, that detect shower particles. The primary cosmic ray energy and the mass of the primary have to be reconstructed from the shower parameters, which in turn are obtained from the densities and the timing in the individual counters. The relation between the shower parameters and the properties of the primary cosmic ray is provided by cascade theory which is nowadays based on montecarlo calculations.

Very generally speaking there are two types of shower data - one related to the shower size (number of all charged particles, mostly electrons + an admixture of GeV muons) at the observation level, and another reflecting secondaries produced in interactions of the original nuclei. A very good example for the second type of data is the number of TeV muons generated in the cascade. Both types of shower data reflect to some extent the mass of the primary. Heavy nuclei, consisting of many nucleons, and heaving large interaction cross-section, dissipate their energy faster than hydrogen. This leads to shower size that develops faster, and is absorbed sooner, in the atmosphere. By studies of the angular dependence of showers, i.e. showers propagating through different atmospheric thickness, one could detect the the fraction of heavy nuclei through its stronger absorption. This is the classical method for investigation of the cosmic ray composition, which has been used to reconstruct the cosmic ray spectrum from the shower size spectra of the Akeno array[9]. A main feature of this approach, which has been realized long ago, is that the attenuation of the showers in the atmosphere is much more consistent with a heavy, rather than hydrogen, primary composition.

Recently the combined results from the Chakaltaya and Akeno experiments (which have to be combined to cover a significant range of atmospheric thicknesses) were interpreted in terms of all particle spectrum in two extreme assumptions[10] for the cosmic ray composition. A phenomenological two component composition model (PTCCM) was constructed for that purpose. Up to $R_0 = 10^{14} \times Z$ V the spectrum and composition were parametrized to agree with the JACEE measurements. This *low energy* component is subject to rigidity dependent escape. At higher energy a second component appears, with a spectral index $\gamma = 2$, changing to 3.1 at 3×10^{15} eV, which is assumed to be either pure hydrogen, or pure iron. The idea is to calculate the corresponding shower size spectra for the two extreme compositions (hopefully bracketing reality) and to adjust the all particle spectrum so that the calculated shower size spectrum matches the experimental one.

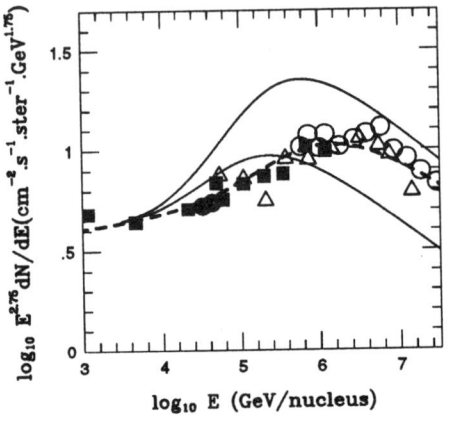

Fig. 6. Spectrum reconstructed from shower sizes(full lines) and from TeV muons (dash).

Fig. 6 shows the all particle spectra that result from this excersize. The upper curve corresponds to the heavy composition, and lower one to the hydrogen one. The comparison of the energy dependence of the calculated and experimental shower size spectra favours the heavy composition.

On the other hand, a comparison of the experimental results on multiple muon groups observed with the MACRO detector in the Gran Sasso laboratory[11] with calculations using PTCCM prefers the light composition with the original normalization, shown on Fig. 6 with a dotted line. There is an obvious disagreement between different types of air shower data. This is not necessarily bad because it is forcing us reexamine much of the basic input that goes into the conversion of measured quantities into primary spectrum. One of the suspected culprits is standard "superposition model" treatment of collisions of heavy nuclei in the atmosphere. A correct treatment[12] can significantly change the derived composition.

There are quite a few other checks in the body of air shower data that must be done before we can with any certainty solve the composition problem at the *knee*. For example, the bracketing compositions of the PTCCM could be significantly restricted by the data on the energy spectrum of single TeV muons, that extend to 10^{13} eV. There are many other relevant bits and pieces of information which have to be treated in a consistant way, and this requires time.

On the other hand there are new generation experiments that are much more sensitive to the composition as they detect simultaneously both types of shower data. A good example is the MACRO-EASTOP[13] combination at Gran Sasso, where MACRO detects the TeV muon group, and the surface array measures the shower size. Since TeV muons and shower size depend in oposite ways on the primary mass, the sensitivity to the composition increases significantly. The aperture of this combination is much smaller than that of either component, but in a few years of operation it should give us significantly better idea of the composition at the *knee*.

Acknowledgements. The author is grateful to Simon Swordy from the CRN group, and to the JACEE collaboration for the data lists. Most of the work is done in constant cooperation and discussion with T.K. Gaisser and many other colleagues. This research is funded in part by NSF.

References.

1. J.J. Engelmann *et al. Astron. Astrophys.* **233**, 96 (1991).
2. D. Müller *et al. Ap. J.* **374**, 356 (1991).
3. K. Asakimori *et al.* (JACEE), *Proc. 22nd Int. Cosmic Ray Conf.* (Dublin Inst. for Advanced Studies, 1991) v. 2, p. 57.
4. D. Müller *et al. Proc. 22nd Int. Cosmic Ray Conf.* (Dublin Inst. for Advanced Studies, 1991) v. 2, p. 25.
5. K. Asakimori *et al.* (JACEE), *Proc. 22nd Int. Cosmic Ray Conf.* (Dublin Inst. for Advanced Studies, 1991) v. 2, p. 97.
6. N.L. Grigorov *et al. Yadernaya Fizika*, **11**, 1058 (1970).
7. T.K. Gaisser, in *Astrophysical Aspects of the Most Energetic Cosmic Rays* (World Scientific, 1991, eds. M. Nagano & F. Takahara) p. 146.
8. B. Peters, *Nuovo Cimento (Suppl)*, **14**, 436 (1959).
9. M. Nagano *et al. J. Phys G*, **10**, 1295 (1984).
10. G. Auriemma *et al, Proc. 22nd Int. Cosmic Ray Conf.* (Dublin Inst. for Advanced Studies, 1991) v. 2, p. 101.
11. R. Bellotti *et al.* (MACRO), *Proc. 22nd Int. Cosmic Ray Conf.* (Dublin Inst. for Advanced Studies, 1991) v. 2, p. 1.
12. J. Engel, T.K. Gaisser, P. Lipari and T. Stanev, *in preparation*.
13. M. Aglietta *et al. Proc. 22nd Int. Cosmic Ray Conf.* (Dublin Inst. for Advanced Studies, 1991) v. 2, p. 61.

THE ARRIVAL DIRECTIONS AND MASS COMPOSITION OF COSMIC RAYS ABOVE 10^{18}eV

A A Watson
Department of Physics, University of Leeds, Leeds LS2 9JT, UK

ABSTRACT

Data on the arrival direction distribution and mass composition of cosmic rays above 10^{18}eV are briefly reviewed. There is no compelling evidence for wide-angle or point-source-like anisotropies and the mass composition above 10^{17}eV is constant to at least 5×10^{18}eV: a mixture of protons and Fe nuclei can describe the data. There is weak evidence that the mean mass increases above 5×10^{18}eV but, as with this and with the arrival direction problem, a vastly greater and better data set is required. This could come from an instrument with an aperture of 10^4km^{2}sr.

INTRODUCTION

Measurable parameters of relevance to the question of the origin of the cosmic rays above 10^{18}eV are the energy spectrum, arrival direction distribution and mass composition of the incoming beam. These three quantities are listed in decreasing order of their currently known precision and become even more uncertain above 10^{19}eV than they are at 10^{18}eV. In an accompanying paper Sokolsky[1] has described the current situation with regard to the energy spectrum as deduced from the experiments known as Fly's Eye, Yakutsk, Haverah Park and AGASA. Over the last ten years differences between the spectra reported from these experiments have been largely resolved: a clear picture has emerged and the spectral shape may now be said to be relatively well understood from 10^{18} to about 5.10^{19}eV. For about two decades up to 10^{18}eV, the differential spectrum has a slope, $\gamma = -3.0$ and there is consistent evidence from all experiments that from 10^{18} to 10^{19}eV the spectrum is significantly steeper, $\gamma \simeq -3.2$, (see references (1, 2) for details). At higher energies the slope becomes significantly flatter and above 10^{19}eV $\gamma \simeq -2.5$, although the uncertainty is quite large. All groups report events with energies above 6.10^{19}eV and there is no evidence for any cut-off in the energy spectrum.

It is worth remarking that the spectra reported by the four groups cited all refer to the ultra energy cosmic rays as seen from the Northern Hemisphere. The only Southern Hemisphere array[3] was of large area but detected only muons and the conversion from what was observed to primary energy is uncertain. What can be said is that there is no gross contradiction between the brief summary just given and the Sydney results. The possibility of differences should not be forgotten: the centre of the Virgo Supercluster is visible most easily from the Northern Hemisphere whilst the centre of our Galaxy can be seen clearly from the Southern Hemisphere.

The features of the energy spectrum listed above have sometimes been described in terms of a model in which the steeper slope between 10^{18} and 10^{19}eV is attributed to a decline, with increasing energy, in the efficiency of acceleration mechanisms within our galaxy. If the acceleration mechanism becomes ineffective when the scale size of the accelerating region is of the order of the Larmor radius of the particle being accelerated, one might expect that, as the energy increases, an increasing fraction of the cosmic rays would be heavy nuclei (conventionally referred to as iron nuclei) assuming that the composition at the source is similar to that found at very much lower energies (~1TeV) where direct measurements are possible. The flattening of the spectrum above 10^{19}eV has been ascribed to the dominance of extragalactic cosmic rays in this region, the slope being identified with the canonical value associated with shock acceleration (believed to be effective at energies between 10^{12} and 10^{15}eV). Whether the cosmic rays of supposed extragalactic origin are light (protons) or heavy (iron) is presently a matter of speculation. At one extreme, if such cosmic rays result from the collapse of loops of cosmic strings, then there is a clear prediction[4] that the primaries are protons: such speculation is not constrained by the data.

It the particles between 10^{18} and 10^{19}eV are galactic, one might ask whether it was reasonable to expect any anisotropies in the arrival direction distributions as a function of energy. At 10^{19}eV in a $3\mu G$ field the Larmor radii for protons and Fe-nuclei are about 3kpc and 100pc respectively. Thus only if the sources were rather close to the earth could one be certain of seeing anisotropies.

DATA AVAILABLE FOR ANISOTROPY SEARCHES AND THEIR INTERPRETATION

The data available above 10^{18}eV for anisotropy searches from all experiments which have so far reported are listed in Table 1.

Table I Data available for anisotropy studies above 10^{18}eV

Volcano Ranch[5] ($\lambda = 35°$N)	794
Sydney[3] ($\lambda = 30.5°$S)	3139
Haverah Park[6] ($\lambda = 54°$N)	8565
*Yakutsk[7] ($\lambda = 62°$N)	17089
*Akeno/AGASA[8] ($\lambda = 36°$N)	2207
*Fly's Eye[9] ($\lambda = 40°$N)	2662
World Total	34,456

Instruments still observing

Recall that the root mean square fractional amplitude of uniformly distributed data is given by $r = 2/\sqrt{n}$, where n is the number of events. Thus, for a wide angle anisotropy which is constant in amplitude at all declinations, with $n = 3 \times 10^4$, we require $r > 2\%$ for even 1% significance. Even if $r = 10\%$ about 2×10^3 events are wanted for 1% significance. Clearly if the direction and amplitude of anisotropy are varying as a function of energy many more events are required. It should also be noted that the Fly's Eye technique is less useful for measuring small amplitude anisotropies as, although the data are accumulated during dark periods at a much higher rate than with other instruments, it is difficult to establish uniform exposures at the level of a few percent across the sky. By contrast all other measurements are with particle arrays and long periods of operation at high efficiency lead naturally to uniform exposures in right ascension.

A feature which might be expected[10] to be present in the data if there are sources in the Galactic Plane is a second harmonic associated with the passage of the Galactic Plane through the aperture of the instruments. A recent assessment[2] was made using data from Haverah Park, Fly's Eye, Yakutsk and Akeno for the energy range $4.10^{18} - 3.10^{19}$ eV. This energy band is chosen to match the energy ranges described from different experiments: it also straddles the region where there is a marked dip in the spectrum (see above) perhaps caused by a galactic/extragalactic transition. For 2105 events from these four experiments the second harmonic amplitude is 8.7 (± 3.1)% in the direction $76 \pm 20°$ and $256 \pm 20°$: the probability that this is not a chance occurrence is only 0.019 and is thus not very convincing. If the amplitude was really 5% then 10^4 events would be required for $p = 10^{-3}$.

An alternative approach to organising the observations has been made by Szabelski, Wdowczyk and Wolfendale[11]. They searched for a dependence of anisotropy on galactic latitude, b (radians), using the relationship

$$I(b) = I_0 [(1 - f_E) + f_E \exp(-b^2)] \tag{1}$$

to describe the data, where f_E is the fraction of cosmic rays which are galactic at an energy E. A recent compilation using this model[8] is shown in Figure 1. It is seen that f_E is positive and increases with energy until above 2×10^{19} eV when f_E changes sign. The measurements shown at the highest energy are from the Haverah Park[6] and Sydney[3] experiments at high galactic latitudes and, if confirmed, provide evidence for an excess of events above the expectation of isotropy. The Fly's Eye data have not been given in a form suitable for this analysis but it is claimed[9] that above 3.2×10^{19} eV there is 'an excess from the Galactic north sky lobe'.

If cosmic ray sources lie at distances very much less that the Larmor radius one might expect point source like features in the arrival direction patterns. Similarly point source features would be expected if the sources were more distant but emitted gamma-rays or neutrons, the mean decay length for neutron of 10^{18} eV being about 9kpc and gamma-ray absorption being negligible above 10^{17} eV. Recently there have been several searches for point-like features at energies above 5.10^{17} eV. The first of these was by the Fly's Eye group[12] who reported evidence for a signal from the x-ray binary system, Cygnus X-3 during the period November 1981 - May 1988. Contemporaneous data were available from the Haverah Park experiment[13] but no similar feature was observed. By contrast the Akeno group[14] claim to have replicated the Fly's Eye result. The fluxes and flux limits from these three experiments are shown in Table 2.

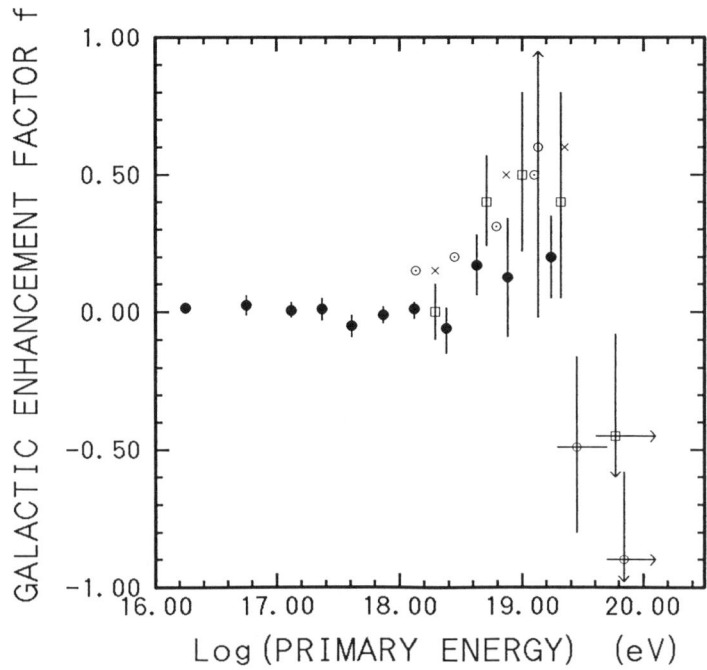

Figure I The galactic enhancement factor f_E as a function of energy for Haverah Park (squares), Sydney (open circles), Yakutsk (crosses) and Akeno (closed circles).

Table II Fluxes and flux limits from Cygnus X-3 at 5.10^{17}eV

Flux $>5.10^{17}$eV

Fly's Eye[12] (Nov '88 - May '88)	$(20 \pm 6) \times 10^{-18}$ cm^{-2} s^{-1}
Haverah Park[13] (Jan '74 - July '87)	$<4 \times 10^{-18}$ (neutrons) or $<8 \times 10^{-18}$ cm^{-2} s^{-1} (gamma-rays)
Akeno[14] (Dec '84 - July '89)	$(18 \pm 7) \times 10^{-18}$ cm^{-2} s^{-1}

The two fluxes associated with the Haverah Park measurements correspond to the assumptions of neutron and gamma-ray primaries respectively as the sensitivity of the water Cerenkov array is slightly mass dependent. Close examination of the Fly's Eye and Akeno reports reveal the following points:-
(a) The two directions of enhancement are not coincident with Cygnus X-3.
(b) The two directions of enhancement are separated by ~5°.
(c) 4.8^h periodic features seen in Fly's Eye data are not seen by the Akeno group.

(d) The dependence of the signal on energy is different in the two experiments.

Furthermore, if the effect is due to gamma-rays (and thus the Haverah Park limit is less strong) it is most surprising that Cygnus X-3 appears to have been quiescent in gamma-rays at 100TeV during the late 1980's. Subsequently, during 1988-90, the Fly's Eye group[15] have doubled their data set but see no signal from Cygnus X-3, while the Akeno group[16] have report sporadic emission, perhaps associated with the radio flares in January and July 1991, at times when Cygnus X-3 is a daytime object for the Fly's Eye group. Clearly further observations will be made and are needed to resolve the problem. A difficulty is that whereas an array such as AGASA at Akeno is operational virtually 100% of the time with a modest aperture, the Fly's Eye instrument has a much greater aperture but can only operate on clear moonless nights and thus has an effective duty cycle of <10%.

A subsequent claim for evidence of point source emission of cosmic rays has been made by the Durham group[17]. Using data listed in catalogues produced by the groups at Volcano Ranch, Haverah Park, Sydney and Yakutsk they have reported evidence of clustering of the arrival directions of the highest energy cosmic rays. In the Northern Hemisphere they identified 14 directions, defined by the arrival direction of events having energy above 3×10^{19}eV around which 5 or more other events ($E \geq 10^{19}$eV) are clustered. Six of the marker directions are found to lie within $6°$ of the Galactic Plane. The claims of the Durham group have been challenged[18] in a study which has used additional data from the Haverah Park array and also a series of Monte Carlo tests. In this study it is concluded that few of the six clusters found close to the Galactic Plane are likely to be true clusters indicative of point-like sources of cosmic rays. Work similar to that of the Durham group has been carried out by Glushkov et al[19] who also find clusters on a $6°$ scale in Yakutsk data. However none of these clusters coincides with those found by the Durham group and the calculations reported by the Yakutsk workers to support their contention that the clustering is other than chance appear dubious[18]. The Durham workers have also analysed other data from some of the large arrays, notably the Sydney group, and argued that it is possible to separate events produced by protons from those produced by iron nuclei. To many workers in the field this seems an inherently improbable possibility given the sparseness of the Sydney data and the fact that the detector was only able to record muons: I would endorse the view expressed recently by Lloyd-Evans[20] on this matter.

EVIDENCE ON MASS COMPOSITION ABOVE 10^{17}eV

Determining the mass composition of cosmic rays beyond energies where direct measurements have been possible ($\sim 10^{14}$eV) with satellites and balloon borne emulsion stacks is exceedingly difficult. In addition to the obvious problem that measurements are made some 10 interaction lengths or so beyond the first interaction of the incoming primary, the experimenter has to contend with systematic uncertainties from unknown particle physics at energies above 10^{16}eV and the probability, based on extrapolation of low energy trends, that the p-Air cross-section will continue to rise with energy so that fluctuation differences expected between iron and proton initiated showers will become smaller than at lower energies.

An approach which has yielded some success in the last decade has been to determine the change in the position of the shower maximum with energy and to measure the fluctuation of this position at fixed energy. Such measurements can be related to changes in the primary mass composition with energy[21]. The most recent and most direct attack on this problem has been made by the Fly's Eye group[22] who have obtained the depth of maximum for 559 well-reconstructed events above 10^{17}eV. It is found that the rate of change of depth of maximum with energy, the elongation rate[21], is 69.4 ± 5gcm^{-2}/decade over the range 1.4 x 10^{17} to 6 x 10^{18}eV.

It has been shown[23] that for reasonable assumptions about the variation of multiplicity of pion production and the rise in cross-section with energy, the elongation rate would be expected to be in the range 67 ± 10gcm^{-2}/decade for a mass composition which does not vary with energy. Hence the Fly's Eye result is quite consistent with little or no change in the mass composition with energy. If, over a decade, the mass changed from pure iron to pure proton then the elongation rate would be about 185gcm^{-2}/decade, depending on the exact nature of the change. Conversely, if the mass composition changed from pure proton to pure iron over the same range, the elongation rate would be negative and approximately equal to -50gcm^{-2}/decade. From the Fly's Eye study of depth of maximum as a function of energy all that can be said is that the change in mass composition is rather small: certainly a change in the ratio proton/iron by a factor of two is probably excluded but a 10% change could not be detected because of uncertainties about particle physics.

From an analysis of the width of the experimental distribution of depth of maximum[22] it is found that the data are not consistent with a pure mass distribution of either iron or protons but rather are consistent with a mixture. This statement refers to an average energy of 3 x 10^{17}eV. A two component model is probably highly simplistic but is all that the data can justify at this stage.

These recent measurements by the Fly's Eye group have allowed a reassessment of much older work done at Haverah Park[24] where the elongation rate was deduced from a complex analysis based on measurements of the rise-time of signals in the 34m^2 deep-water-Cerenkov detectors as a function of distance, zenith angle and energy. The rise-time is sensitive to the position of shower maximum with deeper penetrating showers giving slower rise-times at fixed axial distance, energy and zenith angle. The measurements reported were found to be quite consistent with expectation for a mass composition which is independent of energy from 3.10^{17} to 7.10^{18}eV: the elongation rate was 70 ± 5gcm^{-2}/decade but the complexity of the analysis hampered interpretation and hindered belief. The results from the Fly's Eye group allow further insight[25], particularly when combined with evidence for a change of slope in the energy spectrum near 10^{19}eV. For the seven lowest energy points the elongation rate measured in the Haverah Park experiment is 73 ± 5gcm^{-2}/decade, in excellent agreement with Fly's Eye and thus giving confidence in the Haverah Park method. However, for the three highest energy points the measurements of elongation rate are 26 ± 56, 12 ± 50 and 50 ± 25gcm^{-2}/decade respectively. The first two points cover about 1.5 decades centred on 10^{19}eV and the weighted average of 18 ± 37gcm^{-2}/decade is consistent with a rather rapid increase in the average mass at about this energy. The final datum is consistent with a uniform composition. However the errors are very large and efforts are under way to reduce them so that, for now, we have only a tantalising indication of a mass change.

CONCLUSIONS

The main conclusion which can be drawn from even a cursory reading of the above is that much more high quality data is needed to resolve the problem of arrival direction distributions and the mass composition above 10^{18}eV. Such data can only be acquired through a concerted international effort to build a truly giant array with an aperture of perhaps 10^4km^{2}sr. With such a device approximately 5000 events per year would be recorded above 10^{19}eV where the integral rate[2] is about 0.5km^{-2}y^{-1}sr^{-1}. Thus a factor of 5 more events per year would be collected than the present integrated total from all of the experiments made world-wide. In principle such an array can be constructed: the previous experiments have provided excellent prototyping and the only barriers are the will of sufficient people to do it, and the money (perhaps $50M). However unless this effort is put in - and the money required is less than the $70M overrun on the LEP tunnel at CERN[26] - the experimental situation will not change radically over the next ten years. Such an array would answer definitively questions about the existence or not of primary cosmic rays above 10^{20}eV - of major interest to those studying acceleration mechanisms - and would give a very superior understanding of the arrival direction distribution of ultra high energy cosmic rays. However pretty sky patterns alone would be hard to interpret and provision must be made for mass compositions to be measured in at least 10% of events so that the complex question of the origin of the highest energy cosmic rays can be finally disentangled.

1. P. Sokolsky, This workshop (1991).
2. A. A. Watson, Nuclear Physics B (Proc Suppl) 22B, 116 (1991).
3. M. M. Winn et al, J.Phys.G 12, 675 (1986).
4. P. Bhattacharjee, in Astrophysical Aspects of the Most Energetic Cosmic Rays (World Scientific, Singapore 1991), p. 382.
5. J. Linsley, Proc. 14th Int. Conf. Cos. Ray Phys. 2, 598 (1975).
6. P. V. J. Eames et al, Proc. 19th Int. Cos. Ray Conf. (La Jolla) 2, 254 (1985).
7. T. A. Egorov et al, Proc. 22nd Int. Cos. Ray Conf. (Dublin) 2, 121 (1991).
8. N. Hayashida et al, Proc. 22nd Int. Cos. Ray Conf. (Dublin) 2, 117 (1991).
9. G. L. Cassiday et al, Ap.J. 351, 454 (1990).
10. J. Wdowczyk and A. W. Wolfendale, J.Phys.G 10, 1453 (1984).
11. J. Szabelski, J. Wdowczyk and A. W. Wolfendale, J.Phys.G 12, 1442 (1986).
12. G. L. Cassiday et al, Phys. Rev. Lett. 62, 383 (1988).
13. M. A. Lawrence, D. C. Prosser and A. A. Watson, Phys. Rev. Lett. 63, 1121 (1989).
14. M. Teshima et al, Phys. Rev. Lett. 64, 1628 (1990).
15. R. Cooper et al, in Astrophysics Aspects of the Most Energetic Cosmic Rays (World Scientific, Singapore 1991) p. 34.
16. N. Hayashida et al, Proc. 22nd Int. Cos. Ray Conf. (Dublin) 1, 309 (1991).
17. X. Chi et al, J.Phys.G 18, 539 (1992).
18. M. A. Lawrence, D. C. Prosser and A. A. Watson, Proc. 22nd Int. Cos. Ray Conf. (Dublin) 2, 125 (1991).
    M. A. Lawrence and A. A. Watson, submitted to J.Phys.G (1992).
19. A. V. Glushkov, N. N. Efimov and A. M. Mikhailov, Proc. 22nd Int. Cos. Ray Conf. (Dublin) 2, 113 (1991).
20. J. Lloyd-Evans, Proc. 22nd Int. Cos. Ray Conf. (Dublin) 5, 215 (1991).
21. J. Linsley, Proc. 15th Int. Cos. Ray Conf. (Plovdiv) 12, 89 (1977).
22. G. L. Cassiday et al, Ap.J. 356, 669 (1990).
23. J. Linsley and A. A. Watson, Phys. Rev. Lett. 46, 459 (1981).
24. R Walker and A. A. Watson, J.Phys.G 8, 1131 (1991).
25. A. A. Watson, in Astrophysical Aspects of the Most Energetic Cosmic Rays (World Scientific, Singapore 1991) p. 2.
26. P. Aldhous, Nature 355, 285 (1992).

Ultrahigh-Energy Cosmic Rays from Fanaroff Riley class II Radio Galaxies

Jörg Rachen and Peter L. Biermann
Max Planck Institut für Radioastronomie, Bonn, Germany

February 4, 1992

Abstract

The hot spots of very powerful radio galaxies (Fanaroff Riley class II) are argued to be the sources of the ultrahigh energy component in Cosmic Rays. We present calculations of Cosmic Ray transport in an evolving universe, taking the losses against the microwave background properly into account. As input we use the models for the cosmological radio source evolution derived by radioastronomers (mainly Peacock 1985). The model we adopt for the acceleration in the radio hot spots has been introduced by Biermann and Strittmatter (1987), and Meisenheimer et al. (1989) and is based on first order Fermi theory of particle acceleration at shocks (see, e.g., Drury 1983). As an unknown the actual proportion of energy density in protons enters, which together with structural uncertainties in the hot spots should introduce no more than one order of magnitude in uncertainty: We easily reproduce the observed spectra of high energy cosmic rays. It follows that scattering of charged energetic particles in intergalactic space must be sufficiently small in order to obtain contributions from sources as far away as even the nearest Fanaroff Riley class II radio galaxies. This implies a strong constraint on the turbulent magnetic field in intergalactic space.

1 The origin of the high energy Cosmic Rays

The origin of the Cosmic Ray particles with energies beyond the knee at about 5×10^{15} eV is still a matter of debate. There is no theory which at once explains a) how the particles reach those energies, b) how they attain their spectrum, and c) how the flux can match the flux of those particles with energies below the knee for which we believe we know the origin. For the origin of the particles at energies beyond the knee the leading contender theories are i) from a galactic wind shock, ii) from acceleration by a supernova shock racing through a stellar wind, and iii) from neutron stars. Near 3×10^{18} eV the situation becomes simpler insofar as a spectral change suggests that the origin changes, and the Larmor radii of these particles become so large that a galactic origin can definitely be excluded. It is those particles with extremely high energies which we wish to address in this communication.

The central thesis here will be that these particles originate in the hot spots of radio galaxies. These hot spots are believed to be the sites of very strong shock waves; such a theory has several advantages: a) first of all, it can be tested since we know the space density of radio galaxies, and we can directly calculate what the integrated contribution can be. b) We obtain at once from such a theory i) particle energies, ii) particle flux and spectrum, and iii) a constraint on the particle propagation through the cosmos. In the following we will review the basic argument and then present calculations that describe the properties of Cosmic Rays at such energies.

2 Radiogalaxy hot spots

2.1 Acceleration of particles

The spectra of the radio hot spots of those radio galaxies that have a characteristic double structure (see, e.g., Miley 1980) often show a spectrum that extends into the optical range and then usually cuts off rather abruptly. Similar properties are found for some red quasars and jets. Biermann and Strittmatter (1987) proposed that such spectra could be understood as arising from electrons accelerated in shocks, that are mediated by energetic protons. The role of the protons is crucial here: They provide the messenger for the upstream gas to react to the shock downstream, thereby excite turbulence and so provide the wavefield for the electrons to scatter on both sides of the shock; this repeated scattering is essential for electrons to gain energy on their repeated passages through the shock. A critical assumption is here that this wavefield is of Kolmogorov character; however, from plasma simulations, from the interstellar medium, as well as from the solar wind it is now known that Kolmogorov turbulence appears to be one standard mode of turbulent motion in a plasma outside the dissipation scales. A first order Fermi theory approach was also used by Meisenheimer et al. (1989) to actually fit observed hot spot spectra. The fit is quite good and suggests that the model has merit. For many of the hot spots they discuss the mean free path of scattering can be estimated and so it can actually be checked whether a Kolmogorov spectrum is consistent with the data: It is. From the argument of Biermann and Strittmatter (1987) it then follows that protons exist in the hot spots that reach rather high energies, in magnetic field environments of 10^{-4} Gauss energies up to 10^{21} eV. It can be ascertained that the space available and the competing losses from proton-proton collisions do not normally win over the Synchrotron and photon-interaction losses which limit the proton energy that can be reached. This argument depends, however, on the magnetic field used: For a lower magnetic field, below 100μGauss, the space limitation may introduce a cutoff in the energy distribution of protons and nuclei similar to the loss induced cutoff argued here. For the following arguments the origin of this cutoff is irrelevant, as long as the energy of the cutoff is near 10^{20} eV or higher. We have used for reference an exponential law for the shape of the cutoff, suggested by earlier considerations. The spectra of these energetic protons and heavier nuclei can be estimated from the electron population: These spectra are very close to E^{-2}, as shown by the $\nu^{-0.5}$ Synchrotron spectra observed. Thus, these arguments provide clear evidence that radio hot spots harbor a population of very high energy protons, and, by analogy, heavier nuclei. Such radio hot spots are situated far outside their central galaxy in most cases, and may well be considered to be embedded in intergalactic space. Thus, it it an interesting question to ask to what degree such hot spots might contribute to the observed high energy cosmic rays, since their expected particle energy range is very close to that observed. In the following we limit our calculations to protons, since they are almost certainly the dominant chemical species in radio galaxies.

2.2 Radio source evolution

It is well known that radio sources with steep spectra, i.e. overall radiospectra close to $\nu^{-0.8}$, show a correlation between overall structure and radio luminosity. The powerful sources have the characteristic double lobe structure, where most of the emission is in two symmetrically situated radiolobes on the two sides of a central, usually weak source. These double lobes in turn often show characteristic hot spots that can readily be interpreted as the terminal shock of a powerful gasflow coming out from the central source. These luminous radiosources with the double structure are referred to as Fanaroff Riley class II sources (Fanaroff and Riley 1974), whereas the corresponding weaker sources, where the central emission dominates, are referred to as Fanaroff Riley class I sources. Radio surveys provide then further information on the cosmological evolution of these powerful radio sources. Peacock (1985) has provided a number of models that describe the evolution of the radio source population to explain the radio data.

We will use these radio source evolution models as input to discuss the sources of cosmic rays as a function of cosmological epoch.

3 The evolution of the injected spectrum of energetic particles

3.1 The injection

When observing radio sources, the energy density of the various relativistic components can be estimated from the requirement that the observed radio emission be reproduced in the observed space. The luminosity of Synchrotron emission is clearly proportional to some power of the magnetic field, depending on the spectrum of the emission. The required particle energy density in relativistic electrons then is inversely related to the assumed magnetic field strength. The energy density of the magnetic field is obviously proportional to the square of the field strength. The assumption that the sum is near the minimum in real environments leads then to a unique relation between magnetic field, particle energy density and total energy density. The relative proportion of proton — and heavier nuclei — and electron populations of relativistic particles k_p remains an unknown here; in the interstellar medium this ratio is 100 for particle energies above 1 GeV. In radio lobes and hot spots this number could well be anywhere between unity and more than 1000. This enters below into the fudge factor, the factor given in figure 4.

3.2 The Cosmic Ray evolution

Interaction of the high energy particles with the microwave background is a strong limiting factor in determining for how much time the injected particles can travel in intergalactic space (Greisen 1966, Stecker 1968). The dominant loss processes in this interaction are first of all pion production, due to the extremely strong increase of the proton-photon interaction cross section at pion energies (the Delta resonance), then the corresponding electron-positron pair production (Blumenthal 1970), and last, the adiabatic losses due to the expansion of the universe. Fig. 1 shows the cross sections for the various hadronic interaction channels, all weighted by the inelasticity of the encounter. The figure demonstrates that the first interaction at the $\Delta(1232)$ resonance which produces a pion is dominant for the lowest energies. We note that the corresponding cross section for pair production extends to much lower energies, but is much lower, peaking broadly at an energy of 10 MeV with a cross section times inelasticity just below 1 μbarn. Obviously, we do include the cosmological evolution of the microwave background itself in energy density and temperature. We treat here the proton-photon interaction as a continuous loss process, following Berezinsky and Grigor'eva (1988). They considered this approximation and demonstrated that it does not introduce errors of more than about 10%. Thus, the observed particle energy spectrum is cut off at some energy corresponding to the time that has elapsed since injection, the Greisen cutoff energy. Fig. 2 shows the elapsed time that a particle with an assumed original energy of 10^{21} eV can travel in the universe to reach us here at earth with the energy E_0. It follows that practically no particles can reach us from any interesting distance, such as distances beyond the Virgo cluster, which have an energy upon arrival of above or even very close to 10^{20} eV. This particle energy loss is folded in our calculation with an injected spectrum and the cosmological distribution of sources at different powers corresponding to the observed radio source evolution.

4 The graininess of the universe for Cosmic Ray propagation

Thus we compose total spectra of ultrahigh energy particles at earth out of the contributions from the different radio galaxies that have injected energetic particles in the past. For this calculation it is a priori not relevant whether we have straight line propagation or diffusion, i.e. a random walk through the universe, as long as the graininess of the source distribution does not become dominant. For instance if all energetic particles would stay close to their

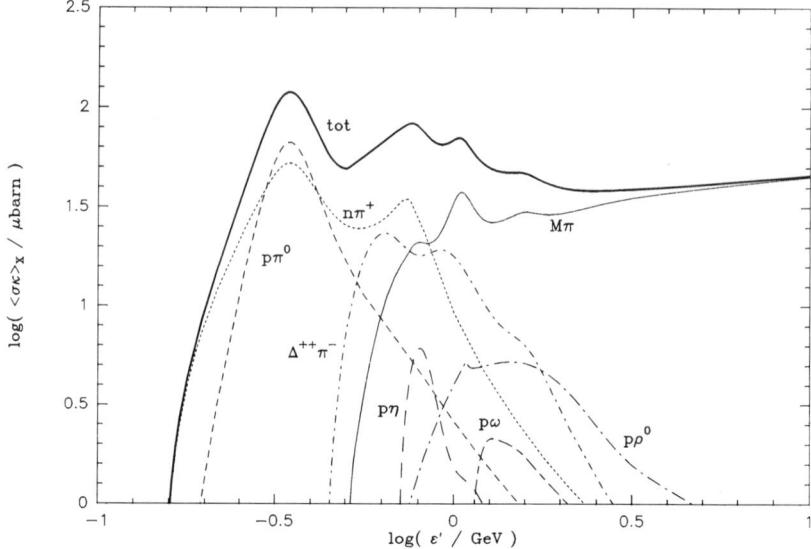

Figure 1: The cross sections of the hadronic proton-photon cross sections in the proton frame, multiplied with the dimensionless inelasticity κ of the interaction. $M\pi$ denotes the sum of all multi-pion channels.

source and form a cosmologically small sphere filled with energetic particles, then outside the spatial sum of all such spheres there would be no contribution whatsoever. Thus, in order to be relevant as a description of real nature, our calculation imposes the condition that the path of energetic particles goes out sufficiently far from the sources, in order to attain a nearly homogeneous distribution. Obviously, this could be translated into a limit on the strength of the intergalactic magnetic field, given a spatial topology. Considering a random walk, the time required to traverse a distance of about half the mean distance between powerful radio galaxies leads to a lower limit to the mean free path of scattering. This lower limit is about $10/h_0$ Mpc for particles of 3×10^{18} eV and about $100/h_0$ Mpc for particles of 10^{20} eV (see Fig. 2).

This means in the quasilinear theory for the scattering of particles in a fluctuating magnetic field, that the gyroradius has to be of this order given a fully turbulent field. In order to gain some insight about the possible strength of the magnetic field implied here, assume first that the turbulent wavefield is saturated, i.e. is of the same energy density as the background field at all wavenumbers. Then the spatial requirement on the energy dependent mean free path implies an upper limit to the strength of the magnetic field of $0.3/h_0$ nGauss and $1/h_0$ nGauss for 3×10^{18} eV and 10^{20} eV, respectively. If the energy density in the turbulent wavefield is a fraction ϵ of the background field, then the implied limit on the magnetic field strength is increased by a factor $1/\epsilon$. One possibility is also, that the turbulence attains maximum intensity at wavenumbers corresponding to 10^{20} eV, and can be characterized by a sawtooth pattern at lower energy; then the mean free path is independent of energy, and the implied upper limit to the magnetic field strength is about $1/h_0$ nGauss. If, one the other hand, the turbulent wavefield is better characterized by convective motions that transport the embedded Cosmic Rays in bulk, then again the mean free path is independent of energy, and the size of these convective cells has to be of order $50/h_0$ Mpc or larger. The magnetic field strength inside such

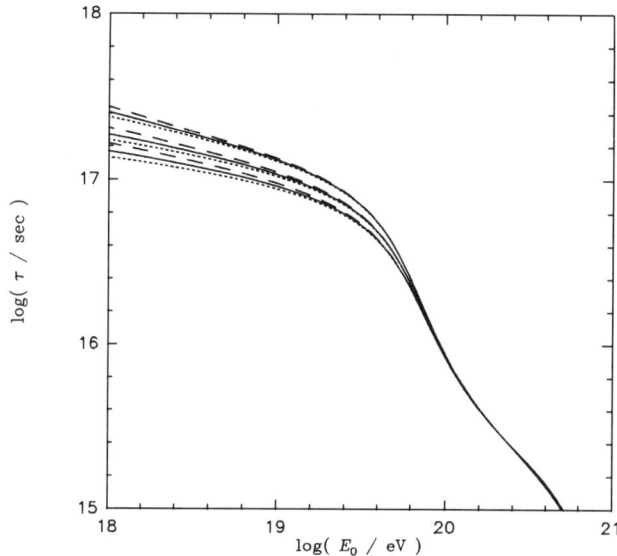

Figure 2: Maximum travel time for a particle arriving at earth with energy E_0, assuming an injection cutoff energy of 10^{21} eV. The different curves correspond to the variation of cosmology; dashed: $q_0 = 0$; solid: $q_0 = \frac{1}{2}$; dotted: $q_0 = 1$; upper triplet: $h_0 = 0.5$; central triplet: $h_0 = 0.75$; lower triplet: $h_0 = 1.0$. We use $H_0 = h_0 \times 100 \,\mathrm{km\, sec^{-1}\, Mpc^{-1}}$.

cells should obviously not exceed the limits just given; a magnetic field of nanoGauss strength would be entirely consistent with all data. It is a tantalizing speculation to ask whether such postulated cells could have any relationship to the overall cell structure believed to exist in the universe.

5 Ultrahigh energy Cosmic Ray spectra

The calculated spectra are displayed in Figs 3 and 4. The first figure gives the contributions from identified radio galaxies, assuming in this case straight line propagation. Therefore we give the spectra here as total flux without specifying which part of 4π the particles actually reach. We see that individual radio galaxies can dominate the total ultrahigh energy Cosmic Ray spectrum, given the simplistic propagation model. In Fig. 4 we show the total spectrum using the radio source evolution model dubbed RLF2 of Peacock (1985), for a range in initial cutoff energies.

The fudge factor given in each case is the factor necessary to fit the Cosmic Ray flux at $10^{19.1}$ eV. This factor has to be compared with the function $\frac{4}{7} k_p (1 + k_p)^{-3/7}$, which is the minimum energy density factor of relativistic protons relative to the case where no protons are assumed; this, however, is the case in the jet and radio lobe energies estimated by Rawlings and Saunders (1991). Keeping k_p as an unknown reduces the uncertainties involved in our estimate to questions like a) is the fraction of energy in protons in the hot spots the same as in the entire lobe, and b) is the fraction of protons which leave the source going into intergalactic space independent of energy? The combination of all such factors, which we assume to be near unity in their effect, with the function involving k_p has to match the fudge factor given in the figures. We see that this is easily done with the possible range for k_p; the k_p-term by itself provides a possible range of 0.42 over 7.9 to 44. for k_p from 1 over 100 to 2000. We note that spectra very

398 Fanaroff Riley Class II Radio Galaxies

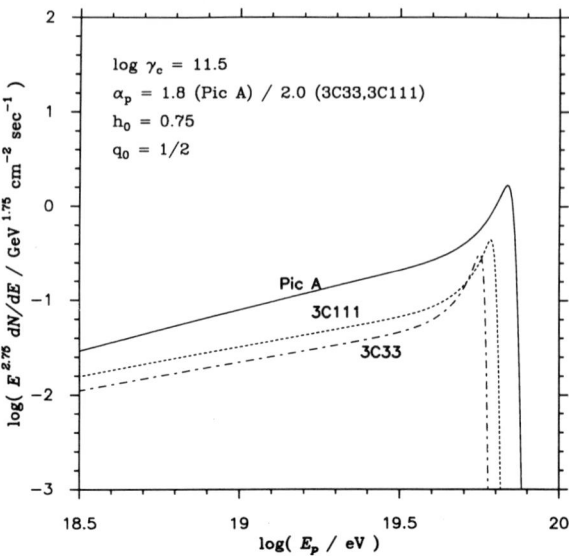

Figure 3: The contribution of identified radio galaxies Pic A, 3C111 and 3C33 to the Cosmic Ray spectrum; the jet luminosities and spectral indices were taken from Meisenheimer et al., 1989, who assumed $k_p \leq 10$.

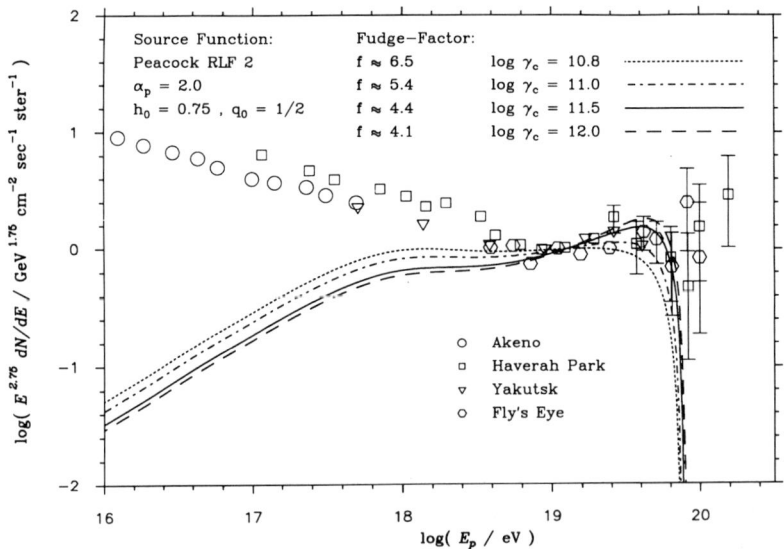

Figure 4: The total Cosmic Ray spectrum for different initial cutoff energies; the fudge factor is quite reasonable for all cases considered.

close to $\alpha_p = 2.0$ or flatter are required; changing the assumed injection spectrum index from 1.9 to 2.0 and 2.1 increases the implied fudge factor from 0.27 to 5.4 to 170; the last of these values is clearly too large and such models have to be excluded probably. We remark that the cutoff energy, assumed here to be the same in all sources, could be as low as 10^{20} eV or even a little less without changing the overall result. An early analytical version of these results was first presented by one of us (P.L.B.) at a conference in honor of M.M. Shapiro in October 1990 in Washington, D.C. It appears possible to explain nearly all data above about 10^{18} eV within this model.

6 Conclusion

We have thus combined knowledge from cosmological radio source evolution with a tested physical model for the acceleration of relativistic nuclei and have found that we can readily explain the entire ultrahigh energy spectrum of cosmic rays. We emphasize that our model requires that the intergalactic diffusion length for the transport of these high energy particles be of order $50/h_0$ Mpc or larger.

Acknowledgments

We wish to thank Drs. T.K. Gaisser, J.R. Jokipii, W.H. Matthaeus, T. Stanev and A.A. Watson for many discussions on the origin of high energy Cosmic Rays and their propagation, and Dr. T. Stanev for giving us his weighted spectrum of Cosmic Rays to compare with our model. One of us, P.L.B., also wishes to acknowledge the generous hospitailty at Bartol on various occasions when he visited.

References

Berezinsky, V.S., Grigor'eva, S.I.: 1988, *Astron. Astrophys.* **199**, 1
Biermann, P.L., Strittmatter, P.A.: 1987, *Astroph. Journal* **322**, 643
Biermann, P.L.: 1991, in *Frontiers in Astrophysics*, eds. R. Silberberg, G. Fazio, M. Rees, *Cambridge University Press*
Blumenthal, G.: 1970, *Phys. Rev. D* **1**, 1596
Drury, L. O'C: 1983, *Rep. Prog. Phys.* **46**, 973
Fanaroff, B.L., Riley, J.M.: 1974, *Monthly Not. Roy. Astr. Soc.* **167**, 31P
Greisen, K.: 1966, *Phys. Rev. Letters* **16**, 748
Meisenheimer, K., Röser, H.-J., Hiltner, P.R., Yates, M.G., Longair, M.S., Chini, R., Perley, R.A.: 1989, *Astron. Astrophys.* **219**, 63
Miley, G.K.: 1980, *Ann. Rev. Astron. Astroph.* **18**, 165
Peacock, J.A.: 1985, *Monthly Not. Roy. Astr. Soc.* **217**, 601
Rawlings, S., Saunders, R.: 1991, *Nature* **349**, 138
Stecker, F.W.: 1968, *Phys. Rev. Letters* **21**, 1016

PARTICLE ACCELERATION UP TO 10^{20} EV

W.-H. Ip and W.I. Axford

Max-Planck-Institut für Aeronomie, W–3411 Katlenburg-Lindau, Germany

ABSTRACT

A scenario for second-stage acceleration of the galactic cosmic rays beyond 10^{14} eV via multiple interaction with SNRs in the galactic disc is investigated. A numerical scheme taking into consideration the size frequency distribution of SNRs, the probability distribution of energy gain/loss during SNR encounter and the compositional abundances of the cosmic ray nuclei is used to illustrate the quantitative aspects of such process.

1. INTRODUCTION

A critical issue in cosmic ray physics concerns the acceleration of charged particles to energies as high as 10^{20} eV. The very successful theory of diffusive shock acceleration can account for the power law behaviour of the energy spectra up to $\sim 10^{14}$ eV [1]. The limitation in the acceleration to higher energy is due to the time available for particle acceleration by a single supernova remnant (SNR) or the finite size of the SNR as discussed by Lagage and Cesarsky [2]. The fact that the power law index of the differential flux ($dj/\alpha E \alpha E^{-\gamma}$) changes from $\gamma \simeq 2.7$ for $E \leq 10^{14}$ eV and $\gamma \simeq 3$ for higher energies suggests that there are different acceleration mechanisms for these two cosmic ray populations.

A number of theoretical models have been proposed to account for such a "bump" in the cosmic ray energy spectrum. These include the ideas of introducing the effect of nuclear fragmentation as a result of interaction with EUV photons in the Galaxy [3], the contribution from other independent sources [4] and galactic wind termination shock acceleration [5]. After considerations of the merits and difficulties of different models, Axford [6] drew the conclusion that the pertinent acceleration of cosmic ray particles beyond the critical energy of 10^{14} eV might actually be closely related to the reacceleration of the cosmic rays via a sequence of shock encounters in the galactic disc. It was suggested that after escape from the source SNRs, the cosmic ray particles with an initial energy of 10^{14} eV, say, would propagate through the Galaxy as guided by the galactic magnetic field configuration. In the absence of effective scattering centres, the particles would mainly interact with large-scale structures in the Galaxy, i.e., the molecular clouds, SNRs and OB associations, etc. and especially with the associated shock waves.

At each encounter with such a structure the cosmic ray particles would be scattered with consequent displacements in their guiding centres and changes in pitch angles. The lifetime of the "free-streaming" cosmic rays is thus to a large extent determined by their random walk motion in the galactic environment, in addition to drift motions in the galactic magnetic field. Berezinsky et al. [7] have considered the latter problem using a simple model of galactic magnetic field in the disc and the galactic corona and found that the storage time of the high energy cosmic rays with $E > 10^{14}$ eV varies as $E^{-0.3}$. Since the magnetic field structure in the Galaxy is still very uncertain it may be appropriate to consider the derived values as lower limits. The upper limit, on the other hand, will be given by the dynamical escape time (t_d) in the leaky box model for particles at 10^{14} eV which is approximately 2×10^5 years. This is in fact the value we shall use in our preliminary consideration. We mention this point here because, as will be discussed later, the nature of the energy spectrum beyond 10^{14} eV depends sensitively on the dynamical escape time and the relevant time between encounters with SNRs.

2. THE GCR II ACCELERATION

We follow the terminology given by Axford [6] in naming the diffusive shock acceleration to 10^{14} eV by one single SNR the first-state (GCR I) acceleration and the subsequent acceleration via multiple SNR encounters the second-stage (GCR II) acceleration. The second-stage process could, in principle, produce particles up to 10^{17} eV for protons and up to 3×10^{18} eV for Fe nuclei. The cosmic ray flux at higher energies (EGCR) is postulated to originate from acceleration external to the Galaxy, namely, the extragalactic radio sources. It has been shown that the hot spots in the radio jets could be favourable sites for particle acceleration up to 10^{20} eV [8,9]. The observed GCR energy spectrum would thus be a combination of the GCR I, GCR II and the EGCR populations. The key information on the individual contributions would have to come from the energy spectra of different species, i.e., protons vs. Fe. For example, the recent report of a change of the proton spectrum from a power law behaviour of $E^{-2.7}$ up to 10^{14} eV and then to $E^{-3.4}$ at higher energies [10] sets a strong constraint on the efficiency of the diffusive shock acceleration and hence indicates the necessity for a second-stage reacceleration process.

For second-stage acceleration via multiple shock encounters to work, the rate of energy gain is quite modest (e.g., $\Delta E/E \approx 0.3\%$ per encounter). However, it is not obvious that such rate can be achieved in a straightforward manner. This is because several complicated factors enter into the acceleration process, namely, the exact nature of the shock drift acceleration mechanism, the size distribution of the SNRs in the Galaxy, the random propagation of cosmic rays in different regions of the Galaxy (and hence the galactic magnetic field configuration). Most of these properties are unfortunately not very well known. Consequently any second-stage GCR acceleration model must necessarily involve uncertainties in the astrophysical parameters chosen. In the following, we shall present some preliminary accounts such that the quantitative aspects of several major elements could be further explored.

It should be noted that, even though the reacceleration by old SNRs is used as an example here, it does not necessarily mean that other processes of similar nature in the galactic disc can be excluded. Astrophysical systems with large electric fields such as strong stellar winds from OB stars and/or rotating neutron stars are all possible sites for the second-stage acceleration. The scenario of multiple shock interaction is first investigated as it provides a convenient extentions of the theory of diffusive shock acceleration.

3. MODEL FOR MULTIPLE SNR INTERACTIONS

As a SNR expands, the interstellar magnetic field lines connected with the shock front are carried outward together with the blast wave. During the passage of a cosmic ray particle through the SNR, it will experience the "motional" electric field given as $\underline{E} = -\underline{V}(r,t) \times \underline{B}(r,t)$ where $\underline{V}(r,t)$ and $\underline{B}(r,t)$ are the plasma flow velocity and magnetic field inside the SNR, respectively. The starting point of our calculation is therefore to find a suitable description of $\underline{V}(r,)$ and $\underline{B}(r,t)$. An analytical model based on the Sedov similarity solution of an SN explosion has been derived by Rakiewicz et al. [11]. In our calculations we have assumed that the interstellar medium has a density equivalent to 3×10^{-3} H-atoms cm^{-3}, which is appropriate for a hot interstellar medium.

The magnetic and electric field structures are illustrated in Fig. 1. It can be seen that the motional electric field has its maximum magnitude near the shock front because of the compression of the interstellar magnetic field and the peak value of the flow velocity therein.

There are several ways for the cosmic ray particles to interact with the SNR. Firstly, in the case in which the gyroradius of the particle is considerably smaller than the size of the SNR and the particle simply enters the SNR from one end and leaves from the other along a magnetic field line, the energy will be reduced as a result of adiabatic cooling inside the expanding SNR. Secondly, if the initial pitch angle is large such that the particle is reflected at encounter with the shock front, an energy increment will result. The absolute energy changes in these two types of encounter geometries are on the order of a few tenths per percent at particle energy $E \sim 10^{16}$ eV. As discussed later, encounters with energy gain generally have higher probability than those with energy loss within a certain energy range. The corresponding acceleration process thus may be essentially considered to be of the nature of the first-order Fermi effect [12].

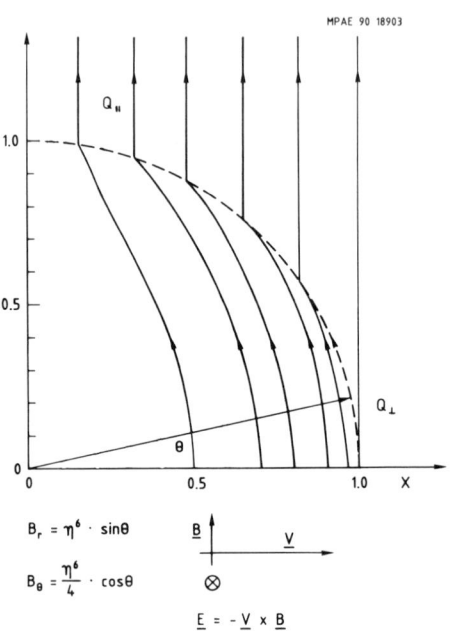

Fig. 1. A sketch of the magnetic field configuration in the vincinity of an expanding SNR during its Sedov phase. The analytical approximation is derived from a similarity solution.

If the particle encounters the shock with guiding centre impact parameter b within one gyroradius (R_g) of the SNR radius, the cosmic ray particles may be temporarily trapped in cycloidal motion across the shock front in the direction of the motional electric field. Relatively large energy increments ($\Delta E/E$ up to 10%) could be achieved for $E \sim 10^{15} - 16^{16}$ eV. A significant increase to the particle energy can therefore be occasionally obtained in the course of the multiple shock encounters. However, such drift acceleration does not operate when R_g is comparable to the radius of the SNR (R_{SNR}). In this case, such "edge-on" encounters will shift from the regime of energy gain to that of energy loss. All these tend to reduce the efficiency of the second-stage acceleration at a certain upper energy limit (E^*). With the maximum SNR radius assumed to be about 160 pc and a galactic magnetic field of 6 μG, we find $E^* = 5 \times 10^{17}$ eV for protons.

During their motion in interstellar space the cosmic ray particles will encounter a wide variety of SNRs and other large-scale structures. In the simplest approximation we may consider the random encounters to be entirely uncorrelated. In other words, each interaction is assumed to be independent of the previous SNR encounter such that all relevant parameters, i.e., pitch angle, gyrophase, impact parameter and size of the SNR, can be chosen randomly. Furthermore, there is no overlapping of the SNRs. Because of the tendency for SNRs to concentrate in clusters [13] it is quite possible that a more realistic model must take into account the localized spatial and age distributions of the SNRs. Taking the life time of the SNR Sedov phase to be 2×10^5 years and making the assumption of random explosion of supernovas in space and time we can construct the cumulative encounter frequency distribution (f_c) for GCR interaction with SNRs.

The next step of our calculations is to compute the probability distribution of energy gain and loss for particles with initial energy E_o. We have considered six cases, e.g., $E_o = 10^{15}$ eV, 10^{16} eV, 3×10^{16} eV, 10^{17} eV and 3×10^{17} eV, with a variety of B_o (3 μG, 6μG and 10 μG). In each case, 10^4 encounters have been simulated with the starting value of the pitch angle, gyrophase and other input parameters chosen at random. The trajectories of the relativistic particles are traced by numerical integration of the equation of motion: $dp/dt = ze(\underline{E} + \underline{V} \times \underline{B}/c)$.

Two basic features can be seen from the cumulative probabilities of the energy changes (f_E): (1) encounters with energy loss tend to become more frequent for increasing particle energies; (2) the multiple shock acceleration works best for large value of B_o. Both effects are related to the finite Larmor radii of the energetic particle.

4. THE SYNTHETIC ENERGY SPECTRA

The numerical values of the cumulative probabilities of energy change (f_E) can be used to simulate the second-stage multiple shock acceleration in the following manner. First, the initial energy is taken to be 10^{14} eV which is taken to be the upper limit for diffusive shock acceleration for protons. For particle energy $E < 10^{15}$ eV, the corresponding value of f_E is assumed to be the same as that for $E = 10^{15}$ eV. A random number (ξ) is generated at each time step (encounter) such that the energy change $\Delta E/E$ is given by assigning $f_E = \xi$. For $E > 10^{15}$ eV, the value $\Delta E/E$ is obtained by interpolation between two curves, f'_E and f''_E, of which $E' < E < E''$. The synthetic energy spectrum is assumed to be divided into N energy bins with bin size δE. At the ith time step, the contribution from the test particle to the jth energy bin $[j = E(t_i)/\delta E]$ is given by the weighting function, $w_i = \exp(-i\Delta t/t_d)\,[= \exp(-i/N_t)]$.

This procedure is repeated until either $i = i_{max}$ or $E_i = E_{max}$. In the above, $N_t = (t_d/\Delta t)$ is the ratio of the dynamical lifetime of the galactic cosmic rays to the average time (Δt) between two shock encounters. As for E_{max}, there are two limiting factors. First, the Larmor radius of the accelerated particles should not be larger than the thickness of the galactic disc. Hence for $r_g < 150$ pc, Z=1 and $B_o = 6\mu$G we have $E_{max} = 5 \times 10^{17}$ eV. The second factor is naturally the efficiency in particle acceleration via shock encounter. In our computations, it is found that, particle energization appears to be impeded at $E > 3 \times 10^{17}$ eV. We consequently take $E_{max} = 3 \times 10^{17}$ eV for protons. For heavy nuclei, $E_{max} = 3 \times 10^{17}$ Z eV.

The simulated cumulative spectra for $B_o = 6\mu G$ are summarized in Fig. 2. As can be seen, while the starting value of the injection rigidity (P_o) does not change the slopes of the energy spectra the power law spectral index (γ) is sensitive to the ratio of the dynamical escape time to the SNR encounter time (i.e., $N_t = t_d/\Delta t$). We have found that for a power law with $\gamma = 3$, N_t is required to be 170 for $B_o = 10\mu G$, and 300 for $B_o = 3\mu G$. The latter value is generally accepted to be the average value of the magnetic field strength in the galactic disc. This also means $t_d \approx N_t\Delta t \approx 3 \times 10^5$ years if $\Delta t \approx 10^3$ years. Such a long storage time of the high energy cosmic rays with energies up to a few times 10^{17} eV is certainly a matter of concern in our model. We have no solution to this problem at the current time. It is to be mentioned that a more realistic model of the large-scaled magnetic field may lead to a longer storage time for the ultra-high energy cosmic rays. Another possible way out is to invoke additional mechanism(s) which could provide more effective acceleration effect at the high energy end. The neutron stellar wind acceleration effects proposed by Bell [14] and Berezhko [15]. are therefore of potential interest in this respect. As with these authors the common point we are making is simply that the GCR beyond 10^{14} eV could originate directly from the same population of cosmic ray particles energized by diffusive shock acceleration.

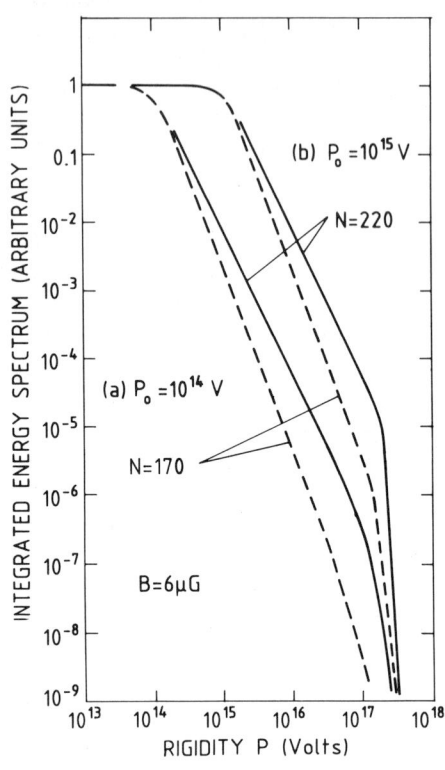

Fig. 2. The intergral energy spectra of the GCR II population derived by using different values of the injection rigidity (P_o) and N_t. The background magnetic field is assumed to be 6 μG in this example.

The final step of our model calculations is to produce a composite differential energy spectrum. This can be done by superimposing the pertinent theoretical curves in Fig. 2 (e.g., $N_t = 220$ and $P_o = 10^{14}$ V) for different nuclei by taking into consideration their Z-dependence and relative abundances. In addition, the power spectra of the nuclei from the EGCR population can also be simulated by assuming a power law distribution with $\gamma = 2.7$. An example of the resulting synthetic spectrum using relative abundances given by Swordy et al. [16] is shown in Fig. 3. This shows that the presence of a "dip" at 3×10^{18} eV in the observed GCR spectrum is consistent with the overlapping of the GCR II and EGCR components with different spectra in this region. On the other hand, the "bump" at approximately 3×10^{15} eV cannot be fitted by the theoretical spectrum [17]. This discrepancy may be partly resolved by adjusting the relative abundances of the heavy nuclei (say, Fe) at the injection energy $E_o = 10^{14}$ eV. However, this point remains to be investigated.

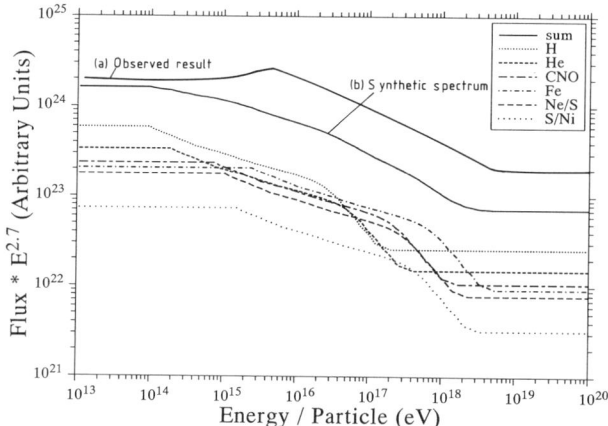

Fig. 3. A synthetic spectrum of the GCR by combining the GCR I component from diffusive shock acceleration, the GCR II component from multiple shock interactions and the EGCR population from extragalactic radio jets. The input parameters are $P_o = 10^{14}$ V and $B_o = 6\mu G$.

REFERENCES

1. L.O.C. Drury, Rep. Prog. Phys. **46**, 973 (1983).
2. P.O. Lagage and C.J. Cesarsky, Astron. Astrophys. **125**, 249 (1983).
3. S. Karakula, S.U. Bujiksjtm, T.N. Stamenov and W. Tkaczyk, Proc. 20th ICRC, Adelaide, 3, 169 (1990).
4. J. Wdowczyk and A.W. Wolfendale, Ann. Rev. Nucl. Part. Sci. **39**, 43 (1989).
5. J.R. Jokipii and G. Morfill, Ap. J. **312**, 170 (1987).
6. V.S. Berezinksy et al., Proc. ICRR Int'l. Symp., Astrophysical Aspects of the Most Energetic Cosmic Rays, Eds., M. Nagano and F. Takahara, (World Scientific, 1991), p. 134.
7. W.I. Axford, Proc. ICRR Int'l. Symp., Astrophysical Aspects of the Most Energetic Cosmic Rays, Eds., M. Nagano and F. Takahara, (World Scientific, 1991), p. 406.
8. P. Biermann and P. Strittmatter, Ap. J. **322**, 643 (1987).
9. W.-H. Ip and W.I. Axford, Proc. ICRR Int'l. Symp. Astrophysical Aspects of the Most Energetic Cosmic Rays, Eds., M. Nagano and F. Takahara, (World Scientific, 1991), p. 273.
10. T. Stanev, these Proceedings.
11. R. Ratkiewicz, J.F. McKenzie and W.I. Axford, these Proceedings.
12. E. Fermi, Astrophys. J., 119, 1 (1954).
13. G.A. Tammann, Supernovae: A Survey of Current Research, Eds. M.J. Rees and R.J. Stoneham, 371 (1981).
14. A.R. Bell, Proc. 22nd Int'l Cosmic Ray Conf., Dublin, Vol. 2, p, 421, (1991).
15. E.G. Berezhko, Proc. 22nd Int'l Cosmic Ray Conf., Dublin, Vol. 2, p, 436, (1991).
16. S.P. Swordy, D. Müller, P. Meyer, et al., Astrophy. J., 349, 625 (1990).
17. T. K. Gaisser, Proc. ICRR Int'l. Symp. Astrophysical Aspects of the Most Energetic Cosmic Rays, Eds., M. Nagano and F. Takahara, (World Scientific, 1991), p. 146.

Electron Acceleration

ELECTRON ACCELERATION IN THE GALAXY: OBSERVATIONS AND THEORY

S. P. Reynolds
North Carolina State University, Raleigh, NC 27695

ABSTRACT

Cosmic-ray electrons form a distinct and important subclass of cosmic rays in general. While they provide almost all the direct evidence we have for the existence of relativistic particles anywhere outside the heliosphere, the mechanisms by which they are produced must differ in crucial ways from those producing ions, and currently remain mysterious. Here I review the observational information on the pool of Galactic cosmic-ray electrons, and obtain the firm observational constraints that must be met by any theories of acceleration. Propagation models are not reviewed in detail. I briefly summarize the state of acceleration theories, and outline the future observational and theoretical directions most likely to prove fruitful in improving our understanding of electron acceleration.

I. OBSERVATIONS OF DISTRIBUTED ELECTRONS

Electrons in the diffuse interstellar medium ('cosmic-ray electrons') make their presence known in several ways. They can of course be directly observed at Earth from balloons or satellites, and elsewhere in the heliosphere from spacecraft. Electrons of Galactic origin are observed from 0.4 to 1000 GeV, though below a few GeV spectral information is uncertain due to solar modulation effects. The nonthermal radiation from cosmic-ray electrons is easily observable as the Galactic synchrotron background between about 10 MHz and 1.4 GHz. Synchrotron radiation at frequency ν is produced by electrons of energy $E(\nu)$ in a magnetic field B, where $E(\nu) = 15 \, [\nu(\text{GHz})/B(\mu\text{g})]^{1/2}$ GeV. Thus, for a mean galactic magnetic field of 3 μgauss, we observe only a single decade of electron energies, from about 1 to 10 GeV. Finally, bremsstrahlung from cosmic-ray electrons contributes to the diffuse Galactic gamma ray background, and can with some effort be partially deconvolved from the contribution due to the decay of pions produced in interactions between cosmic-ray protons and background gas, since the latter cuts off below gamma-ray energies of about 150 MeV. Gamma ray energies below this are produced primarily by electrons with energies of below 300 MeV or so, thus not overlapping with directly observed cosmic-ray electrons. We note that the energy range accessible at Earth is far larger than that studied by either of the electromagnetic channels.

A. Cosmic-ray electrons at Earth

In the last decade, observers of cosmic-ray electrons have come to general agreement, at least to within a factor of two. The figure (Golden et al. 1984) shows the most recent data from several groups for energies between 5 and 1000 GeV. Golden et al. (1984) quote the best power-law fit to their data between 4.5 and 62.5 GeV as

$$N(E) = (1.5 \pm 0.2) \times 10^{-17} E^{-(3.15 \pm 0.20)} \text{ cm}^{-3} \text{ erg}^{-1}$$

for electrons only. This spectrum is consistent with later observations (e.g., Tang 1984; Basini et al. 1991). If it extends down to $E_l = 0.4$ GeV, and if positrons contribute roughly another 10% (Basini et al. 1991), we infer a total energy density in electrons of $u_e = 0.04$ eV cm^{-3} with an uncertainty of perhaps a factor of 2, from uncertainties in the spectral shape and the value of E_l, primarily. Of course, the spectrum of Galactic electrons undoubtedly extends all the way down to $m_e c^2$ and below, though the shape may be uncertain.

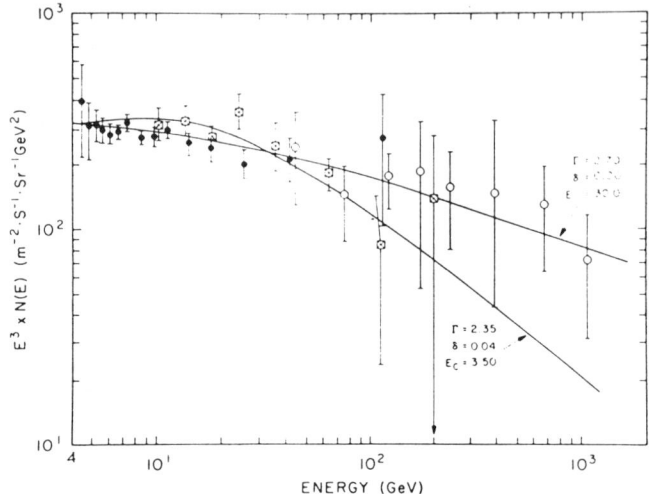

The diffuse Galactic gamma rays indicate that the spectrum does not change suddenly just below 0.4 GeV. However, the energy index p ($N(E) \propto E^{-p}$) quoted above cannot hold to much lower energies; the lower-energy electron spectrum, while badly confused by solar modulation, seems at least consistent with p = 2.3 to 2.7 (see below). If the spectrum below 0.4 GeV continues back to $m_e c^2$ with a slope of 2.3, the total energy in relativistic electrons is of order 1 eV cm^{-3} – comparable to that in relativistic ($E > 1$ GeV) protons! (The deficiency of a factor of about 30 in cosmic-ray electrons at a given energy compared to protons may simply reflect the kinematics of electrons becoming relativistic at lower energies.) For the purposes of energetics, we shall use the number above, recalling that it refers to the population of electrons directly observable at Earth, and is certainly a lower limit to the total energy density in suprathermal electrons to be found throughout the Galaxy. It should be noted that the positron fraction is probably energy-dependent, and may rise to values as large as 20% above 20 GeV (Müller and Tang 1987).

B. Galactic synchrotron background

A great deal of work has been done on this subject in the past decade (see Salter and Brown [1988] for a review). Some workers have derived the spatial structure (volume emissivities) from all-sky surveys at one frequency (e.g., Beuermann et al. 1985), while others have studied the spectrum averaged over substantial portions of the sky (e.g., Lawson et al. 1987). Reich and Reich (1987) have exhibited a spectral-index map of the galactic background between 408 and 1420 MHz; unfortunately for the study of electron spectra, this frequency range corresponds to only a factor of 1.9 in electron energies.

Probably the most elaborate discussion deriving volume emissivities is that of Beuermann et al. (1985), who find that the data require two components, a thick and thin disk, of comparable emissivity but different spatial extent. The thick disk is about 20 kpc in radius, with a full thickness of 2.3 kpc near the Galactic center, flaring to about 6.3 kpc at a Galactocentric radius of 20 kpc, for a total volume V of about 5×10^{67} cm^3. The thin disk flares as well, with a similar radius but about one-tenth the thickness, thus amounting simply to a small emissivity enhancement near the Galactic plane; the thick disk dominates the total energetics. Evidence for spiral structure is seen in both components.

The spectrum of the synchrotron background appears to steepen with increasing energy in most directions (Lawson et al. 1987). In addition it flattens as one moves away from the plane, and toward the anticenter, counter to one's naïve expectations. The large structures known as Galactic Loops (of which Loop I is also commonly called the North Polar Spur) seem to have steeper spectra than nearby regions above 408 MHz (Reich and Reich 1988). The spectral indices α ($S_\nu \propto \nu^\alpha$) range from -0.3 to -1.0 between 408 and 1420 MHz, but at lower frequencies, toward the Galactic poles and the anticenter, appear to converge at about -0.6 (Webber 1983; Cane 1979), implying $p = -2\alpha + 1 = 2.2$. This may be the closest to an intrinsic value we can obtain, though it is slightly flatter than the values deduced from propagation models applied to directly observed cosmic rays (see below).

Beuermann et al. (1985) find that the emissivity drops with Galactocentric radius, with a scale distance of about 4 kpc for that radial variation; at 408 MHz, for radii 7 – 10 kpc, they quote

$$j_\nu = (1.3 - 2.6) \times 10^{-40} \text{ erg cm}^{-3} \text{ s}^{-1} \text{ sr}^{-1} \text{ Hz}^{-1},$$

including both thin and thick components. For a mean local magnetic field of 3 μgauss, and $\alpha = -0.6$, the derived energy density is $u_e = (0.05 - 0.10)$ eV cm^{-3}, roughly twice the value derived from measurements at Earth. The shortfall may simply represent the effect of Galactocentric gradients, since the emissivities derived by Beuermann et al. actually vary by over an order of magnitude between Galactocentric radii of 0 to 20 kpc.

C. Low-energy γ-rays from electron bremsstrahlung

The COS-B data on the diffuse gamma-ray background have been used to derive information on cosmic-ray electrons by Bloemen et al. (1984, 1986) and by Gualandris and Strong (1984). The quality of the information is substantially below that of the synchrotron or direct observations, but the above-cited workers agree that both the total observed intensities and the Galactocentric gradients derived by Beuermann et al. (1985) are consistent with the gamma-ray data, implying continuity of the basic acceleration processes across the divide at around 400 MeV separating electrons observable directly and through bremsstrahlung.

II. THEORIES OF ELECTRON ACCELERATION

A. Observational constraints on sources

The primary constraint on possible sources of relativistic electrons throughout the Galaxy is that of energetics. We may express the constraint in terms of a luminosity, $L_{\text{in}} = u_e V t_c$ (erg s^{-1}), where t_c is a characteristic lifetime against energy losses, leakage, convection away, etc.:

$$L_{\text{in}} = 6 \times 10^{39} \left(\frac{u_e}{.05 \text{ eV cm}^{-3}} \right) \left(\frac{V}{5 \times 10^{67} \text{ cm}^3} \right) \left(\frac{t_c}{2 \times 10^7 \text{ yr}} \right)^{-1} \text{ erg s}^{-1}.$$

Here the nominal value of t_c is the ^{10}Be age for cosmic ray nuclei (Garcia-Muñoz, Mason, and Simpson 1977). This luminosity, to within a factor of 3 or so, is insensitive to detailed models for electron propagation (leaky boxes, winds, etc.). It is roughly 30 times smaller than that required to produce cosmic-ray nuclei; but it is very likely that putative sources of cosmic-ray electrons will exhibit them through synchrotron radiation, making them difficult to hide.

The required input spectrum cannot be extracted without more model-dependence. Simplest leaky-box models (Silverberg and Ramaty 1973) give results from $E^{-2.3}$ (Webber 1983) to $E^{-2.7}$ (Tang 1984). More detailed models, such as winds or diffusion (e.g., Owens and Jokipii 1977; Lerche and Schlickeiser 1980, erratum 1981), produce similar inferences. Better electron fluxes above 100 GeV will be necessary to resolve these uncertainties, but the experimental difficulties are daunting.

A rough lower limit to the total number of cosmic-ray electrons in the Galaxy (with energies greater than some E_l) can be obtained by integrating over a spectrum with $p = 2.2$, the flattest value consistent with observations of both cosmic-ray electrons and the synchrotron background:

$$n_e = 1.4 \times 10^{-11} \left(\frac{E_l}{400 \text{ MeV}}\right)^{-1.2} \Rightarrow N_{\text{tot}} = 7 \times 10^{56} \left(\frac{V}{5 \times 10^{67} \text{ cm}^3}\right) \text{ electrons.}$$

If this spectrum extends down to $m_e c^2$, the number increases to $N_{\text{tot}} = 2 \times 10^{60}$ electrons, replenished on a time-scale t_c, for a production rate of 10^{42} s^{-1}.

Beuermann et al. (1985) point out that their derived Galactocentric gradients in synchrotron emissivity imply, for roughly constant magnetic field, flatter gradients in electron density than are observed for most Population I objects, such as supernova remnants (SNRs), H II regions, or pulsars, which collectively represent practically every possible source of cosmic rays. If substantiated, this result will require either new sources of cosmic-ray electrons, or peculiar propagation effects. Beuermann et al. suggest a galactic fountain picture in which electrons circulate through a Galactic halo before raining down at larger radii than they began.

B. A catalogue of possible sources

Given these requirements, in roughly decreasing order of stringency, we can enumerate possible sources of cosmic-ray electrons. Stellar winds from early-type stars (Cesarsky and Montmerle 1983) or perhaps from evolved stars have been mentioned as candidates. For supergiants of types O, Of, or WR, we can estimate liberally

$$L(KE) \sim 10^{36} \left(\frac{M}{10^{-6} M_\odot \text{ yr}^{-1}}\right) \left(\frac{v}{2000 \text{ km s}^{-1}}\right) \text{ erg s}^{-1}$$

per star, but with fewer than about 10^4 of these extremely massive stars in the Galaxy, each would need to be essentially perfectly efficient at turning stellar-wind mechanical energy into cosmic-ray electrons. (Cesarsky and Montmerle [1983] used a rather more elaborate analysis to conclude that early-type stars could contribute no more than 5% – 20% of the power of supernovæ.) Similarly, red supergiants may have comparable mass-loss rates, but characteristic wind speeds of only 20 km s^{-1} (Dupree 1986), so that shocks may not form at all in the interstellar medium. Even if they do, or if some other means is found for turning this mechanical energy into relativistic electrons, there are still too few stars by roughly a factor of 10 – 100 (Mihalas and Binney 1981) to provide the required energy.

Classical novæ occur at rates roughly five orders of magnitude greater than the supernova rate. However, only one example of a classical nova is known to be an extended nonthermal radio source: the old nova GK Persei (Nova Persei 1901; Reynolds and Chevalier 1984a). While it may be more efficient than even Cas A at turning outburst energy into relativistic electrons and magnetic field, the absence of any other examples (Bode et al. 1987) suggests that the phenomenon is not generally true of novæ. Furthermore, since the typical nova outburst releases roughly 10^{45} erg, every nova would need to be roughly 10 times more efficient than the typical supernova to contribute a comparable amount of energy toward particle acceleration.

Pulsars could be imagined to be important sources of relativistic electrons, either in isolation or as the power sources of Crablike supernova remnants (Weiler 1985, Reynolds and Chevalier 1984b). Solitary pulsars, however, are probably born as relatively slow rotators (Chevalier and Emmering 1986), with initial periods P_0 of order 10 – 100 ms. Their rotational energy supply is then

$$E_{\rm rot} = 2 \times 10^{48} \left(\frac{I}{10^{45} \text{ g cm}^2} \right) \left(\frac{P_0}{100 \text{ ms}} \right)^{-2} \text{ erg,}$$

so that even at one pulsar born per 20 yr, the luminosity does not reach the required value. Furthermore, one expects equal numbers of positrons, since pair cascades are thought to be responsible for the bulk of the low-energy particles (e.g., Cheng et al. 1986). As the generators of Crablike supernova remnants, pulsars might be preferentially faster rotators; but the fact that fewer than 10% of Galactic supernova remnants are Crablike suggests that most pulsars do not produce radio emission at Crablike levels for a very long period. Furthermore, Crablike remnants are distinguished for having a spectrum far too flat to explain cosmic-ray electrons (p = 1 to 1.5; Weiler 1985). And, like isolated pulsars, they are thought to generate equal fractions of electrons and positrons. We conclude that the best hope for producing the observed cosmic-ray electrons remains shell supernova remnants.

C. Shell supernova remnants

About 160 supernova remnants (SNRs) have been identified by their radio emission in the Galaxy (Green 1988a). Of these, at most 10% are powered primarily by pulsars; a few others contain both Crablike and shell components, but the latter dominates in all such cases (Weiler 1985). Radio luminosities are small, $L_r \sim 10^{33} - 10^{35}$ erg s^{-1}, with few objects at the bright end. (The total radio luminosity of the Milky Way is very roughly 10^{38} erg s^{-1}; Beuermann et al. 1985). Synchrotron spectral indices range from -0.3 to -0.8, heavily concentrated between about -0.45 and -0.6. There is no tendency for older, larger, fainter remnants to have spectral indices closer to the steeper end (nearer the presumed injection spectrum for cosmic-ray electrons derived above); if anything, older remnants may have flatter spectra (Green 1988b). The supernova rate has been controversial recently; the fairly standard figures of Tammann

(1982) giving about 5 SN/century in our Galaxy have more recently been questioned by van den Bergh et al. (1987), who derive a rate closer to one or two per century (all SN types). Defining the mean interval between SNs as t_{SN} (years), we can thus estimate a total energy input rate of

$$L(SN) = 3 \times 10^{41} \left(\frac{t_{SN}}{100 \text{ yr}}\right) \left(\frac{E_{SN}}{10^{51} \text{ erg}}\right) \text{ erg s}^{-1}.$$

– certainly ample to provide for both cosmic-ray electrons and nuclei, as has so frequently been noted in the past, though the lower SN rate does raise the required efficiency to several tens of percent, and if substantial numbers of supernovæ have explosion energies of considerably less than 10^{51} erg, there may be difficulties.

Minimum (equipartition) energies in magnetic field and relativistic electrons can be derived for SNRs using standard formulæ (e.g., Pacholczyk 1970), to obtain $E_{min} = 10^{47.5} - 10^{49}$ erg, of which roughly half is in electrons. These values imply equipartition magnetic fields in the range $7 - 300$ μgauss. There is, of course, little evidence that SNRs as radio sources are anywhere close to equipartition; these results are crude guidelines and lower limits for the nonthermal energy in SNRs. If SNRs are also required to produce 30 times the electron energy density in relativistic ions, and the magnetic field is in equipartition with the total relativistic-particle population, the values for E_{min} rise by a factor of $30^{4/7} \sim 7$. (But the electrons remain at 1/30 of the ions, so $7 \times 10^{46} - 3 \times 10^{48}$ erg). The equipartition magnetic fields would in this case rise by about a factor of 10.

A quick summary of SNR evolution can be found in Reynolds (1988). Here it is sufficient to note that truly radiative, high-compression shocks are found only in denser portions of SNR shells, where the blast wave encounters dense clouds with concomitant low shock velocities; elsewhere, the bulk evolutionary phase is probably still adiabatic. The significance of low compression ratios in young, bright SNRs like Tycho and Kepler is that they guarantee that new electrons must be produced; compression of ambient cosmic-ray electrons and magnetic field is insufficient (Fulbright and Reynolds 1990). Remnants with sharp-edged radio shells can be argued to be producing new electrons via diffusive shock acceleration (Blandford, this volume).

D. Shock acceleration of electrons

While diffusive shock acceleration in general has been extensively studied and reviewed over the last decade (e.g., Blandford and Eichler 1987; Jones and Ellison 1991), the acceleration of electrons in a proton-dominated shock has received little attention. The difficulties of injecting thermal electrons in a shock smoothed by the pressure of backstreaming accelerated ions have been noted frequently (e.g., Blandford 1988): on the scale of the thermal electron gyroradius, the shock is not a discontinuity but a slow change in fluid properties, only adiabatically compressing electrons, and waves of sufficiently short wavelength to scatter thermal electrons may be difficult to produce. Even if electrons can somehow be injected, their nonthermal distribution may no longer agree with the simplest test-particle predictions, when nonlinear effects are taken into account. It is ironic that the prediction by the simplest test-particle theories of a spectral index of -0.5 from shock-accelerated electrons should be taken to confirm the operation of this process, when more careful study of shock acceleration has revealed that compression ratios in efficient shocks may be much larger than 4, implying asymptotic electron distributions with slopes much flatter than those observed! We need more attention to the spectra of electrons accelerated in a realistic shock model.

One calculation of the spectrum of electrons accelerated in a shock modified by the pressure of accelerated ions has been done by Ellison and Reynolds (1991), using a Monte Carlo simulation. They find that for electron energies near 1 GeV, which produce observed SNR radio emission, the electron spectra (a) have a very different spectrum than protons at those energies; (b) have a mean spectral index of $p \sim 2.2$, consistent with observations of young supernova remnants; and (c) are slightly positively curved (flattening to higher energies). These curved spectra actually imply synchrotron spectra in better agreement with observations than straight power-laws, for the remnants of Tycho's and Kepler's SNRs (Reynolds and Ellison 1991). To strengthen this conclusion, and to test the prediction in other SNRs, considerably better total flux density measurements of SNRs are needed. In addition, more sophisticated simulations are necessary to include effects of time-dependence and varying shock obliquity.

The problem of electron injection deserves attention in its own right (Melrose, this volume). Heliospheric observations of low-energy suprathermal electrons, and the turbulence in which they scatter, may contain important information. The possibility of thermal electron scattering by whistlers needs careful consideration, and has begun (e.g., Baring 1991). Radio observations of SNRs may help with this issue, since the observation of strong azimuthal modulation of radio intensity around shells of young SNRs probably reflects obliquity-dependence of the injection process (Fulbright and Reynolds 1990). Possible contributions of radioactive-decay positrons to the radio emission of remnants of Type I supernovæ should be examined (Lingenfelter, private communication).

III. OUTSTANDING QUESTIONS

In addition to the particular deficits in our understanding of diffusive shock acceleration of electrons described above, several other issues concerning the Galactic population of cosmic-ray electrons deserve consideration. If the oldest, largest SNRs exhibit electrons closest to being contributed to the Galactic cosmic-ray population, the disagreement between their radio spectral index (−0.3 to −0.5; Green 1988b) and the inferred injection spectrum for the electrons radiating the Galactic synchrotron background (~ -0.6 to -0.8) should be a concern. The origin of 1000 GeV electrons is a problem. If they arrive at Earth after diffusing through a medium with diffusion coefficient D, losing energy to synchrotron radiation in a mean magnetic field of 3 μgauss, they will lose half their energy in a time $t_{\rm rad} \sim 4 \times 10^5$ yr, and can random walk only a distance $l = (Dt_{\rm rad})^{1/2} \sim 330$ pc for $D \sim 10^{29}$ cm^2 s^{-1}. This may simply represent the residue from the last local supernova explosion. But if future experiments detect even higher-energy electrons, the problem may become acute.

Another issue related to the problem of electron injection concerns the observation in several SNRs of 'bipolar' X-ray emission coincident with bright regions of radio emission. While the latter may owe its existence to obliquity dependence of nonthermal particle acceleration, the role the shock obliquity might play in thermal-gas heating is not nearly so clear. This effect may in fact contain important clues about possible obliquity-dependent plasma instabilities (e.g., the Buneman instability; Cargill and Papadopoulos 1988) which might collisionlessly heat electrons to produce the X-ray emission, and also inject some fraction to be accelerated. Finally, a fundamental difficulty concerns the considerable disparity between a SN rate of one per 50 – 100 yr and the inferred SNR production rate of one per 250 or more yr (e.g., Clark and Caswell 1976). Can radio-quiet objects make cosmic-ray electrons? Perhaps many SNRs occur in H II regions or other confused environments where they cannot be recognized as such. In general, the study of cosmic-ray electrons will be of great interest in the future, with applications to understanding of nonthermal electron populations everywhere in the Universe.

REFERENCES

Baring, M.G. 1991, *Proc. ICRC 22* (Dublin), 2, 348.
Basini, G. et al. 1991, *Proc. ICRC 22* (Dublin), 2, 137.
Beuermann, K., Kanbach, G., and Berkhuijsen, E.M. 1985, AA, 153, 17.
Blandford, R.D. 1988, in *Supernova Remnants and the Interstellar Medium*, ed. R.S. Roger and T.L. Landecker (Cambridge: Cambridge University Press), p. 309.
Blandford, R.D., and Eichler, D. 1987, Phys.Repts., 154, 1.
Bloemen, H. et al. 1984, AA, 135, 12; 1986, AA, 154, 25.
Bode, M.F., Seaquist, E.R., and Evans, A. 1987, MNRAS, 228, 217.
Cane, H.V. 1979, MNRAS, 189, 465.
Cargill, P.J., and Papadopoulos, K. 1988, ApJ, 329, L29.
Cesarsky, C.J., and Montmerle, T. 1983, SpSciRev, 36, 173.
Cheng, K.S., Ho, C., and Ruderman, M. 1986, ApJ, 300, 500.
Chevalier, R.A., and Emmering, R.T. 1986, ApJ, 304, 140.
Clark, D.H., and Caswell, J.L. 1976, MNRAS, 174, 267.
Dupree, A.K. 1986, ARAA, 24, 377.
Ellison, D.C., and Reynolds, S.P. 1991, ApJ, 382, 242.
Fulbright, M.S., and Reynolds, S.P. 1990, ApJ, 357, 591.
Garcia-Muñoz, M., Mason, G.M., and Simpson, J.A. 1977, ApJ, 217, 859.
Golden, R.L., Mauger, B.G., Badhwar, G.D., Daniel, R.R., Lacy, J.L., Stephens, S.A., and Zipse, J.E. 1984, ApJ, 287, 622.
Green, D.A. 1988a, ApSpSci, 148, 3.
——. 1988b, in *Genesis and Propagation of Cosmic Rays*, ed. M.M. Shapiro and J.P. Wefel (Dordrecht: Reidel), p. 205.
Gualandris, F., and Strong, A.W. 1984, AA, 140, 357.
Jones, F.C., and Ellison, D.C. 1991, SpSciRev, 58, 259.
Lawson, K.D., Mayer, C.J., Osborne, J.L, and Parkinson, M.L, 1987, MNRAS, 225, 307.
Lerche, I., and Schlickeiser, R. 1980, ApJ, 239, 1089; Erratum, 1981, ApJ, 246, 360.
Mihalas, D., and Binney, J. 1981, *Galactic Astronomy* (San Francisco: Freeman), p. 227.
Müller, D., and Tang, K.-K. 1987, ApJ, 312, 183.
Owens, A.J., and Jokipii, J.R. 1977, ApJ, 215, 685.
Pacholczyk, A.G. 1970, *Radio Astrophysics* (San Francisco: Freeman).
Reich, P. and Reich, W. 1988, AASupp, 74, 7.
Reynolds, S.P. 1988, in *Galactic and Extragalactic Radio Astronomy*, ed. G.L. Verschuur and K.I. Kellermann (Berlin: Springer). p. 439.
Reynolds, S.P., and Chevalier, R.A. 1984a, ApJ, 281, L33.
Reynolds, S.P., and Chevalier, R.A. 1984b, ApJ, 278, 630.
Reynolds, S.P., and Ellison, D.C. 1991, in *Proc. ICRC 22* (Dublin), 2, 404.
Salter, C.J., and Brown, R.L. 1988, in *Galactic and Extragalactic Radio Astronomy*, ed. G.L. Verschuur and K.I. Kellermann (Berlin: Springer), p. 1.
Silverberg, R.F., and Ramaty, R. 1973, Nature Phys.Sci., 243, 134.
Tammann, G. 1982, in *Supernovæ: A Survey of Current Research*, ed. M.J. Rees and R.J. Stoneham (Dordrecht: Reidel), p. 371.
Tang, K.-K. 1984, ApJ, 278, 881.
van den Bergh, S., McClure, R.D., and Evans, R. 1987, ApJ, 323, 44.
Webber, W.R. 1983, in *Composition and Origin of Cosmic Rays*, ed. M.M. Shapiro (Dordrecht: Reidel), p. 83.
Weiler, K.W. 1985, in *The Crab Nebula and Related Supernova Remnants*, ed. M.C. Kafatos and R.B.C. Henry (Cambridge: Cambridge University Press). p. 227.

COSMIC RAYS AND SHOCK WAVES IN ACTIVE GALAXIES

Andrew S. Wilson
Space Telescope Science Institute, 3700 San Martin Drive,
Baltimore, MD 21218
and
Astronomy Department, University of Maryland
College Park, MD 20742

ABSTRACT

Studies of non-thermal radio emission in the inner ($\simeq$kpc scale) regions of active galaxies reveal clear morphological evidence for shock waves. It is argued that the shocks are driven by collimated outflow of plasma from the active galactic nucleus. I describe a model in which the synchrotron radio emission arises from ambient relativistic electrons and magnetic fields which have been compressed in a radiative shock wave. No active particle acceleration at the shock front is required in this picture. The prospects for diagnosing active acceleration from radio continuum and optical emission-line observations are briefly discussed.

1. INTRODUCTION

One of the greatest challenges to any acceleration mechanism of cosmic rays is the huge energies (up to $\simeq 10^{60}$ erg) inferred for the population of relativistic particles in the lobes of radio galaxies and quasars. Although it is widely believed that cosmic rays are accelerated by first order Fermi acceleration in shock waves[1], direct observational evidence for this process in active galaxies is limited. Morphological signatures of shocks, such as sharp-edged, linear synchrotron features, are rarely seen in radio galaxy jets and lobes, a notable exception being feature A of the jet of M87[6]. In this paper, I should like to discuss a class of active galaxies in which the non-thermal radio emission, although of much lower intrinsic power than is characteristic of radio galaxies and radio-loud quasars, does seem to be intimately associated with shock waves driven by "jets" and "lobes" ejected from the active nucleus.

In Seyfert galaxy nuclei and steep spectrum core radio galaxies, the radio emission is generally confined to the kpc scale, which is well within the host galaxy. These sources are usually interpreted[9] in terms of the interaction of a low energy jet or quasi-continuous ejection from the nucleus with the interstellar gas of the galaxy. I shall first describe a few examples of these objects and then discuss whether the observed synchrotron emission results from passive compression of ambient cosmic rays and magnetic fields in a radiative shock wave, or whether Fermi acceleration of cosmic rays at the shock front is necessary.

2. RADIO OBSERVATIONS

Figure 1 shows VLA radio maps of the Seyfert galaxy NGC 1068. The brightest

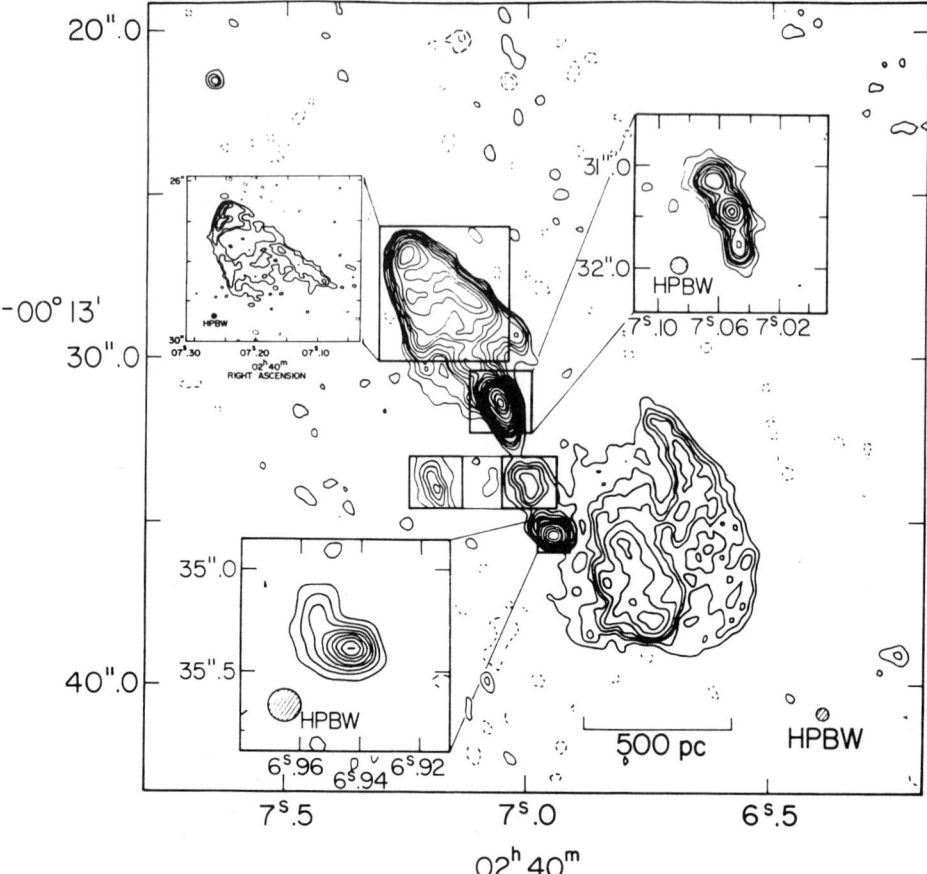

Figure 1 - Contour maps of the radio emission of the inner region of NGC 1068 (from refs 10 and 11). The NE lobe may represent radio emission from a radiative bow shock (see text).

emission coincides with the Seyfert nucleus but for our present purposes, the most interesting feature is the "arrow-shaped" lobe to the NE of the nucleus. The true three dimensional shape of this structure is presumably conical, with the apex of the cone some 6″ (400 pc) from the nucleus. The chief properties of the lobe[11] are i) a limb-brightened morphology with a very sharp outer edge (see inset in Fig. 1); ii) a very large jump in synchrotron emissivity across the boundary of the lobe; iii) linear polarization up to $\simeq$ 30% at 2 cm; and iv) a projected magnetic field running essentially parallel to the lobe edges in the limb-brightened regions. The morphology is obviously suggestive of a bow shock.

There is an intimate relation between the radio emission of NGC 1068 and the thermal gas near 10^4K, as studied in, for example, the [OIII]λ5007, Hα or [NII]$\lambda\lambda$6548, 6584 emission lines. The most complete study[3] reveals high velocity (up to $\simeq$ 1500 km s^{-1}) gas, which is tightly aligned with the linear radio structure. According to Cecil, Bland and Tully (1990), this high velocity emission constitutes most of the mass and kinetic energy of the narrow line region. There is also lower velocity ($<$ 60 km s^{-1}) emission associated with the NE radio lobe. Although the optical data are limited by atmospheric seeing, this low velocity emission appears to be enhanced along at least the NW edge of the limb-brightened lobe.

Figure 2 is a VLA map of another classical Seyfert galaxy, NGC 3516. The nucleus is the southernmost of the bright double source SW of the map center. The radio emission is seen to extend NE of the nucleus, with a bright feature at the end, some 20″ (3.6 kpc) from the nucleus. The feature is elongated *transverse* to the source axis and has a sharp "outer" (NE) edge. Images in the [OIII]λ5007 and Hα + [NII]$\lambda\lambda$6548, 6584 emission lines[5] are dominated by a bright bi-polar nebulosity within 10″ (1.8 kpc) of the nucleus. There is, however, also an elongated emission-line feature which runs roughly parallel to the elongated radio feature, but $\simeq$ 1.5″ (270 pc) closer to the nucleus.

Lastly, Figure 3 is a VLA radio image of NGC 5548. The northern lobe exhibits a limb-brightened structure with a sharp outer edge. Although the nature of this morphology is more ambiguous than that of the features seen in NGC 1068 and NGC 3516, the lobe can at least be considered a candidate for radio emission from a bow shock.

3. SHOCK WAVE MODELS

The data suggest that the radio emission of classical radio galaxies and radio-loud quasars is associated with jets and cocoons of old jet material[7]. On the other hand, the maps presented above are more suggestive of radio emission from bow-shocks. The key factor that renders bow-shocks visible within inter*stellar* gas and invisible in inter*galactic* gas may be the radiative, rather than adiabatic, nature of the shock in the denser environment of the former medium. A cartoon of a possible model involving a radiative shock is illustrated in Figure 4. Interstellar gas is heated at the shock to a high temperature and then cools down at roughly constant pressure (e.g. ref. 2), forming a dense shell. If there are no other sources of energy, the gas will eventually reach a very low temperature (cf. ref. 4); in the case of Seyfert galaxies, however, the nucleus is

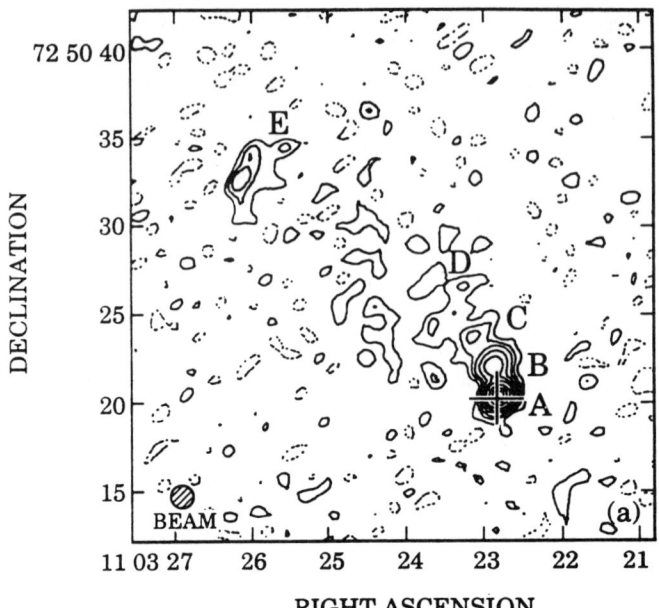

Figure 2 - Contour map of the radio emission from NGC 3516 (from ref. 5). Feature A coincides with the optical nucleus, while feature E may represent radio emission from a shock wave.

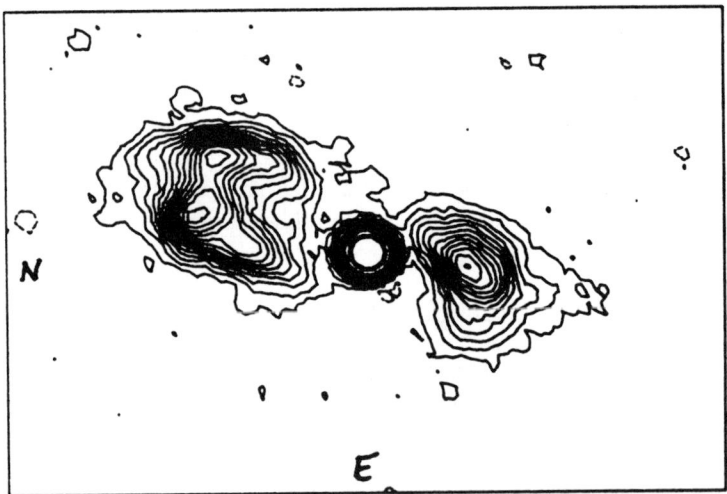

Figure 3 - Contour map of the radio emission from NGC 5548 (from ref. 12). The limb-brightened north lobe may be associated with a shock wave.

a luminous source of ionizing photons, which are likely to prevent the shell cooling to much below 10^4K. The magnetic pressure increases as χ^2, where χ is the ratio of the density in the cooling gas to the density in the pre-shock gas, so the compression may be limited by magnetic forces. Any cosmic rays and magnetic field in the ambient medium are compressed along with the thermal gas, boosting the synchrotron *emissivity* by a factor $\simeq \chi^{3.3}$. Because the *volume* of the gas decreases by only $\simeq \chi^{-1}$, large increases in synchrotron power are expected.

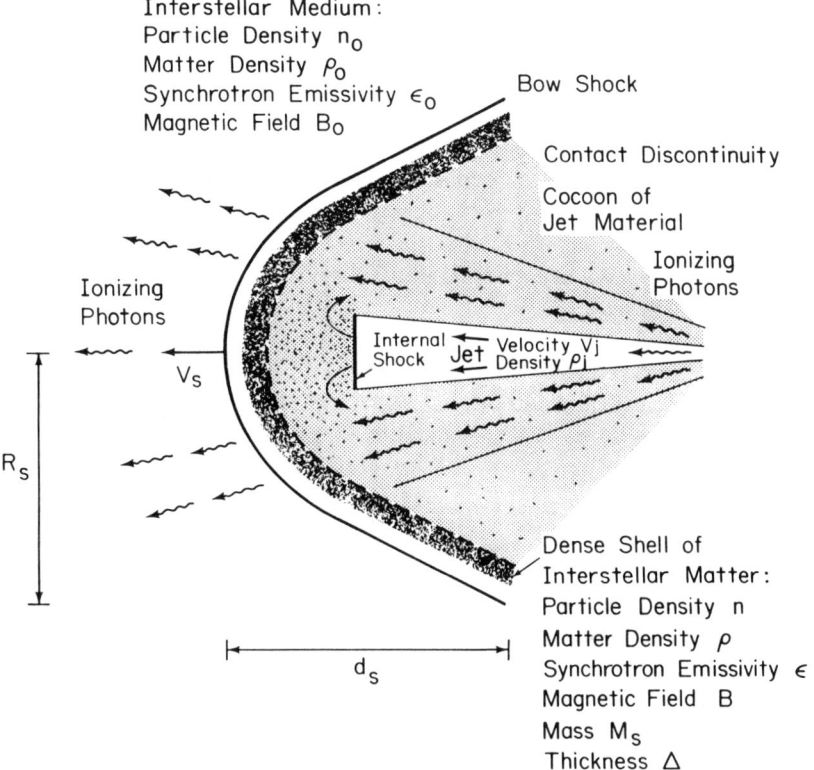

Figure 4 - Schematic diagram of the interaction of a jet with the interstellar medium of the host galaxy. A dense shell of interstellar gas forms downstream of the radiative bow shock wave (from ref. 11).

A model of this type has been developed for the NE lobe of NGC 1068[11]. In this galaxy, there is radio emission associated with the starburst disk of the galaxy (this radio emission is too faint to see in Figure 1). We assumed that the NE radio lobe is plowing into this disk, which is probably quite thick, being "puffed up" by the starburst activity; the tip of the NE radio lobe projects only 440 pc from the nucleus, so this assumption is plausible. The enhancement in synchrotron emissivity across the shock could then be inferred directly from observations, allowing χ to be estimated. We found

$\chi = 21$ at the tip of the lobe and $\chi = 13.5$ at the side; such lower compression at the side is expected because of the lower effective shock velocity there. We assumed that the magnetic field in the pre-shock medium takes its synchrotron equipartition value. The velocity of the bow-shock can then be calculated for various assumed pre-shock densities. More recently, optical observations[3] have isolated and measured the flux of Hα and [NII] emission along the limb-brightened edges of the NE lobe. Presuming this emission reflects the gas in the cooling zone, the pre-shock density is found to be $\simeq 7$ cm^{-3} and the shock velocity 190 km s^{-1} (cf. Sect. Vfiii of ref. 11).

This model successfully accounts for most of the observed results on the NE lobe. It also "explains" the spectral index of the radio lobe emission: because the power-law index of the cosmic ray energy spectrum is unchanged in the compression, we expect the spectral index of the disk and lobe emission to be identical. This does appear to be the case. The strongest challenge to the model is the high electron density (100 - 200 cm^{-3}) in the postshock cooling zones. This density is much higher than those of $\simeq 0.2$ cm^{-3} inferred from Faraday rotation arguments. A possible explanation is that the pre-shock field is completely random in direction, so that the field in the cooled gas is also random but confined to a plane parallel to the shock plane. When viewing the edge of the lobe (where the line of sight is parallel to the plane of the shock), high synchrotron polarization, but little Faraday depolarization would be observed. The optical data[3] confirm the existence of dense gas along the edge of the lobe, but much higher resolution optical observations (e.g. with HST) are required to check whether this gas is co-spatial with the synchrotron emission, as expected in this model.

Too little information is available to check whether this type of model may apply to other active galaxies with kpc-scale radio emission. If the *primary* site of acceleration of the cosmic ray electrons is supernova remnants in the disk, a correlation between disk and nuclear radio powers would be expected. Such a correlation would be a strong argument that the radio emission of the active nucleus results from passive compresion of disk fields and cosmic rays. The alternative picture is that cosmic rays are accelerated by Fermi acceleration at the bow shock; in this case the nuclear radio emission is not causally related to that of the disk.

In NGC 3516, the radio emission associated with the shock (Fig. 2) lies upstream of the emission-line gas[5]. One interpretation is that the radio emission is associated with the shock front and the emission-line gas is only seen after the shocked gas has cooled to 10^4K (cf. ref. 8).

4. CONCLUDING REMARKS

From the perspective of this conference, the most pertinent question is whether the shock waves inferred to exist around active galactic nuclei are accelerators of cosmic rays. I have argued that a very conservative model, in which the synchrotron-emitting relativistic electrons represent compressed, ambient, disk cosmic rays, can account for the radio emission of the "bow shock" in NGC 1068. No active acceleration at the shock front is required, but such is also not ruled out. A combination of high resolution emission-line and radio continuum imaging can define the shock properties and check whether this model applies to other objects. If the predicted radio powers are below those observed, Fermi acceleration at the bow shock would presumably be necessary. Both the efficiency with which the shock power is converted into cosmic ray electrons and the electron energy spectrum could be determined from the radio and optical observations.

This research was supported in part by NASA grant NAGW-2689 to the University of Maryland.

REFERENCES

1. Bell, A. R., MNRAS, 182, 147 (1978).
2. Blondin, J. M. & Cioffi, D. F., ApJ, 345, 853 (1989).
3. Cecil, G., Bland, J. & Tully, R. B., ApJ, 355, 70. (1990).
4. Hollenbach, D. J. & McKee, C. F. ApJ Suppl., 41, 555 (1979).
5. Miyaji, T., Wilson, A. S. & Pérez-Fournon, I., ApJ, 385, 137 (1992).
6. Owen, F. N., Hardee, P. E. & Cornwell, T. J., ApJ, 340, 698 (1989).
7. Scheuer, P. A. G., MNRAS, 166, 513 (1974).
8. Whittle, M. et al., MNRAS, 222, 189 (1986).
9. Wilson, A. S. & Willis, A. G., ApJ, 240, 429 (1980).
10. Wilson, A. S. & Ulvestad, J. S., ApJ, 275, 8 (1982).
11. Wilson, A. S. & Ulvestad, J. S., ApJ, 319, 105 (1987).
12. Wrobel, J. M., Weiss, L. E. & Unger, S. W., Bull. A. A. S. 22, 1316 (1990).

The Quest for Particle Acceleration in Extragalactic Sources

Lawrence Rudnick, Debora Katz-Stone, Martha Anderson
Department of Astronomy, University of Minnesota
116 Church St. S.E., Minneapolis, MN 55455

ABSTRACT

In this paper, we discuss the types of observations used to study the acceleration of relativistic electrons in the diffuse plasma of extragalactic radio sources. The goal of this work is to identify the conditions under which such acceleration is important and which acceleration processes dominate. We identify some of the pitfalls in such analyses and suggest directions for future work.

Why is it important to study particle acceleration in extragalactic sources?

For astronomers interested in these very powerful radio sources, particle acceleration is one of the key processes to be understood. Other coupled questions include the confinement and transport of the relativistic particles and fields, and the original energy source for both the relativistic and underlying plasmas.

More generally, radio galaxies and quasars also offer some unique opportunities for studying particle acceleration under physical conditions different from those in the solar or galactic systems. Relativistic shocks are the clearest example - they are certainly present in the superluminal nuclear jets studied with VLBI. Relativistic shocks have also been suggested[1] as an explanation for the energy and spectra of "hot spots", where the jets are presumed to dump their energy.

Observations of extragalactic synchrotron sources also represent a large averaging over time and space, much greater than any temporal or spatial scales important to particle acceleration. This allows simplified modelling, *e.g.*, the use of steady-state particle distributions.

Finally, extragalactic sources seem to be the last great hope for particles with energies above 10^{19} eV. Although the synchrotron radiation does not provide a direct test for these particles, it helps us understand the laboratories where they may be produced.

What is the evidence for particle acceleration in extragalactic sources?

One of the early indications that relativistic electrons were being energized *in situ*, far from their nuclear source, was the sub-adiabatic dropoff in brightness of jets.[2] But the most dramatic evidence for *in-situ* acceleration has been the observation of optical and even X-ray synchrotron emission from jets and hot spots[3,4,5], where the life-time of the emitting particles is much shorter than the transit time from the nucleus. Figure 1 shows the synchrotron spectrum of the southern hot spot of 3C33[6,7] - one such case.

The detailed argument for the necessity of *in situ* particle acceleration proceeds as follows. Assuming that a relativistic electron of energy γ_{jet} travels in a jet of magnetic field B_{jet}, we can calculate its lifetime against losses, τ, as follows (modified from reference [8]):

$$\tau = \frac{1.66 \times 10^{11}}{\gamma_{jet}} \times \left(\left(\frac{0.33\, u_{rad}}{10 eV\, m^{-3}}\right) + \left(\frac{B_{jet}}{nT}\right)^2 \right)^{-1} \text{years} \qquad (1)$$

u_{rad} is the energy density of the microwave background, and this provides an upper limit to the age because of inverse Compton scattering, even if $B_{jet} = 0$. In the best established cases, the electrons cannot reach the hot spot in this limit, even travelling at the speed of light, indicating the need for local (re–)acceleration.

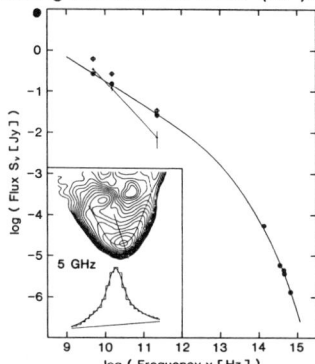

Fig. 1. The broadband synchrotron spectrum and structure of the southern terminus (hot spot) of radio galaxy 3C33.

Since γ_{jet} is often not observed directly when optical hot spot emission is seen, $\gamma_{hotspot}$ is usually substituted directly for γ_{jet} in eq. (1). However, $\gamma_{hotspot}$ can be $\gg \gamma_{jet}$ from adiabatic compression alone. It is therefore not clear, for optical hotspots, whether further energization, *e.g.*, Fermi-type, is necessary.

How are the observations used to probe particle acceleration?

The main tool for probing particle acceleration mechanisms is the broad-band spectrum of the synchrotron emission. The first step is actually to model the synchrotron losses, since they are always important in extragalactic sources, where we find magnetic field values of 1-100 μGauss and ages of 10^6-10^8 years. Modelling these losses has enjoyed great popular success; an example from the radio galaxy 3C337 is shown in Figure 2[9].

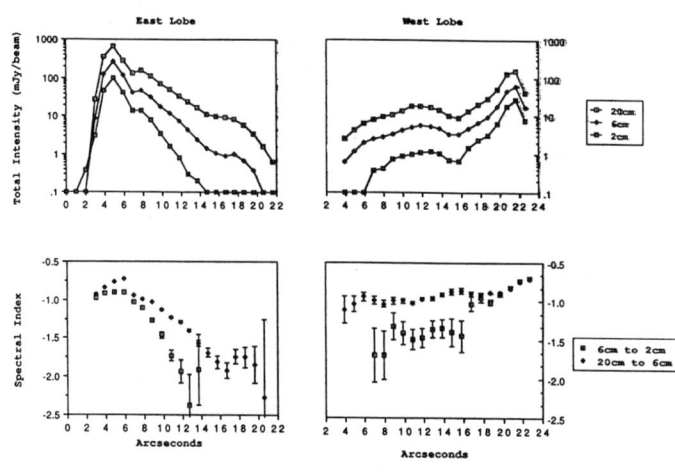

Fig. 2. Cuts in brightness (top) and spectral index (bottom) of the radio galaxy 3C337. Hot spots are at the peaks (outer edges).

In this figure, the source is brightest and the spectral index is flattest (hardest) at the outer edges, where the compact "hot spots" are found. As the plasmas flow back towards the nucleus (towards the center of Figure 2), the intensity falls off and the spectrum becomes steeper, as expected from synchrotron losses.

In seeking specific evidence for particle acceleration, the procedure is thus to model the observed spectrum as one resulting from a distant injection of a power-law distribution of electrons, which then gets modified by losses <u>and gains</u> (particle acceleration). Observations of the brightness structures in the source can then be used to identify what types of hydrodynamic features are serving as the acceleration sites. A schematic of this process is shown in Figure 3. In the standard analysis, one would follow the dark arrows in reverse, starting from the observed spectrum, inferring the relativistic particle distribution and linking it to the observed hydrodynamic features.

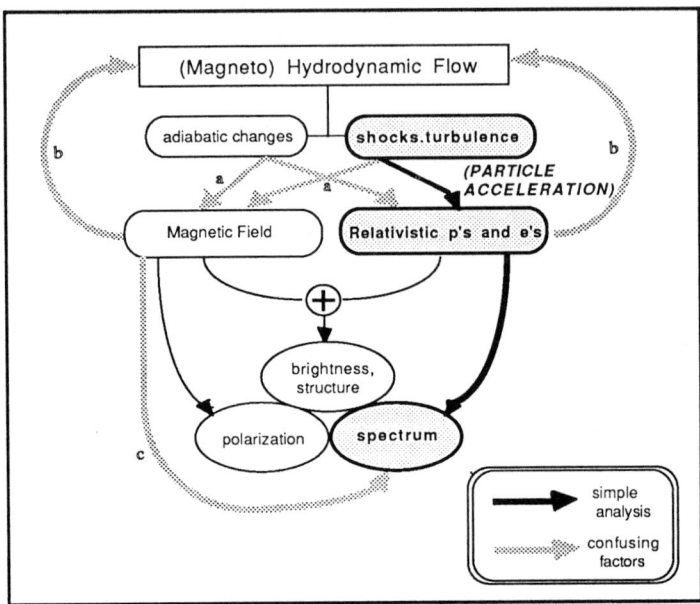

Fig. 3. Schematic of inferred relationships between physical processes and observations of synchrotron sources.

Complications in the real world....

The gray arrows in Figure 3 identify some of the difficulties in this work. First, <u>the observed brightness structures are **not** simply connected to hydrodynamic (non-relativistic) features</u>. Between the hydrodynamic features and the synchrotron brightness are effects (a) on the particles and fields, *e.g.*, shear which increases the magnetic field and the brightness, but not the underlying pressure. At present, one can simply compare observed structures to the non-relativistic fluid numerical simulations[10]. These simulations are starting to include evolved magnetic fields, but not yet any particles.

An example of a hot spot structure we would like to match with a specific hydrodynamic feature is shown in Fig. 4. 3C33 South, shown at lower resolution in Figure 1, is seen here with a resolution of ≈150 pc, one of the highest available. It has features which subtend an angle of only 10^{-3} at the nucleus and thus results from a highly collimated (invisible) flow. In some ways, this structure is reminiscent of features seen in the simulations of jet terminations, but it is impossible at this stage to identify it unambiguously, and so to determine what type of hydrodynamic feature is responsible for the particle acceleration.

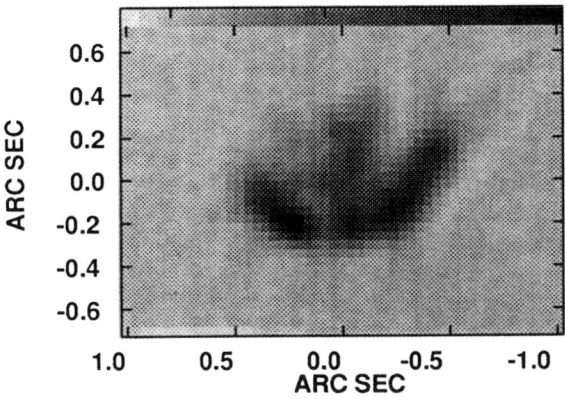

Fig. 4. High resolution map at λ2cm of south hot spot of 3C33.

The (b) arrows in Figure 3 indicate a second major problem - if enough momentum and energy are pumped into the relativistic particles and fields, they can actually modify the flow. In this case, the relativistic and non-relativistic plasmas cannot be considered separable model parameters, but must be solved for self-consistently.

Perhaps the most insidious source of confusion is illustrated with arrow (c). In this case, a change in the magnetic field strength, whether due to adiabatic or other processes, can cause an <u>apparent</u> change in the spectrum. This is simply because, when observing at a fixed frequency, ν_{obs}, the actual energy of the electrons being observed depends on the local magnetic field value as:

$$\nu_{obs} \approx k\, \mathbf{B} \cdot E^2 \qquad (2)$$

(with k as a constant). If the slope s of the electron distribution is itself a function of energy, as is the case for synchrotron losses,

$$\frac{dN(E)}{dE} = E^{-s(E)} \qquad (3)$$

then when the magnetic field changes, but the observing frequency remains fixed, the observed spectral index, α_{obs}, will change, as follows:

$$\alpha_{obs}(\mathbf{B}) \equiv \frac{d\,\ln(I)}{d\,\ln(\nu)} = \frac{s(E)-1}{2} = \frac{s\left(\sqrt{\frac{\nu_{obs}}{k\mathbf{B}}}\right) - 1}{2} \qquad (4)$$

This B-field relationship is well understood, but seldom applied in the analysis of spatial variations in α_{obs}. Most observed variations in α_{obs} are interpreted as due to particle energy losses or gains, and not as simply due to a changing B. The potential importance of this B field confusion can be seen in Figure 5, which shows a slice down one lobe of Cygnus A (data from reference[11]). The lines show the change in both brightness (which depends on B) and the "break frequency" (*i.e.*, where losses become important), which depends on both the B field and the physical age of the electrons. How much of the spectral change is due to changes in B, and how much is due to spectral aging, is difficult to answer. In a few, but not all, places in the lobes of Cygnus A one can show that particle losses must dominate.[11]

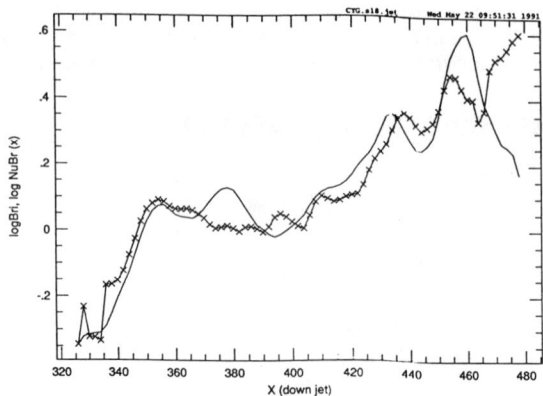

FIGURE 5. Slices through radio galaxy Cygnus A. The solid line is the brightness, and the X-d line shows the break frequency, as determined in reference 11.

Another important *caveat* is provided by the optical and X-ray synchrotron radiation from M87[5]. The current observations show no steepening of the spectrum between the bright regions in the jet, where acceleration sites might be present, and the more diffuse regions, where no acceleration is expected. However, the radiating electrons cannot live long enough to move between the bright and diffuse regions, so a change in spectrum should be seen. Is this telling us that there is no *in situ* particle acceleration in the jet, but simply a replenishment by fresh electrons travelling along the jet in a region of $B \approx 0$?[5]

Where do we go next?

Continued work is necessary along a number of different lines to understand the role of particle acceleration in extragalactic sources. These include:

•More sophisticated modelling of 3-D hydrodynamic flows, including tracer B field and particle distributions which retain their history of gains and losses; (included here is the need for the development of clear observational signatures of various hydrodynamic features);

•Further modelling of a range of particle injection and acceleration processes, to use as input to the simulations above[12,13];

•More optical and X-ray data to identify the injection sites of high energy electrons;

•Detailed studies of the brightness and spectral behavior in the radio to understand the interplay between adiabatic and other gain and loss mechanisms.

On this last point, we have developed a new tool for examining the synchrotron and particle distribution throughout a source. Changes in the shape of the spectra are a key tool in identifying non-adiabatic processes. but such changes are hard to detect observationally. We are therefore suggesting use of the "color-color" diagram[14], as in Figure 6, where the spectral index between one pair of frequencies is plotted against the spectral index at another frequency pair, for each point in a source. The locus of such points is a curve which fixes - with virtually continuous sampling - the shape of the underlying electron distribution (*caveat lector*). One can then compare the observed distribution with various theoretical models of relativistic electron injection, gains and losses. Using this diagram, we can vastly improve on current methods for distinguishing between different models, and can examine the entire source in one diagram.

Our results on Cygnus A are striking. None of the simple theoretical models fit the observations, nor do simple age or B-field mixtures. We are presently in the process of "inverting" the color-color plot; our hope is that we will obtain a much more realistic picture of the electron distribution to re-examine the role of particle acceleration. We are also planning observations of other sources, to see if this unexpected particle distribution is universal.

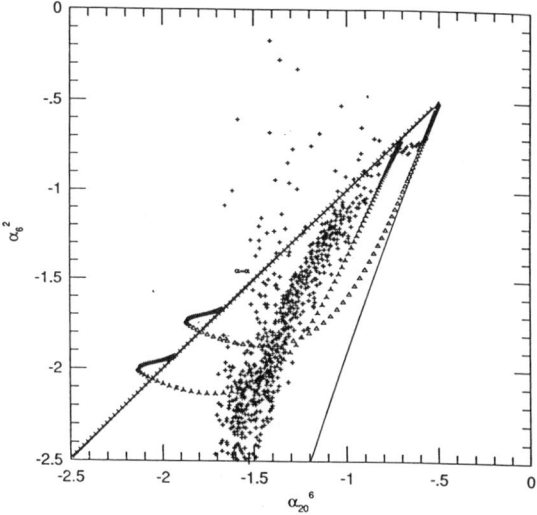

Fig. 6. The "color-color" diagram. The hatched curve $\alpha = \alpha$ represents pure power laws. The solid curve is a loss spectrum with pitch angle scattering. The two curves with symbols are losses with no scattering. The data points are from Cygnus A.

Acknowledgements

We gratefully acknowledge the use of data on Cygnus A (reference 11) as well as fruitful conversations with Leahy, Carilli, Colin Lonsdale and Tom Jones. These observations are from the Very Large Array, operated by NRAO through AUI's contract with the National Science Foundation. This work is supported in part by the NSF through grants AST-8720285 and AST-9100486.

References

[1] Begelman, M.C., and Kirk, J.G., *Ap. J.*, **353**, 66, (1990).
[2] see general discussion in Eilek, J.A. and Hughes, P. in *Beams and Jets in Astrophysics*, ed. P. Hughes, Cambridge Univ. Press, Ch. 9, (1991).
[3] Simkin, S.M., *Ap.J.*, **234**, 56 (1978).
[4] Rudnick, L., Saslaw, W.C., Crane, P., Tyson, J.A., *Ap. J.*, **246**, 647, (1981).
[5] Biretta, J.A., Stern, C.P. and Harris, D.E., *Astron. J.*, **101**, 1632, (1991).
[6] Meisenheimer, K. et al., *Astron. and Ap.*, **219**, 63, (1989).
[7] Rudnick, L. *Ap. J.*, **325**, 189, (1988).
[8] Longair, M.S., *High Energy Astrophysics*, Cambridge Univ. Press, (1981).
[9] Pedelty, J.A., Rudnick, L., McCarthy, P.J., Spinrad, H., *Astron. J.*, **98**, 1232, (1989).
[10] anearly reference: Smith, M.D., Norman, M.L., Winkler, K.-H., Smarr, L., *MNRAS*, **214**, 67, (1984).
[11] Carilli, C.L., Perley, R.A., Dreher, J.W., Leahy, J.P., *Ap. J.*, **383**, 554, (1991).
[12] Jones, T.W., and Kang, H., *these proceedings*.
[13] Kang, H. and Jones, T.W., *MNRAS*, **249**, 4 39 (1991).
[14] Katz-Stone, D., Rudnick, L., Anderson, M.C., in preparation, (1992).

ELECTRON ACCELERATION BY YOUNG SUPERNOVA REMNANT BLAST WAVES

R. D. Blandford
Caltech, Pasadena, CA 91125.

ABSTRACT

Some general considerations regarding relativistic particle acceleration by young supernova remnants are reviewed. Recent radio observations of supernova remnants apparently locate the bounding shock and exhibit large electron density gradients which verify the presence of strong particle scattering. The radio "rim" in Tycho's remnant has been found to contain a predominantly radial magnetic field. This may be attributable to an instability of the shock surface and a progress report on an investigation of the stability of strong shocks in partially ionized media is presented.

INTRODUCTION

Supernova remnants younger than $\sim 10^3$ yr are believed to form strong, adiabatic shocks which propagate roughly radially with speeds $\gtrsim 300$km s^{-1}into the interstellar medium. [1] These shock waves are held responsible for the acceleration of the majority of Galactic cosmic rays. However the manner in which this occurs is still not well understood. In this short report, I shall briefly describe a couple of ways in which recent radio observations might serve as diagnostics of high Mach number collisionless shocks.

I will start with some general comments. Firstly, despite their luminosity, supernova remnants are not particularly efficient when it comes to relativistic electron acceleration. The observed synchrotron radio power of a remnant like Cas A, is $\sim 10^{-4}$ of the maximum (equipartition) power that might be associated with the pressure within the remnant. This contrasts with the assumption that is commonly made in analyses of the most powerful extragalactic radio sources. Perhaps only relativistic shocks are highly efficient at accelerating electrons (*cf.*the discussion of the Crab Nebula [2]). Secondly, the gross morphological properties of supernova remnant radio emission can be explained qualitatively in terms of shock acceleration, post-shock magnetic evolution, and, possibly supplementary acceleration within the main body of the remnant. Quantitative theory that distinguishes Cas A from Tycho etc remains elusive though. The gross features of the cosmic ray nucleon spectrum up to energies $\sim 10^{14}$ eV can likewise be explained by shock acceleration. If these explanations are taken at face value, then shocks transmit no more than a few percent of their momentum flux in the form of relativistic electrons, in contrast to $\gtrsim 30$ percent as ions.

Thirdly, supernova shocks must heat thermal electrons fairly quickly, because the bremsstrahlung X-ray emission in sources like Tycho is seen to extend all the way up to the bounding shock front as delineated by radio observations. Although it is not required that the post shock electron temperature be as large as the Rankine-Hugoniot value, detailed emission models that attempt to accommodate the continuum and the line observations do suggest that a temperature $\gtrsim 1/3$ the Rankine-Hugoniot value is needed. This is in accord with analysis of slower, interplanetary shocks. [3]

A central issue is to understand the relative efficiency of quasi-parallel and quasi-perpendicular shocks for electron acceleration. Contrasting theoretical arguments can be given. On the one hand, injection into the Fermi process is kinematically easier for a quasi-parallel shock. [4] On the other, electrons will be confined more readily to the vicinity of a quasi-perpendicular shock as numerical simulations [5] and observations of the earth's bow shock [6] bear out. Shock drift acceleration [7] also requires a quasi-perpendicular shock. A subset of observed remnants are described as bipolar and exhibit a barrel morphology. This has been interpreted by proposing that the interstellar field has a uniform component. If so, the preponderance of "barrel" as opposed to "filled remnant" is suggestive of preferential quasi-perpendicular acceleration. [8]

RADIO INTENSITY GRADIENTS

High angular resolution observations of certain young supernova remnants exhibit scale-heights that are a small fraction of the remnant radius. [9] If we define the scale height H in terms of the intensity I_ν by

$$H = \min[|\nabla(\ln I_\nu)|^{-1}]$$

then, typically, $H \sim 10^{17}$ cm. 5 GHz synchrotron emission in a typical field $B \sim 10^{-5}$ G requires ~ 10 GeV electrons with Larmor radii $r_L \sim 3 \times 10^{12}$ cm, far too small to be resolved. Now, the pitch angle scattering rate calculated under the most primitive quasi-linear theory [10] is

$$\nu_C \sim \frac{c}{r_L}\left(\frac{\delta B}{B}\right)^2 \sim 10^{-2}\left(\frac{\delta B}{B}\right)^2 \text{ Hz}$$

The associated spatial diffusion coefficient is $D \sim r_L c/3 \sim 3 \times 10^{22}(B/\delta B)^2$ cm^2 s^{-1}. The predicted scale height is then

$$H \sim \frac{D}{V} \sim 10^{14}\left(\frac{B\cos\theta_{Bn}}{\delta B}\right)^2 \text{ cm}$$

for a shock speed $V \sim 3000$ km s^{-1}. Observations (see below) suggest that $\theta_{Bn} \lesssim 45°$. We therefore conclude that a resonant wave turbulent amplitude $(\delta B/B) \gtrsim 0.03$ is necessary to accomodate the observations.

Now contrast this with the conditions inferred in the general interstellar medium at the solar radius. The mean 10 GeV cosmic ray drift speed is $< v >_{CR} \sim 30$ km s^{-1}, of order the Alfvén speed. (This is derivable from both the escape grammage and the anisotropy.) The cosmic ray scale height in the galaxy appears to be ~ 1 kpc. Therefore $(\delta B/B)_{ISM} \sim 0.003$ at 10 Gev, at least an order of magnitude smaller than the value derived above at a young SNR shock front. This is *prima facie* evidence that Alfvén waves are accelerated at supernova remnant shock fronts.

A more detailed calculation (Achterberg, Blandford and Reynolds 1991, in preparation) confirms this general conclusion using recent VLA maps of Tycho, SN1006 and W49B. Observations are underway to improve the radio limit on H_{SNR}. If, as commonly assumed in particle acceleration theory, $(\delta B/B)$ is roughly constant with wavelength, then we predict that $H_{SNR} \propto \gamma \propto \nu^{1/2}$. Observations at high frequency should establish the tightest upper bound on H as the linear resolution of an interferometer of fixed baseline scales $\propto \nu^{-1}$. Unfortunately, it will not be possible to resolve H_{SNR} if $(\delta B/B) \sim 1$ as is supposed in some versions of shock acceleration theory. [10] High angular resolution X-ray observations using the ROSAT satellite, will also be important for locating shock fronts.

SHOCK "RIMS"

One of the most intriguing aspects of recent radio observations [9] is the discovery of thin, limb-brightened, tangential features, called "rims", along the periphery of Tycho's supernova remnant. What is most surprising, and quite unexpected, is that the radio emission is significantly polarized and, under the assumption that we are observing synchrotron radiation, indicates a radial magnetic field. If the magnetic field is isotropic ahead of the shock, then we would have expected that the field would have been preferentially strengthened along the direction of the projected shock surface. Setting aside the possibility that an alternative emission mechanism is at work, a possible explanation of these observations is that the shock is dynamically unstable and that corrugation of its surface is followed by the magnetic field. If such an instability were to have a large enough amplitude, then it might produce a net radial field.

STABILITY OF SHOCKS IN PARTIALLY IONIZED PLASMA

Motivated by this observation, Achterberg, Reynolds and I are investigating the stability of supernova shocks propagating into a partially ionized plasma. The simplest case to consider is a planar, perpendicular shock in which the magnetic field is weak enough that its dynamical effects can be ignored. The inflowing gas is modeled as two cold fluids, one neutral, the other ionized. We suppose that the shock is strong and fast enough to ensure collisional ionization of the neutral component on a length scale $\sim 10^{15}$ cm, some 4-5 orders of magnitude larger than the ion Larmor radius, which is a measure of the thickness of the collisionless subshock in the ionized component.

A cold, neutral atom crossing this shock will then move with a speed $\sim 3V/4$ relative to the shocked, ionised plasma, where V is the shock speed relative to the upstream gas. For the envisaged speeds of several thousand km s^{-1}, the neutral atom will eventually either undergo charge exchange when the relative velocity between the ion and the neutral is $\lesssim c/137$ or direct collisional ionisation when this inequality is reversed. The post shock electron temperature is uncertain. If it is low, then ionisation by electron impact is not likely to dominate. However, if the electrons achieve equipartition with the ions, then their greater speeds may ensure that they dominate the ionization rate. The combined cross sections can be represented by the approximate expression [11]

$$\sigma \sim 2 \times 10^{-16} + 2 \times 10^{-15} v_8^{-2} \text{cm}^2; \qquad 2 \lesssim v_8 \lesssim 5$$

where v_8 is the relative speed in units of 1000 km s^{-1}.

When an atom undergoes charge exchange or collisional ionisation, it will be swept up by the magnetic field and start to gyrate in the frame defined by the magnetic field. This will happen in a few ion Larmor radii. If the magnetic field is quasi-parallel, the freshly-created ion can stream along the field. However, scattering by plasma waves and magnetic inhomogeneity should remove any residual drift between the new and the old ion population. The momentum transfer will be relatively small in most charge exchange reactions, and so the newly recombined atom will not be given a large recoil and will, itself, soon be ionized. Elastic collisions between the incoming atoms and shock electrons and ions are can also occur prior to ionization. However, these mostly also involve relatively little momentum transfer.

For these reasons we make the approximation that the streaming atoms join the ion fluid immediately after ionisation and that the two fluid model model is adequate for exploring shock dynamics. We adopt a constant post-shock collision rate

$$\alpha = \frac{<\sigma v>}{m_p}$$

which is a fair approximation for the range of shock velocities of interest. Typically $\alpha = 5-10 \times 10^{16}$ cm^3 s^{-1} g^{-1}. Monte Carlo simulations could undoubtedly test the validity of this assumption.

The strong shock in the ionised component is described by the non-magnetic Rankine-Hugoniot jump conditions. Specifically, in the frame of the shock, the post shock ion velocity, henceforth v, is reduced to $V/4$ and the post-shock pressure is $p = 3\rho v^2$, where ρ is the ion density. In other words, in the frame of the shocked ions, the relative kinetic energy per unit mass of the incoming neutrals, $9v^2/2$, is converted completely into internal energy.

Now, we have made the *ansatz* that an incoming neutral immediately joins the shocked fluid as soon as it is ionised. As the shock is strong, it will have just the same kinetic energy per particle as the ions and contribute an equivalent internal energy per particle. It will automatically satisfy the same jump conditions as the ions. There will therefore be no acceleration of the post ion shock fluid as it is augmented by freshly ionized atoms (*i.e.* $v = V$ throughout). A direct solution of the equations of mass, momentum and energy conservation verifies this.

If we denote the density of neutral atoms, ρ_n, then the equation of mass conservation becomes

$$\frac{d}{dx}(\rho v) = -V\frac{d}{dx}\rho_n = \alpha\rho\rho_n$$

where the distance coordinate x advances in the direction of the fluid velocity. Now let the combined density of ions and neutrals upstream be ρ_0 so that $\rho = 4(\rho_0 - \rho_n)$. We then obtain

$$\rho = \frac{4\rho_0}{1 + e^{-\xi}}$$

$$\rho_n = \frac{\rho_0}{1 + e^{\xi}}$$

where

$$\xi = \frac{4\alpha\rho_0 x}{V} > \xi_0$$

is a dimensionless length coordinate. The ion pressure is given by

$$p = \frac{3\rho_0 V^2}{4(1+e^{-\xi})}$$

The neutral pressure is ignored throughout. In integrating these equations, we have chosen the origin of x and ξ so as to remove a constant from the argument of the exponentials. These relations are only appropriate downstream from the location of the shock, which we place at $\xi_0 = \ln(\rho/\rho_n)_-$, the natural logarithm of the ion-neutral density ratio ahead of the shock. It is convenient that one density profile suffices for all such ratios.

We now consider the fate of small perturbations about this equilibrium flow. Our technique is quite similar to that described by Wardle.[12] We assume that there is a small displacement to the subshock by an amount $\zeta(y,t) \propto e^{iky+st}$, where y is a coordinate lying in the shock front, k is a wave vector and s is the growth rate. We then derive five first order equations describing the downstream variations of the five perturbations $\delta V_{x,y}, \delta p, \delta \rho, \delta \rho_n$. These must be solved subject to the condition that the disturbance is local to the shock. Upstream, the flow is undisturbed and the modified Rankine-Hugoniot conditions at the perturbed subshock provide initial conditions. Far downstream, where the gas is fully ionized and flows with constant velocity, it turns out that there are four spatially decaying modes and one spatially growing mode. A general disturbance can be expanded as a sum over these five eigenfunctions. We now assume values for ξ_0, k and solve the coupled evolution equations for different values of the growth rate. We compute the complex coefficient of the single spatially growing mode, call it $C(s)$ as an analytic function of the complex variable s. (Analyticity provides a convenient check upon the accuracy of the numerical computations.) The temporally growing modes that we seek are given by zeros of $C(s)$, for $\Re(s) > 0$.

Somewhat surprisingly, it turns out that this simple model exhibits dynamical instability, although only under quite restrictive conditions. The unstable modes are present only if the upstream ionized fraction, is less than ~ 1 percent. Under these conditions all $k \lesssim k_{max}$ are unstable where k_{max} diminishes from $\sim \alpha \rho_0/V$ as the ionization fraction increases. The maximum growth rate is $\Re(s) \sim 0.08 \alpha \rho_0$ for $k \to 0$. The associated wave frequency is $\Im(s) \sim 0.9 \alpha \rho_0$ so that the unstable long wavelength modes are actually overstable.

Physically, what seems to be happening is that a local accumulation of ions downstream leads to an increase in the rate at which the neutrals are ionized and a reduction in the ion bulk speed. This effect is counterbalanced by the increase in ion pressure which causes the region to expand and thus is a stabilizing effect. It seems that the two effects are of comparable importance, but that there is net stability except when the initial ionization fraction is low.

The ionization fraction and the velocity of young supernova remnant shock fronts can be probed by observing the Balmer lines.[13] Two components are seen, a narrow line from excited neutrals prior to ionization, and a broad line from once-ionized and accelerated ions following charge exchange. The interpretation of these observations is now rather more complicated than was once thought[14] but it does seem possible to conclude that the pre-shock medium is more than ~ 1 percent ionized, especially in view of the presence of precursor cosmic rays. Therefore, we must conclude that the shock is formally stable to this particular type of instability.

However, it is apparent from the analysis so far that the flow is not far from marginal stability to long wavelength modes. Adding extra features to the physical model, most notably an improved prescription for ion neutral coupling, buoyancy, magnetic stress and cosmic ray pressure [15] [16] [11] may well lead to generic instability. Further calculations are necessary to see if this is indeed the case.

ACKNOWLEDGEMENTS

I thank my collaborators, A. Achterberg and S.Reynolds together with L. Drury, S. Phinney, M. Voit and G. Zank, for helpful discussions. Support under NASA grant NAGW2372 and NSF grant AST89-17765 is gratefully acknowledged.

REFERENCES

1. S. P. Reynolds, *Particle Acceleration in Cosmic Plasmas*, ed. G. Zank & T. Gaisser. New York: American Institute of Physics (1992, in press).
2. A. Harding, *Particle Acceleration in Cosmic Plasmas*, ed. G. Zank & T. Gaisser. New York: American Institute of Physics (1992, in press).
3. C. F. McKee & B. T. Draine, *Science* **252**, 397 (1991).
4. J. P. Edmiston, C. F. Kennel & D. Eichler, *Geophys. Res. Lett.* **9**, 531 (1982).
5. K. B. Quest, *J. Geophys. Res.* **93**, 9649 (1988).
6. R. G. Stone & B. T. Tsurutani, Eds., *Collisionless Shocks in the Heliosphere*Washington D. C.:American Geophysical Union (1985).
7. G. M. Webb, W. I. Axford & T. Terasawa, *Ap. J.* **270**, 537 (1983).
8. M. S. Fulbright & S. P. Reynolds, *Ap. J.* **357**, 591 (1990).
9. J. R. Dickel, W. J. M. van Breughel & R. G. Strom, *Astronom. J.* **101**, 2151 (1991).
10. R. D. Blandford & D. Eichler, *Phys. Rep.* **154**, 1 (1987).
11. R. Ptak & R. E. Stoner, *Ap. J.* **185**, 121 (1973).
12. M. Wardle, *MNRAS* **246**, 98 (1990).
13. R. A. Chevalier, & J. C. Raymond, *Ap. J. Lett.* **225**, L27 (1978).
14. R. C. Smith, R. P. Kirshner, W. P. Blair & P. F. Winkler, *Ap. J.* **375**, 662 (1991).
15. L. O'C. Drury & S. A. E. G. Falle, *MNRAS* **223**, 353 (1986).
16. G. P. Zank, W. I. Axford & J. F. McKenzie, *Astron. Astrophys.* **233**, 275 (1990).

DIFFUSIVE ACCELERATION OF ELECTRONS IN SUPERNOVA 1987A

Lewis Ball
Research Centre for Theoretical Astrophysics,
University of Sydney, N.S.W. 2006, Australia

John Kirk
Max-Planck-Institut für Kernphysik, Heidelberg, D-6900, Germany

ABSTRACT

The reappearance of radio emission from SN 1987A in July 1990 is discussed. We propose a model involving synchrotron radiation from electrons which are accelerated by the expanding supernova shock wave. From the observed delay in the switch-on between 843 MHz and 4.8 GHz we derive the mean free path of the electrons and the time at which the acceleration process must have started, presumably as a result of the shock encountering a density jump in the circumstellar material.

INTRODUCTION

Supernova 1987A reappeared at radio frequencies in July 1990 after being undetectable for roughly three years and 4 months [1]. Other type II supernovae have also exhibited a substantial delay before becoming visible at radio wavelengths, and this has been attributed to the time taken by the supernova shock front to penetrate the thick absorbing screen expelled in the stellar wind of the progenitor [2]. However, in the case of SN 1987A, this explanation is inadequate. Not only does the spectrum show no sign of the low frequency turnover characteristic of absorption, but the shock front was visible at radio wavelengths as early as two days after explosion [3].

In this paper we advance an explanation of the radio emission based on the diffusive acceleration of electrons at the outer shock front of the supernova [4]. In order to keep the model as simple as possible, we assume the shock front remains spherically symmetric and propagates at constant speed. After a certain time t_a, electrons begin to be accelerated. From our interpretation of the observations, this must have occurred about two years after explosion. It is conceivable that the shock front crossed into a region of higher ambient density at around that time, perhaps as speculated by Chevalier [5], upon reaching the reverse shock front thought to terminate the wind from the blue giant phase of the progenitor. Whatever the reason, we assume that subsequent to this time there is a constant input of electrons into the diffusive acceleration mechanism. As these electrons gain energy, they begin to emit synchrotron radiation in the compressed magnetic field swept up with the stellar wind material. In our model, we assume this field is toroidal and proportional to the inverse of the radius r. Extrapolating estimates of the magnetic field at the site of the radio flare on day 2[4, 6, 7] we arrive at a value of the order of 10^{-7} T (1 mG). In the nature of the diffusive process, there is a steady leakage of accelerated particles swept out of the immediate vicinity of the shock front by the downstream flow of plasma. These particles continue to emit synchrotron radiation, but suffer adiabatic losses in the overall expansion. As a consequence, the emission at a fixed frequency first rises as particles are accelerated up to the required energy, and then falls as adiabatic losses set in. To calculate the time dependence we neglect the effects of the finite light travel time across the spherical remnant. Since the shock front expands at roughly $c/10$, this means we are unable to model features in the light curve which have a time scale of less than or about one tenth of the time since explosion. However, even when the observed light curves are averaged to this time resolution there remains an almost linear rise at both 843 MHz and 4.8 GHz which seems to flatten after about 350 days. Furthermore, the emission at 843 MHz systematically precedes that at 4.8 GHz. If we extrapolate back to zero flux the time difference at switch-on is roughly 60 days. This is the most important datum for the model; it determines the rate at which electrons are accelerated, leading to an estimate of their mean free path and of the time at which they began to be diffusively accelerated. Another important property of the emission is its spectrum. During the period when SN 1987A was detectable at 843 MHz and not at 4.8 GHz, the spectrum must have fallen off more steeply than $\nu^{-1.6}$. After switch on at 4.8 GHz the spectrum quickly flattened from a power-law of spectral index $\alpha \sim 1.3$ at switch-on, to one lying between $\alpha = 0.8$ and $\alpha = 1$. Diffusive acceleration predicts a rapid flattening of the spectrum to a value $\alpha = 3/[2(\rho - 1)]$, where ρ is the compression ratio of the shock. Thus, observations indicate $\rho \approx 2.5$ rather than the value 4 expected of a strong shock front in a gas whose ratio of specific heats is 5/3. We do not enter into a discussion of how the shock can have weakened. Perhaps the most likely explanation lies in the back reaction effect of cosmic rays (other than electrons) which the shock has presumably been accelerating. Arguments along these lines have been used by Ellison and Reynolds[8] in applying diffusive acceleration to older supernova remnants. A possible alternative explanation is that the acceleration occurs at a weaker shock associated with the supernova blast wave, perhaps a reverse shock set up in the following ejecta.

THE MODEL

Time dependent, diffusive acceleration in self-similar, spherically symmetric expanding flows has been considered by several authors. The solutions depend critically on the parameter $\epsilon = \kappa/(r_s v_s)$, where κ is the diffusion coefficient, v_s is the velocity (possibly time-dependent) of the shock front and r_s is the radius of the shock. If ϵ is independent of time, the transport equation separates and analytic solutions are available for certain assumptions regarding the dependence of v_s on time[9]. Unfortunately, these solutions are unwieldy unless ϵ can be treated as a small quantity. Solutions which use ϵ as a small parameter and which have a momentum dependence which separates from the spatial and temporal dependence have been discussed by Krymsky and Petukhov[10] and by Drury[11]. These, however, are not useful in the present context in which we wish to describe how the shape of the particle spectrum changes with time.

Physically, ϵ represents the ratio of the size of the region around the shock front within which acceleration occurs (essentially the length scale of the precursor) to the radius of the shock front. Equivalently, it is the ratio of the acceleration time scale to the expansion timescale r/v_s. Adiabatic losses occur on the latter timescale and clearly compete with acceleration at the shock front if ϵ is not small. Fortunately, in our problem observations indicate that ϵ is indeed small. Thus, whilst a particle is resident close to the shock, its adiabatic losses are negligible compared to its acceleration rate. Therefore, to a first approximation, the problem of acceleration is identical to that at a plane shock front. However, once a particle ventures several diffusion lengths ($\sim \kappa/v_s$) downstream of the shock, the probability of return becomes extremely small and adiabatic losses dominate its future. In view of this, we split our treatment into two, taking the time-dependent acceleration at a plane shock and matching it onto a diffusion-free flow in which particles experience only adiabatic losses. The philosophy of this approach is similar to that of the "onion-skin" models introduced by Bogdan and Völk[12] in that once particles leave the shock they are assumed frozen into the shell of plasma in which they find themselves. However, our approach to the computation is different, since we are interested in the synchrotron spectrum as a function of time, and not in the cosmic ray distribution at large times.

Consider, then, a plane shock front into which plasma flows along the normal at speed v_1 and exits also along the normal at speed v_2. This shock, of compression ratio $\rho = v_1/v_2$, accelerates a population of test particles which diffuse in the background plasma with spatial diffusion coefficients $\kappa_{1,2}$. Test particles enter the acceleration process at the rate Q with momentum p_0. The distribution function of such particles can be found as a function of momentum, time and space[13,14]. However, it suffices for our purposes to use a simple spatially averaged model of the acceleration process (cf. Bogdan and Völk[12]): a test particle is presumed to undergo continuous acceleration, whilst in the vicinity of the shock, such that its momentum p increases at a rate given by $p\Delta/t_c$, where $\Delta = 4(v_1 - v_2)/3v$ is the average fractional momentum gain per shock crossing/recrossing (with v the test particle's velocity) and $t_c = 4(\kappa_1/v_1 + \kappa_2/v_2)/v$ is the average time taken to perform such a cycle[11,15]. In addition, these particles escape from the shock region at a rate which is just the escape probability per cycle $P_{esc} = 4v_2/v$ divided by the average cycle time. The equation determining $N(p,t)$ (the number of particles in the acceleration region with momentum between p and $p + dp$, divided by dp) is then:

$$\frac{\partial N}{\partial t} + \frac{\partial}{\partial p}\left(\frac{p\Delta}{t_c}N\right) + \frac{P_{esc}}{t_c}N = Q\delta(p - p_0). \qquad (1)$$

Assuming κ is independent of p, the solution satisfying the boundary condition

$N(p, t_\mathrm{a}) = 0$ is

$$N(p,t) = \frac{t_\mathrm{c} Q}{p_0 \Delta} \left(\frac{p}{p_0}\right)^{-2\alpha-1} \left[H\left(p-p_0\right) - H\left(p - p_0 e^{(t-t_\mathrm{a})\Delta/t_\mathrm{c}}\right)\right] \quad (2)$$

where $\alpha = P_\mathrm{esc}/(2\Delta)$ and $H(x)$ is a Heaviside function equal to zero for $x < 0$ and unity for $x > 0$. The number of particles *leaving* the acceleration region per second is $N(p,t) P_\mathrm{esc}/t_\mathrm{c}$ and, since plasma leaves this region at speed v_2, the distribution function of particles advected with it is

$$f_\mathrm{s}(p,t) = \frac{1}{4\pi p^2} N(p,t) \frac{P_\mathrm{esc}}{A t_\mathrm{c} v_2} \quad (3)$$

where A is the area through which the plasma leaves the acceleration region.

We now move to the rest frame of the upstream plasma. Assume the shock front to be moving out from the site of the explosion at constant speed v_s and to be spherical in shape. At time t_a electrons begin to be accelerated, perhaps as a result of a sudden increase in external density. The rate Q at which they are picked up may be expected to remain constant provided the external density falls off after the jump as r^{-2}. The shock front has a constant compression ratio ρ and it is assumed that the downstream plasma moves in the radial direction with (constant) speed $v_\mathrm{d} = v_\mathrm{s}(\rho-1)/\rho$. Particles which have left the shock are frozen in to the downstream plasma, so that the equation satisfied by the distribution function is:

$$\frac{\partial f}{\partial t} + v_\mathrm{d} \frac{\partial f}{\partial r} - \frac{2}{3}\frac{v_\mathrm{d}}{r} p \frac{\partial f}{\partial p} = 0. \quad (4)$$

This equation is easily solved using the Lagrangian (comoving) coordinate $R = r - v_\mathrm{d} t$, where $t = 0$ is the time of explosion. One finds the general solution to be an arbitrary function of R and of the combination $p r^{2/3}$. The boundary condition to be applied is that the distribution at the position of the shock front is given by Eq. (3). Noting that the radius at which the fluid element labelled by R passed through the shock is ρR, one finds the solution:

$$f(R,p,t) = f_\mathrm{s}(xp, \rho R/v_\mathrm{s}) \quad (5)$$

where

$$x = [r/(\rho R)]^{2/3} \quad (6)$$

is the factor by which adiabatic losses reduce the particle momentum. To simplify the calculation of synchrotron emission from this distribution, the emissivity of a single particle can be approximated by

$$j_\nu(p) = a_0 p^2 B^2 \delta(\nu - a_1 p^2 B) \quad (7)$$

with a_0 and a_1 constants, and where the magnetic field B is assumed to vary as r^{-1}. Integrating over the entire downstream electron population then leads to the predicted emission from particles behind the shock. According to the frequency, different regions contribute to the emission, but these are easily located when $\rho > 7/4$, which covers the cases of interest to us. Introducing dimensionless variables according to

$$\hat{t} = t/t_0$$
$$\hat{\nu} = \nu/(a_1 B_0 p_0^2)$$

where B_0 is the magnetic field immediately behind the shock at the (arbitrary) time t_0, i.e., $B = B_0 v_s t_0/r$, we find that the effects of electrons of momentum close to that of injection can be ignored provided $\hat{\nu} > 1/\hat{t}$. Then the emission occurs up to a cut-off given by

$$\hat{\nu}_{\max}(\hat{t}) = \hat{t}^{-1} \exp[2(\hat{t} - \hat{t}_a)/\rho\eta] \tag{8}$$

where $\eta = t_c/(\Delta \rho t_0)$ is a dimensionless parameter roughly equal to the acceleration timescale in units of t_0.

For frequencies $\hat{t}^{-1} < \hat{\nu} < \hat{\nu}_{\max}$ contributions to the emission arise not only from particles which have left the shock front but also from those still engaged in the acceleration process. These particles are described by Eq. (2) and radiate primarily in the magnetic field just downstream of the shock front. With these simplifications it is possible to obtain the predicted total flux density at Earth in closed form:

$$F(\nu, t) = C(\hat{\nu}\hat{t})^{-\alpha} \left\{ \left(\frac{\rho}{\rho - 1}\right)^\alpha \left[B_{y_1}\left(\alpha, 1 + \frac{4\alpha}{3}\right) - B_{y_2}\left(\alpha, 1 + \frac{4\alpha}{3}\right)\right] + \frac{\eta}{2\alpha \hat{t}} \rho^{-4\alpha/3} \right\} \tag{9}$$

where $B_y(a,b)$ is the incomplete beta function [16]. The constant C is given by

$$C = \frac{a_0 Q B_0 t_0 \alpha}{4\pi D^2 a_1} \rho^{1 + \frac{4\alpha}{3}} \tag{10}$$

where D is the distance to SN 1987A, and

$$y_1 = (\rho - 1)\hat{t}/[\rho \hat{R}_1 + (\rho - 1)\hat{t}] \tag{11}$$
$$y_2 = (\rho - 1)/\rho \tag{12}$$

with $\hat{R}_1$ the solution of the equation

$$\hat{\nu}(\rho \hat{R}_1)^{-4/3}[\hat{R}_1 + \hat{t}(\rho - 1)/\rho]^{7/3} = \exp[2(\hat{R}_1 - \hat{t}_a/\rho)/\eta] . \tag{13}$$

The term in square brackets in Eq. (9) is the contribution of particles which have left the acceleration region. Clearly, for small η the emission is in general dominated by such particles. However, close to the cut-off $\hat{\nu}_{\max}$ the contribution of

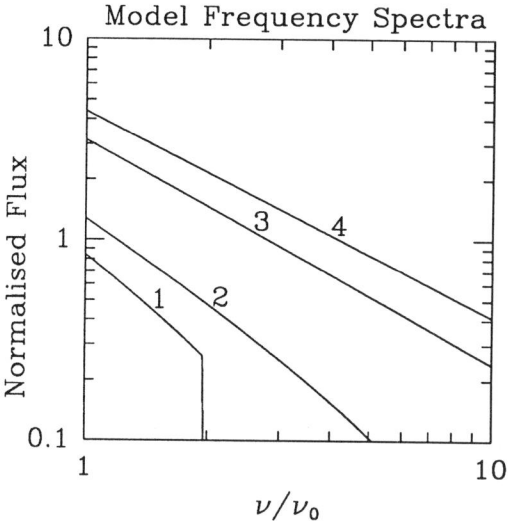

Figure 1: The predicted emission spectra at fixed times denoted by the labels: $1 \to \hat{t} = 1.02$, $2 \to \hat{t} = 1.05$, $3 \to \hat{t} = 1.25$, $4 \to \hat{t} = 2.0$.

particles in the shock can be comparable, since then $y_1 \approx y_2$ and the two beta functions almost cancel. The basic features of the predicted emission are illustrated in Figures 1 and 2. Both figures are plotted with t_0 and the parameters ρ, η and $\hat{t}_a$ chosen as discussed below. A power law of index α is obtained extending from $\hat{\nu} = 1/\hat{t}$ up to close to $\hat{\nu}_{\max}$. At a given frequency, the emission first grows rapidly, reaches a maximum and decays because of adiabatic losses.

COMPARISON WITH OBSERVATIONS

The emission from SN 1987A at 843 MHz appears to precede that at 4.8 GHz by about 60 days. In our model, we assume the start of emission at a given frequency $\hat{\nu}$ occurs when $\hat{\nu} = \hat{\nu}_{\max}(\hat{t})$. Using a normalisation $t_0 = 1170$ days, so that switch-on at the lower frequency occurs at $\hat{t} = 1$, we find, using (8), a value $\eta = 0.057/\rho$. The smallness of this quantity guarantees the validity of the approximation in which the acceleration is calculated as if the shock front were planar. From η we can find values of the average diffusion coefficient of the particles (assuming $\kappa = \kappa_1 = \kappa_2$):

$$\kappa = 3.0 \times 10^{20} \text{ m}^2 \text{ s}^{-1} \tag{14}$$

in which we have assumed $\rho = 5/2$ (see below) and [4] $v_s = 30\,000$ km s^{-1}. The corresponding mean free path of the electrons is then roughly 3×10^{12} m. In addition, the quantity $\hat{t}_a$, which enters only in the factor $\exp(\hat{t}_a/\rho\eta)$ scaling the dimensionless frequency in Eq. (13), is determined by the magnetic field B_0 and

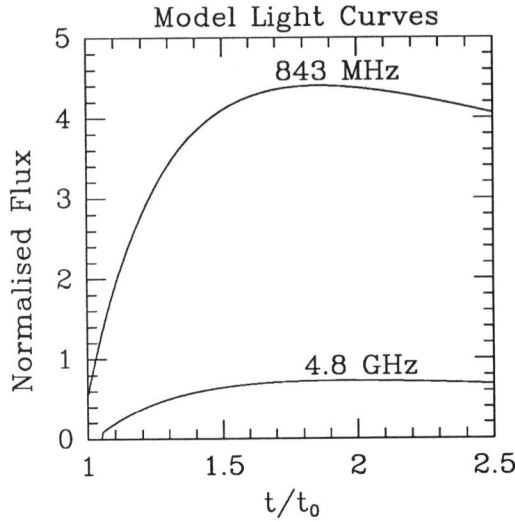

Figure 2: The predicted light curves at 843 MHz and 4.8 GHz.

the injection momentum p_0:

$$t_a = \left[721 + 34 \ln\left(\gamma_0^2 B_{-7}\right)\right] \text{ days} \qquad (15)$$

where B_{-7} is the value of B_0 expressed in units of 10^{-7} T (milligauss) and $\gamma_0 = p_0/mc$ is the Lorentz factor of injected electrons (provided these are relativistic). Thus, in this picture particle acceleration started about two years after the explosion.

Since our model predicts emission with a power law index $\alpha = 3/[2(\rho-1)]$ soon after switch on, the observed spectral index of about 1 indicates $\rho \approx 2.5$. The only remaining free parameter in the model is the injection rate Q. This can be determined by normalising the predicted flux density to that observed at a single frequency on a given date. Using the value of 10 mJy observed at 843 MHz on day 1386[1] implies that the supernova shock is picking up electrons at the rate of about $10^{45}\,\text{s}^{-1}$, which corresponds to a density in front of the shock of electrons which are to be injected of about $2 \times 10^5\,\text{m}^{-3}$. This is five orders of magnitude smaller than the electron density observed in the circumstellar ring surrounding SN 1987A [17]. On the other hand, the required electron pick-up rate, or equivalently the required electron density, is about five percent of the total number of electrons which would be encountered per second by the shock front if it were moving in the uncompressed blue giant wind. Taking[18] a mass-loss rate of $\dot{M} = 10^{-5}\,M_\odot/\text{yr}$ and a wind speed of $v_w = 550\,\text{km s}^{-1}$, gives a preshock density of $\sim 4 \times 10^6\,\text{m}^{-3}$ and an encounter rate of $\sim 2 \times 10^{46}\,\text{s}^{-1}$. It is unlikely that the "injection efficiency" of the acceleration mechanism is as high as 5%, so the uncompressed blue giant wind alone is unlikely to be sufficiently dense to account for the observed emission. However, since we expect that the ambient density jumped to a value significantly above that in the blue giant wind at about day 720, the required injection 'efficiency' is significantly less than 5%.

DISCUSSION

The model presented above seems to fit quite well with the overall features of the radio emission. The spectral fit is good, provided the shock responsible for the acceleration has a compression ratio of about 2.5. The observed flux density requires that something of the order of one percent of the encountered electrons is injected into the diffusive acceleration process at the shock front. If our interpretation of the radio emission is correct, the implication is that the supernova shock wave ran into a jump in the ambient density about 2 years after the supernova explosion. This density jump may well be the termination shock of the blue giant wind as it is slowed down by the more dense and more slowly expanding material from progenitor's red giant phase. Regardless of its origin, this density jump has not yet been detected in any other way. Another implication of the model is that the delay between the appearance of emission at 843 MHz and at 4.8 GHz is a direct observation of the acceleration time. This is probably the first time the consequences of the acceleration process itself have been directly observed in a synchrotron source. The model also makes some predictions. If the injection rate and compression ratio remain constant, the flux density at both 843 MHz and at 4.8 GHz should peak at about day 2200 (early 1993).

Several important simplifications have been made in order to obtain a result in closed form. Perhaps the most important is the neglect of the effects of the finite light travel time across the remnant. However, we have also treated synchrotron radiation in a rather crude manner. Both of these shortcomings should be easy to remove by numerical means.

ACKNOWLEDGMENTS

We thank D. B. Melrose, G. Bicknell, J. Kuijpers, R. N. Manchester, L. Staveley-Smith, and A. J. Turtle for numerous helpful discussions.

REFERENCES

1. L. Staveley-Smith, R. N. Manchester, M. J. Kesteven, D. Campbell-Wilson, D. F. Crawford, A. J. Turtle, J. E. Reynolds, A. K. Tzioumis, N. E. B. Killeen, and D. L. Jauncey, Nature, *355*, 147, 1992.

2. R. A. Chevalier, Astrophys. J., *259*, 302, 1982.

3. A. J. Turtle, D. Campbell-Wilson, J. D. Bunton, D. L. Jauncey, M. J. Kesteven, R. N. Manchester, R. P. Norris, M. C. Storey, and J. E. Reynolds, Nature, *327*, 38, 1987.

4. L. T. Ball and J. G. Kirk, *Radio Supernova 1987A*, Proc. Astron. Soc. Aust., in press, 1992.

5. R. A. Chevalier, *The radio evolution of SN 1987A*, Nature, in press, 1992.

6. M. C. Storey and R. N. Manchester, Nature, *329*, 421, 1987.

7. J. G. Kirk and M. Wassmann, *Electron acceleration at a supernova blast wave. A model for the prompt radio flare from SN 1987A,* Astron. Astrophys., in press, 1992.

8. D. C. Ellison and S. P. Reynolds, Astrophys. J., *382*, 242, 1991.

9. V. L. Prishchep and V. S. Ptuskin, Sov. Astron., *25*, 446, 1981.

10. G. F. Krymsky and S. I. Petukhov, Sov. Astron. Lett., *6*, 124, 1980.

11. L. O'C. Drury, Rep. Prog. Phys., *46*, 973, 1983.

12. T. J. Bogdan and H. J. Völk, Astron. Astrophys., *122*, 129, 1983.

13. I. N. Toptygin, Space Sc. Rev., *26*, 157, 1980.

14. L. O'C. Drury, Monthly Notices Roy. Astron. Soc., *251*, 340, 1991.

15. A. R. Bell, Monthly Notices Roy. Astron. Soc., *182*, 147, 1978.

16. M. Abramowitz and I. E. Stegun, Handbook of Mathematical Functions (National Bureau of Standards, Washington, 1972).

17. C. Fransson, A. Cassatella, R. Gilmozzi, R. P. Kirshner, N. Panagia, G. Sonneborn, and W. Wamsteker, Astrophys. J., *336*, 429, 1989.

18. R. A. Chevalier and C. Fransson, Nature, *328*, 44, 1987.

SIMULATION OF ELECTRON ACCELERATION AT COLLISIONLESS SHOCKS

D. KRAUSS-VARBAN

Department of Electrical and Computer Engineering and California Space Institute, University of California at San Diego, La Jolla, CA 92093-0407

ABSTRACT

The motion of suprathermal electrons through the quasi-perpendicular, curved bow shock is considered, using a combined test-particle and hybrid-code approach. Whistler-wave generation and pitch-angle scattering of the derived distributions is studied with an implicit full-particle code. The results are compared to published ISEE observations at the Earth's bow shock.

I. INTRODUCTION

Interplanetary shocks, planetary bow shocks and shocks in astrophysical settings can accelerate charged particles in a variety of ways. Shocks are often invoked to explain evidence of high energy tails, cosmic rays, and accelerated particle beams.[1-4] The Earth's bow shock can be viewed as a laboratory that allows us to test ideas and compare results of analytical studies and simulations with highly accurate measurements. In the case of electrons, evidence of acceleration was first accumulated in the 1960s[5,6]. Later, observations by the ISEE satellites documented that energetic electrons originate near the point of tangency of the interplanetary magnetic field and the bow shock, and are emitted upstream in thin sheets at energies up to > 100 keV.[7-9] The pertinent acceleration mechanism was unclear until studies gave an explanation via a "fast" Fermi process.[10,11] When viewed in the deHoffmann-Teller frame (HTF), in which the motional solar wind electric field vanishes, the process is a simple mirror reflection in the shock magnetic field gradient. Equivalently, the process can also be viewed as gradient drift acceleration in the normal incidence frame (NIF), in which the plasma flow is along the shock normal. There it can be shown that the electrons undergo a drift due to the magnetic field gradient, antiparallel to the motional electric field.[10-14]

The high turbulence level in shocks can also lead to acceleration due to wave-particle interactions, *e.g.*, with lower-hybrid waves.[15,16] However, recent elaborate analytical[17] and test-particle calculations in one[18,19] and two dimensions[20] have shown qualitatively as well as quantitatively that gradient drift acceleration alone can explain observations quite well. From this one may conclude that scattering processes usually play a secondary role, only. Still, a mystery persists in the fact that the largest fluxes of energetic electrons are commonly observed just downstream of quasi-perpendicular shocks, in the vicinity of the overshoot of the magnetic field.[21,22] Mirror reflected particles are apparently restricted to the upstream, and it is not immediately obvious how the gradient drift acceleration process extends to the downstream. It has been claimed that the observed downstream fluxes are larger than what would be expected from simple adiabatic mapping.[22] A second interesting feature of the observations is a nearly isotropic velocity distribution by the time the plasma has reached the overshoot[22]. Obviously, at least pitch-angle scattering of the transmitted electrons (if not stochastic acceleration) has to be invoked to explain this finding. We will address both of these issues in this paper.

A one-dimensional, self-consistent hybrid code[23] with kinetic ions and a massless electron fluid is used. The solar wind electrons are modelled in two parts: a thermal population (core electrons, fluid in simulation) and an energetic population (halo electrons, test particles). The basic approach and procedure for a curved bow shock field-geometry has been described previously.[20] A more complete report of the present simulations results will be presented elsewhere[24]. Parameters applicable to the Earth's bow shock are: $\omega_i/\Omega_{ci} = 5000$, where ω_i and Ω_{ci} are the upstream ion plasma and ion cyclotron frequency, respectively, and $\beta_i = \beta_e = 0.5$ for both the upstream ion and core electron plasma beta. The values correspond to a core energy of about 10 eV, the halo electrons have 6 times the core thermal energy.[15] We are not concerned with relativistic energies. The shock Mach number is $M_A \simeq 6.3$. We model the halo population with a κ-distribution $f_\kappa(v) = c_\kappa \{1 + v^2/[(\kappa - 3/2)v_{th}^2]\}^{-\kappa-1}$, where $c_\kappa = [\pi(\kappa - 3/2)v_{th}^2]^{-3/2} \kappa! / \Gamma(\kappa - 1/2)$. We use an index of $\kappa = 6$, that fits the observations quite well for energies up to several keV.[25,26] Observed distributions cannot be well described by a single power law. For energies above several keV a distribution with index 3-4 gives a better fit.[26]

II. TWO-DIMENSIONAL DRIFTS AND UPSTREAM ELECTRON ENERGIZATION

Let x be in the shock normal direction **n**, and let the upstream magnetic field be contained in the x-z plane making an angle θ_{Bn} with **n**. It can be shown that the drift ℓ_z along z is much longer than that in the y direction.[20] For 1-D adiabatic theory[10,11] to be applicable, $\ell_z \simeq \tau v_1/\cos\theta_{Bn}$ should be *much* smaller than the radius of curvature of the shock, since only angles θ_{Bn} in the direct vicinity of 90° lead to significant final (parallel) energies, $\langle E_\parallel \rangle \simeq 2m_e v_1^2/\cos^2\theta_{Bn}$. Here, m_e is the electron mass, v_1 is the solar wind velocity in the NIF, and $\tau \sim 2\text{-}3\ \Omega_{ci}^{-1}$ is the typical reflection time scale[18,20]. For typical shocks, the above can be rewritten as $\ell_z \sim 5 \cdot \sqrt{(E_\parallel/\text{keV})}\ R_E$, where R_E is the Earth's radius. The radius of curvature of the Earth's bow shock may vary from less than 20 R_E to more than 100 R_E. Typically, one would expect considerably reduced fluxes at energies larger than a few keV.

Indeed, single electrons when drifting along z experience an average, smaller θ_{Bn} in the shock layer which reduces their final *parallel* energy. Yet, our computations show reflected electron fluxes of the order of observed fluxes:[20] The differential flux of reflected electrons agrees well with observations made by the ISEE spacecraft[27] and PROGNOZ 10[16,28] in the 4 keV to 15 keV energy range.

Consistent with the halo characteristics, one can demonstrate that a seed population with $\kappa \sim 4$ produces reasonably good results at very high energies $> 15\,\text{keV}$, but would result in fluxes that are too large at smaller energies.

Why doesn't the curvature inhibit fluxes at high energy? First, it turns out that the high energy tail of reflected electrons originates from large *perpendicular* energies, has shorter than average reflection times, and thus sees less of the curvature than expected. Moreover, there is a flux concentration due to the fact that the drift length depends on the angle θ_{Bn} at which an electron enters the shock. The result is a very spiky high energy flux profile with respect to θ_{Bn}.

III. DOWNSTREAM (TRANSMITTED) ELECTRONS

What are the predictions using a test-particle approach for the halo electrons? Close to $\theta_{Bn} = 90°$, nearly all electrons are situated within the loss-cone and penetrate the shock. Without pitch-angle scattering, energization (in the NIF) takes place in the perpendicular velocity component only, proportional to the magnetic field jump, whereas the parallel velocity component is virtually unaffected.[13] Since the electrons are tied to the magnetic field lines, there is a corresponding compression of the distribution associated with the transmission. Away from $\theta_{Bn} = 90°$, fewer electrons transmit, there is a finite energy contribution from the cross-shock potential, and (ignoring the potential) the gain in perpendicular energy is eventually balanced by a loss in parallel energy.

Fig. 1 shows the distribution of halo electrons versus total energy in the vicinity of the overshoot, at four different angles θ_{Bn}. Circles denote the upstream distribution, for comparison. The radius of shock curvature is $5000\,c/\omega_{pi}$ (approximately 50 R_E). Below $\sim 10^{-1}$ keV, the curves would join with the heated core population. Apparently, in addition to a general compression, the halo electrons have been heated significantly. The results close to $90°$ are consistent with a density increase *and* heating by a factor between 3 and 4. In contrast to reflected electrons, the results are not very sensitive to θ_{Bn}, i.e., the spatial mobility of the electrons has significantly smoothed any dependence on θ_{Bn}.

Fig. 2 shows contour plots of the velocity distributions with respect to the magnetic field, in the ramp and overshoot regions, averaged over $85° \leq \theta_{Bn} < 90°$. The two most striking features are the loss-cone in the ramp, and the large anisotropy especially at the overshoot. Such anisotropies are not observed. Thus, it is evident that very efficient pitch-angle scattering must take place while the electrons reach the downstream.

In the above treatment, wave-particle interactions on electron scales are not self-consistently included. Let us consider the wave generation and pitch-angle scattering separately from the electron dynamics that lead to the unstable distributions. We present here initial results of such a study. In the parameter region under consideration, use of an implicit code is advantageous since we are not interested in ω_{pe} (electron plasma frequency) time scales. We have $\omega_{pe}/\Omega_{ce} \sim 100$, where Ω_{ce} is the electron gyrofrequency. So, we only demand a time step $\Delta t \sim 1/40\, \Omega_{ce}^{-1} = 2.5\, \omega_{pe}^{-1}$, even when resolving the electron gyromotion very well. For typical resonance wavelengths $\lambda \simeq 5...10\, c/\omega_{pe}$ we may choose a cell size of $\Delta x \sim 1\, c/\omega_{pe}$.

The simulations are executed with the one-dimensional version of the implicit electromagnetic particle code CELEST1D.[29] We choose a box size of 100 c/ω_{pe}, 500 particles per cell and a propagation angle of 30° with respect to the ambient magnetic field $\mathbf{B}_o$. In the first simulation we have loaded a core population with a velocity anisotropy of $v_\perp/v_\parallel = 3$. This type of anisotropy would be expected at the overshoot, if no scattering were to take place. A 5% halo population has an anisotropy of 2 and a 45° loss-cone similar to the ramp distribution in Fig. 2. Runs with a halo anisotropy of 3 or a core anisotropy of 2 gave qualitatively similar results. In Fig. 3 we can see that electromagnetic whistler waves are generated at the expected wavelength. The isotropization of the electrons is displayed in Fig. 4, which shows scatter plots of the phase space parallel and perpendicular to $\mathbf{B}_o$. At time $t = 25\, \Omega_{ce}^{-1}$ the loss-cone of the halo population has filled up considerably, and by $t = 50\, \Omega_{ce}^{-1}$ most of the anisotropy has vanished. Thus, the isotropization takes place on a time scale very much faster than the drift time scale through the shock of $\sim \Omega_{ci}^{-1}$.

How much do the halo electrons contribute to the waves? In Fig. 5 we have plotted results from a similar simulation as above, except that the core electrons are isotropic at $t = 0$. It is evident that now even at relatively late time $t = 62.5\, \Omega_{ce}^{-1}$, neither the loss-cone nor the anisotropy of the halo electrons are affected significantly. Thus, we conclude that the halo electrons are isotropized in the whistler waves generated by the core population.

IV. SUMMARY

In agreement with observations,[22] we obtain downstream fluxes of energetic electrons much larger than upstream, that are most pronounced in the vicinity of the overshoot. At a curved shock, the great mobility of the electrons can lead to enhanced fluxes away from $\theta_{Bn} = 90°$, resulting in local values larger than expected from 1-D adiabatic mapping. From implicit, full-particle simulations we conclude that during the passage through the shock the halo electrons are pitch-angle scattered in the whistler waves generated by the core population. We suggest that this mechanism is responsible for the observed isotropy[22] of the energetic electrons once they have reached the overshoot.

ACKNOWLEDGMENTS

The author would like to thank D. Burgess, H. X. Vu and J. U. Brackbill for many helpful discussions. The CELEST1D code used in the implicit calculations was originally developed by J. U. Brackbill and H. X. Vu at LANL. This work was supported by the National Aeronautics and Space Administration, research grant NAGW-2618. The computational facilities were provided by the NSF San Diego Supercomputer Center.

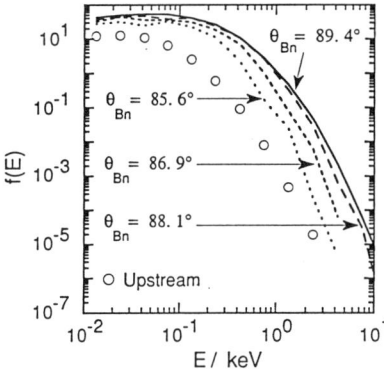

Fig. 1. Energy distribution of halo electrons at the overshoot for several values of θ_{Bn}, as indicated. Circles denote upstream distribution.

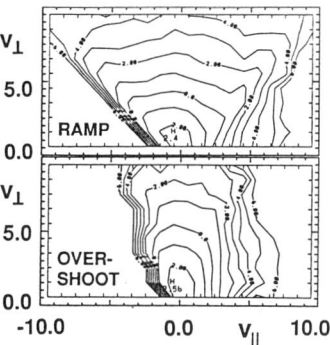

Fig. 2. Contour plots of halo velocity distribution in shock ramp and overshoot. Normalized with upstream thermal speed, one contour per decade; negative $v_\parallel$ points upstream.

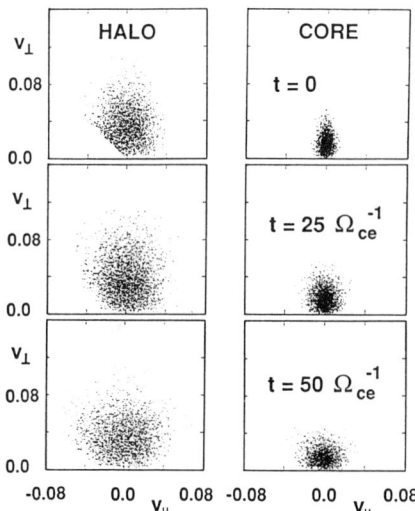

Fig. 4. Pitch-angle scattering and isotropization of the halo and core populations in self-consistently generated whistler-wave turbulence. Velocities are normalized with the speed of light.

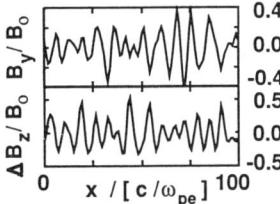

Fig. 3. Plot of the perturbation in the two transverse magnetic field components for whistler waves generated by a temperature anisotropy in a full particle simulation, at $t = 25\Omega_{ce}^{-1}$.

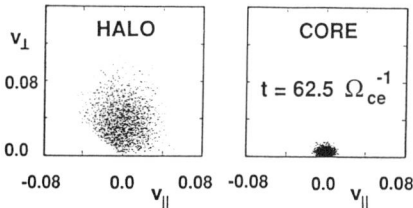

Fig. 5. Same as Fig. 4, except here the core population was isotropic at $t = 0$. The conclusion is that the halo electrons scatter in the whistler waves generated by the core population.

REFERENCES

1. Toptyghin, I. N., Space Sci. Rev., **26**, 157 (1980).
2. Drury, L. O., Rep. Prog. Phys., **46**, 973 (1983).
3. Armstrong, T. P., M. E. Pesses, and R. B. Decker, in *Collisionless Shocks in the Heliosphere: Reviews of Current Research, Geophys. Monogr. Ser.*, vol. 35, edited by B. T. Tsurutani and R. G. Stone, pp. 271–285, AGU, Washington, D. C. (1985).
4. Scholer, M., in *Geophys. Monogr. Ser.*, vol. 35, pp. 287–301, AGU, Washington, D. C. (1985).
5. Fan, C. Y., G. Gloeckler, and J. A. Simpson, Phys. Rev. Lett., **13**, 149 (1964).
6. Anderson, K. A., J. Geophys. Res., **74**, 95 (1969).
7. Anderson, K. A., Nuovo Cimento Soc. Ital. Fis., **2C** (N6), 747 (1979).
8. Anderson, K. A., R. P. Lin, F. Martel, C. S. Lin, G. K. Parks, and H. Rème, Geophys. Res. Lett., **6**, 401 (1979).
9. Parks, G. K., E. Greenstadt, C. S. Wu, C. S. Lin, A. St.-Marc, R. P. Lin, K. A. Anderson, C. Gurgiolo, B. Mauk, H. Rème, R. Anderson, and T. Eastman, J. Geophys. Res., **86**, 4343 (1981).
10. Wu, C. S., J. Geophys. Res., **89**, 8857 (1984).
11. Leroy, M. M., and A. Mangeney, Ann. Geophys., **2**, 449 (1984).
12. Webb, G. M., W. I. Axford, and T. Terasawa, Astrophys. J., **270**, 537 (1983).
13. Krauss-Varban, D., and C. S. Wu, J. Geophys. Res., **94**, 15,367 (1989).
14. Vandas, M., Bull. Astron. Inst. Czech., **40**, 189 (1989).
15. Wu, C. S., J. D. Gaffey, Jr., and B. Liberman, J. Plasma Phys., **25**, 391 (1981).
16. Galeev, A. A., S. Fisher, S. I. Klimov, K. Kudela, V. N. Lutsenko, Z. Němeček, M. N. Nozdrachev, J. Šafránková, P. Tříska, O. Vaisberg, and G. Zastenker, Adv. Space Res., **6**, 45 (1986).
17. Vandas, M., Bull. Astron. Inst. Czech., **40**, 175 (1989).
18. Krauss-Varban, D., D. Burgess, and C. S. Wu, J. Geophys. Res., **94**, 15,089 (1989).
19. Vandas, M., Bull. Astron. Inst. Czech., **42**, 170 (1991).
20. Krauss-Varban, D., and D. Burgess, J. Geophys. Res., **96**, 143 (1991).
21. Anderson, K. A., J. Geophys., **40**, 701 (1974).
22. Gosling, J. T., M. F. Thomsen, S. J. Bame, and C. T. Russell, J. Geophys. Res., **94**, 10,011 (1989).
23. Winske, D., and M. M. Leroy, in *Computer Simulation of Space Plasmas—Selected Lectures From the First ISSS*, edited by H. Matsumoto and T. Sato, pp. 255–278, Kluvier Academic, Hingham, Mass. (1984).
24. Krauss-Varban, D., in preparation for J. Geophys. Res., (1992).
25. Feldman, W. C., R. C. Anderson, S. J. Bame, S. P. Gary, J. T. Gosling, D. J. McComas, M. F. Thomsen, G. Paschmann, and M. M. Hoppe, J. Geophys. Res., **88**, 96 (1983).
26. Lin, R. P., Solar Phys., **100**, 537 (1985).
27. Anderson, K. A., J. Geophys. Res., **86**, 4445 (1981).
28. Vandas, M., S. Fisher, V. N. Lutsenko, K. Kudela, J. Šafránková, and Z. Němeček, Bull. Astron. Inst. Czech., **39**, 308 (1988).
29. Vu, H. X., and J. Brackbill, Comp. Phys. Commun., in press (1992).

WHISTLER WAVES AND ELECTRON-WHISTLER SCATTERING IN ASTROPHYSICAL PLASMAS

Matthew G. Baring
Department of Physics, Box 8202,
North Carolina State University,
Raleigh, NC 27695

ABSTRACT

In shocked astrophysical plasmas, whistler waves can act as scattering centres for electrons in the turbulent fluid, resonantly interacting with electrons of much lower speeds than Alfvén waves scatter with. They therefore show great potential to energize electrons from shocked thermal gas and thereby effect electron injection in the usual diffusive shock acceleration process. In this context, whistlers as a source of electron scattering have received only limited attention in the literature, so here a study of the basic properties of whistler waves in warm plasmas and electron-whistler scattering is presented. Including the effects of finite plasma temperature T is a significant advance on previous research. The whistler dispersion relation is found in the limit of small wave damping, and it is observed that the range of whistler frequencies depends strongly on T: for high temperatures, whistlers can have low frequencies and phase velocities below the Alfvén speed. The spatial diffusion coefficient for electron-whistler scattering from quasi-linear theory is also presented.

INTRODUCTION

A major problem in the theory of diffusive shock acceleration is the mechanism by which electrons can be accelerated out of the ambient thermal gas into an energetic population of cosmic rays. For near-thermal protons, this mechanism is scattering by Alfvén turbulence in the fluid flow, which is very effective in reversing the protons so that they can cross the shock front many times, reaching very high energies (e.g. see Bell, 1978). However, electrons cannot interact resonantly with Alfvén waves unless their energies are already highly suprathermal. Therefore another mechanism is needed to provide the diffusive scattering of near-thermal electrons, and thus effect the injection of electrons into a suprathermal population that can participate in the usual shock acceleration mechanism mediated by Alfvén wave scattering.

Whistler waves are a natural candidate for scatterers of thermal electrons. This dispersive mode of helicon-like waves can resonantly interact with electrons with near-thermal speeds: this is to be expected because whistlers are strongly damped only at frequencies corresponding to resonant interaction with thermal electrons. In this paper, a preliminary investigation of the properties of whistlers relevant to shock acceleration is made. The dispersive features of whistlers in warm plasmas is analyzed, and the introduction of temperatures T typical of shocked gas in supernova remnants produces a surprising reduction of the whistler phase velocity. The treatment here focusses on a parameter regime different from the solar flare applications of Steinacker and Miller (these proceedings). The spatial diffusion coefficient for electron-whistler scatterings at near-thermal electron velocities, as obtained from quasi-linear theory, is presented assuming canonical power-law wave spectra and isotropy.

WHISTLER WAVES AND ELECTRON-WHISTLER SCATTERING

Whistlers and Alfvén waves are transverse, circularly-polarized electromagnetic eigenmodes of magnetized media. Their freqencies are below the electron cyclotron frequency Ω_e, which is normally much smaller than the plasma frequency ω_e. Alfvén waves exist in both left-handed and right-handed states of polarization while whistlers can only be right-handed (e.g. see Ichimaru, 1973). The dispersion relation for the right-handed states varies continuously from the domain of Alfvén waves to that of whistlers. It can be found simply from the dielectric response for the magnetized plasma (see Ichimaru, 1973):

$$\frac{k^2}{\omega^2} = \epsilon_r(k,\omega) = 1 + \frac{\omega_e^2}{k\omega\sqrt{2\theta_e}}\mathcal{Z}(z_e) + \frac{\omega_p^2}{k\omega\sqrt{2\theta_p}}\mathcal{Z}(z_p) \quad . \tag{1}$$

This relation is evaluated for waves of frequency ω and wavenumber k propagating parallel to the field. Here $\mathcal{Z}(\zeta)$ is the non-relativistic *plasma dispersion function*, defined by Fried and Conte (1961), with $z_e = (\omega - \Omega_e)/(k\sqrt{2\theta_e})$ and $z_p = (\omega + \Omega_p)/(k\sqrt{2\theta_p})$. Also, $\theta_e = k_B T_e/m_e$, $\theta_p = k_B T_p/m_p$, with $\omega_j = \sqrt{4\pi e^2 n_j/m_j}$ as the plasma frequency and $\Omega_j = eB/m_j$ as the cyclotron frequency of species j. Here also k_B is Boltzmann's constant and $c = 1$.

This form for the response function assumes that the waves propagate almost along the magnetic field lines: if $\arccos\eta$ is the angle of propagation of the wave with respect to the magnetic field, then Eq. (1) is valid for an isothermal ($T_e = T_p$) electron-proton plasma at whistler frequencies only when $1 - \eta^2 \ll \Omega_e/(\omega_e\sqrt{\theta_e})$ (for a discussion of this constraint and deviations of the response function from eq. (1), see Baring 1992, in preparation). In this paper, $\eta = 1$ will be assumed. Analytic forms for eq. (1) can be obtained when the arguments z_e and z_p of the plasma dispersion functions (which essentially are ratios of phase velocities to thermal velocities) are large and therefore the damping is small. The criterion for wave damping to be small is that the imaginary part of ω be negligible, which is satisfied if $\mathrm{Im}\,\epsilon_r(k,\omega) \ll \mathrm{Re}\,\epsilon_r(k,\omega)$.

When wave damping is small, the dispersion relation (1) simplifies because of the asymptotic form $\mathcal{Z}(\zeta) \sim -1/(\zeta\sqrt{\pi})$. It is then easily demonstrated that there are two distinct wave modes. When $k \ll 2\omega_p$, the propagation modes are Alfvén waves with $\omega = (\Omega_p/\omega_p)k$. For larger values of k, the modes are whistlers with the dispersion relation (for $\eta \neq 1$ see e.g. Melrose, 1980)

$$\frac{\omega}{\Omega_e} = \frac{k^2}{\omega_e^2} \quad , \qquad \frac{\Phi}{2\sqrt{\theta_e}}\sqrt{\frac{m_e}{m_p}} \lesssim \frac{k}{\omega_e} \lesssim \left(\frac{\Phi^2}{4\theta_e}\right)^{1/4} \quad . \tag{2}$$

Here $\Phi = \Omega_e/\omega_e$, and is of the order of 10^{-3} for microgauss magnetic fields and particle densities of 1 cm^{-3}, typical conditions for supernova remnants and the earth's bow shock. In the cold plasma limit, the whistlers satisfy $2\omega_p \ll k \ll \omega_e$, however large temperatures $\theta_e \gtrsim \Phi^{-2}$ alter this range of k. This effect is displayed in Fig. 1, where the solution of Eq. (1) is given for temperatures that are typical of shocked plasmas in supernova remnants, as well as for the case of a cold plasma. The Alfvén (non-dispersive and low k) and whistler modes (high k) are separated by a region of anomalous dispersion where wave damping is large. The breadth and position of this region is determined by the proton temperature largely independently of T_e; a more accurate determination of the branch of anomalous dispersion requires solution in complex frequency space.

It is easily shown that maximizing T_p and minimizing T_e will lead to a large range of k for whistlers. This range is an important ingredient for the calculation of the diffusion coefficients for electron-whistler scattering below. Further, a central feature is that not only does a significant range of whistlers exist at temperatures $\theta_e \sim 10^{-3}$, but that the phase velocities drop below the Alfvén speed. This startling behaviour, not normally associated with whistlers, may break down when the waves propagate obliquely to the field lines. From the point of view of shock acceleration models, this feature can dramatically affect the way whistlers can turn electrons around in shocked flows. High frequency whistlers rapidly propagate through the fluid flow, colliding with electrons in a stochastic fashion, leading to relatively inefficient excitation of electrons out the the thermal gas and to long scattering mean-free-paths. On the other hand, for high temperature conditions, the whistlers are slow, and therefore are effectively anchored in the fluid, acting much like Alfvén waves (i.e. non-stochastically), with much shorter mean-free-paths. The detailed determination of the whistler dispersive properties clearly may be crucial to any shock acceleration model that uses electron-whistler scattering to effect injection.

The physically important and informative quantity in a model of diffusive shock acceleration is the mean free path λ of the particle scattering. To calculate this for electron-whistler resonant interactions requires knowledge of the pitch angle and spatial diffusion coefficients, which can be derived from standard quasi-linear wave-particle scattering theory (e.g. see Melrose, 1980). The resonance condition for an electron of momentum $p_e = \gamma_e \beta_e m_e$ colliding with a wave of freqency ω is $s\Omega_e = \omega - kp_e\eta\mu/m_e$, where s is an integer and μ is the pitch angle of the electron. An estimate for the spatial diffusion coefficient using the dispersive properties calculated here was obtained by Baring (1991), assuming an isotropic energy distribution of whistlers of the form $W_w(k,\eta) = \{\mathcal{E}_w k_l^2/(4\pi^2)\}(k/k_l)^{-\alpha}$, with $k_l < k < k_u$ and $\alpha < 0$. The assumption of isotropy is crude, since efficient generation may usually require anisotropic conditions (Melrose, 1974). Further, the variation of the whistler frequency range with wave propagation angle η, may be much more complicated than is assumed in Baring (1991). However, the estimate obtained by Baring using the formalism of Hasselmann and Wibberenz (1970) has order of magnitude accuracy, and if σ_T is the Thomson cross-section, it is given by

$$\lambda = \frac{16\theta_e^2}{\sigma_T n_e} \frac{\alpha - 1}{\alpha} \frac{m_e}{\mathcal{E}_w} \left[(2m_e/m_p)\Phi\sqrt{\theta_e}\right]^{-1-\alpha/2}. \quad (3)$$

The mean free path λ can be calculated for $B = 3\mu$G and $n_e = 1$ cm^{-3}, conditions typical of shocked plasmas in supernova remnants. When $\theta \sim 10^{-4}$

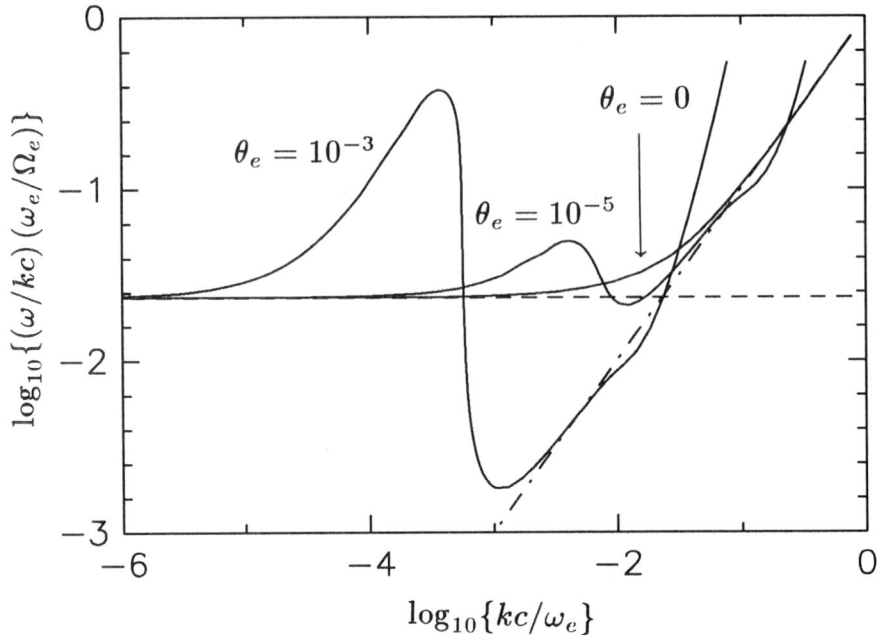

Fig. 1: The Alfvén-whistler dispersion relation (solid lines) in Eq. (1) for a magnetized plasma with isothermal ($T_p = T_e$), isotropic electron and proton distributions, shown for three different temperatures $\theta_e = k_B T_e/m_e c^2$. The dashed line represents the Alfvén dispersion relation and the dash-dot line corresponds to whistlers. The vertical axis is just the (scaled) phase velocity of the waves. Here $\Phi = \Omega_e/\omega_e = 10^{-3}$ and the wave damping is assumed to be small.

and $\alpha = -2$, $\lambda \sim 10^{21}(m_e \theta_e)/\mathcal{E}_W$ cm, and rapidly becomes larger when $\alpha \to 0$. In supernova remnant shocks, electrons can achieve relativistic energies on distance scales of a tenth of a parsec or so, indicating that $\mathcal{E}_W/(m_e \theta_e) \gg 10^4$: the whistler population is then much greater than can be supplied at equipartition with the ambient thermal plasma. It can therefore be deduced that a very efficient generation of whistlers would be needed to effect electron injection in shock acceleration via electron-whistler resonant scattering.

REFERENCES

Baring, M. G.: 1991 in *Proc. 22nd ICRC*, Vol. II, p. 348
Bell, A. R.: 1978 *Mon. Not. R. astr. Soc.* **182**, 147
Fried, B. D., and Conte, S. D.: 1961 *The Plasma Dispersion Function* (Academic Press, New York)
Hasselmann, K., and Wibberenz, G.: 1970 *Astrophys. J.* **162**, 1049
Ichimaru, S.: 1973 *Basic Principles of Plasma Physics*, (W. A. Benjamin, Reading)
Melrose, D. B.: 1974 *Solar Phys.* **37**, 353
Melrose, D. B.: 1980 *Plasma Astrophysics, I and II*, (Gordon and Breach, New York)
Miller, J. A.: 1990 *Ph.D. Thesis*, University of Maryland

DIFFUSIVE TRANSPORT IN CLUSTER-CENTER RADIO SOURCES

Jean A. Eilek

New Mexico Tech, Socorro NM 87801

ABSTRACT

Unlike most jet-fed radio sources, those attached to dominant central galaxies in clusters are thought to be maintained by diffusion of relativistic electrons through the turbulent, magnetized cluster gas. However, severe synchrotron losses do not allow the electrons to diffuse far enough to account for these sources, unless some *in situ* reacceleration process is taking place. In this paper I consider the possibility that second-order turbulent acceleration can account for these sources.

CLUSTER-CENTER RADIO SOURCES ARE DIFFERENT

Most extragalactic radio sources have been fairly successfully described by models in which the extended radio lobes are supplied with mass and energy by jets arising from the galactic nucleus. However, recent work on radio sources associated with dominant cluster-center galaxies shows that these sources are different. Their morphology is diffuse rather than linear, and their radio spectra are steeper than is typical of jet-fed sources. Small jets ($\lesssim \sim$ kpc in length) can be found in the cores of these sources, but cannot be followed through the extent of the source. (Examples are the halo source in M87, Owen 1992; Abell 2052, Ge 1991; or sources in Burns 1990). These sources are only found in cluster centers where radiative cooling of the gas is important. In addition to the gas being at a higher pressure in these cores, we know the gas is also turbulent and magnetized (*e.g.*, Owen, Eilek & Keel 1990, 1992; Ge 1992; Baum 1992).

It is tempting to speculate, then, that the jets in these sources are disrupted close to their origins, by the unusual ambient conditions; and that the relativistic electrons then propagate by diffusion into the cluster gas. This simple model has a severe problem, however: strong synchrotron losses due to the local magnetic field limit the distance the electrons can travel by diffusion. "Reasonable" estimates of diffusion rates (I give some numbers later in this paper) suggest the electrons cannot get farther than ~ 1 kpc before suffering severe energy losses. The sources are, however, a factor of 10 - 30 times larger than this. Thus, some *in situ* reacceleration mechanism is needed if this picture is to work. The most likely candidate seems to be second-order, turbulent acceleration. Unlike the more typical, jet-fed radio sources, these sources show no evidence for shocks, or indeed for transonic flow. However, we know the medium into which they propagate is turbulent; turbulent velocities comparable to the local sound speed have been detected on $\sim$ kpc scales through emission-line cloud motions, and turbulent dynamos have been suggested to be the origin of the magnetic fields in the cluster. In this paper I present preliminary work on the combined effects of turbulent spatial and momentum diffusion on the electron spectrum in these objects.

COMBINED SPATIAL AND MOMENTUM DIFFUSION

The rate at which relativistic particles propagate through a turbulent medium is controlled by the level and spectrum of the turbulence. Diffusion along the magnetic field is controlled by resonant pitch-angle scattering, which depends on the Alfven wave turbulent spectrum. Writing the general diffusion coefficient as D_{zz}, parallel diffusion in the quasilinear limit gives (*e.g.*, Melrose 1980),

$$D_{zz}(p) \simeq \frac{p^2 c^3}{6\pi^2 e^2} \frac{1}{W_A(k_{res})} \qquad (1)$$

where the Alfven energy density per wavenumber at $k = k_{res}(p) \simeq eB/pc$ is $W_A(k_{res})$. (In this and the following expressions, I have approximated terms depending on pitch angle, assuming a non-pathological pitch angle distribution.) Diffusion across field lines, or in a highly tangled magnetic field, is controlled by the turbulent power at low-wavenumbers;

$$D_{zz} \simeq \frac{c}{4B^2} W_A(k_o) \qquad (2)$$

(Jokipii 1966) which is independent of the particle energy, and depends on the power at the longest wavelengths; $k_o \to 0$ formally. Stochastic particle acceleration can be treated as a diffusion process in momentum space; this diffusion coefficient, $D_{pp}(p)$, also depends on the turbulence. This acceleration can come from Alfven resonant acceleration, giving (*e.g.* Eilek & Hughes 1991)

$$D_{pp}(p) \simeq \frac{2\pi e^2 v_A^2}{c^3} W_A(k_{res}) \qquad (3)$$

where v_A is the Alfven wave speed. Alternatively, the acceleration can be due to Fermi-type acceleration (such as magnetic pumping or magnetosonic wave acceleration). One form of the diffusion coefficient, which describes acceleration due to Landau-resonant interactions with magnetosonic waves, is

$$D_{pp}(p) \simeq \frac{p^2 v_{MS}^2}{cB^2} k_o^2 W_{MS}(k_o) \qquad (4)$$

where $W_{MS}(k)$ is the magnetosonic wave spectrum. Other models for Fermi acceleration give other functional forms for $D_{pp}(p)$ (*e.g.*, Eilek & Hughes 1991; Borovsky 1986); all share the momentum dependence $D_{pp}(p) \propto p^2$.

Thus, the same turbulence which allows the particles to diffuse can energize them while they are diffusing. One might expect that turbulent energization could be strong enough to maintain the electron spectrum against synchrotron losses, so that energetic electrons could propogate to distances at least $\sim$ tens of kpc, the characteristic size of the cluster-center sources. In addition, if the spatial diffusion coefficient depends on the electron energy – as it does in the parallel-diffusion case – the electron spectrum produced by the acceleration process will be modified, as a function of position and time, by the diffusion itself.

ANALYTIC METHODS

I have explored this problem by looking at analytic solutions for an electron spectrum undergoing both spatial and momentum diffusion, while undergoing synchrotron losses. I have simplified the problem by restricting the diffusion to one spatial dimension, and by assuming D_{zz} and D_{pp} are independent of position (this assumes the turbulent spectra and magnetic field are constant). Details of the calculation will be presented elsewhere (Eilek 1992); here I summarize the approach and general conclusions which can be drawn. (Schlickheiser, Seivers & Thiemann, 1987, also considered turbulent transport in clusters, but compared their results to data only after integrating over the cluster volume. Here I explicitly retain the spatial information.

Let the electron transport be described by a spatial diffusion coefficient, $D_{zz}(p)$, and by a momentum diffusion coefficient, $D_{pp}(p)$. The specific forms of $D_{zz}(p)$ and $D_{pp}(p)$ depend on the specific transport mechanism. Synchrotron losses at momentum p are $dp/dt = Sp^2$, where S contains fundamental constants. The time evolution of the electron distribution function, $F(z,p,t)$, is given by

$$\frac{\partial F}{\partial t} - \frac{\partial}{\partial z}\left[D_{zz}(p)\frac{\partial F}{\partial z}\right] = \frac{1}{p^2}\frac{\partial}{\partial p}\left[D_{pp}(p)p^2\frac{\partial F}{\partial p} + Sp^4 F\right] \quad (5)$$

One can immediately see important scaling parameters from this equation. Momentum can be scaled by p_c, the momentum where acceleration balances synchrotron losses. Picking a power law, $D_{pp}(p) = D_o p^r$ gives $p_c = (D_o/S)^{1/(3-r)}$. Thus, in the absence of diffusion, one expects the spectrum to have a peak or a turnover at $p \sim p_c$ (e.g. Borovsky & Eilek 1986). Using this, time is usefully scaled by the synchrotron lifetime at p_c, namely, $t_{sy} = 1/Sp_c$. In addition, a useful variable transform scales z to the length scale "stretched" by diffusion: $y^2 = z^2/D_{zz}(p)$.

I have investigated solutions to (5) by separating the combined distribution function as $F_\omega(z,p,t) = g_\omega(z,t)f_\omega(p)$, with a separation eigenvalue ω:

$$\frac{\partial g_\omega}{\partial t} - \frac{\partial^2 g_\omega}{\partial y^2} = -\omega g_\omega \quad (6)$$

and

$$\frac{1}{p^2}\frac{\partial}{\partial p}\left[D_{pp}(p)p^2\frac{\partial f_\omega}{\partial p} + Sp^4 f_\omega\right] = -\omega f_\omega \quad (7)$$

This approach has the advantage that solutions to (6) and (7) are similar to known solutions. Solutions to (6) resemble the usual diffusion solutions, (which contain the term $e^{-y^2/4t}$), modified by a term depending on ω. Well-behaved series solutions have been found to (7) for values $0 \leq \omega t_{sy} \leq 9/4$. These solutions have high-p and low-p asymptotes very similar to standard diffusive-acceleration solutions; they turn out not to be strongly sensitive to ω.

These eigensolutions (with a continuous "power spectrum" $a(\omega)$ of eigenvalues) must then be combined as

$$F(z,p,t) = \int a(\omega) g_\omega(z,t) f_\omega(p) d\omega \qquad (8)$$

The spectrum $a(\omega)$ is properly specified by considering initial or boundary conditions; however one can gain insight by considering "interesting" $a(\omega)$ forms, without solving a particular boundary value problem. In particular, since the $f_\omega(p)$ solutions turn out not to be very sensitive to ω, it makes sense to try simple forms for $a(\omega)$, and to work with a "mean" $\tilde{f}(p)$ function, weighted by the $a(\omega)$ spectrum. The restrictions on allowed frequencies lead one to work with spectra which have power at low ω. Picking such spectra, it turns out that the integral in (8) can be approximated as

$$F(z,p,t) \simeq \frac{F_o}{t^{3/2}} \exp\left[-\frac{z^2}{4D_{zz}(p)t}\right] \tilde{f}(p) \qquad (9)$$

where F_o is a normalizing constant. This expression provides a good approximation to the high-p and low-p behavior of the solution; it probably does not do justice to the details around $p \sim p_c$.

RESULTS: CONSEQUENCES FOR CLUSTER-CENTER SOURCES

The simple expression for $F(z,p,t)$, (9), is extremely useful: it can be interpreted as the "intrinsic" diffusive acceleration spectrum, $\tilde{f}(p)$, modified by a propagator, $P(z,p,t) = t^{-3/2} e^{-K(z,p,t)}$, with $K(z,p,t) = z^2/[4D_{zz}(p)t]$. Thus, the $\tilde{f}(p)$ spectrum which turbulent acceleration supports can be thought of as propagating out, away from the central source, at a speed $\sim (D_{zz}/t)^{1/2}$. The nature of the particle distribution at point (z,t), then, depends on the nature of the diffusion coefficient $D_{zz}(p)$.

If the magnetic field is tangled on scales small compared to the length scale of the system being studied, the cross-field diffusion coefficient (equation 2) will be relevant. In this case, the intrinsic $\tilde{f}(p)$ distribution will be maintained throughout the source, with an amplitude which decays with time. This solution is probably not the one to consider for cluster-center sources, however; the evidence from Faraday rotation (*e.g.*, Ge 1991) is that the magnetic field is *ordered* on scales comparable to the size of the radio sources themselves ($\sim$ tens of kpc). The diffusion when the field has large-scale order is more likely to be described by the parallel-diffusion case. In this case, the dependence of D_{zz} on p means that the particle spectrum found at some (z,t) will be quite sensitive to the effects of the propagator.

The effect of the propagator depends on the Alfven wave spectrum. To be specific, assume the wave spectrum can be described by a power law: $W_A(k) \propto$

k^{-m}. This gives $D_{zz}(p) \propto p^{(2-m)}$, and a propagator exponent $K(z,p,t) \propto z^2 p^{(m-2)}/t$. Thus, steep wave spectra (with $m > 2$) will deplete high electron energies, leading to a steeper particle spectrum; flat wave spectra (with $m < 2$) will deplete low electron energies, leading to a flatter wave spectrum. To illustrate this effect, I have chosen a "typical" $\tilde{f}(p)$ solution, $\tilde{f}(p) \propto p^{-0.8}$ for $p < p_c$, and $\tilde{f}(p) \propto p^{-2}$ for $p < p_c$. I have written the propagator factor $K(z,p,t) = K_o(p/p_c)^{(m-2)}$ and picked $m = 3/2$ ("flat") and $m = 5/2$ ("steep"). The momentum distribution functions are shown in the figures, for several K_o values, for the flat and steep wave spectrum cases.

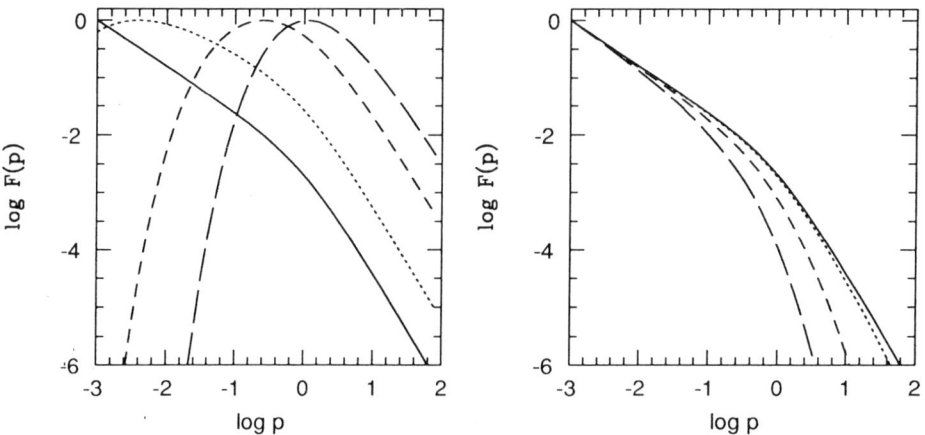

Figure 1. Momentum distribution functions resulting from an intrinsic spectrum, $\tilde{f}(p)$, modified by propagator, $P(z,p,t)$, as described in the text above. Left, for a "flat" wave spectrum; right, for a "steep" wave spectrum. The solid lines show the intrinsic $\tilde{f}(p)$; the dotted line is for $K_o = .1$, the short-dashed line for $K_o = 1$ and the long-dashed line for $K_o = 3$.

This illustration points out the dramatic effect of the propagator, especially in the case of flat wave spectra. At a given point in space, the $F(p)$ solutions evolve towards smaller K_o values, approaching the intrinsic $\tilde{f}(p)$ as $t \to \infty$. Alternatively, at a given time, the solutions move to higher K_o going away from the central source; thus, for a steep wave spectrum, the $F(p)$ solutions are steeper, further from the source; but for a flat wave spectrum, the solutions are flatter, due to the faster propagation of high-p electrons.

DO THE NUMBERS WORK OUT?

Finally, one must ask whether these solutions can be applied quantitatively to cluster-center sources. That is, do the numbers work? This question can be addressed by estimating physical parameters in cluster cores. Spatial diffusion must be able to maintain sources at a size $\sim$ tens of kpc, in a magnetic field $\sim 10\mu$G, over the lifetime of the source. Current "conventional wisdom" puts the lifetime of these sources to be at least $\sim 10^8$ years, and probably larger, while the synchrotron lifetime of electrons radiating in the radio band is $t_{sy} \sim 5\times10^7 B_{10\mu G}^{-3/2}$ years. Thus, the sources are very likely older than the radiative lifetime of single electrons. In addition, momentum diffusion must maintain electrons at energies observable in the radio band: $p_c \gtrsim 10^3 m_e c$ is needed to radiate at ~ 1 GHz in the local magnetic field. In terms of the solutions presented in this paper, these conditions translate to two constraints on the turbulence parameters (which determine the D_{zz} and D_{pp} coefficients):

(a) the momentum diffusion coefficient must satisfy $D_{pp}(p_c) \simeq S p_c^3$, for p_c given above;

(b) the spatial diffusion coefficient must satisfy $K(z,p_c,t) \simeq 1$ for the length scales and ages given above; this translates to $D_{zz}(p) \simeq z^2/4t$.

At this point, quantitative answers require specific assumptions about the mechanisms supporting D_{zz} and D_{pp}; a detailed discussion will be presented in Eilek (1992). In brief, one finds from (a) that momentum diffusion may – just – be able to maintain p_c at the needed values, if the turbulent wave energy is close to the background magnetic field energy. This is not a new result, but rather has been recognized as an important piece of the radio-galaxy "puzzle" for quite some time. On the other hand, one finds from (b) that spatial diffusion, *acting over the source lifetime rather than the shorter t_{sy}*, can transport electrons with $p \sim p_c$ out to distances $\sim$ tens of kpc. This suggests that turbulent transport may account for cluster-center radio sources. This is the new result of this calculation, and requires the combined actions of spatial and momentum diffusion. However, simple diffusion of electrons from a central source, even with turbulent acceleration, does not seem able to account for cluster-wide radio haloes, such as in Coma. Such large sources would seem to require extended sources of seed electrons, rather than simply one central source.

This work has been partially supported by NASA grant NAGW-1591.

Baum, S. A., 1992, to appear in A. Fabian and M. M. Colless, *Clusters and Superclusters of Galaxies* (Cambridge: NATO/ASI School).
Borovsky, J. E., 1986, *Phys. Fluids.*, 29, 3245.
Borovsky, J. E. and Eilek, J. A., 1986, *Ap. J.*, 308, 929.
Burns, J. O., 1991, *A.J.*, 99, 14.
Eilek, J. A., 1992, in preparation.
Eilek, J. A. and Hughes, P. E., 1991, in P. Hughes, ed., *Astrophysical Jets* (Cambridge: Cambridge University Press), 428.
Ge, J.-P., 1991, Ph.D. thesis, New Mexico Tech.
Jokipii, J. R., 1966, *Ap. J.*, 146, 480.
Melrose, D. B., 1980, *Plasma Astrophysics* (New York: Gordon & Breach).
Owen, F. N., 1992, in preparation.
Owen, F. N., Eilek, J. A. and Keel, W. C., 199, *Ap. J.*, 362, 449.
Owen, F. N., Eilek, J. A. and Keel, W. C., 1992, in preparation.
Schlickheiser, R., Seivers, A. and Thiemann, H., 1987, *Astr. Ap.*, 182, 21.

JOVIAN ELECTRON TRANSPORT TO THE POLAR HELIOSPHERE: AN ANALOGY TO MAGNETOSPHERIC RECIRCULATION

J. F. Cooper
National Space Science Data Center/Hughes STX Corporation,
NASA/Goddard Space Flight Center, Greenbelt, MD 20771

D. N. Baker
Laboratory for Extraterrestrial Physics,
NASA/Goddard Space Flight Center, Greenbelt, MD 20771

ABSTRACT

The theory of magnetospheric recirculation for cyclic electron energization and transport may apply in part to heliospheric transport of jovian electrons if enhanced cross-IMF propagation occurs at heliospheric altitudes near and below the solar wind transition region. Low altitude, ecliptic-to-polar transport would short-circuit conventional interplanetary diffusion, facilitate rapid access to the polar heliosphere with minimal adiabatic energy losses, and provide a seed population for acceleration to 10^{2-3} MeV energies at the solar wind termination shock and in the heliomagnetotail.

INTRODUCTION

In the earth's magnetosphere recent measurements of energetic electrons at geosynchronous orbit [1] have provided experimental confirmation for the magnetospheric recirculation theory first proposed for the Jovian magnetosphere [2,3,4] and later recast for the earth's magnetosphere [5]. The theory states that electrons may be accelerated to high energies during cyclic transport between the inner and outer magnetospheres. This would occur whenever there exist non-adiabatic processes of sufficient strength to scatter low altitude trapped particles in pitch angle so they cross field lines and thereby return a small fraction of an inwardly diffusing, accelerated particle population to the source particle reservoir in the outer magnetosphere. The recirculation model requires the substorm generation of a spectrally-soft electron component with subsequent inward radial diffusion in the conventional manner by violation of the third adiabatic invariant (i.e., the drift shell L value). This process increases the electron momentum transverse to the local magnetic field direction and converts an initially isotropic distribution into a more equatorially-trapped, "pancake"-shaped pitch angle distribution as the electrons diffuse towards the inner magnetosphere. In the deep magnetosphere strong pitch angle scattering moves the electron mirror points to low altitudes where cross-field diffusion transports the electrons without significant energy loss back onto field lines connecting to the outer magnetosphere. The expected low altitude diffusion is consistent with the known intensity of ULF turbulence in the high-latitude topside ionosphere. The lack of energy loss arises in part from the relatively low field gradients across field lines near low altitude ionospheric footpoints.

LOW-ALTITUDE HELIOSPHERIC TRANSPORT

A heliospheric analog for magnetospheric recirculation would resemble the magnetospheric equivalent in the sense that latitudinal transport between the ecliptic and the polar heliosphere may be enhanced by the non-dipolar magnetic fields and increasing Alfvenic turbulence to be expected below heliospheric radii of 30 solar radii (0.14 AU). Interplanetary radio scintillation (IPS) observations [6] indicate that the solar wind bulk plasma velocity is accelerating from fifty kilometers per second to average terminal values of 300-400 kilometers per second in the transition region at 10 to 30 solar radii. The same observations indicate that enhanced plasma turbulence, perhaps due to dissipation of non-linear Alfven waves, occurs in the same region as the bulk plasma acceleration and may also enhance cross-field propagation in this region in a manner analogous to ULF waves in the low altitude magnetosphere of Earth.

Solar energetic particle studies provide information on particle propagation in the solar corona out to several solar radii. This provides lower limits on ecliptic-to-polar particle transport times in the low altitude heliosphere. Observations indicate that solar energetic particles can often traverse large coronal distances without significant losses in intensity or energy. A recent study of flare particle time profiles at three different spacecraft locations and heliographic longitudes [7] shows that 0.5 MeV electrons in some flares are found at the same intensities up to about sixty degrees in heliocentric longitude separation from the flare site. The coronal propagation times for electrons are only a few minutes in these cases as compared to hours for protons from the same flares or for electrons and protons from more poorly connected flares. In the latter case the electron propagation time ranges up to 16 hours. Time for propagation from the corona to the Earth along the best connected IMF spiral line is of order 0.2 to 4 hours, so the total time for electron transport to Earth from a flare site is no more than one day.

At heliospheric altitudes of 0.5 to 1 AU [8] and 1 to 5 AU [9] the measurements of 10-MeV Jovian electrons by the Mariner-10 and Pioneer 10/11 spacecraft, respectively, provide information on longitudinal and latitudinal transport parameters which set upper limits on the ecliptic-to-polar transport times. At 1-5 AU the electron diffusion coefficients D_n for diffusion parallel to the IMF, cross-field in heliolongitude, and cross-field in heliolatitude were found to be 5×10^{22}, 7×10^{20}, and 5×10^{19} cm^2/sec, respectively. If the propagation time T is scaled with respect to a propagation distance scale L by the relation $T = L^2/4D$, the value of T is about 2.5 days for electrons to propagate from Jupiter over a distance of 14 AU along the archimedean spiral IMF to 1 AU. In comparison the cross-field values of T for diffusion over sixty degrees in heliolongitude separation are about one day to three weeks at heliocentric radii of 1 to 5 AU, respectively.

Vertical diffusion from the ecliptic plane to 60 degrees heliolatitude in the same radial range requires propagation times from two weeks at 1 AU to almost one year at 5 AU.

It may often be the case, therefore, that Jovian electrons preferentially will propagate faster in longitude and latitude via entry into the low-altitude heliosphere, where coronal transport times may apply, than by diffusing across the IMF at 0.5 to 5 AU. In fact, the comparison [8] of jovian electron increases (at times of best IMF connection to Jupiter) at 0.5 and 1 AU showed occasional evidence for increasing intensities at lower heliospheric altitudes. Furthermore, jovian electron studies with ISEE-3 measurements [10] showed spectral shifts attributed to adiabatic cooling effects. These shifts were interpreted in terms of shorter times for propagation from Jupiter to near Earth (ISEE-3 orbited the L1 Sun-Earth lagrangian point) after 1981 than before. Such observations might be manifestations of low altitude transport which enhances longitudinal and latitudinal propagation in the inner heliosphere, particularly during high and variable solar activity in the 1980-1984 period.

Future Ulysses observations after the February 1992 Jupiter flyby will study the latitudinal profiles of jovian electrons in the inner heliosphere within radial distances of 2 to 5 AU from the Sun. If the jovian electron increases above 40-60 degrees heliolatitude are significant and in excess of predictions from interplanetary diffusion models, this may be sign of enhanced poleward transport at low heliospheric altitudes. A future solar probe mission might provide direct measurements of jovian electrons and transport parameters down into the solar wind transition region. In the case of earth's magnetosphere the analogous low-altitude transport in magnetic recirculation will be tested with electron measurements by the polar-orbiting SAMPEX satellite, planned for launch in June of 1992.

ACCESS TO THE OUTER HELIOSPHERE AND HELIOMAGNETOTAIL

At some yet unknown distance beyond the 50 AU heliocentric distance reached by Pioneer 10 the solar wind reaches a termination shock (i.e, transition to subsonic flow) which bounds the "plasmaspheric" region of the heliosphere, in analogy to the plasmapause in earth's magnetosphere. It is expected that the solar wind transitions through the "heliosheath" region into the interstellar plasma and fields outside the heliopause [11]. Since the average IMF direction out to the termination shock has the geometry of a tightly wound archimedean spiral (a 400 km/sec solar wind flow gives about seventeen windings out to 100 AU) the heliolatitude has a potentially large effect on transport of charged particles along the IMF field lines between the inner and outer heliospheres. Whereas in the heliographic equatorial plane the path length parallel to the average IMF is thousands of AU to reach the termination shock, the pathlength to this shock through the polar IMF fields would be of the order of a few hundred AU or less, approaching the radial distance of order 100 AU at the solar rotational pole.

The heliosheath region between the termination shock and the heliopause transitions the radial solar wind flow to the tailward flow direction defined to be that of the inflowing interstellar plasma. If at least some significant fraction of IMF field lines are continuous across the termination shock, then these field lines will be carried with the plasma back into the heliomagnetotail in the same manner as the polar field lines of the earth's polar magnetosphere are swept back into the terrestrial magnetotail region. If north-south polarity is maintained all along the IMF field lines and back into the heliomagnetotail, the heliospheric neutral sheet must extend into the heliomagnetotail to separate north and south lobes of opposite polarity in analogy to the neutral sheet in the Earth's magnetotail.

ELECTRON ACCELERATION IN THE HELIOMAGNETOTAIL

The heliomagnetotail region could provide high energy electron acceleration via the magnetic reconnection process which is known to accelerate electrons up to about one MeV in the earth's magnetotail at 10-20 earth radii [12]. Maximum energies have been calculated from theoretical simulations [13] for particle acceleration associated with non-steadystate reconnection. In the neutral sheets of the near-earth magnetotail and the inner heliosphere the maximum energies are 2 MeV and a few hundred keV, respectively. In the heliotail a magnetic field of one nanotelsa, density of 0.01/cc for ambient plasma, and cross-tail dimension of 200 AU would give a maximum energy of about 700 MeV. This value could reach the GeV range for higher magnetic fields or lower plasma densities in the heliomagnetotail region. In comparison electron acceleration at the solar wind termination shock would be limited to about 300 MeV for a source electron population below 30 MeV at Jupiter [14]. In either case the strong adiabatic losses associated with radial diffusion in the heliosphere (which slow MeV jovian electrons to keV energies) might be reduced by more rapid access to the polar heliosphere via low-altitude transport in the inner heliosphere.

REFERENCES

1. D. N. Baker, et al., Geophys. Res. Lett., **16**, 559 (1989).
2. A. Nishida, J. Geophys. Res., **81**, 1771 (1976).
3. D. D. Sentman, et al., Geophys. Res. Lett., **2**, 465 (1975).
4. M. Fujimoto and A. Nishida, J. Geophys. Res., **95**, 3841 (1990).
5. M. Fujimoto and A. Nishida, J. Geophys. Res., **95**, 4265 (1990).
6. M. Tokumaru, et al., J. Geomag. Geoelectr., **43**, 619 (1991).
7. G. Wibberenz, et al., Solar Phys., **124**, 353 (1989).
8. J. H. Eraker and J. A. Simpson, Ap. J., **232**, L131 (1979).
9. D. L. Chenette, J. Geophys. Res., **85**, 2243 (1980).
10. D. Moses, Ap. J., **313**, 471 (1987).
11. S. T. Suess, Rev. Geophys., **28**, 97 (1990).
12. D. N. Baker, in Magnetic Reconnection in Space and Laboratory Plasmas, Geophys. Monogr. **30**, ed. E. W. Hones, Jr., (American Geophysical Union, Washington, D.C., 1984), p. 193.
13. M. L. Goldstein, et al., Geophys. Res. Lett., **13**, 205 (1986).
14. H. Moraal, et al., Ap. J., **367**, 191 (1991).

DIRECT ELECTRIC FIELD HEATING AND ACCELERATION OF ELECTRONS IN SOLAR FLARES

Gordon D. Holman
NASA/Goddard Space Flight Center, Greenbelt, MD 20771

Stephen G. Benka[*]
Naval Research Laboratory, Washington, DC 20375-5000

ABSTRACT

We show that the observed properties of solar flare X-ray and microwave emission can be explained through the Joule heating and electric field acceleration of runaway electrons in current channels. The global properties of the flaring region required for this are presented. We have fit a hybrid thermal/nonthermal electron distribution, consisting of hot, isothermal electrons with a nonthermal tail of runaway electrons, to high-resolution hard X-ray and microwave spectra and have obtained excellent fits to both. The hybrid model relaxes the electron number and energy flux requirements for the hard X-ray emission over those of a purely nonthermal model. The model also provides explanations for several previously unexplained aspects of the high-resolution microwave spectra. The fit parameters can be related to physical properties (such as the electric field strength in the current channels) of the acceleration region.

The rapid rise and fall of microwave, hard X-ray and γ-ray emissions on a time scale of seconds to minutes is characteristic of impulsive solar flares. When all three emissions are observed, their time profiles are often remarkably similar. The microwave emission is understood to be gyrosynchrotron radiation from mildly relativistic electrons spiraling around ambient magnetic field lines. The hard X-ray and part of the γ-ray emission are understood to be bremsstrahlung radiation from energetic electrons interacting with ambient ions, especially in the lower, denser parts of the solar atmosphere. These emissions provide our best diagnostics of impulsive electron acceleration in flares.

Most flare models involve electric currents. These currents and their associated electric fields are either aligned with the magnetic field or are in reconnection regions where a component of the magnetic field reverses direction. In the presence of these electric fields, part of the thermal electron distribution will not be collisionally confined to the bulk current, but will run away and be freely accelerated by the electric field. This electric field acceleration of runaway electrons provides the most direct means of producing and accelerating energetic electrons in flares. Our recent flare research has been concerned with applying this acceleration process to the interpretation of flare X-ray and microwave data to determine the physical conditions required for this mechanism to be important. Our results are summarized here.

Electron runaway occurs in a current-carrying plasma because the collisional drag on a particle decreases with increasing velocity [1,2]. For electrons with velocities above a critical velocity v_c, the force exerted by the electric field exceeds the collisional drag and the electrons are accelerated out of the thermal bulk. Electrons with velocities below v_c remain thermalized and the electric field drives Joule heating of the plasma. Hence electron heating and acceleration naturally occur together in the acceleration region. Comparing heating and acceleration in flares can provide information about the physical conditions in the acceleration region [3]. The properties and evolution of hot flare plasma can be determined from observations of its thermal bremsstrahlung emission at soft X-ray wavelengths. The hottest plasma may also be observed at the lowest hard X-ray energies, or at microwave frequencies through its thermal gyrosynchrotron radiation.

---

[*] NAS/NRC Research Associate

One of the most important constraints on the direct electric field acceleration of electrons in flares is the flux (and, therefore, current) of electrons required for a nonthermal hard X-ray burst [4,5,3]. Taking the hard X-ray emission observed above photon energies of 25 keV to arise entirely from thick-target nonthermal bremsstrahlung, the electron flux required to generate this emission is typically 10^{35} electrons/sec or higher (a current of 10^{16} amps). Taking a maximum possible current channel width of 100,000 km (the radius of the sun is 700,000 km), the induction magnetic field associated with the runaway electrons alone is, from Ampere's law, one-million gauss. Since the maximum magnetic field strengths observed on the sun are on the order of 1,000 gauss, the *total* electron flux in a single current channel is limited to 10^{32} electrons/sec (10^{13} amps). Hence, if the accelerated electrons remain in the current sheets, ~10,000 or more oppositely directed current channels (i.e., current / return current pairs) are required. It is feasible to accelerate the electrons in a single current channel if they escape the channel on a short scale length so that the total current does not become too large. However, acceleration to the required energies requires a resistivity in the channel that is much greater than the classical Coulomb resistivity [6].

In addition to the high electron flux, the nonthermal hypothesis for the hard X-ray emission generally also implies that most of the energy released in the flare goes initially into accelerated electrons. These requirements are relaxed if part of the hard X-ray emission is from thermal electrons. Unfortunately, most X-ray observations have been unable to distinguish these possibilities. High-spectral-resolution observations of one flare have clearly demonstrated the presence of emission from hot, thermal plasma at hard X-ray energies below 40 keV [7]. This thermal component, which first appeared around the time of the peak hard X-ray emission, may have resulted from the heating of plasma by the nonthermal electrons. In this case the high electron flux and total energy are still required. Alternatively, however, the hot plasma may have resulted from the Joule heating of plasma in and around current channels in which runaway electrons are also accelerated. In this case the tail of nonthermal electrons only extends down to electron energies ~40 keV and most of the released energy goes into heating rather than particle acceleration.

Using an electron distribution consisting of an isothermal component and a nonthermal tail of runaways, we have fit representative spectra from the flare observed by Lin *et al.* [7]. The results are shown in Figure 1 (from Ref. 8). The top frame in the figure shows the fit to a spectrum from early in the flare, before the presence of a thermal component is apparent. The bottom frame shows the fit to a spectrum from later in the flare, when the thermal component is clearly present. The spectral fit is characterized by five free parameters: the temperature (T) and emission measure (E.M.) of the thermal plasma; the energy of an electron with critical velocity v_c ($\mathcal{E}_{critical}$); the maximum energy attained by a particle with initial energy $\mathcal{E}_{critical}$ ($\mathcal{E}_{cut-off}$); the area of the thick-target interaction region (A). Also shown in the figure are the ratio of the electric field strength to the Dreicer field strength (ε), where the Dreicer field is the electric field strength for which all of the thermal electrons are in the runaway regime, and the electric field strength (E) and density (n or N) in the current channel when the length of the channel (L) is taken to be 30,000 km. The density is obtained on the assumption that the resistivity in the channels is classical. The distribution function for the runaway electrons was based upon the results of Fuchs *et al.* [9], with an exponential function included above $\mathcal{E}_{cut-off}$ [8,10].

The parameters derived from the spectral fits demonstrate that the X-ray emission can be produced with a sub-Dreicer electric field and classical resistivity with a reasonable density in the flaring region. This is generally true for typical flare parameters [3,6]. The small area of the interaction region is consistent with the thickness of the current channels which, from Ampere's law, is ~1—10 meters. The temperature and emission measure derived for the thermal component in the later spectrum are consistent with those obtained by Lin *et al.* [7]. Our best fit to the earlier spectrum, however, provides an interesting alternative to the nonthermal interpretation of Lin *et al.* and Lin and Schwartz [11]. We find that hot plasma is present here as well, but with a higher temperature and a lower emission measure. This would occur if early in the flare the heating is confined to a volume in the immediate vicinity of the current channels, while later in the flare the heat is distributed to a

Fig. 1. Fits to flare X-ray spectra

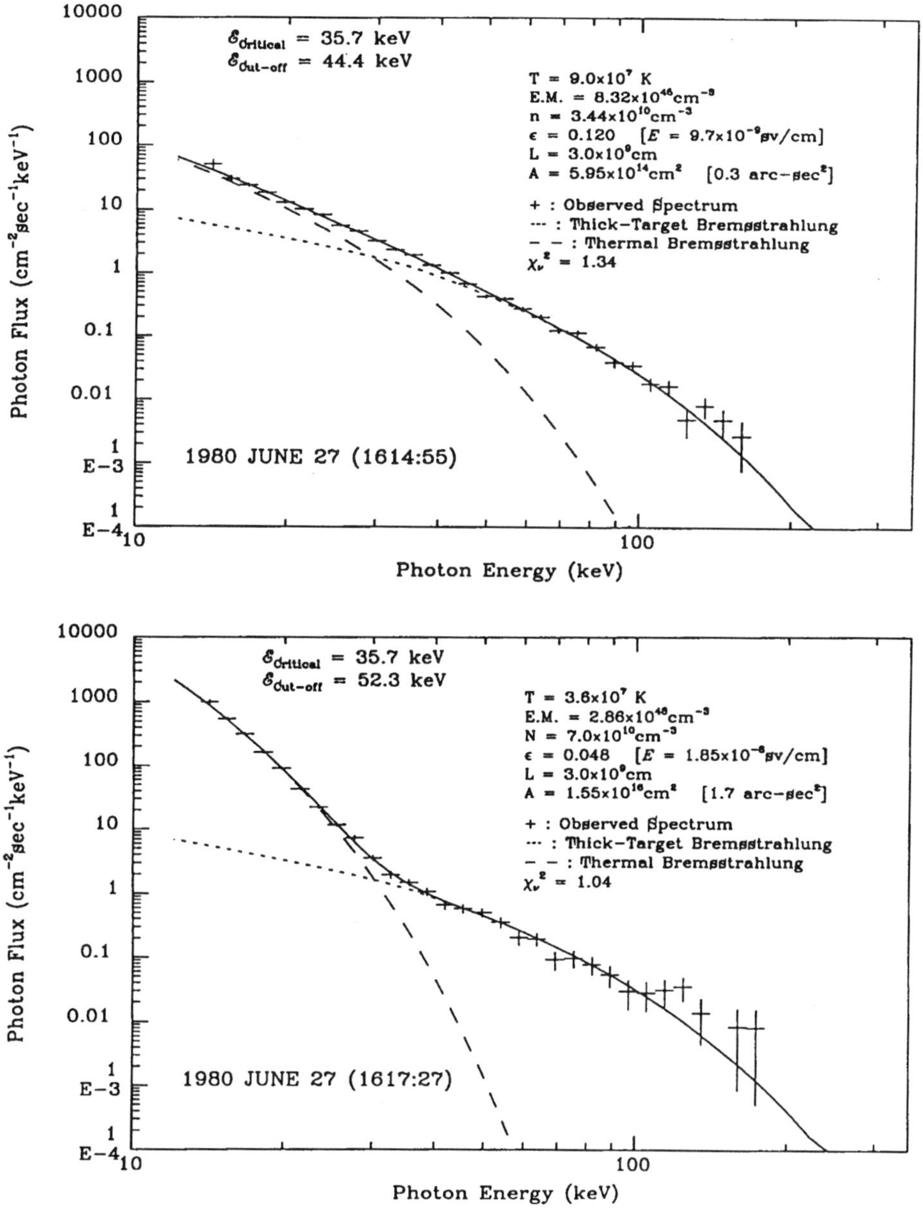

1980 JUNE 27 (1614:55)

$\mathcal{E}_{\text{critical}} = 35.7$ keV
$\mathcal{E}_{\text{cut-off}} = 44.4$ keV

$T = 9.0 \times 10^7$ K
E.M. $= 8.32 \times 10^{46}$ cm^{-3}
$n = 3.44 \times 10^{10}$ cm^{-3}
$\epsilon = 0.120$ [$E = 9.7 \times 10^{-9}$ sv/cm]
$L = 3.0 \times 10^9$ cm
$A = 5.95 \times 10^{14}$ cm^2 [0.3 arc-sec^2]

+ : Observed Spectrum
··· : Thick-Target Bremsstrahlung
− − : Thermal Bremsstrahlung
$\chi_\nu^2 = 1.34$

1980 JUNE 27 (1617:27)

$\mathcal{E}_{\text{critical}} = 35.7$ keV
$\mathcal{E}_{\text{cut-off}} = 52.3$ keV

$T = 3.6 \times 10^7$ K
E.M. $= 2.86 \times 10^{46}$ cm^{-3}
$N = 7.0 \times 10^{10}$ cm^{-3}
$\epsilon = 0.048$ [$E = 1.85 \times 10^{-8}$ sv/cm]
$L = 3.0 \times 10^9$ cm
$A = 1.55 \times 10^{16}$ cm^2 [1.7 arc-sec^2]

+ : Observed Spectrum
··· : Thick-Target Bremsstrahlung
− − : Thermal Bremsstrahlung
$\chi_\nu^2 = 1.04$

larger volume. (The plasma density may also be higher later in the flare, as indicated by our derived densities.)

The traditional nonthermal interpretation of the Lin et al. spectra, after removing the 35 million degree thermal component and assuming that the nonthermal component extends down to the lowest observed energy channel (13 keV), requires a flux in accelerated particles of 10^{35} - 10^{36}

electrons/sec. Our hybrid thermal/nonthermal interpretation, on the other hand, requires $10^{33} - 10^{34}$ electrons/sec, about two orders of magnitude smaller. Likewise, the energy flux in accelerated electrons is ~ 30 times smaller than that required by the nonthermal interpretation. In the hybrid thermal/nonthermal model, most of the released energy goes directly into heating the thermal plasma rather than accelerating particles. This will always be the case as long as the electric field strength is much less than the Dreicer field ($\varepsilon << 1$).

We have also applied the hybrid model to flare microwave emission [8,12]. High-resolution spectra from the Owens Valley Radio Observatory have typically shown flare microwave spectra to be somewhat complex [13,14]. A typical flare spectrum and our fit to that spectrum are shown in Figure 2. Since the nonthermal component of the microwave emission most likely arises from electrons that have diffused out of the current channels, we have represented the tail of accelerated particles as an isotropic power-law distribution with spectral index δ ($f(\mathcal{E}) \sim \mathcal{E}^{-\delta}$). The gyrosynchrotron radio emission depends upon the magnetic field properties and source geometry, as well as the electron distribution function. The spectral fit was done for a homogeneous source with uniform thickness along the line of sight. It is characterized by eight parameters: the temperature (T) and density (n_{th}) of the thermal plasma; the spectral index δ; ε; the magnetic field strength (B) and angle (θ) to the direction of propagation of the observed radiation; the source thickness (D) and area (A). The critical energy at which the transition from predominantly thermal to predominantly nonthermal electrons occurs (E_{cr}) is also given. "TNT" refers to the computed, hybrid thermal/nonthermal microwave spectrum (solid line). The dotted line shows the computed emission from nonthermal electrons alone, and the dashed line from thermal electrons alone.

Fig. 2. Fit to a flare microwave spectrum

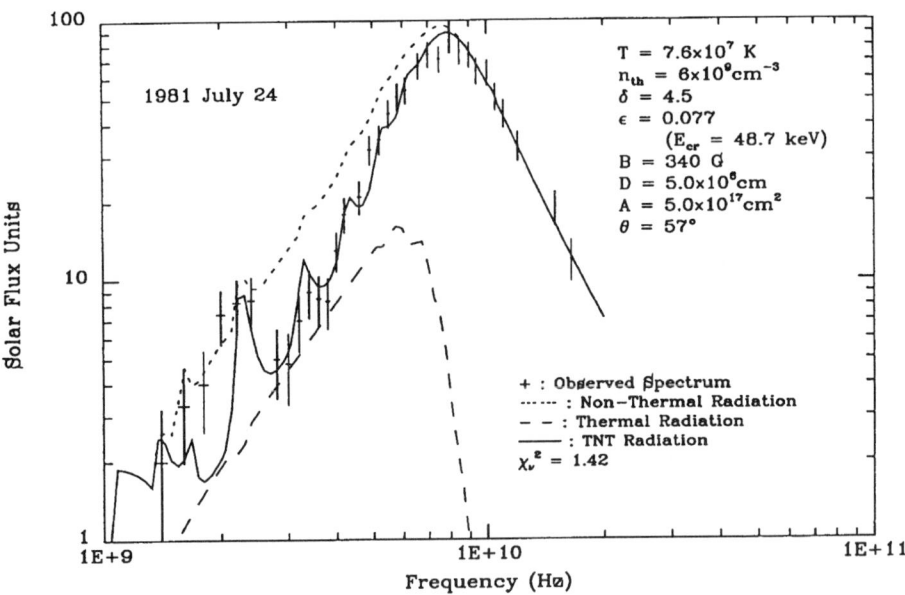

The most notable features of our fit to the microwave spectrum are the reproduction of the secondary components near 2 and 3 GHz, and the steep positive slope between 4 GHz and the flux peak at 8 GHz. Stäli, Gary, and Hurford [13] have observed one or more such secondary components in ~80% of their observed flares. These components arise naturally in the hybrid thermal/nonthermal model, occuring between the low harmonics of the electron gyrofrequency, where the opacity of the thermal component is too low to suppress the nonthermal radiation. The steep positive slope of the (optically thick) low-frequency spectrum is also explained with the hybrid thermal/nonthermal distribution. A thermal electron distribution alone produces a positive, optically thick (Rayleigh-Jeans) slope of 2, while the self-absorbed emission from nonthermal electrons alone has a slope of ~2.5—3. The data in Figure 2, however, have a slope of ~4 between 4 and 7 GHz. Positive slopes >3 are frequently observed [13,14]. These steep slopes result from the combination of the emissivity of the nonthermal particles and the opacity of the thermal plasma.

The fit to the microwave spectrum has yielded interesting results for the source dimensions: a small, 50 km thickness and a fairly small source area. This may indicate that the emission is predominantly from the leg of a coronal loop, where mirroring of the particles could enhance their density and their mean pitch angle to the magnetic field. This small effective source area may explain the success of the simple, homogeneous source model that we have used.

An additional facet of flare microwave spectra that needs to be explained is the observation that the frequency of the peak emission does not change significantly as the microwave flux rises and falls with time during the flare. This is contrary to what is expected from a simple, self-absorbed thermal or nonthermal source model. This behavior can be explained in the hybrid thermal/nonthermal model, however, if the evolution of the emission is primarily due to the evolution of ε, the ratio of the electric field strength to the Dreicer field. In this case the primary and secondary components evolve together, as observed, and there is little change in the peak frequencies.

We have found that a hybrid thermal/nonthermal model relaxes the particle flux and total energy requirements for accelerated electrons in a solar flare. It is consistent with theoretical expectations that most of the energy released into the flare plasma will go directly into heating rather than accelerated particles (cf. Ref. 15). The model also provides natural explanations for several previously unexplained features of flare microwave spectra. When the results are interpreted in terms of simultaneous Joule heating and direct electric field acceleration of runaway electrons, the coupling between the two components is apparent and the model provides physical information about the acceleration region. This interpretation is consistent with most global models for the energy release mechanism in flares. It requires either multiple current channels, consisting primarily of current/return current pairs so that most of the current is neutralized, or a single current channel with anomalous resistivity in the channel and rapid escape of the accelerated electrons out of the channel. We note the similarity of the need for multiple current channels to current ideas about coronal heating (e.g., Ref. 16) and to the results of *in situ* observations of the currents associated with the earth's aurora [17].

Future theoretical work of particular interest to this research would be a better determination of the runaway electron distribution function and runaway rate. Existing calculations and numerical simulations have generally been done for a spatially infinite, homogeneous current. Recent numerical computations have shown that the distribution function and runaway rate can be significantly different from these results in a finite-length current channel [18]. This is particularly important for determining the shape of the distribution function above $\varepsilon_{cut-off}$. A comparison of the computed emission from the runaway tail with high-resolution hard X-ray and γ-ray spectra can be used to determine if another acceleration mechanism (a stochastic mechanism, for example) is required at the higher electron energies. More studies of the distribution function in a finite-length current channel are needed for this to be done with confidence. Computations that account for the finite width of the current channel and include the magnetic field are also needed (particularly in a sheet geometry). These calculations would also provide a better starting point for studying instabilities in the current channels. (Moghaddam-Taaheri and Goertz [19] have done a recent study of the stability of and microwave emission from runaway electrons confined within a spatially infinite current channel.)

Observationally, the most immediate need is for more high-resolution X-ray spectra. The Lin et al. [7] spectra, obtained for a single flare, are still the only available high-resolution flare spectra. Also desirable are simultaneous high-resolution microwave spectra, so that the properties of the plasma and accelerated electrons deduced for each can be compared for the same flare. Such observations are possible now, but depend upon the availability and success of balloon flights for the X-ray observations. The hard X-ray observations are limited to energies above ~ 20 keV, so that plasma with a temperature below ~ 20 million degrees is not detectable. Thanks to Yohkoh, the Japanese satellite dedicated to solar observations, and the Gamma Ray Observatory (GRO), lower resolution hard X-ray and γ-ray spectra are now being obtained. Yohkoh is also obtaining rudimentary imaging of hard X-ray bursts in four energy bands between 15 keV and 100 keV. (Microwave imaging is available with the VLA and the Owens Valley Radio Observatory.) These low-resolution spectra are valuable, but there is greater ambiguity in their interpretation.

A NASA program that is in the developmental stages and is well suited to this line of research is the High Energy Solar Physics Mission (HESP). The primary HESP instrument, to be flown on an earth-orbiting satellite at the time of the next solar maximum, is designed to obtain high-resolution X-ray spectra from 2 keV up to γ-ray energies of 20 MeV. Obtaining high-quality spectra from a single instrument over this wide range of photon energies will remove most of the uncertainty which has plagued previous attempts to determine the distribution of electrons that are impulsively heated and accelerated during a flare. The instrument is also designed to obtain, using pairs of grids and a Fourier transform technique, imaging of the emission as a function of photon energy. This high-resolution imaging spectrometer will provide valuable new information about the spatial relationship between the thermal and nonthermal hard X-ray and γ-ray emissions.

ACKNOWLEDGEMENTS

This work was supported in part by NASA RTOP 170-38-53-16. We thank Brian Dennis for his comments on the manuscript.

REFERENCES

1. H. Dreicer, Phys. Rev. 117, 329 (1960).
2. H. Knoepfel, and D. A. Spong, Nucl. Fusion 19, 785 (1979).
3. G. D. Holman, Ap. J. 293, 584 (1985).
4. P. Hoyng, Astr. Ap. 55, 23 (1977).
5. D. S. Spicer, Adv. Space Res. 2, No. 11, 135 (1983).
6. G. D. Holman, M. R. Kundu, and S. R. Kane, Ap. J. 345, 1050 (1989).
7. R. P. Lin, R. A. Schwartz, R. M. Pelling, and K. C. Hurley, Ap. J. Lett. 251, L109 (1981).
8. S. G. Benka, Ph. D. Dissertation, Univ. of North Carolina, Chapel Hill (1991).
9. V. Fuchs, M. M. Shoucri, J. Teichman, and A. Bers, Phys. Fluids 31, 2221 (1988).
10. S. G. Benka and G. D. Holman, in preparation (1992b).
11. R. P. Lin and R. A. Schwartz, Ap. J. 312, 462 (1987).
12. S. G. Benka and G. D. Holman, Ap. J. 390, in press (1992a).
13. M. Stäli, D. E. Gary, and G. J. Hurford, Solar Phys. 120, 351 (1989).
14. M. Stäli, D. E. Gary, and G. J. Hurford, Solar Phys. 125, 343 (1990).
15. D. F. Smith, Solar Phys. 66, 135 (1980).
16. E. N. Parker, Ap. J. 330, 474 (1988).
17. R. A. Hoffman, M. Sugiura, and N. C. Maynard, Adv. Space Res. 5, No. 4, 109 (1985).
18. P. MacNeice and N. N. Ljepojevic, in preparation (1992).
19. E. Moghaddam-Taaheri and C. K. Goertz, Ap. J. 352, 361 (1990).

CONSTRAINTS ON ELECTRON ACCELERATION IN THE CRAB NEBULA

A. K. Harding
Code 665, NASA/Goddard Space Flight Center, Greenbelt, MD 20771

O. C. DeJager
Dept. of Physics, Potchefstroom University, Potchefstroom 2520, South Africa

ABSTRACT

Using the radio through hard X-ray images of the Crab nebula to derive the spatial dependence of the electron spectrum and the magnetic field distribution from MHD flow models, we have rederived the high-energy spectrum of inverse Compton scattered gamma rays. We find agreement with the observed spectrum at TeV energies, but it is clear that the inverse Compton flux does not contribute significantly to the unpulsed nebular emission observed by COS-B from 50 to 500 MeV, which is consistent with a smooth continuation of the spectrum in hard X-rays. The emission at these energies must therefore be due to synchrotron radiation by electrons of at least PeV energies. It appears that the emission in the high-energy γ-ray range, sensitive to the highest energy electrons in the nebula, can put interesting constraints on the acceleration mechanism.

INTRODUCTION

The synchrotron nature of the Crab Nebula from radio through X-ray energies has been established through 1) the power law spectrum between 10^7 Hz and 10 MeV, 2) strong linear polarization, and 3) the observed decrease in scale size of the nebular images with increasing frequency. The scale size is roughly proportional to the lifetime of electrons radiating at a frequency ν times the radial velocity of the bulk motion[1]. This picture requires a continuous injection of relativistic particles near the center of the nebula, presumably from the pulsar, whose spin-down power is about five times the radiated power of the nebula. Although the overall energetics makes sense, the mechanism by which the pulsar transfers its spin-down energy to relativistic particles is still not understood.

Gould[2] predicted a flux of TeV γ-rays from inverse Compton (IC) scattering of the synchrotron radiation by the relativistic electrons. Although the observations to confirm this prediction have been inconclusive for several decades, the development by the Whipple group[3,4] of an imaging technique to eliminate background hardron-initiated showers has finally produced a solid ($\approx 20\sigma$) detection of unpulsed γ-rays from the Crab nebula with a spectral index of 2.4 ± 0.3. Since the Crab nebula is optically thin to absorption and scattering, the calculation of the IC spectrum is straightforward with the aid of the detailed maps and photon spectra, provided that the field distribution is known. Previous calculations of the IC spectrum of the Crab[2,5,6,7] have assumed constant or unphysical nebular field distributions. We have recently reexamined[8] the synchrotron self-Compton (SSC) model of the Crab nebula, incorporating more realistic magnetic field distributions derived from MHD flow models[9]. We show that the optimal field distribution for the Crab nebula as derived by Kennel and

Coroniti[9] provides a TeV flux and spectral index which is consistent with the observations and also gives the required $\bar{B} \approx 3 \times 10^{-4}$ G calculated from the spectral break between radio and optical[10]. However, the extension of this IC spectrum below GeV energies cannot account for the 50 - 500 MeV unpulsed flux detected by COS-B (ref. 11). We have fit this emission with an extension of the synchrotron spectrum, assuming the same nebular field distribution used to compute the IC spectrum, and find that an electron spectrum extending to $\sim 10^{16}$ eV is required to produce this radiation.

SYNCHROTRON SELF-COMPTON MODEL

Gould[2] has calculated the IC spectrum of the Crab nebula assuming a spherical model and has shown that it may be observable at VHE energies. Rieke and Weekes[5] have repeated these calculations for a prolate spheroid, and obtained an IC flux which is nearly ten times smaller than Gould's values, which is mainly due to a more accurate treatment of the Thomson limit. In both approaches a constant field strength as well as a delta function approximation for the cross section were used. Grindlay and Hoffman[6], using a more accurate estimate for the IC cross section,[12] showed that the delta function approximation overestimates the true flux by a factor of about 3. They used a $1/r$ dependence for the field distribution in the nebula which resulted in a differential spectral index of 4 at TeV energies, which is much steeper than is now observed.

The treatment of the field distribution is most crucial in SSC estimates, since the scattered flux depends on the field strength B as $B^{-(\alpha+1)/2}$ (given an electron spectral index of α). Since the earlier calculations of the IC flux of the Crab, more sophisticated models have been developed for the field distribution in the nebula, based on the pulsar wind model first suggested by Rees and Gunn[13]. They argued that the spindown power of the pulsar would be transported in the form of a relativistic MHD wind and showed that a shock would form in the pulsar wind due to the confinement by the nebula expanding at $v \ll c$. They estimated r_s as the distance from the pulsar where the ram pressure in the pulsar wind equals the total nebular pressure, giving $r_s \approx 10''$. In the MHD models investigated by Kennel and Coroniti[9] the spindown luminosity L of the pulsar upstream of the shock is divided between particle and electromagnetic energy as $L = 4\pi n_1 \gamma_1 u_1 r_s^2 mc^3 \times (1 + \sigma)$, where n_1 is the proper density and u_1 is the radial three speed (such that $\gamma_1^2 = 1 + u_1^2$). The only free parameter in their model is $\sigma \equiv B_1^2/4\pi n_1 u_1 \gamma_1 mc^2$, or the ratio of magnetic to particle energy density just ahead of the shock. The best estimate of σ, obtained by matching the the wind flow solutions to the observed velocity of ≈ 2000 km.s^{-1} at the outer boundary of the nebula, gives a value of $\sigma \approx 0.003$. For such a small σ the magnetic field strength at r_s would be a minimum, but would build up as $B \propto r$ for $r > r_s$ until equipartition is reached, thereafter decreasing until the nebular radius r_N is encountered.

In our recent calculation[8], we have used the field distribution derived by Kennel and Coroniti to obtain a more accurate estimate of the IC spectrum of the Crab Nebula. The synchrotron emissivity was assumed to be spherically symmetric, having a Gaussian shape with a standard deviation of λ_ν in the radial direction: $\epsilon_i(\nu, r) = \kappa_\nu \exp(-r^2/2\lambda_\nu^2)$. We obtained functional estimates for the variation of the scale length λ_ν between 10^7 Hz and 500 MeV by using maps at radio, optical and X-ray frequencies. The emissivity integrated over a 4π solid angle (i.e. $4\pi\epsilon$) was then normalized to the total nebular flux, $4\pi d^2 K_i \nu^{-\Gamma_i}$ erg.s^{-1}Hz^{-1}, where $d = 2$ kpc is the assumed distance. The index i refers to each of the four identified synchrotron power law components. These components are separated by the three spectral breaks at $\approx 1.8 \times 10^{13}$ Hz, $\approx 9 \times 10^{15}$ Hz, and $\approx 3.6 \times 10^{19}$ Hz. Two dust components are also expected to act as targets for IC scattering by relativistic electrons and they were given the same size as the radio emitting nebula.

With the one dimensional specification of B (with r), we were able to obtain the electron spectra at each r by inverting the synchrotron photon spectra from the formula

$$4\pi d^2 K_i \nu^{-\Gamma_i} C_\nu \lambda_\nu^{-3} r^2 \exp[-(r/\lambda_\nu)^2/2] dr = \int_{E_{min}}^{E_{max}} P(\nu,r) N(E,r) dr dE. \quad (1)$$

where $P(\nu,r)$ is the power radiated by a single electron with energy E and $N(E,r)dr$ is the total number of electrons per unit energy in a spherical shell $(r, r+dr)$. The minimum (E_{min}) and maximum (E_{max}) electron energies depend on the field strength $B(r)$ such that the corresponding characteristic frequencies are $\nu_{min} = 10^7$ Hz and $\nu_{max} = 10^{23}$ Hz respectively. Analytical expressions have been obtained for $N(E,r)$ for the field distributions of interest. These functions are given in ref. 8. To calculate the IC spectrum we also need to know the spatially resolved photon density. We determine the photon energy density $U(\nu,r)$ from the volume emissivity $\epsilon(\nu,r)$ using the equation for radiative transfer in the optically thin (for both absorption and scattering) case.

With the above information we were able to estimate the IC spectrum at Earth using the production rate $d^2N_{\nu'}/dtd\nu'$ calculated originally by Jones[12]:

$$J(\nu') = \int_{\nu_{min}}^{\nu_{max}} \int_{E_{min}}^{E_{max}} \int_{r_s}^{r_N} U(\nu,r) N(E,r) \left(\frac{\gamma^2 h\nu}{1 + \frac{\gamma h\nu}{mc^2}}\right) \frac{d^2N_{\nu'}}{dtd\nu'} dr dE d\nu. \quad (2)$$

This spectrum has been added to the calculated synchrotron spectrum (which reproduces the observed low energy γ-ray spectrum up to a high energy cutoff, given the field strength at the shock, and an assumed maximum energy $E_{max} = 10^{16}$ eV. The resulting spectrum is shown in Figure 1 for different field distributions. It is clear that the optimal estimate of $\sigma = 0.003$ by Kennel and Coroniti[9] provides an IC spectrum which is consistent with the observed TeV intensity and differential spectral index at 1 TeV. The range $\sigma = 0.001$ to 0.003, which is consistent with the Kennel and Coroniti model, is also consistent with the error in the TeV spectrum. The range of field strengths inferred at the shock from σ is then between ≈ 5 and 8×10^{-5} G.

These results and those of previous studies[2,5,6] indicate that the IC spectrum is much flatter and much lower in flux than the observed COS-B unpulsed spectrum[11] between 50 MeV and 500 MeV, so that it would be difficult to associate the COS-B flux with an inverse Compton component in the nebula. Interpreting this spectrum as synchrotron radiation at the pulsar wind shock allows us to estimate the maximum electron energy at the shock. In Figure 2 we have calculated the synchrotron component at high energies for $B_s = 8 \times 10^{-5}$ G ($\sigma = 0.003$) for various values of E_{max}. The spectrum for $E_{max} = 10^{16}$ eV agrees best with the COS-B spectrum and is consistent with the spectrum

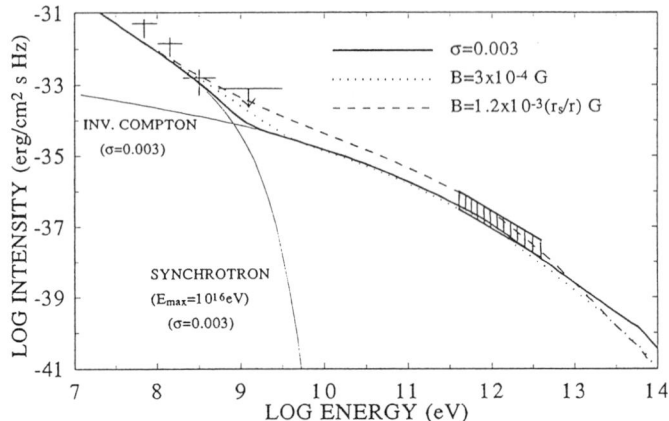

Figure 1. Differential γ-ray spectrum of the Crab Nebula computed from various field distributions. Unpulsed detections from COS-B (ref. 11) and Whipple[4] are also shown.

extrapolated from the low energy γ-ray range. Reducing E_{max} by a factor of about 3 (e.g. due to variability) can turn off the detection of unpulsed γ-rays (from the shock) between 50 MeV and 1 GeV, but the emission above 1 GeV is expected to be unaffected due to the IC scattering of radio and optical photons by much lower energy electrons (GeV to TeV) in the nebula.

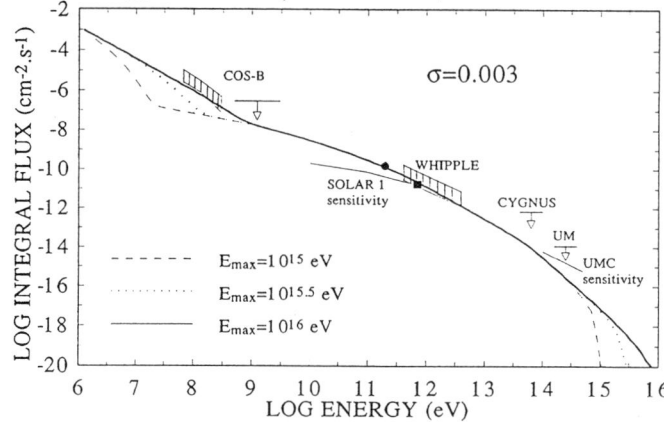

Figure 2. Integral γ-ray spectrum of the Crab Nebula for $\sigma = .003$ and various maximum electron energies E_{max} (from ref. 8).

ELECTRON ACCELERATION IN THE NEBULA

We have shown that inverse Compton scattering of synchrotron photons in the nebula cannot account for the unpulsed 50 to 500 MeV emission observed by COS-B. If the synchrotron spectrum of the nebula extends to these energies, and if our adopted MHD flow model giving a magnetic field downstream of the shock $B_s = 8 \times 10^{-5}$ G is correct, then electrons of energy $E_{max} \sim 10^{16}$ eV are required to produce this synchotron emission. It would be difficult to accelerate these UHE electrons inside the pulsar magnetosphere where curvature radiation would limit their energies to less than $\approx 3 \times 10^{12}$ eV. The electrons responsible for the soft X-ray to γ-ray emission must somehow be accelerated outside the pulsar magnetosphere, the most likely mechanism being shock acceleration at r_s.

Most of the literature on shock acceleration (see ref. 14, for review) concerns non-relativistic, parallel shocks. Pulsar wind shocks are relativistic, closer to a perpendicular configuration (magnetic field perpendicular to the shock normal), and most likely superluminal. However, the upper limit on the maximum energy to which a particle can be accelerated is the same for parallel and oblique (including perpendicular) shocks, because this energy depends on the product, $\kappa_\| \kappa_\perp \approx r_L v$, of the diffusion coefficients parallel and perpendicular to the magnetic field[15], where r_L is the particle Larmor radius. The maximum electron energy in all cases is the "shock drift" acceleration limit, or the energy an electron would acquire on drifting through the potential drop along the shock:

$$E_{\max} \approx eV_{sh} \approx e(\mathbf{v} \times \mathbf{B})r_s/c = eB_s r_s, \quad (3)$$

where the upstream velocity, $v \approx c$. For $B_s = 8 \times 10^{-5}$ G and $r_s = 2.4 \times 10^{17}$ cm, $E_{\max} \approx 5 \times 10^{15}$ eV.

One can show that the potential drop across the shock as given above is approximately equal to the vacuum potential drop across the pulsar polar cap when $\sigma \geq 1$. In the corotating pulsar magnetosphere, $\mathbf{E} \times \mathbf{B} = 0$ nearly everywhere, so that magnetic field lines are equipotentials. The potential drop across field lines at the neutron star surface between the magnetic pole and the edge of the polar cap is

$$V_{pc} \approx \frac{1}{2}(\frac{2\pi}{cP})^2 B_o r_o^3 = 3 \times 10^{16} \text{ eV}. \quad (4)$$

for the Crab pulsar with period $P = 2\pi/\Omega = 33$ ms, radius $r_o = 10^6$ cm and surface field $B_o = 5 \times 10^{12}$ G. We can interpret this as implying that the potential drop across field lines in the open magnetosphere is preserved across field lines in the pulsar wind, where particles do not cross field lines, and transported intact as long as equipartitition is maintained. However, the MHD flow model of the nebula indicates that equipartition is not maintained out to the shock ($\sigma \ll 1$) and therefore the potential across the shock will be less than the full polar cap potential by a factor $\sim \sigma^{1/2}$.

Even if a potential drop across the shock of $\approx 10^{16}$ eV can be maintained, the question still remains whether enough turbulence will be developed by the electron-positron wind to give the particle scattering necessary to attain the highest energies. This problem also holds for drift acceleration which is expected to be the dominant acceleration mechanism in perpendicular shocks. If the turbulence generated is too small to account for the observations, then another particle acceleration mechanism would be required. Arons et al.[16] discuss a mechanism which accelerates only *positrons* at pulsar wind shocks, which may provide the necessary turbulence. But, the fact that the nebular emission at radio to optical frequencies shows strong linear polarization but weak circular polarization would indicate that roughly equal numbers of positrons and electrons are accelerated, at least in the population that radiates these frequencies.

The maximum energy in any model is limited by the Larmor radius of the electron and/or positron which cannot greatly exceed either the shock radius or the scale length of the magnetic field (both $\approx r_s$), since an accelerated particle with $r_L > r_s$ would escape from the shock and move into the nebula. For relativistic flow velocities, this limit turns out to be the same as the "shock drift" limit. Synchrotron losses may further limit the maximum energy of the accelerating particles. Equating the synchrotron loss timescale to a minimum acceleration timescale given by $t_{\rm acc}^{\min} \simeq 1/\nu_c$, where $\nu_c = eB/2\pi\gamma mc$, we obtain a maximum energy, $(E_{\max})_{\rm syn} = 2 \times 10^{15}$ eV$/\sin\psi_s$, where ψ_s is the particle pitch angle. Even for isotropic pitch angles, the maximum energy is only a factor of 2 below the "shock drift" limit. However, it is clearly below the energy required to explain the COS-B emission up to 500 MeV. It is possible that the pitch angles upstream of the shock in the relativistic flow will not have isotropic pitch angles, which would increase the maximum energy, but only up to the "shock drift" limit.

CONCLUSION

If we consider the observed unpulsed flux up to 500 MeV from the Crab nebula as a solid result, and we assume that this emission is synchrotron radiation at the pulsar wind shock, then there seems to be a significant discrepancy between the maximum electron/positron energy required and theoretical upper limits to acceleration of such particles in the shock. It is possible that this high-energy γ-ray emission originates from a higher field region nearer the pulsar, but the smooth continuation of the spectrum from X-rays, which are clearly extended, would argue against this. It is possible that we do not understand the geometry of the pulsar wind flow. If the emission between 50 and 500 MeV is indeed due to the highest energy particles accelerated in the nebula, then we would expect the emission in this range could be variable on a timescale of months. If the EGRET detector on Compton GRO is able to confirm the presence of unpulsed emission in this energy range, then we may obtain a more accurate fit to the spectrum and thus to the required maximum electron energy.

REFERENCES

1. N. Aschenbach and W. Brinkmann, Astron. Astrophys., **41**, 147 (1975).
2. R. J. Gould, Phys. Rev. Lett., **15**, 577 (1965).
3. T. C. Weekes et al. Ap. J., **342** 379 (1989).
4. G. Vacanti et al., Ap. J., **377**, 467 (1991).
5. G. H. Rieke and T. C. Weekes, Ap. J., **155**, 429 (1969).
6. J. E. Grindlay and J. A. Hoffman, Astrophys. Lett., **8**, 209 (1971).
7. A. A. Stepanyan, in *Proc. Vulcano Workshop 1990* (Italian Phys. Soc.: Bologna), 377 (1990).
8. O. C. DeJager and A. K. Harding, Ap. J., in press (1992).
9. C. F. Kennel and F. V. Coroniti, Ap. J., **283**, 694 (1984).
10. P. L. Marsden et al., Astron. Astrophys., **278**, L29 (1984).
11. J. Clear et al., Astron. Astrophys., **174**, 85 (1987).
12. F. C. Jones, Phys. Rev., **167**, 1159 (1968).
13. M. J. Rees and J. E. Gunn, M.N.R.A.S., **167**, 1 (1974).
14. F. C. Jones and D. C. Ellsion, Space Sci. Rev., **58**, 259 (1991).
15. D. Eichler, Ap. J., **244**, 711 (1981).
16. J. Arons et al., this proceeding (1992).

A MODEL FOR THE RADIO FLARE FROM SN 1987A

J. G. Kirk, M. Wassmann
Max-Planck-Institut für Kernphysik, W-6900 Heidelberg, Germany

ABSTRACT

We present an acceleration mechanism based on multiple reflection of electrons by a shock front and apply it to explain the radio flare observed from SN1987A two days after explosion. The radiation mechanism is assumed to be synchrotron radiation from accelerated electrons which pass through the supernova shock front after having bounced off it many times during its early expansion into the progenitor's stellar wind. Multiple reflection occurs because the wind contains a magnetic field wound into the form of a spiral by stellar rotation: the spiral becomes ever more tightly wound as the shock progresses outwards and reflected electrons are unable to escape radially. For parameters appropriate to SN1987A, the shock becomes superluminal and reflections cease after a time of the order of days. The subsequent decay of the flare is a result of adiabatic losses in the expanding downstream plasma. The predicted spectrum is a power law $I \propto \nu^{-1}$ above the synchrotron self-absorption frequency, consistent with observation.

INTRODUCTION

About two days after the increase of the optical brightness of SN1987A, radio emission was detected at Australian observatories [11]. The emission was of relatively short duration (a few days), had a nonthermal spectrum and emerged from a region of at least the same size as that engulfed by the blast wave at that time [7]. There have been several proposed explanations of this phenomenon [10,2,1,3]. In our model [8] we assume the emission mechanism is synchrotron radiation, and present a simple way of accelerating electrons at the blast wave. The basic process is as follows: charged particles are tied to the spiral magnetic field lines in the stellar wind before the explosion and are reflected when hit by the magnetic compression of the blast wave. This results in a modest increase in energy, and can happen only to particles with approximately the same speed as the blast wave. Other particles have their pitch angles in the loss-cone, and are overtaken by the blast wave. In particular, ions of the pre-SN stellar wind are overtaken. A single reflection is not an effective means of accelerating particles, but the key to our mechanism is that a reflected electron will *always* be caught again by the spherically expanding shock front. This is because the shock moves out at almost constant speed, but the magnetic field line to which the electron is tied becomes more and more tightly wound [9], restricting the radial progress of the particle. Thus, electrons can suffer repeated reflections, gaining energy at each one; they can 'surf' along the outer edge of the blast wave. The process must stop before the time t_L at which the speed of the point of intersection of the blast wave and a magnetic field line reaches the speed of light. Surfing particles gain most of their energy just before this happens. Afterwards there is no escape and they are overtaken by the shock. Given the speed of the shock and the structure of the spiral (determined by the stellar rotation rate and the stellar wind speed) it is simple to find t_L. Typical values for the SN1987A lead to $t_L \sim$ days, and our calculations show that particles are continually accelerated until this time. Consequently, a few days after explosion a population of accelerated electrons can be expected to be injected into the plasma downstream of the blast wave. These electrons will be scattered in the presumably turbulent downstream medium and will radiate synchrotron radiation, suffering at the same time adiabatic losses from the overall expansion.

An interesting question in our model is the number of electrons which are injected into the acceleration mechanism. The stellar wind upstream of the shock contains a population of electrons which are likely to be hot. Provided their energy is high enough that ionization losses can be overcome ($\gtrsim 15$ keV), they are swept up and accelerated. Alternatively, injection may occur by leakage of electrons through the shock front from the downstream region, or may be caused by the strong neutrino burst from the core collapse, as suggested by Bisnovaty-Kogan[3].

ACCELERATION MECHANISM

Before explosion, the stellar wind of the progenitor was of low density $n < 10^6$ cm^{-3} at radial distances greater than about 10^{14} cm [10], since the radio emission would otherwise have been absorbed. Taking values typical of OB stars in our galaxy, indicates that the progenitor probably had a wind speed w at explosion of about 500 km s^{-1} and a mass loss rate of some $10^{-5} M_\odot$ yr^{-1},[4] which implies about the same density as the upper limit derived from the radio emission. The existence of the dense *ring*[6] suggests that the star must have been rotating at a substantial rate. Assuming an angular velocity Ω of about 10^{-6} s^{-1} (or, equivalently, a period of 73 days) which corresponds to about 10% of the break-up rate Ω_{max} ($= 10^{-5} (R_*/3 \times 10^{12}$ cm$)^{-3/2} (M_*/20 M_\odot)^{1/2}$ s^{-1}, with R_* the stellar radius and M_* the stellar mass), and a surface magnetic field of 300 G derived from synchrotron self-absorption arguments implies that the wind was super-Alfvénic and that the geometry must have been close to that of an Archimedian spiral[9].

Into this medium, the supernova launched a strong shock front at a speed $v_s \simeq 25000$ km s^{-1} (Chevalier & Fransson 1987), compressing the plasma and the azimuthal component of the magnetic field by a factor of roughly 4. If we presume that this shock front was spherical, then we can compute the time at which the front went superluminal:

$$t_L \approx 2.8 \left(\frac{v_s}{25\,000 \text{ km s}^{-1}}\right)^{-2} \left(\frac{w}{500 \text{ km s}^{-1}}\right) \left(\frac{\Omega}{10^{-6} \text{ s}^{-1}}\right)^{-1} \text{ days}. \quad (1)$$

Consider the motion of a charged particle which is in the region upstream of the blast wave and is being caught up by it. We assume that the upstream region is not turbulent, so that the trajectory is well described by an equation of motion containing only the spiral magnetic field. Proceeding on the assumption that

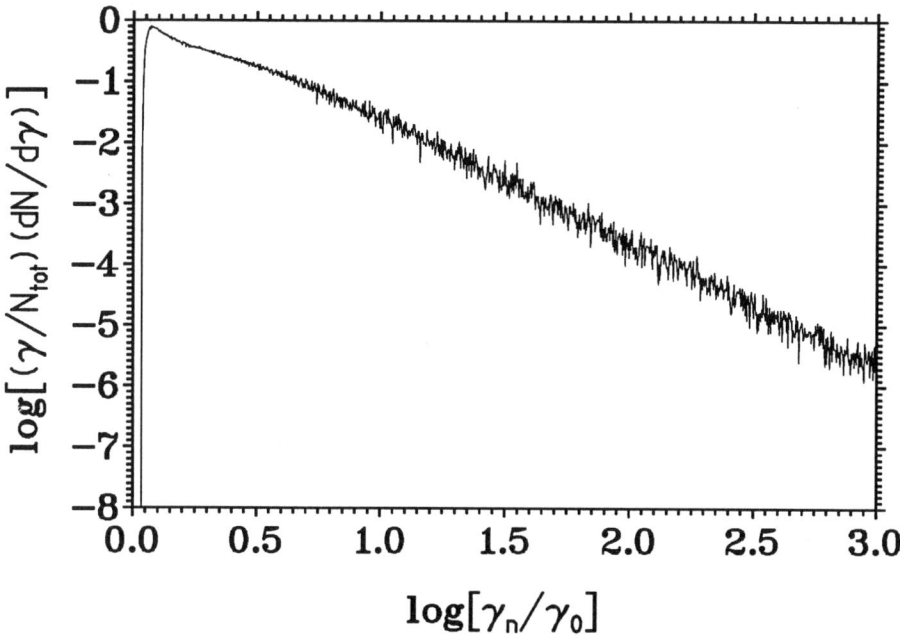

Figure 1: The energy spectrum for monoenergetic injection. The particles injected in this case already possess high Lorentz factor ($= 10^3$). The total number of injected particles is $N_{\rm tot}$ and $dN/d\gamma$ is the differential particle spectrum. At high energy a power law with index $dN/d\gamma \propto \gamma^{-3}$ is found, in agreement with the analytic result.

the magnetic moment is conserved, we can easily compute the effect of a shock encounter on a particle if we assume the shock front to be locally plane and if use is made of Lorentz transformations to and from the frame (the *dHT frame*) in which the electric field vanishes [5]. The dHT frame is reached from that in which the supernova progenitor was at rest (the *star frame*) by a Lorentz transformation in the direction of the upstream magnetic field with speed equal to the speed of the point of intersection of the shock front and a field line $v_{\rm int}$. Such a transformation is possible only for $v_{\rm int} < c$, or, equivalently, $t < t_{\rm L}$. Although a single reflection results in only modest acceleration, we can solve the equation of motion of the particle after reflection to determine the point at which it is again caught by the shock and in this way compute the effects of multiple encounters.

Approximating the Parker spiral by a *tightly wound* spiral, enables the problem to be formulated as a set of two first order difference equations:

$$v_{\|n+1} = \frac{2v_{\text{int}n} - (1 + v_{\text{int}n}^2/c^2) \cdot v_{\|n}}{(1 + v_{\text{int}n}^2/c^2) - 2v_{\text{int}n} \cdot v_{\|n}/c^2} \quad (2)$$

$$v_{\text{int}n+1} = 2v_{\|n+1} - v_{\text{int}n} \quad (3)$$

where $v_{\text{int}n}$ refers to the speed of the intersection point at the reflection labelled n, and $v_{\|n}$ is the speed of a particle along the field line *before* the reflection labelled n, the first reflection being labelled by $n = 0$. In the limit that a particle surfs on the wave, an analytic solution to these equations can be found, together with an estimate of the total energy a particle gains before $t = t_L$. This quantity turns out[8] to be inversely proportional to the cosine δ_0 of the pitch angle in the dHT frame which the particle possessed when it first encountered the shock.

$$\gamma_{\max} \propto 1/\delta_0 \quad (4)$$

where $\gamma_{\max}$ is the final value of the Lorentz factor of a particle attained after many reflections. An isotropic distribution of seed particles leads to an injection rate which is proportional to the quantity δ_0, i.e., $dN/d\delta_0 \propto \delta_0$ and thus gives a spectrum of accelerated particles which is a power law of index -3:

$$\frac{dN}{d\gamma_{\max}} \propto \frac{d\delta_0}{d\gamma_{\max}} \delta_0 \quad (5)$$

$$\propto \gamma_{\max}^{-3} \quad (6)$$

A test of this result is presented in Fig. 1 which shows the numerically determined spectrum of particles injected at relatively high energy $\gamma = 10^3$ (corresponding to electrons of 500 MeV). These particles are faster than the shock front in the tightly wound part of the spiral and undergo injection there. The result accurately reproduces the slope given by (6) at high energies. At low energies, the analytic approach is inadequate because the number of reflections is too small to be considered a continuous process. In order to extend the calculation to include particles injected at lower energy, it is necessary to abandon the approximation of a tightly wound spiral, making a numerical treatment essential. We have performed such a generalisation and present results and a detailed application to SN 1987A elsewhere[8].

REFERENCES

1. Benz, A.O., Spicer, D.S., 1990, Astron. & Astrophys. 228, L13

2. Bisnovaty-Kogan, G.S., 1987, ESO Workshop on SN1987A, page 347

3. Bisnovaty-Kogan, G.S., Illarionov A.F., Slysh V.I. 1990 in "Plasma Astrophysics" ESA-SP-311, page 289.

4. Chevalier, R.A., Fransson, C., 1987, Nature 328, 44

5. de Hoffmann, F., Teller, E., 1950 Phys. Rev. 80, 692
6. Jakobsen P., Albrecht R., Barbieri C. et al 1991, ApJ 369, L63
7. Jauncey, D.L., Kemball, A., Bartel, N., et al, 1988, Nature 334, 412
8. Kirk, J.G., Wassmann, M. 1992 Astron. & Astrophys. 254, 000
9. Parker, E.N., 1958, ApJ 128, 664
10. Storey, M.C., Manchester, R.N., 1987, Nature 329, 421
11. Turtle, A.J., Campbell-Wilson, D., Bunton, J.D., et al, 1987, Nature 327, 38

Observation Priorities and Perspectives

Ground Based Experiments - Observational Priorities

J. A. Goodman
Department of Physics
University of Maryland
College Park, Maryland 20742-4111

Abstract

The study of cosmic ray sources above 30 GeV is beyond the reach of current space based detectors. The study of these most energetic particles is limited to indirect ground based experiments which study the secondary particles from interactions in the atmosphere. The experimental techniques used to observe these particles differ depending on the primary energy range they detect. Significant work is being performed in the energy range of 10^{11}eV to 10^{15}eV to look for neutral radiation produced at the source of high energy cosmic ray acceleration. At the highest energies the question of anisotropy and the end of the spectrum are currently being studied. In this paper, a review the current experiments is presented as well as a discussion of several of the next generation experiments.

Overview

Since their discovery in the early part of this century the source of very energetic cosmic rays has puzzled observers. Charged particles which make up the bulk of high energy radiation spiral toward earth in the magnetic fields which are omnipresent in the interstellar medium. This motion erases virtually all information about the particle's point of origin. The integral flux of these energetic particles drops as $E^{-1.7}$ below several times 10^{15}eV and as $E^{-3.2}$ above that energy. The high energies involved require that detectors be massive which, coupled with the small fluxes, makes detectors which directly observe these particles impractical to fly.

Ground based detectors[1] use a variety of techniques to observe the air showers produced when energetic particles interact in the atmosphere. In the range near 1 TeV (10^{12}), air showers do not directly reach the ground. These showers are detected by observing the Čerenkov light produced by the cascade electrons in the atmosphere. Large mirrors collect the light on dark moonless nights. In this paper, the Whipple Čerenkov telescope and their next generation upgrade called Granite is described.

At energies above 50 TeV, tens of thousands of shower particles reach the ground at mountain altitude. The cascade propagates very near the speed of light so that fast timing of the shower front can be used to determine the primary's direction. Experiments in this energy interval use arrays of scintillation counters spread out over areas of $\sim 10^5 m^2$ to measure the particle density and arrival times. Two such arrays, the CYGNUS experiment and the CASA array, are operational in the western United States and are described briefly here, along with a description of a new technique of using swimming pools in the CYGNUS array to enhance angular resolution.

At the highest energies, above 10^{18}eV (1 EeV), the scarce flux makes even direct observation of the secondary particles difficult. In this energy range a detector called the Fly's Eye is used to observe the nitrogen florescence in the night sky produced by the cascade in the atmosphere. This detector, while restricted in viewing time to dark conditions, gains by having an enormous sensitive area. This detector and the upgrade called HiRes are described.

VHE Observations

Whipple – The Whipple detector [2] consists of an array of 109 photomultipliers at the focus of a 10m diameter mirror. Čerenkov light from electrons and positrons in the atmospheric cascade is recorded by this "camera". The Čerenkov measurement integrates light from the entire cascade which makes it a calorimetric measurement. This results in the signal recorded being proportional to the primary particle energy. The light from the shower falls in a cone of radius of $\approx$100 m. This yields an effective area of about $3 \times 10^4 m^2$. The angular acceptance of the telescope is about 3.75°. The trigger rate for this device is about 5 events per second when more than 50 photoelectrons are required.

The shape of the light signal as recorded by the pattern of PMT hits has been shown by simulations to be sensitive to the type of primary particle. Primary photons from a source produce a highly elliptical pattern whose major axis passes through the center of the field of view. Background proton induced showers are less regularly shaped and do not point as clearly toward the center of the field. By cutting on a parameter, azwidth, which combines both of these characteristics, simulations predict that strong discrimination between photons and protons can be obtained.

VHE Results – Using this azwidth cut, the Whipple camera has been used to observe the Crab Nebula. They have observed a highly significant signal which was considerably strengthened by the azwidth cut which removes 97% of the background. This signal is observed to be non-pulsed and steady on a time scale of years. This is the only consistent observation of a steady source above 30 GeV. The origin of this photon signal is thought to arise from inverse Compton scattering of energetic electrons with photons in the region surrounding the pulsar. Because of the predicted cutoff of the electron spectrum, it is expected that the photon spectrum should exhibit a significant steepening above 10 TeV. Currently the Whipple experiment reports an observed integral spectrum of the form $E^{-1.4}$.

As of this writing the collaboration has not reported any other sources using this technique. Previously, they had reported episodic emission from several objects such has Her X-1. These data were taken using an earlier version of the camera which had 37 elements. These data have been subsequently reanalyzed using an azwidth cut. The cut has either reduced or eliminated the significance of the older observations. In the spring of 1986 an observation of Her X-1 was reported by Whipple, Haleakala and CYGNUS collaborations[3][4] [5]. This Whipple observation also disappears when the azwidth cut is applied.

One of the characteristics of this type of detector is its low duty cycle. The requirement of clear, dark, moonless observing periods restricts significantly the opportunity to scan the sky for episodic emission which has been regularly reported by other experiments in this energy range.

UHE Air Shower Experiments

CYGNUS – Since the first reports of directional radiation from point sources, more questions about the nature and origin of this radiation have been raised than answered[6]. The CYGNUS experiment[7] was one of the first experiments specifically designed to look for sources of Ultra High Energy (UHE) radiation above 3×10^{13}eV. This experiment was designed to have good angular resolution when reconstructing the direction of Extensive Air Showers (EAS) and unambiguous detection of muons accompanying these showers.

The CYGNUS experiment consists of an air shower array and detectors to sample the muon content of these showers. It is located at the end station of the LAMPF accelerator at Los Alamos National Laboratory at an altitude of 2134m, corresponding to an atmospheric depth of 800 gm/cm^2. An air shower array covering $\sim 8 \times 10^4$ m^2 area is operating with 204 scintillation counters, each of $\sim$ 1m^2. The E225 [8] detector is used as a muon detector, as is an array of scintillation counters buried in tunnels beneath the array.

A typical trigger requirement is 20 counters with signal greater than 0.1 minimum ionizing particle. This trigger allows showers generated by primaries of E>50 TeV to be detected with good efficiency. The hardware rate is about 5 hz. At this time, no cuts are made and all data is written to tape. The array has operated with over 90% on-time.

The ratio of a signal from a point source to that from the isotropic background of cosmic ray hadrons is increased by improving angular resolution. The angular resolution of this array was determined to be about 0.75° for small showers and about 0.45° for large showers. This resolution has been measured by observing the shadowing of background cosmic ray protons by the sun and moon.

CASA – CASA[9] is an air shower array located in the Utah desert (at the site of the Fly's Eye) which covers 2.3×10^5m^2. It is made up of 1089 stations each with 1.5m^2 of scintillation counter and local electronics. This array is operated in conjunction with an array of 2560m^2 of buried scintillation counters for muon detection built and operated by the University of Michigan. The CASA experiment, which has recently become fully operational, has a data rate of more than 20 Hz. It has an angular resolution comparable to the CYGNUS array and also verified its pointing accuracy with the moon and sun shadow as well as with a set of four Čerenkov telescopes.

UHE Results – Neither CYGNUS nor CASA has reported any significant evidence for continuous emission from any northern hemisphere source candidate.

The lack of reported signal suggests that either sources such as CYG X-3 which were reported in the early 1980's are no longer producing gamma rays or that the original observations were incorrect. The level of the upper limits placed on the flux from CYG X-3 are more than an order of magnitude below the flux reported by both Kiel and Haverah Park [10] [11]. The only reported observation of UHE emission was a episode in 1986 where the CYGNUS array reported a significant observation of showers from the direction of HER X-1 which showed both an anomalously high muon content for gamma induced showers and a period shifted from the reported x-ray period. The same shifted period was reported in observations made within two months by both the Whipple and Haleakala groups. The significance of this observation (particularly the muon anomaly) requires further confirmation by simultaneous observations. CASA and CYGNUS now have this capability since they observe the same region of the sky at the same time.

EHE Observations – The End of the Spectrum?

The Fly's Eye – The Fly's Eye experiment [12] of the University of Utah has been recording showers at energies above 10^{17}ev since 1981. The Fly's Eye detector measures nitrogen florescence in the atmosphere to determine the track and energy deposition of showers as they cascade in the atmosphere. The detector, while having a limited on-time because of the need for dark, clear nights, makes up for this with an effective aperture of up to almost 20 Km. The current operation uses two eyes for stereo reconstruction of some showers, but the vast majority of showers have only monaural reconstruction. The pointing accuracy of these events is $\pm 2°$ in the direction perpendicular to the projection of the track on the sphere of the sky and $\pm 9°$ along the other direction.

EHE Results – The Fly's Eye has observed events whose energy is very near 10^{20}eV (100 EeV). Their data along with data from Haverah Park show that there may be is some evidence for a slight flattening of the spectrum above 10 EeV, but this issue is not yet settled. Also their data show no firm evidence for the predicted cutoff due to interaction of particles with the 3° background radiation (the Griesen cutoff[13]).

What We Know and What We Don't

- Are there continuous sources around 1 TeV?
  - Yes, at least one: The Crab Nebula as seen by Whipple.

- Are there other sources in this energy range?
  - Maybe, GRO may tell us where to look. (e.g. Geminga)

- Is there continuous emission above 100 TeV?
  - Probably not! CYGNUS and CASA do not see any.

- Are there episodic sources above 1 TeV?
  – Probably! Many experiments have reported bursts, but it is difficult to tell until we can get a better duty cycle.

- What is happening at the "end" of the spectrum?
  – There is not enough data yet to know.

Future Plans Near 1 TeV

MILAGRO – The CYGNUS collaboration is proposing[14] a new water Čerenkov gamma ray observatory to cover the energy range above 1 TeV thus covering a largely unexplored region of the spectrum. This detector would be located near Los Alamos at an elevation of 2650m in an existing pond. The air shower would be detected by 500 photomultiplier tubes spaced three meters apart, located one meter below the light-tight cover on the water's surface; 150 tubes near the bottom of the pond are optically isolated from the top layer and would measure the muon content of the showers. The pond measures 60m by 80m at the water's surface and is about 8m deep. Simulations show that MILAGRO will trigger on a large number of events below 1 TeV and that its angular resolution will be significantly better than existing air shower arrays. This detector will observe, with 5σ significance, the Crab signal reported by Whipple in less than one year. This detector has the advantages of an open aperture and 24 hour a day operation, making possible the search for episodic and known sources of TeV gamma rays.

Granite – The Whipple[15] group is actively building a second 10m camera similar to the existing one. This will allow stereo reconstruction of events. The combination of the two telescopes will further improve the photon signal to background by an order of magnitude. This factor will allow detection of the crab signal at 5σ within two hours.

Future Plans Near 100 TeV

Swimming Pools – The CYGNUS collaboration has deployed five backyard-style swimming pools of radius 3.7 m (42 m^2 area) and depth 2.3m which are the enclosures for an array of water Čerenkov detectors. There is 2 m of water above the faces of the seven fast Burle PMTs (< 2.5 ns FWHM).

The pools are being run in coincidence with the CYGNUS air shower array; preliminary results on air-shower timing show that the angular resolution obtainable with these pools is substantially better than with the CYGNUS array alone. This will provide enhanced signal to background for the study of point sources.

Future Plans Near 10 EeV

HiRes – The Fly's Eye collaboration[16] is in the process of developing a new detector similar to the existing Fly's Eye. In this detector, three stations will be located about 15 Km apart and will provide stereo viewing of air showers. Each PMT will view 1° instead of the current 5°. This detector when complete will be able to observe more than 200 events per year above 10 EeV. This is an order of magnitude improvement over the existing detector. In several years of operation HiRes should be capable of seeing the Greisen cutoff as well as making some measurements of anisotropy at the highest energies.

References

[1] L. J. Rosenberg: Univer. of Chicago preprint EFI 90-53 August 1990 to be published.

[2] M. J. Lang, et. al., Nuclear Physics B - (Proc. Suppl.) P165 14A (1990)

[3] L. Resvanis, et. al.,Ap. J., **328**, L9, 1988.

[4] R. C. Lamb, et al., Ap. J. **328**, L13(1988).

[5] B.L. Dingus et. al., Phys. Rev. Letters,**61**,1906 ,1988.

[6] J. A. Goodman, Nuclear Physics B - (Proc. Suppl.) P84 14A (1990)

[7] Alexandreas, et. al., Nuclear Instr. and Methods (1991)

[8] R. C. Allen, et. al., Nuclear Instr. and Methods (1991) and Phys. Rev. Lett.,**55**, 2401, 1985.

[9] R. A. Ong, Nuclear Physics B - (Proc. Suppl.) P273 14A (1990)

[10] M. Samorski and W. Stamm, Ap. J. 268, 117 (1983)

[11] J. Lloyd-Evans, et. al., Nature, 305, 784, (1983)

[12] R. M. Baltrusaitis, et al., Ap. J. (Lett.),**293**, L69, (1985b)

[13] K. Greisen, Phys. Rev. Letters, **16**, 748 (1966)

[14] D. Berley et. al., High Energy Gamma Ray Astronomy AIP Conf Proc 220 (1990)

[15] C. W. Akerlof, et. al., Nuclear Physics B - (Proc. Suppl.) P237 14A (1990)

[16] L. Borodovsky et. al., Proc. 22nd ICRC Dublin, **Vol 2** P688 (1991)

PERSPECTIVES OF FUTURE OBSERVATIONS ON PARTICLE ACCELERATION IN THE HELIOSPHERE

Eberhard Möbius
Department of Physics and Institute for the Study of Earth, Oceans and Space
The University of New Hampshire, Durham, NH 03824

ABSTRACT

With sophisticated composition instruments covering a wide energy range on spacecraft, the quantitative study of acceleration processes close to the Earth, on the sun and in the outer heliosphere has become a feasible task. The paper outlines this development into the foreseeable future with a few examples. A major laboratory for in-situ studies of acceleration processes is the Earth's bow shock. Significant progress has been made to simulate shock processes and their effects on particle acceleration which now allows a quantitative comparison with data from ISEE and AMPTE. A next natural step is the extension to simultaneous observations immediately upstream and downstream of the shock with full composition and distribution function information. This will become available with the CLUSTER mission. A good understanding of shock acceleration from bow shock studies may also cross-fertilize the investigation of the anomalous cosmic rays and the heliospheric termination shock. Improved data sets on this topic will be obtained with SAMPEX and ACE - in particular, by directly measuring the charge state - which will complement ideally the data sampled by the Pioneer and Voyager probes right now. Detailed charge state measurement will also be of key importance for the understanding of the acceleration processes in solar flares. Instruments to be flown on SOHO and ACE will make a significant step towards this goal.

INTRODUCTION

A variety of situations and phenomena involving particle acceleration have been covered at this conference, ranging from the Earth's magnetosphere to the most distant objects in the universe. The acceleration processes in the heliosphere have been reviewed by Lee[1]. Since this field still comprises a broad range of diverse applications, this paper on future perspectives of particle acceleration will, rather than presenting a comprehensive overview, concentrate on three specific topics. Each of them has been selected, because new diagnostic tools, available through a combination of new instruments and spacecraft mission concepts, should shed new light on them.

• First the acceleration of ions at the Earth's bow shock will be discussed for which the capabilities of simultaneous measurements at different locations and of the full 3 D distribution function resolved in species with the CLUSTER mission will give the utmost detailed diagnostics. This is the most easily accessible example of shock acceleration which may be extrapolated to other situations in the cosmos.

• The second example of shock acceleration is the production of the anomalous component of cosmic rays (ACR) at the heliospheric termination shock. A good cross-fertilization may be expected with the first topic and we look forward to a unique constellation of spacecraft to study the large scale structure of this shock within the next 2 decades with the outward moving armada of Pioneer 10/11 and Voyager 1/2. Furthermore, the opportunity to study directly the charge state of these particles will be at hand shortly with the instrumentation on SAMPEX.

• Finally, the selective processes in the acceleration during small compact flares with unusual ion composition are still not understood. The existing models of the processes suggest that a detailed study of the charge states may help to resolve the remaining

© 1992 American Institute of Physics

questions. The WIND, SOHO and ACE instrumentation will make this possible and will also cover the difficult energy range between ≈ 10 keV/nucleon and ≈ 300 keV/nucleon which so far has been largely unexplored.

In combination these studies will hopefully link the detailed plasma physics which is possible in the immediate neighborhood of the Earth with the growing information from other acceleration sites in the heliosphere through increasingly capable simulations of the acceleration processes. This will set a solid base for extrapolations to the large scale acceleration processes in the heliosphere and likewise to other places in the universe.

ACCELERATION OF IONS AT THE EARTH'S BOW SHOCK

The closest laboratory for shock acceleration in space is the Earth's bow shock. It has been demonstrated that in-situ acceleration at the shock and leakage of energetic particles out of the magnetosphere contribute to these particle populations. For the purpose of this paper we will concentrate solely on the aspect of in-situ acceleration. For reviews of the shock acceleration see, for example Scholer [2] and Forman and Webb [3]. An overview of the current status of the observations has been presented by Ipavich [4] at this conference. In general two different distributions of accelerated ions are observed upstream of the bow shock depending on the orientation of the magnetic field with respect to the shock normal:
- a directed beam of reflected ions in the case of a quasi-perpendicular shock
- diffuse ions under the condition of a parallel or quasi-parallel shock

Most recently the behavior of the energetic ion distributions and their interaction with the shock was revealed in numerical hybrid simulations. In particular, an intimate link between the shock formation and the acceleration of the ions has been uncovered. These findings are described in various papers [5, 6, 7, 8] in this volume. One of the most important results of these simulations is the fact that a considerable fraction of the incoming solar wind ions - out of the tail of the ion distribution - are reflected at the shock for all magnetic field conditions. Particularly in the quasi-parallel case these reflected ions create and amplify waves upstream of the shock. The ions become bunched in the first wave crest upstream until after a while it becomes the new shock. This situation is sketched in Fig. 1. This so-called shock re-formation [9] exhibits a cyclic behavior with a well-defined re-formation length scale.

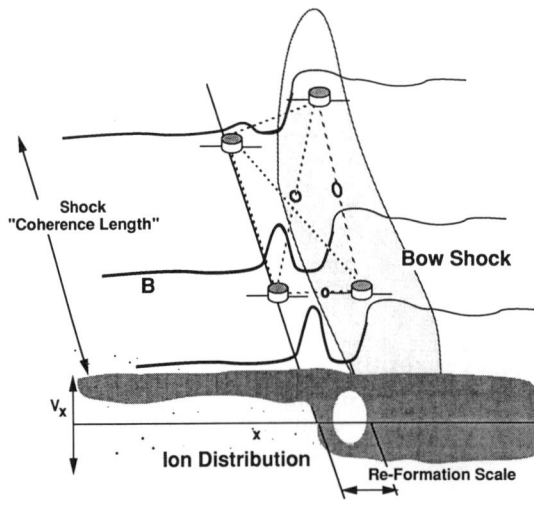

Fig. 1: Perspective view of the situation at the Earth's bow shock with a potential configuration of CLUSTER. The re-formation of the shock and a potential coherence length have been adopted in the sketch of the magnetic field structure. The insert shows the typical phase space distribution of solar wind ions and the accelerated component.

The resulting time-variable magnetic and electric field structure is responsible for a significant acceleration of ions during their first reflection[10]. The statistics of this first acceleration basically leads to energy spectra of the ions which are consistent with observations[11]. The longer an ion stays with the shock, before it returns into the upstream region, the higher its final energy. In this sense a typical length scale along the shock may determine the slope of the energy spectra. Such a length scale may also explain the observed scaling of upstream ion spectra with energy/charge, since the energy gain of the ions is directly connected with electric fields in the shock plane[10]. Both 2 and 3-dimensional simulations are underway which can delineate this scale length.

This new understanding of the shock physics enables us to ask more detailed questions about the observations:
• Firstly, these findings may be validated by a detailed comparison between observations of the ion velocity distributions, and measurements of the magnetic and electric fields. A first attempt in this direction can be undertaken with the existing AMPTE IRM data set which contains 3 dimensional ion distributions up to 40 keV/Q, energy spectra of the most abundant species from 10 - 230 keV/Q, as well as magnetic and electric field data. An initial comparison of simulated and measured energy spectra for protons and alpha particles has been successful[11].
• Secondly, a detailed study of the spatial and temporal structure of the shock and its immediate neighborhood up- and downstream will become possible with the CLUSTER Mission[12]. CLUSTER will comprise a tetrahedron configuration of 4 identically equipped spacecraft, which will allow unambiguous determination of the temporal and spatial evolution of plasma structures in all 3 spatial dimensions for the first time. A potential configuration of the 4 spacecraft for the bow shock is included in Fig. 1. Simultaneous measurements upstream and downstream as well as in 2 dimensions along the shock front will be possible.

This configuration of the spacecraft may enable the direct determination of the two important length scales involved in the shock formation and their dependence on solar wind conditions. From the ongoing shock modeling efforts essential constraints for the CLUSTER operations can be derived:
• According to the simulations the shock re-formation process occurs on a typical length scale of 300 - 500 km perpendicular to the shock front. The separation of at least two spacecraft a few 100 km along the shock normal, i.e., along their orbital trajectory, will provide a good baseline for simultaneous observations of the old shock front and the new wave structure upstream including the corresponding ion distributions.
• As discussed above the typical coherence length of the shock structure along its plane may limit the energy gain of ions due to electric field acceleration during one shock encounter and thus determine the spectral slope in energy/charge. The direct measurement of the coherence length may give a clue on the absolute calibration of the acceleration efficiency under various conditions. The scale length involved may be roughly estimated from the arguments that all electric fields in the shock are of the order of the induced electric field $-v_{sw} \times B$ and that the e-folding energies of upstream ion distributions are about 20-30 keV/Q. With $v_{sw} \approx 500$ km/s and the magnetic field in the shock $B \approx 20$ nT a typical scale length of 2-3000 km in the shock plane is estimated.

A configuration of the CLUSTER spacecraft which could measure both crucial scales simultaneously would be the optimum for the study of the shock structure, the ion acceleration, and the comparison with simulations. Even with a separation strategy which targets both scales consecutively these topics may be covered.

The CLUSTER spacecraft will contain a full complement of plasma diagnostic instruments including the unique ability to measure with one spin (≈ 4 sec) resolution, 3-D ion distributions[13] for H^+, He^{2+} and He^+ from 0 to 40 keV/Q. This package contains a diagnostic power comparable to the current hybrid simulations. In particular, 3-D distributions of individual species will be available with the resolution of a fast plasma instrument for the first time. The CODIF (COmposition and DIstribution Function analyzer) combines a hemispherical top-hat electrostatic analyzer, post-acceleration of the incoming ions to at least 25 keV/Q, and a time-of-flight analysis. This combination provides the energy/charge, mass/charge and directional distribution of the ions. The instrument has two geometrical factors differing by a factor of 100 to cope with the enormous dynamic range of fluxes. On board computation of the moments of the distribution function allows the transmission of the most important parameters with full time resolution at all times.

The combination of 3-dimensional distribution functions, composition, and spatial coverage with CLUSTER along with the rapidly developing simulations will allow a giant step to be made in the understanding of shock acceleration processes. This will also allow a safer extrapolation to shocks at more remote places in the universe and foster the understanding of other shocks in the heliosphere, such as interplanetary traveling shocks, corotating shocks, and the heliospheric termination shock.

ACCELERATION OF IONS AT THE SOLAR WIND TERMINATION SHOCK

The anomalous component of cosmic rays (ACR) has been discovered in the early 70's as a component in the energy spectrum between solar cosmic rays at low energies and galactic cosmic rays at high energies with an enhanced abundance of elements with high ionization potential[1]. The most likely source for these particles is interstellar neutral gas which penetrates the heliosphere where it is then ionized. The resulting ions are transported to the termination shock and finally accelerated. These views have gained support by a series of independent observations and modeling efforts. From a phase shift with respect to the solar cycle it could be deduced that the ions are most likely singly ionized[14]. A model which includes acceleration at the termination shock via both Fermi and shock-drift processes and the transport to the inner solar system by diffusion and drift in the large scale magnetic field[15] describes most of the observations correctly. The ionized interstellar gas, or "pick-up" ions, which are swept outwards with the solar wind have recently been identified[16].

Yet the ACR offer some open questions:
• The current acceleration models become efficient only at energies much higher than the potential injection energy of the pick-up ions. In particular, the termination shock is expected to be an almost exclusively perpendicular shock for which almost all particles will be swept downstream after their first encounter. The injection of the ions is still a mystery.
• Linked to this is the question: how does the ion distribution look at the shock?
• What is the 3-dimensional distribution of the particles in the heliosphere? Are certain regions of the shock more efficient for acceleration? To date the observations have been limited to regions close to the ecliptic plane.
• That all ACR are singly ionized, still awaits direct confirmation.

For most of these questions the perspectives in the near future are extremely good. A unique constellation of spacecraft, the rapid evolution of simulations of the bow shock, as well as new instrumentation will pave the way.

Partial answers to the question on the injection efficiency may be expected from a comparison of the observed pick-up ion distributions, their appearance at the Earth's bow shock, and shock simulations including the injection of pick-up ions. Observations are available from the AMPTE and Ulysses spacecraft. However, the extrapolation to the termination shock may still have to bridge too large a gap, since the conditions at this shock are still unknown.

With Voyager-1 and -2 and Pioneer-10 and -11 this final question may be brought closer to an answer within the next 10 to 20 years, depending on which prediction for the shock distance is correct. Figure 2 shows the current configuration of these interplanetary spacecraft. Even long before the shock encounter a valuable data set is being collected. With Ulysses passing over the polar region of the sun, the 3-dimensional spatial distribution of cosmic rays can be studied for the first time. This will establish a full set of radial, longitudinal, and latitudinal gradients and thus may eliminate existing uncertainties about the transport of ACR into the inner heliosphere. The final encounter of the shock may reveal the magnetic structure of the shock and the local ACR distributions, but unfortunately not the seed population for the acceleration process, since the instrumentation is not sensitive enough. This complete set of information may still have to await an Interstellar Probe Mission.

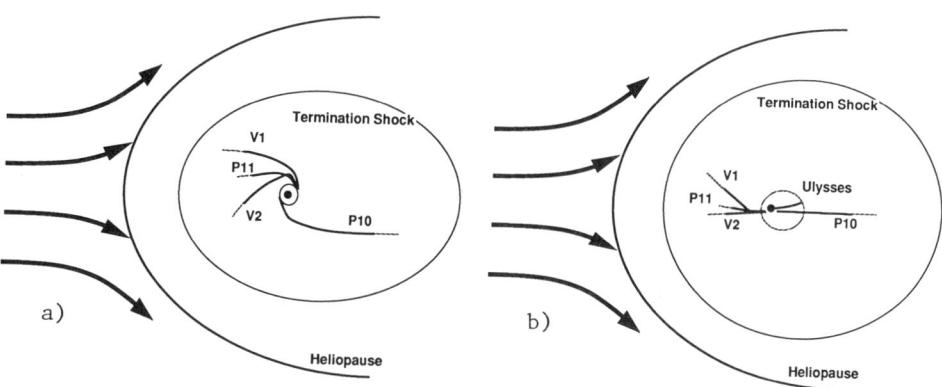

Fig. 2: Past (full line) and future trajectories of the Pioneer 10, 11, Voyager 1, 2, and Ulysses spacecraft in (a) and perpendicular to (b) the ecliptic together with the heliospheric boundary structures.

A direct charge state measurement of the ACR is planned with the SAMPEX (Solar, Anomalous, Magnetospheric Particle EXplorer) spacecraft[17] to be launched by mid 1992. The technique is the determination of the cut-off rigidity of ACR ions in the Earth's magnetic field. The key element is the SAMPEX/HILT sensor which combines a drift chamber and a position-sensitive proportional counter at the entrance with a matrix of solid state detectors at the bottom for the determination of the arrival direction, which is necessary to retrace the trajectory in the magnetic field. The energy and nuclear charge of the particles are determined by a dE/dx and total energy measurement in these detectors. With the high collecting power instrumentation on ACE[18] (Advanced Composition Explorer) the detailed element and isotope composition of the ACR will be obtained in the more distant future.

ACCELERATION PROCESSES IN SOLAR FLARES

An overview on the current understanding of energetic particles from solar flares and possible acceleration mechanisms has been given by Reames[19]. In general solar flare events may be divided into two groups:
 • Large flare events, accompanied by strong shock waves and mass ejections, show energetic particle populations with a composition similar to that in the solar corona.
 • Small impulsive events generally exhibit large overabundances of heavy ions, such as iron. In addition, many of these events show enhancements of ^{3}He over ^{4}He by several orders of magnitude. The particles from these events find their way to the Earth along field lines which are well connected to the flare site and thus very likely reflect material accelerated in the flare.

This last section will concentrate on these ^{3}He-rich events. See, e.g., review by Ramaty et al.[20]. They are always enriched in Fe. They also correlate well with electron events and type III radio bursts[21] indicating impulsive flare events. Still under debate is the nature of the fractionation process which leads to the extreme ^{3}He enrichment. A compilation of the proposed mechanisms is given in Table 1. All mechanisms cover the range of observed ^{3}He enrichment factors. However, in order to maintain reasonable abundances of heavy ions in the energetic particle population they vary in the predicted charge states for the ions. Most discriminating would be detailed observations of charge states of Fe and Si. So far results with moderate statistics and resolution are available[26] which seem to favor the models based on ion cyclotron waves. With the particularly poor statistics of Si[27] - all events observed over the full ISEE mission had to be integrated - and the poor resolution of Fe charge states no decision between the individual models involving ion cyclotron waves is possible to date.

Table 1: Predictions of selective heating models for ^{3}He-rich solar flares

Model	Predicted Ionic Charge States			
	^{3}He	O	Si	Fe
Turbulent heating Ion Sound Waves (Ibragimov et al.[22])	^{3}He^{2+}	O$^{5+ - 6+}$	Si$^{6+ - 8+}$	Fe$^{7+ - 9+}$
Resonant Heating Electrostatic Ion Cyclotron Waves (Fisk[23])	^{3}He^{2+}	O^{5+}	Si^{7+}	Fe$^{\approx 17+}$
Resonant Heating Electromagnetic Ion Cyclotron Waves (Temerin and Roth[24])	^{3}He^{2+}	O$^{5+ - 7+}$	Si^{7+-10+}	Fe^{17-22+}
Non-Resonant Heating (Hydrogen) Electrostatic Ion Cyclotron Waves (Varvoglis and Papadopoulos[25])	^{3}He^{2+}	O^{5+}	Si$^{\approx 14+}$	Fe^{20-22+}

A significant improvement can be expected with the upcoming SAMPEX, WIND, SOHO and ACE missions. A compilation of the energy coverage of mass and ionic charge resolving measurements as of today and for each of these missions is given in Fig. 3. In particular, SOHO will carry a time-of-flight instrument for the suprathermal ion energy range between 20 and 1000 keV/Q which will allow mass and charge composition measurements of solar energetic particles with energies < several 100 keV/nucleon for the first time. The CELIAS STOF sensor combines an electrostatic analyzer for the selection of ions according to energy/charge with a time-of-flight section to determine the energy/mass and the measurement of the total energy in a solid state detector[28]. This allows the extension of studies into a regime where the injection into the acceleration and local transport processes in the flares become important. The charge resolution will be about $\Delta Q/Q \approx 0.1$. The corresponding instrument on WIND has a lower energy limit ($\approx$ 230 keV/Charge) and thus will be limited in the observation of flare particles. The Suprathermal Energy Particle Charge Analyzer (SEPICA) on ACE will cover the energy range from $\approx$ 0.2 to 5 MeV/nucleon with separate sections for large geometrical factor (0.4 cm^{2}sr) with moderate charge resolution ($\Delta Q/Q \approx 0.3$) and high charge resolution ($\Delta Q/Q \approx 0.1$) with moderate geometrical factor (0.04 cm^{2}sr). The nuclear charge and energy of ions are determined with a dE/dx - E measurement in a proportional counter solid state detector combination and E/Q from a position measurement in the proportional counter after collimation in a slit collimator and deflection between two plates with a potential difference of 30 kV.

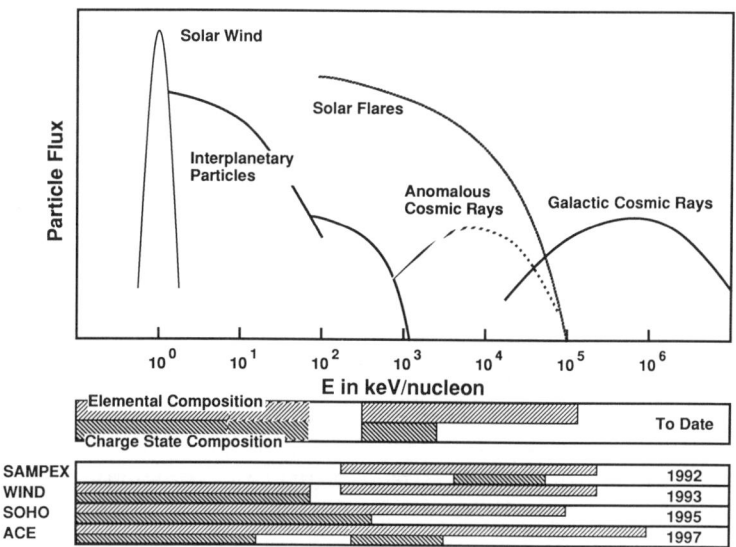

Fig. 3: Energetic particle distributions in the heliosphere and coverage with mass and ionic charge resolving instruments up to now and by instruments on future missions.

With the presently scheduled launches of WIND (1993), SOHO (1995), and ACE (1997) a considerable overlap of the data sets may be possible, which will allow us for the first time the full energy range of solar flare particles with mass and ionic charge resolution to be covered. This will certainly lead to a boost in understanding of the injection, acceleration, and transport processes involved.

CONCLUSIONS

In the near future the combination of unique constellations of spacecraft, new improved instrumentation, and powerful plasma simulations will enhance the understanding of acceleration processes in the near Earth environment, on the sun, and in the outer heliosphere. The multi-spacecraft mission CLUSTER will allow the most detailed view of the bow shock, while the armada of spacecraft passing through various regions of the inner and outer heliosphere will provide the large scale view needed to understand the heliospheric boundary. Improvements in ionic charge resolution and collecting power will hopefully give a clue on still puzzling fractionation processes. The resulting improved understanding of the physical process can then lead to a fruitful cross-fertilization between the studies of the diverse acceleration sites in the heliosphere.

ACKNOWLEDGEMENTS

The author gratefully acknowledges support of this work through NASA grants NAG5-1548, NAGW-2579, and NASA contract NA5-31459. The presentation has greatly benefitted from discussions with L.M. Kistler, M.A. Lee and M. Scholer.

REFERENCES

1. M.A. Lee, this volume (1992)
2. M. Scholer, in Collisionless Shocks in the Heliosphere, ed. R.G Stone and B.T. Tsurutani, Geophys. Monogr. 35, 287 (1985)
3. M.A. Forman and G.M. Webb, in Collisionless Shocks in the Heliosphere, ed. R.G Stone and B.T. Tsurutani, Geophys. Monogr. 35, 91 (1985)
4. F.M. Ipavich, this volume (1992)
5. D.C. Ellison, this volume (1992)
6. D. Burgess, this volume (1992)
7. K.B. Quest, this volume (1992)
8. M. Scholer, this volume (1992)
9. D. Burgess, Geophys. Res. Lett., **16**, 345 (1989)
10. H. Kucharek and M. Scholer, J. Geophys. Res., **96**, 21195 (1991)
11. K.-H. Trattner and M. Scholer, Geophys. Res. Lett., **18**, 1817 (1991)
12. R. Schmidt and L. M. Goldstein, ESA SP-1103, 7 (1988)
13. H. Reme et al., ESA SP-1103, 65 (1988)
14. J.R. Jokipii, this volume (1992)
15. B. Klecker, D. Hovestadt, G. Gloeckler and C.Y. Fan, Geophys. Res. Lett., **7**, 1033 (1980)
16. E. Möbius et al., Nature, **318**, 426 (1985)
17. G.M. Mason et al., in AIP Conf. Proc. No. 203, Particle Astrophysics: the NASA Cosmic Ray Program for the 1990's and Beyond, eds.: W.V. Jones, F.J. Kerr and J.F. Ormes (1990), p. 44
18. E. C. Stone, Phase A Study of an Advanced Composition Explorer, CALTECH, Pasadena (1989)
19. D. Reames, this volume (1992)
20. R. Ramaty et al., in Solar Flares, A Monograph from the Skylab Solar Workshop II, ed. P.A. Sturrock (Colorado Assoc. Univ. Press, Boulder, 1980), p. 117
21. D.V. Reames, T.T. von Rosenvinge and R.P. Lin, Astrophys. J., **292**, 716 (1985)
22. J.A. Ibragimov, G.E. Kocharov and L.G. Kocharov, Dokl. Akad. Nauk. SSSR, No. **588** (1978)
23. L.A. Fisk, Astrophys. J., **224**, 1048 (1978)
24. H. Varvoglis and K. Papadopoulos, Astrophys. J., **270**, L95 (1983)
25. M. Temerin and I. Roth, Astrophys. J., subm. (1991)
26. B. Klecker et al., Astrophys. J., **281**, 458 (1984)
27. A. Luhn, B. Klecker, D. Hovestadt and E. Möbius, Astrophys. J., **317**, 951 (1987)
28. D. Hovestadt et al., ESA SP-1104, 69 (1988)

AIP Conference Proceedings

		L.C. Number	ISBN
No. 195	Dense Z-Pinches (Laguna Beach, CA, 1989)	89-46212	0-88318-396-X
No. 196	Heavy Quark Physics (Ithaca, NY, 1989)	89-81583	0-88318-644-6
No. 197	Drops and Bubbles (Monterey, CA, 1988)	89-46360	0-88318-392-7
No. 198	Astrophysics in Antarctica (Newark, DE, 1989)	89-46421	0-88318-398-6
No. 199	Surface Conditioning of Vacuum Systems (Los Angeles, CA, 1989)	89-82542	0-88318-756-6
No. 200	High T_c Superconducting Thin Films: Processing, Characterization, and Applications (Boston, MA, 1989)	90-80006	0-88318-759-0
No. 201	QED Structure Functions (Ann Arbor, MI, 1989)	90-80229	0-88318-671-3
No. 202	NASA Workshop on Physics From a Lunar Base (Stanford, CA, 1989)	90-55073	0-88318-646-2
No. 203	Particle Astrophysics: The NASA Cosmic Ray Program for the 1990s and Beyond (Greenbelt, MD, 1989)	90-55077	0-88318-763-9
No. 204	Aspects of Electron-Molecule Scattering and Photoionization (New Haven, CT, 1989)	90-55175	0-88318-764-7
No. 205	The Physics of Electronic and Atomic Collisions (XVI International Conference) (New York, NY, 1989)	90-53183	0-88318-390-0
No. 206	Atomic Processes in Plasmas (Gaithersburg, MD, 1989)	90-55265	0-88318-769-8
No. 207	Astrophysics from the Moon (Annapolis, MD, 1990)	90-55582	0-88318-770-1
No. 208	Current Topics in Shock Waves (Bethlehem, PA, 1989)	90-55617	0-88318-776-0
No. 209	Computing for High Luminosity and High Intensity Facilities (Santa Fe, NM, 1990)	90-55634	0-88318-786-8
No. 210	Production and Neutralization of Negative Ions and Beams (Brookhaven, NY, 1990)	90-55316	0-88318-786-8
No. 211	High-Energy Astrophysics in the 21st Century (Taos, NM, 1989)	90-55644	0-88318-803-1
No. 212	Accelerator Instrumentation (Brookhaven, NY, 1989)	90-55838	0-88318-645-4

No. 213	Frontiers in Condensed Matter Theory (New York, NY, 1989)	90-6421	0-88318-771-X 0-88318-772-8 (pbk.)
No. 214	Beam Dynamics Issues of High-Luminosity Asymmetric Collider Rings (Berkeley, CA, 1990)	90-55857	0-88318-767-1
No. 215	X-Ray and Inner-Shell Processes (Knoxville, TN, 1990)	90-84700	0-88318-790-6
No. 216	Spectral Line Shapes, Vol. 6 (Austin, TX, 1990)	90-06278	0-88318-791-4
No. 217	Space Nuclear Power Systems (Albuquerque, NM, 1991)	90-56220	0-88318-838-4
No. 218	Positron Beams for Solids and Surfaces (London, Canada, 1990)	90-56407	0-88318-842-2
No. 219	Superconductivity and Its Applications (Buffalo, NY, 1990)	91-55020	0-88318-835-X
No. 220	High Energy Gamma-Ray Astronomy (Ann Arbor, MI, 1990)	91-70876	0-88318-812-0
No. 221	Particle Production Near Threshold (Nashville, IN, 1990)	91-55134	0-88318-829-5
No. 222	After the First Three Minutes (College Park, MD, 1990)	91-55214	0-88318-828-7
No. 223	Polarized Collider Workshop (University Park, PA, 1990)	91-71303	0-88318-826-0
No. 224	LAMPF Workshop on (π, K) Physics (Los Alamos, NM, 1990)	91-71304	0-88318-825-2
No. 225	Half Collision Resonance Phenomena in Molecules (Caracas, Venezuela, 1990)	91-55210	0-88318-840-6
No. 226	The Living Cell in Four Dimensions (Gif sur Yvette, France, 1990)	91-55209	0-88318-794-9
No. 227	Advanced Processing and Characterization Technologies (Clearwater, FL, 1991)	91-55194	0-88318-910-0
No. 228	Anomalous Nuclear Effects in Deuterium/Solid Systems (Provo, UT, 1990)	91-55245	0-88318-833-3
No. 229	Accelerator Instrumentation (Batavia, IL, 1990)	91-55347	0-88318-832-1
No. 230	Nonlinear Dynamics and Particle Acceleration (Tsukuba, Japan, 1990)	91-55348	0-88318-824-4
No. 231	Boron-Rich Solids (Albuquerque, NM, 1990)	91-53024	0-88318-793-4

No.	Title	LCCN	ISBN
No. 232	Gamma-Ray Line Astrophysics (Paris-Saclay, France, 1990)	91-55492	0-88318-875-9
No. 233	Atomic Physics 12 (Ann Arbor, MI, 1990)	91-55595	088318-811-2
No. 234	Amorphous Silicon Materials and Solar Cells (Denver, CO, 1991)	91-55575	088318-831-7
No. 235	Physics and Chemistry of MCT and Novel IR Detector Materials (San Francisco, CA, 1990)	91-55493	0-88318-931-3
No. 236	Vacuum Design of Synchrotron Light Sources (Argonne, IL, 1990)	91-55527	0-88318-873-2
No. 237	Kent M. Terwilliger Memorial Symposium (Ann Arbor, MI, 1989)	91-55576	0-88318-788-4
No. 238	Capture Gamma-Ray Spectroscopy (Pacific Grove, CA, 1990)	91-57923	0-88318-830-9
No. 239	Advances in Biomolecular Simulations (Obernai, France, 1991)	91-58106	0-88318-940-2
No. 240	Joint Soviet-American Workshop on the Physics of Semiconductor Lasers (Leningrad, USSR, 1991)	91-58537	0-88318-936-4
No. 241	Scanned Probe Microscopy (Santa Barbara, CA, 1991)	91-76758	0-88318-816-3
No. 242	Strong, Weak, and Electromagnetic Interactions in Nuclei, Atoms, and Astrophysics: A Workshop in Honor of Stewart D. Bloom's Retirement (Livermore, CA, 1991)	91-76876	0-88318-943-7
No. 243	Intersections Between Particle and Nuclear Physics (Tucson, AZ, 1991)	91-77580	0-88318-950-X
No. 244	Radio Frequency Power in Plasmas (Charleston, SC, 1991)	91-77853	0-88318-937-2
No. 245	Basic Space Science (Bangalore, India, 1991)	91-78379	0-88318-951-8
No. 246	Space Nuclear Power Systems (Albuquerque, NM, 1992)	91-58793	1-56396-027-3 1-56396-026-5 (pbk.)
No. 247	Global Warming: Physics and Facts (Washington, DC, 1991)	91-78423	0-88318-932-1
No. 248	Computer-Aided Statistical Physics (Taipei, Taiwan, 1991)	91-78378	0-88318-942-9

No. 249	The Physics of Particle Accelerators (Upton, NY, 1989, 1990)	92-52843	0-88318-789-2
No. 250	Towards a Unified Picture of Nuclear Dynamics (Nikko, Japan, 1991)	92-70143	0-88318-951-8
No. 251	Superconductivity and its Applications (Buffalo, NY, 1991)	92-52726	1-56396-016-8
No. 252	Accelerator Instrumentation (Newport News, VA, 1991)	92-70356	0-88318-934-8
No. 253	High-Brightness Beams for Advanced Accelerator Applications (College Park, MD, 1991)	92-52705	0-88318-947-X
No. 254	Testing the AGN Paradigm (College Park, MD, 1991)	92-52780	1-56396-009-5
No. 255	Advanced Beam Dynamics Workshop on Effects of Errors in Accelerators, Their Diagnosis and Corrections (Corpus Christi, TX, 1991)	92-52842	1-56396-006-0
No. 256	Slow Dynamics in Condensed Matter (Fukuoka, Japan, 1991)	92-53120	0-88318-938-0
No. 257	Atomic Processes in Plasmas (Portland, ME, 1991)	91-08105	0-88318-939-9
No. 258	Synchrotron Radiation and Dynamic Phenomena (Grenoble, France, 1991)	92-53790	1-56396-008-7
No. 259	Future Directions in Nuclear Physics with 4π Gamma Detection Systems of the New Generation (Strasbourg, France, 1991)	92-53222	0-88318-952-6
No. 260	Computational Quantum Physics (Nashville, TN, 1991)	92-71777	0-88318-933-X
No. 261	Rare and Exclusive B&K Decays and Novel Flavor Factories (Santa Monica, CA, 1991)	92-71873	1-56396-055-9
No. 262	Molecular Electronics—Science and Technology (St. Thomas, Virgin Islands, 1991)	92-72210	1-56396-041-9
No. 263	Stress-Induced Phenomena in Metallization: First International Workshop (Ithaca, NY, 1991)	92-72292	1-56396-082-6
No. 264	Particle Acceleration in Cosmic Plasmas (Newark, DE, 1991)	92-73316	0-88318-948-8
No. 265	Gamma-Ray Bursts (Huntsville, AL, 1991)	92-73456	1-56396-018-4
No. 266	Group Theory in Physics (Cocoyoc, Morelos, Mexico, 1991)	92-73457	1-56396-101-6